CIRC   {CASTL}   M6/487272
TK/7872/S8/N18/1973
C. 1
NATO ADVANCED STUDY INSTITUTE,
SUPERCONDUCTING MACHINES AND

D0793375

TK
7872   NATO Advanced Study Insti-
S8   tute, Entreves, Italy, 1973.
N18   Superconducting machines
1973   and devices

NOV 1 2 1982
DEC 1 2 1990

Superconducting Machines and Devices

Large Systems Applications

NATO ADVANCED STUDY INSTITUTES SERIES

A series of edited volumes comprising multifaceted studies of contemporary scientific issues by some of the best scientific minds in the world, assembled in cooperation with NATO Scientific Affairs Division.

Series B: Physics

Volume 1 – Superconducting Machines and Devices
edited by S. Foner and B. B. Schwartz

Volume 2 – Elementary Excitations in Solids, Molecules, and Atoms
(Parts A and B)
edited by J. Devreese

Volume 3 – Photon Correlation and Light Beating Spectroscopy
edited by H. Z. Cummins and E. R. Pike

The series is published by an international board of publishers in conjunction with NATO Scientific Affairs Division

A	Life Sciences	Plenum Publishing Corporation
B	Physics	New York and London
C	Mathematical and Physical Sciences	D. Reidel Publishing Company Dordrecht and Boston
D	Behavioral and Social Sciences	Sijthoff International Publishing Company Leiden
E	Applied Sciences	Noordhoff International Publishing Leiden

Superconducting Machines and Devices

Large Systems Applications

Edited by

Simon Foner and Brian B. Schwartz

Francis Bitter National Magnet Laboratory
M. I. T.
Cambridge, Massachusetts

SELKIRK COLLEGE LIBRARY

DISCARDED

OCT 0 6 2021

DATE

PLENUM PRESS • NEW YORK AND LONDON
Published in cooperation with NATO Scientific Affairs Division

Library of Congress Cataloging in Publication Data

NATO Advanced Study Institute, Entreves, Italy, 1973.
  Superconducting machines and devices.

  (The NATO Advanced Study Institutes series (Series B: Physics, v. 1))
  Lectures presented at the NATO Advanced Study Institute, Entreves, Italy, Sept.
5-14, 1973.
  Includes bibliographical references.
  1. Superconductors–Congresses. 2. Superconductivity–Congresses. I. Foner, Simon,
ed. II. Schwartz, Brian B., 1938-   ed. III. Title. IV. Series.
TK7872.S8N18 1973   621.39   74-624
ISBN 0-306-35701-1

The Francis Bitter National Magnet Laboratory is sponsored by
the National Science Foundation

Lectures presented at the NATO Advanced Study Institute, Entreves, Italy,
September 5-14, 1973

© 1974 Plenum Press, New York
A Division of Plenum Publishing Corporation
227 West 17th Street, New York, N.Y. 10011

United Kingdom edition published by Plenum Press, London
A Division of Plenum Publishing Company, Ltd.
4a Lower John Street, London W1R 3PD, England

All rights reserved

No part of this book may be reproduced, stored in a retrieval system, or transmitted,
in any form or by any means, electronic, mechanical, photocopying, microfilming,
recording, or otherwise, without written permission from the Publisher

Printed in the United States of America

SELKIRK COLLEGE
CASTLEGAR. B. C.

PREFACE

This book presents detailed discussions of several of the large scale applications of superconductivity which will have major economic impact on technical developments in the industrial world. The world-wide concern with energy problems makes this work particularly timely. Some of the large scale devices and systems such as superconducting generators, motors, power transmission, large magnets, high speed ground transportation and industrial processing clearly speak directly to improved efficiencies of generation and utilization of energy. The articles treat each subject in depth. The text is suitable for advanced undergraduate or graduate engineering or applied science courses. The text should also be of immediate use to practicing engineers and scientists in applied superconductivity. The unique summaries of national efforts in applied superconductivity will also be valuable to industrial and government planners.

The book is based on a NATO Advanced Study Institute entitled, "Large Scale Applications of Superconductivity and Magnetism" which was held September 5 to 14 in the Hotel des Alpes, Entrèves, Valle d'Aosta, Northern Italy. This Study Institute represented a departure from other NATO Advanced Study Institutes in that it was very strongly directed toward engineering applications rather than purely scientifically oriented interests. The planning of this Institute developed over several years and would not have been possible without continued interest by several key NATO Scientific Affairs Division scientists. It started when one of us (S. F.) met with Dr. H. Arnth-Jensen of the NATO Scientific Affairs Division at a Conference on High Magnetic Fields in Nottingham in 1969. In early 1971, Dr. Arnth-Jensen in several communications asked whether it would be reasonable to consider an Institute which dealt with the problems of applications of high magnetic fields to practical systems, and whether it would be appropriate to develop an international forum on this matter. We were faced with a number of unusual problems concerning this Institute because we recognized that in order for it to be successful, it would require involvement of the industrial engineering community. Furthermore, scheduling would depend on completion of tests of some devices going on-line. After Dr. Arnth-Jensen left NATO, we continued

correspondence with Dr. E. Kovach, and later Dr. T. Kester of the
NATO Scientific Affairs Division. The developments of some of the large
superconducting systems were sufficiently far advanced that at the Applied
Superconductivity Conference in Annapolis, Maryland in May 1972 we received assurances from several of the proposed principal speakers and
members of our International Advisory Committee that this Institute would
be timely and valuable.

The lectures at the Institute provided reviews of the major technical developments in large applied superconductivity programs. In all
cases the lecturers were outstanding scientists and engineers in the fore-
front of their respective fields. The first seven chapters in this book are
review articles devoted to these main lectures. The review articles are
extensive and go beyond the lectures presented at the Institute. These
chapters cover an introduction to applied superconductivity, practical
materials, large magnet systems, dc and ac machines, magnetic levitation and power transmission. We believe that these chapters present an
excellent basis for developing a research knowledge of the field of applied
superconductivity. The next three chapters are shorter and are based on
seminars on the hydrogen economy, superconducting computer devices and
magnetic separation. They supplement the main chapters by focusing on
specific problem areas. The last seven chapters of this book present a
new innovation we introduced at the Institute; a series of reviews of national efforts of applied superconductivity programs of France, Germany,
Italy, Japan, Switzerland, the United Kingdom and the United States. These
give a good up-to-date summary of the worldwide developments in applied
superconductivity and should lead to better national planning as well as to
international cooperation. As with earlier chapters, the national reviews
were presented by technologists in the most favorable positions to know
and write about their own countries. A subject index is included to permit
rapid access to all the material.

The leadoff lecture given by Professor L. N. Cooper on the Theory of
Superconductivity does not appear in this volume. For a good summary
we refer the reader to his Nobel Prize acceptance speech which appeared
in Physics Today 26, 31 (1973). The Institute also included a number of
informal seminars and technology assessment sessions. In order to maintain our very tight publication schedule, we have not included transcrip-
tions of any of the informal seminars on technology assessments unless
they were immediately incorporated in the texts on that subject.

The development of the NATO Institute which resulted in the present volume involved planning over a period of three years. We have
been very fortunate in having an effective advisory committee which

helped us with the planning. Professor J.I. Budnick served as Assistant
to the Directors. Members of the International Advisory Committee
included A.D. Appleton, H. Brechna, B. Birmingham, G. Bogner, P.P.
Craig, G.R. Fox, R.L. Garwin, R.A. Hein, J. Horowitz, J.K. Hulm,
R. Pauthenet, C. Rizzuto, G. Sacerdoti, R. Stevenson and K. Yasukochi.
In addition we had further assistance from R.H. Kropschot, E.M. Purcell
and C.H. Rosner. Throughout the development of the Institute, Dr. Appleton
has been extremely helpful to us. In addition to making many suggestions
at various stages, he also assisted us in looking for a suitable location for
the Institute. We wish to thank Drs. H. Arnth-Jensen, E. Kovach and
T. Kester for their continued interest, and the NATO Science Founcil for
their support of the Advanced Study Institute. We also wish to thank the
General Electric Company, The National Science Foundation, and The
National Research Council of Italy (C.N.R.) for their support.

In addition to the lectures, the NATO Institute had approximately
95 participants from 20 countries. Members of the host country, Italy,
and the Local Chairman, Professor Carlo Rizzuto did many things to
make our Institute a success. We would like to thank Professor Rizzuto
and his associates at the University of Genova who comprised the Local
Committee, for searching out several ideal locations for the Institute and
for their help with all phases of the operation of the Institute.
Mr. Ballabenni and the staff at the Hotel des Alpes helped ensure a well-
run Institute and accommodations. We would like to thank Mr. Turchet of
the Courmayeur Tourist Office, and the Aosta Valley Regional Tourist
Office for help in arrangement of travel plans, evening entertainments,
and two bus tours of the beautiful Aosta Valley and the Italian Alps. We
also were fortunate to have two weeks of exceptionally clear, delightful
weather during the Institute, as predicted by the Local Tourist Bureau.

We received the utmost **cooperation** from all of the lecturers and
students at the Institute and wish to thank them. The manuscripts were
prepared by the time of the Institute and enabled us to meet our tight
publication schedule. We apologize to all the lecturers for having been
so demanding in keeping to our extremely short publication schedule.
Without the continued dedication of each lecturer before, during, and
after the Institute, we would not have succeeded. In addition to our personal thanks, we hope that the Institute and the present volume justifies
their efforts.

We would especially like to thank Edward J. McNiff, Jr., Hernan
C. Praddaude and Richard Frankel for their assistance with the Institute.
We also wish to thank Mary Filoso, Nancy Galvin, Jo Dean Matthews, and
Elisabeth Taylor for typing and correcting the manuscripts. Mary Filoso

contributed to the Institute from our early correspondence to the final stages. Her continued dedication both to the Institute and this book guaranteed success.

Simon Foner
Brian B. Schwartz

Cambridge, Massachusetts
December, 1973

CONTENTS

CHAPTER 1:   LARGE SCALE APPLICATIONS OF
               SUPERCONDUCTIVITY
               J. Powell

CHAPTER 2:   PRACTICAL SUPERCONDUCTING
               MATERIALS
               D. Dew-Hughes

CHAPTER 3:   SUPERCONDUCTING MAGNETS
              H. Brechna

CHAPTER 4:   SUPERCONDUCTING DC MACHINES
               A.D. Appleton

CHAPTER 5:   APPLICATIONS OF SUPERCONDUCTIVITY
               TO AC ROTATING MACHINES
               J.L. Smith, Jr. and T.A. Keim

CHAPTER 6:   HIGH SPEED MAGNETICALLY LEVITATED
                 AND PROPELLED MASS GROUND TRANS-
                 PORTATION
                 Y. Iwasa

CHAPTER 7:   TRANSMISSION OF ELECTRICAL ENERGY
                BY SUPERCONDUCTING CABLES
                    G. Bogner

CHAPTER 8:   THE USE OF HYDROGEN AS AN ENERGY
                CARRIER
                    C. Marchetti

CHAPTER 9:   TUNNEL JUNCTIONS FOR COMPUTER
                APPLICATIONS
                    J. Matisoo

CHAPTER 13:   PROGRAMS ON LARGE SCALE APPLICA-
               TIONS OF SUPERCONDUCTING IN ITALY
               C. Rizzuto

CHAPTER 14:   PROGRAMS ON LARGE SCALE APPLICA-
               TIONS OF SUPERCONDUCTIVITY IN JAPAN
               K. Yasukochi and T. Ogasawara

Chapter 1

LARGE SCALE APPLICATIONS OF SUPERCONDUCTIVITY

J. Powell

Brookhaven National Laboratory

Upton, Long Island, New York

I. INTRODUCTION

Many large scale industrial applications of superconductivity have been proposed. These can be classified in various ways; for example, one could categorize applications in terms of the superconductors characteristic behavior, i.e., Type 1, Type 2, ac, dc, etc. Perhaps the most useful classification, however, is in terms of end use. Table 1 lists the major applications presently visualized as to end use, together with an estimate of the date such applications could make a significant commercial impact. There are, of course, many other possible industrial applications: instruments, computers, electron microscopes, etc. These applications will probably be significantly less important in terms of overall use and economic impact.

An interesting aspect of the applications in Table 1 is that several alternative commercial methods not using superconductivity now commercially exist for each area, and advanced methods, again not using superconductivity, are being developed. Thus, if superconductivity is to be applied in a given area it, and the system it is part of, must offer significantly better performance and/or costs than alternative methods.

Many factors will affect whether or not the above applications ultimately prove feasible. In certain applications, such as some fusion concepts for example, the superconductor requirements are beyond the present state of the art. Thus, whether or not superconductors are used in these applications depends in large part on advances in superconductor technology.

In other areas, such as MHD, superconductor requirements are within present capabilities, but other system components, like the MHD

TABLE 1.   Large Scale Applications of Superconductivity

	Approximate Year for Earliest Commercial Use
Energy Production	
Fusion	2000
MHD	1985
Energy Storage	
Magnetic	1980
Energy Transformation	
AC Generators and Motors	1980
DC Generators and Motors	1980
Transformers	?
Energy Transmission	
DC Cables	1990
AC Cables	1990
Transportation	
High Speed Ground Transport (Passengers and/or Freight)	1980
Space Craft Launch	1980
Industrial Processing	
Separation (Ore, Recycling)	1980
Water Filtration	1980
Effect on Chemical, Metallurgical Reactions	?

duct, are not. If these components cannot be developed, superconductors will not be used for these applications.

In still other areas, like magnetic energy storage, all system components are within present capabilities, but the economic practicality of the application is not certain.

For some applications technical feasibility seems assured, but use may hinge on intangible factors such as esthetics. This will probably be the case with superconducting transmission lines, for example. Underground superconducting transmission lines will always be more expensive than overhead lines and can only compete with underground transmission methods. The decision whether or not to go underground will depend in many cases not on cost, but on how strongly people object to overhead lines. Economic practicality will then hinge on the relative costs of superconducting vs. conventional underground lines.

Finally, in some cases political and social factors may be overriding. Magnetically levitated trains could be in this category, for example. Such a system might turn out to be cheaper and more efficient than air transport, but difficulties in acquiring the necessary right of way, plus lobbying by air and auto transport groups, could prevent it from being built. Conversely, national prestige may dictate that systems be built that are not economically justifiable.

Questions on the reliability and safety of superconducting systems will become much more important for large scale applications. The failure rates will have to be extremely low to protect both lives and property. For example, if magnetically levitated trains are to match the fatality rate of airlines, i.e., one fatality per 10^8 passenger miles, the failure rate of the superconducting lift magnets must be less than one failure per 300,000 hours of vehicle operation at cruising speed. Similar stringent safety requirements will be necessary for fusion. A magnet rupture might damage part of the reactor blanket and disperse some radioactive structural material. Such failures can be made virtually impossible, however, with suitable designs and operating stresses. Many of the applications of superconductivity will involve very large capital investments, hundreds of millions of dollars in a MHD or fusion plant, for example. Even if lives are not at stake, failure rates must be very low to prevent loss of such large investments. In some cases, meeting these very strict requirements of reliability and safety may impose serious technical and/or economic penalities.

These are some of the many factors that will ultimately decide whether or not a given application will be successful. Some factors are quantitative, but are not well known, and some are qualitative and very intangible. Predictions of ultimate success for any given application are thus very subjective. Table 2 attempts to summarize a personal assessment of the chances for success of the various applications, based on technical factors (field capability, degree of losses, stability); economics; problems of other system components; and intangible factors.

TABLE 2. Assessment of Large Scale Superconductor Applications

Application	Capability of Current Superconductors and Cryogenic System to Meet Desired Application			Projected Cost of Superconductor & Crogenic System	Probability of Significant Application
	Field	Losses	Stability		
Fusion (dependent on confinement method)	G to P	G to P	G to P	G to P	VG
MHD	G	G	G	VG	F to G
Magnetic Storage	G	G	G	F to G	F to G
AC Gen. and Motors	G	G	G	G	G
DC Gen. and Motors	G	G	G	G	G
Transformers	G	P	F	P	P
DC Transmission	G	G	G	F	F to G
AC Transmission	G	F	G	F	F to G
High Speed Train	G	F	G	G	G
Ore Separation	G	G	G	G	G

VG - Very Good

G - Good

F - Fair

P - Poor

Perhaps the most intangible factor of all is the attitude of people who will make the decision whether or not to apply superconductivity for a given industrial purpose. Many times the benefits of a new concept are fully realized and appreciated, but there is a hesitation to make major commitments for fear of failure. In this regard, the first major commercial application of superconductivity will have the toughest time. Once decision makers see superconductivity operating successfully on a large scale in industry, subsequent applications will be much easier to sell. This breakthrough could happen in a few years, and probably will be either trains or generators.

II.   ENERGY GENERATION-FUSION

To mankind, the single most important use of superconductivity will probably be for fusion power. While there are alternative long term large scale sources of energy, i.e., nuclear fission, solar, and geothermal, fusion seems to offer the optimum combination of low cost, universal availability, and minimum environmental problems, including essentially no radioactive wastes.

It is not certain of course that fusion reactors will be technologically practical, and that even if practical, that they will need superconductivity. For example, the effort going into the laser-pellet approach to fusion is comparable to that for the magnetic confinement approach. However, the results with magnetically confined plasmas are very encouraging, and the technological problems may be easier than those for laser-pellet systems.

Figure 1 illustrates the magnitude of the power that will be required in the U.S. and the world after 2000 A.D. Commercial fusion reactors are expected to be operational in 2000 A.D., but several decades will elapse before they could supply most of this demand.

The electric demand triples [1] if the U.S. or world converts to a synthetic fuel economy based on electrolytic hydrogen. This conversion may be delayed until 2100 AD by using coal derived fuels but the environmental effects of the very large amounts of coal mining that are required will be severe.

The U.S. electric demand for 2000-2020 is based on detailed projections [2] for a large number of energy consuming sectors, taking into account population increases, etc. The U.S. growth rate after 2020 is taken as 1% per year which is probably low. World demand is taken as a smoothly increasing multiplier of U.S. demand, increasing from a factor of 3 in 2000 AD to 5.5 in 2100 AD. These projections are probably low if anything, but serve to illustrate the magnitude of the application.

To those involved with superconductivity, the most interesting aspect of Fig. 1 is the right hand scale, which shows the cumulative amount of

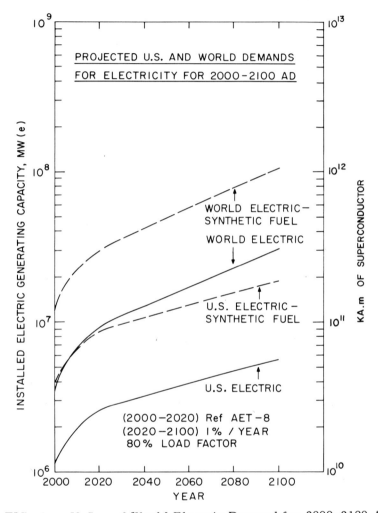

FIG. 1.   U.S. and World Electric Demand for 2000-2100 AD.

superconductor needed in fusion reactors if all power demand is to be met by fusion. This estimate assumes DT tokamaks with B_0 = 5T and first wall fluxes of 1 MW(th)/m^2 and it could be conservative; other reactor types such as stellarators and theta pinches, would require more super- conductor; further, containment may be poorer than expected, or other fuel cycles may be more desirable, which will require higher fields and more superconductor.

Even the conservative estimate is staggering - on the order of 10^{12} kAm by 2100 AD if the world goes to a full fusion economy with synthetic fuel production. This is approximately six orders of magnitude greater than the present annual production of superconductors. Large reductions in superconductor cost can be expected with such increases in production levels. A detailed earlier study [3] found that the future superconductor costs in a fusion economy should be quite low and strongly dominated by material costs; for example, Nb_3Sn conductor should cost only \$0.50/kAm   (1971 dollars) at 10T including stabilizer. The true cost will probably be even lower since no improvements in processing technology and current density were assumed and unit costs for labor, etc. were quite conservative.

On this basis a full fusion economy for the world in 2100 AD with synthetic fuel production requires a total capital investment of approximately 500 billion dollars for fusion reactor superconductors. This tremendous sum should be put in perspective, however. The average annual investment is only \$5 billion/year for the interval 2000-2100 and the total investment is only $\sim 10^{-4}$ of the total world GNP for the same interval. To put it another way, the superconductor in a fusion reactor will only cost about \$5/KW(e), which will be only 1 to 2% of the total plant cost. The total cost of the magnet structure will be approximately an order of magnitude greater than the superconductor cost, i.e., $\sim$ 10 to 20% of the total plant cost.

Thus application to fusion depends more on how well superconductors can meet the technical requirements of fusion reactors than on cost. The reliability, safety, allowable mechanical stress, etc, of superconductors will be much more important than cost, and these factors will determine what superconductor should be used.

There is a further implication for fusion. The DT fuel cycle is generally favored, at least for the first generation of fusion reactors, because of the large cross section, low temperature and high energy for the DT reaction. The required magnetic fields are thus minimized and it appears almost certain that present superconductors, i.e., NbTi and Nb_3Sn, will be adequate. However, use of the DT cycle will raise other technological problems that may delay the commercial application of fusion:

1. Potential severe radiation damage to the blanket structure by 14 MeV neutrons.

2. Necessity of a tritium breeding blanket with lithium as a major component; materials problems may be difficult.

In addition, the 14 MeV neutrons inevitably activate the blanket structure, which could lead to a radioactive waste problem (though this can be minimized by using an aluminum structure [4]), and further, most (~80%) of the DT reaction energy is carried away by the neutrons. The neutron energy ends up as heat and is converted to electricity via a conventional thermal cycle. If one uses a fuel cycle in which the fusion energy is deposited in the hot plasma, however, direct conversion devices may substantially increase overall conversion efficiency.

There are fuel cycles, such as DHe^3 (where He^3 is bred by DD reactions) and pB^{11}, in which almost all of the fusion reaction energy appears as charged particles. Reactors running on these fuels will not have the problems caused by the 14 MeV neutrons from DT fuel, and could be used for direct conversion. However, these fuels require higher magnetic field strengths, on the order of 3 times, than are needed with DT fuel. For example, in tokamaks, the toroidal field coils would see a maximum field of between 20 and 30 Tesla. The superconductor and structure problems would be difficult, of course, but could conceivably present less problems than the development of a practical tritium breeding for a DT reactor. Thus, even if the magnets for DHe^3 or pB^{11} reactors were much more expensive than those for DT reactors, they might be worth it. This possibility should be explored as it not only could lead to a more desirable fusion reactor but perhaps one could get there faster.

Turning now to what magnet problems can be expected for fusion reactors, Table 3 summarizes the principal features of the four main reactor types being considered in the world-tokamaks, stellarators, mirror machines, and θ pinches - for the DT fuel cycle [5, 6, 7, 8, 9, 10, 11, 12].

For purposes of comparison, the four reactor types are normalized to a nominal power output of 1000 MW(e) and a first wall flux of 1 MW(th)/m^2. Other power levels and first wall fluxes could result in more efficient reactors; however, these nominal values are consistent with the desired plant generation level in present power systems, and the neutron fluxes that will probably be allowed for the blanket structure. The other major fusion reactor approach, the laser-pellet, does not require a magnetic field to work, though it might be useful for direct conversion. That is, if the ignited pellet expands in a magnetic field some of its energy could appear directly as electricity without going through a thermal cycle [13]. This could be important for DD and pB^{11} pellets but would not be especially important for DT pellets.

TABLE 3. Comparison of Fusion Power Reactors

Parameters: 1000 MW(e)
1 MW(th)/m^2
40% Thermal Cycle Efficiency

Reactor	Magnet Dimensions	Pulse Length	Duty Cycle	Plasma Pressure (atm)	Maximum Magnetic Pressure (atm)
Tokamak	minor R = 6.6 m major R = 13.8 m	?	Depends on Fueling Mode	5.2	250
Stellarator	minor R = 6.6 m major R = 16.8 m	DC	DC	5	280
θ Pinch	minor R = 0.7 m major R = 210 m	0.1	1%	900	900
Yin-Yang Mirror	Mean Radius of Yin Yang Fan = 17 m	DC	DC	–	1600

In the first type, the tokamak, Fig. 2 [6], the plasma is a fat torus with an aspect ratio (major/minor diameter) between 3 and 4. The plasma is confined by a toroidal field that is established by a set of pancake shaped superconducting toroidal field (TF) coils. Since the toroidal field varies as R^{-1}, where R is the distance from the axis of the toroid, a superconducting turn in the TF coil experiences a maximum field on the inside of the turn and a minimum field on the outside. The ratio of maximum/minimum field is approximately 3/1.

Another set of superconducting coils (OH or vortex coils) is rapidly energized to establish a plasma ring current by transformer action. This current ohmically heats the plasma to 1-2 KeV to aid in its ignition, and also stabilizes it. The plasma ring current produces a poloidal magnetic field which is orthogonal to the toroidal confinement field. At the TF super-conductor, the magnitude of this poloidal field is only a few kG, which is small compared to the maximum toroidal field. However, the poloidal field has a rapid rise rate which may affect the stability of both the TF and OH superconductors. During the initial plasma heating, the OH super-conductor will experience a $\dot{B}$ of $\sim 10^5$ G sec^{-1} and the TF superconductor $\sim 10^4$ G sec^{-1}. There will be a much smaller $\dot{B}$ during the plasma burn, as the changing magnetic field of the vortex coil compensates for the L/R drop of the plasma ring current. The burn could last for a few seconds without refueling and many minutes with refueling. It is not yet established that reflueling will be feasible, and this will have an important effect on the design of tokamak reactors.

In tokamak power reactor designs, the OH coils are usually windings on an iron core transformer which threads the "throat" (i.e., the center) of the torus that is defined by the set of TF coils. The field change is then limited by the properties of the iron; generally, the iron swings from $\sim +2T$ to $\sim -2T$ when the plasma ring current is established. An air core transformer could also be used, though much larger amounts of energy would have to be supplied to establish the current. Also, the leakage fields from the OH coils would be large, causing a much greater $\dot{B}$ at the TF coils.

Additionally, there are vertical field (VF) coils producing a weak vertical field that keeps the plasma ring stationary, and divertor coils that shape the main magnetic field in order to divert the plasma leaking through the confinement field away from the first wall and into a suitable trap. These coils, while very important to reactor operation, are not of prime interest in terms of superconductivity. Accordingly, we do not consider them further.

Tokamak power reactors operating on DT fuel must be large because of their thick blanket and neutron shield (typically 2 meters). Practical TF coils must then be at least 10 meters in diameter, since they enclose the blanket and shield and the plasma, and the smallest practical reactors have major diameters of ~ 20 meters. The DT fuel cycle not

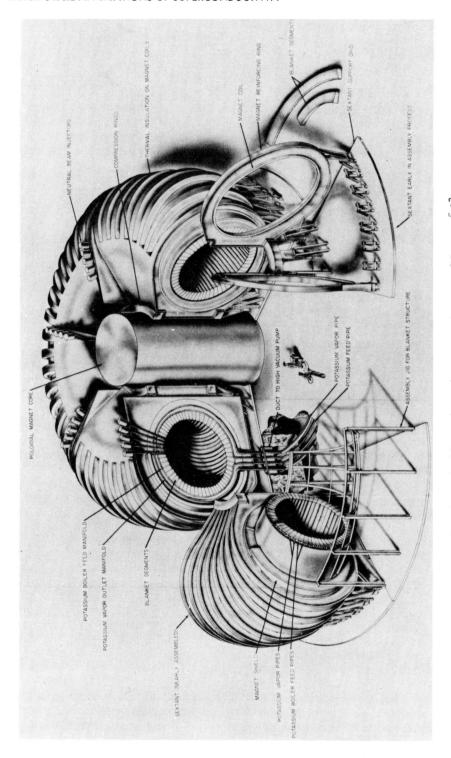

FIG. 2.   ORNL Tokamak Fusion Reactor Magnet [6]

only dictates a large reactor, but considerably increases the maximum magnetic field in the TF coil, due to the 1/R variation of field.

It is not certain what toroidal field strength is needed for a practical reactor. A lower limit can be calculated from the Kruskal-Shafranov condition ($q = 1$, $\beta_\theta = A$) but in present experiments confinement fields are typically 4 to 5 times greater than the theoretical minimum. Thus a practical reactor with the conditions of Table 3 would require a B_0 (field on magnetic axis of the plasma) of $\sim 3.6T$, and the maximum field in the TF superconductor would be ~ 7.5 T.

The large size and strong field yield a very large magnetic stored energy in the TF coils for the reactor of Table 3, amounting to approximately 80,000 MJ.

Even from this drastically simplified description of the magnet components of a tokamak power reactor, it is clear that the magnet system will be quite complex. It is also clear that many details of the system are not yet well defined, and must await more experimental results, as for example, the plasma confinement field strength. This complexity and lack of definition are generally true for all fusion reactor magnet systems.

However, we can roughly sketch out what the superconductor requirements will probably be and from this determine whether the current state of the art seems adequate. If it does not, we can determine what improvements are necessary.

This assessment must reflect the realities of operating power plants. Like other plant components, the superconducting magnets must meet certain criteria of reliability and safety. There are three areas of concern with regard to magnets:

1. Quench of the magnet without damage but with a subsequent cooldown to re-establish superconductivity.

2. Quench of the magnet with sufficient damage that it must be replaced.

3. Destruction of the magnet structure, with consequent damage to blanket and reactor containment building.

The first type of occurrence would keep the fusion reactor turned off for several days after the quench while the magnet was cooled down. An operating power plant could afford a quench of this type about once a year. The second type of occurrence would require a shutdown of many weeks while the damaged magnet was replaced. A power plant could probably afford an occurrence of this type once during its lifetime, i.e., once per 30 years. The third type of occurrence could be potentially serious. If the magnet structure breaks, it seems possible that the blanket may be partially destroyed, and some portions of radioactive blanket materials could be scattered outside the containment building. An occurrence of this type should be extremely rare, i.e., less than once per 10,000 years of reactor operation. This not only affects the choice of superconductor, but also limits the allowable stress in the magnet structure.

The superconductor requirements for the four fusion reactor types are summarized in Table 4. These are estimated assuming a DT fuel cycle and a first wall flux of 1 MW(th)/m^2. The tokamak TF coils seem well within present commercial superconductor capabilities. The maximum field appears to be on the order of 8 T, so that multi-filament NbTi can be used. If the TF windings are shielded with copper or better, high purity aluminum, the rate of field change during the OH pulse is reduced to a few kG/sec. Cryogenic stability appears mandatory for the TF coils, because of the very large stored energy and the need for an extremely safe system.

The OH windings see a relatively low maximum field, $\sim$2T if an iron core is used, but the maximum rate of field change is much greater, though probably achievable with conductors developed for pulsed synchrotrons, which have operated successfully at $\sim$10 T/sec. At present there are three possible approaches: a flat transposed braid of small diameter ($\sim$8 mil) multifilament NbTi wire, which is impregnated with an indium-thallium metallic filling; a transposed cable of bundles of small diameter multifilament NbTi wires, made rigid with Staybrite solder; and a mechanically tight winding of commercial large size ($\sim$100 mil) rectangular multifilament NbTi conductor wound on grooved high purity aluminum sheets, which act as heat drains to remove hysteretic heat developed in the superconductor. The latter has somewhat greater heating, but is less sensitive to mechanical movement of the conductor, since the heat removal capability is greater. The conductors as developed are not cryogenically stabilized, since current density must be very high for synchrotron application. Losses are small enough to be acceptable for OH coils. More stabilizer could probably be used in OH coils, since current density is not limiting.

Like the tokamak, the stellarator is a low β toroidal reactor. The TF coils will be similar to those in the tokamak. Unlike the tokamak, stability is achieved with a superconducting multipole (typically sextopole) helical winding that twists around the torus. Stellarators do not have an OH pulse and the superconductor sees a constant dc field. Also, stellarators require a higher maximum field than tokamaks, since helical windings are placed between the blanket/shield region and the TF coil and the R^{-1} effect increases maximum field. NbTi will probably be adequate but may have to operate below 4.2 K to get the required field capability. Construction of a large helical winding inside the TF coils will be very difficult mechanically. Cryogenic stabilization appears mandatory for both coils.

TABLE 4.   Superconductor Requirements for Fusion Reactors

Fusion Reactor	Superconductor					Status
	Stable	Form	Material	(B)max	(dB/dT)max	
Tokamak						
TF Coils	Cryogenic	Multifilament	NbTi	~8T	~0.3T sec^{-1}	Commercial
OH Coils	?	Multifilament	NbTi	~2T	~100T sec^{-1}	Development (a),(b),(c)
Stellarator						
TF Coils	Cryogenic	Multifilament	NbTi	~10T	0	Commercial (~2 K operation)
Helical Wind.	Ditto	Ditto	NbTi	~10T	0	Ditto
θ Pinch						
Storage Coils	?	?	NbTi	~6T	~1000T sec^{-1}	Research
Mirror						
Minimum B	Cryogenic	Multifilament	Nb$_3$Sn	~15T	0	Development

(a) McInturff, A. D., "Superconductors for Pulsed Magnets," p. 395 in Proc. 1972 App. Supercond. Conf., Annapolis, Md. (1972).

(b) Gilbert, W., et al., "Coupling in Superconducting Braids and Cables," p. 486, ibid.

(c) Allinger, J., et al., "A Superconducting 8° Bending Magnet System," p. 293, ibid.

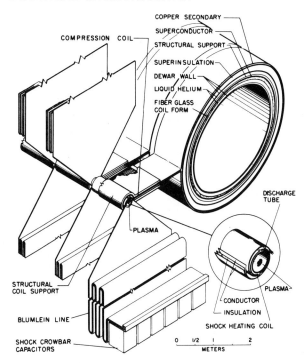

FIG. 3.   LASL θ Pinch Fusion Reactor Magnet [9].

The θ pinch reactor, Fig. 3, uses a room temperature coil to mag-netically compress a shock-heated plasma in ~10 milliseconds to the ignition temperature. The plasma burn continues for ~100 milliseconds, and the reactor then shuts off for ~10 seconds. The magnetic compres-sion energy comes from a set of rotating superconducting coils during the compression, remains in the compression coil until the burn is com-plete, and then is returned to the superconducting coils. Makeup energy is supplied to compensate for the joule losses in the room temperature compression coil. The prime problem with the magnetic energy storage is not the field, which is well within NbTi capability, but in the very low loss allowed per cycle in the superconducting storage. Approximately 3 to 4 parts in 10^5 of the energy in superconducting storage coil could appear as heat. Much more, and the overall plant efficiency would be unacceptably lowered. It may be difficult to meet this requirement with a stable superconductor pulsed at ~1000 kG/sec.

The mirror reactors, Fig. 4, are expected to be high β devices, so that the confinement field would normally be quite low, on the order of 5T, well within the capability of NbTi. Mirror reactors are power amplifiers, however, and depend critically on the leak rate through the mirrors. It is probable that the mirror ratio will be pushed as high as possible to reduce leak rate, which scales as the log of mirror ratio. Thus, the mirror will probably require Nb_3Sn conductor - again, cryogen-ically stable for safety.

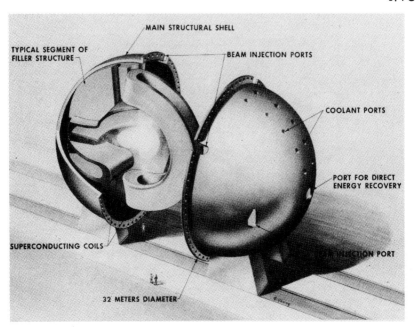

FIG. 4.   LLL Mirror Fusion Reactor Magnet [12].

We turn now to the problems of magnet structure. There are two basic ways to restrain the large magnetic forces on the superconductor - cold reinforcement and warm reinforcement.

On the conventional cold reinforcement approach the superconductor is imbedded in a massive cold (~4 K) steel structure, which counters all magnetic forces. The only force carried to the outside warm world is the weight of the structure. The structure is made massive enough that the maximum stress in it is within acceptable limits. Ordinary steels, though strong, are brittle at 4 K. It is very doubtful that large brittle magnet structures are safe (they have not been used for bubble chamber magnets, for example), so that a ductile, relatively expensive stainless steel (e.g., 304, 21-6-9) is required.

With cold reinforcement, the major part of the fusion reactor magnet costs appears to be the cost of the stainless steel. Other cost elements like the superconductor and magnet assembly costs appear to be a small part (20-30%) of the magnet cost. Various designs of fusion magnets have attempted to minimize the amount of stainless steel by allowing very large structural stresses (e.g., 100,000 psi in some designs) or by shaping the magnet coils to reduce stresses due to bending moments resulting from non-uniform fields. Files [14] D shaped TF coil for the 1/R field in tokamaks is an example of this approach.

There is as yet no clear idea of what limit on structural stress will be allowed for fusion magnet structures. Such a determination must await detailed analytical and experimental studies. The reliability and safety requirements for fusion reactor magnets will be much more stringent than these for any magnets built to date, because of the radio-active hazards and large plant investment. My feeling is that the allow-able maximum structural stress in fusion reactor magnets will probably be in the range of 20-40 kpsi. By way of comparison, the operating stress in PWR reactor pressure vessel walls is on the order of 20 kpsi.

Moreover, the allowable operating stress in magnet structures may well be lower   for brittle superconductors than for ductile super-conductors. Ductile superconductors (e.g., NbTi) are only useful to ~8 Tesla, since current density rapidly drops in higher fields, so that brittle superconductors (e.g., Nb_3Sn) must be used to generate fields >8 Tesla. Brittle superconductors fail in tension and it has only been possible to use them by bonding a stiff ductile reinforcement (e.g., stainless steel) to the superconductor. The steel thermally contracts more than the superconductor, putting the latter in compression. Laboratory magnets have successfully operated with such conductors, and short samples sustain up to ~100 kpsi before failure.

Large engineering structures inevitably have stress concentration points and large fusion magnets should be no exception. Such stress risers are not dangerous with ductile materials, since the material then locally deforms plastically and distributes its load to the adjacent structure Such local deformation would cause a brittle conductor like Nb3Sn to fail, however. As a result, the safety factor that the magnet designer applies may have to be substantially greater with Nb3Sn than with NbTi. Nuclear safety codes generally specify a safety factor of about 2 for primary stresses; that is, the design stress is 50% of the uniaxial failure stress. With a brittle superconductor, a safety factor of 4 may well be required, which would limit design stress in the magnet to ~25 kpsi.

With cold reinforcement, such low design stresses lead to very expensive magnet structures. If one abandons the idea that all magnetic forces have to be contained by a structure operating at 4K, however, magnet structure costs can be lowered by factors of 3-4, and important operational advantages are gained. An illustration of this concept, which I term "warm reinforcement" [15], is shown in Fig. 5 for a tokamak. The magnetic forces on the superconductor are carried directly to a room temperature structure through a high compressive strength, low thermal conductivity thermal barrier. This thermal barrier is a layered structure of epoxy fiberglass and stainless steel sheets, wrapped around the superconducting coil, and completely sup-porting it on all points.

A schematic cross section of the magnet winding and thermal barrier is shown in Fig. 6. The two intermediate heat sinks at 20K and 77K reduce refrigeration cost. The barrier thickness is typically

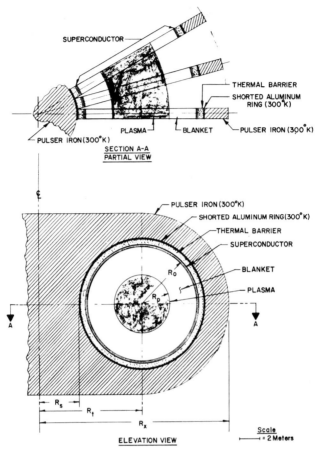

FIG. 5.   Tokamak Fusion Reactor Magnet with Warm Reinforcement

0.5 meter, while the extra cost (barrier materials, refrigerator, and capitalized operating cost) for a 1 MW(th)/m^2 fusion reactor is about \$5/KW(e). The savings in magnet structure cost are much greater than this extra cost, and range up to \$100/KW(e), depending on field and magnet diameter. Warm reinforcement only pays for large magnets with low surface to volume ratios, since refrigeration costs scale with surface area, while structure costs scale with volume. Also, low refrigeration unit costs are necessary, and these can be achieved for large refrigeration loads since refrigerator cost typically scales as the 0.6 power of capacity.

Besides overall cost savings and low structural stress, warm reinforcement has operational advantages. First, aluminum windings can be incorporated in the warm reinforcement structure, so that magnetic energy can be inductively transferred and deposited at room temperature if a superconductor quench occurs. Besides being safer, this allows much quicker magnet recooling after a quench. Second, in the case of tokamaks, the warm reinforcement structure can be the pulser steel used to carry the flux change that establishes the plasma ring current. In this case, the cost of the reinforcement structure is zero, since the pulser steel is required anyway; in addition, more flux change can be carried in the throat of the torus, since much more pulser steel is available. This will permit much larger tokamak pulse times, up to several hours. Third, the warm reinforcement magnet structure can be designed so that magnet rupture is virtually impossible.

With warm reinforcement it should be economically practical to build magnets for DD and pB^{11} fuel cycles, which would greatly reduce radio-active inventory in the reactor, significantly increase its electrical con-version efficiency, and permit a much simpler blanket. Magnets for such fuel cycles will certainly require Nb_3Sn conductor, and perhaps still higher field conductors.

Further reductions in magnet cost are possible if high strength rock, e.g., granites, gneisses and competent limestones and sandstones, are used for the warm reinforcement structure, instead of ordinary steel. This type of construction is described in the section on magnetic energy storage. Such underground siting provides additional containment and safety in the event of accidents.

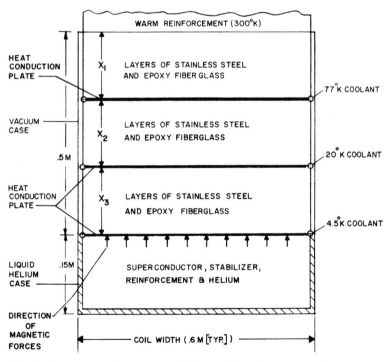

FIG. 6.   Cross Section of Warm Reinforcement Magnet

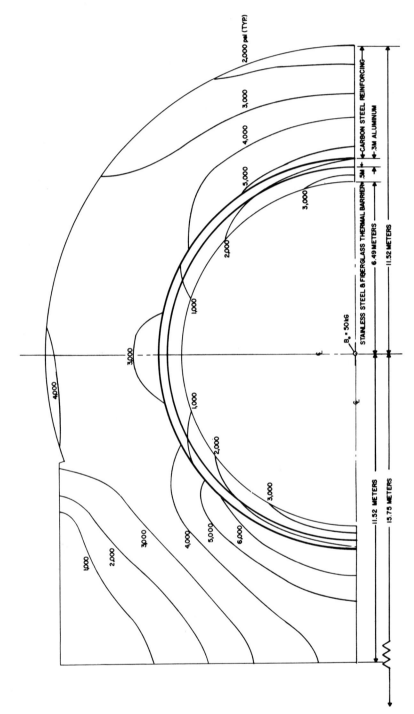

FIG. 7.   Stresses in Warm Reinforcement Magnet Structure

The magnitude of stresses in a fusion magnet structure can vary quite widely if non-uniform magnetic confinement fields are used. Figure 7 shows stress contours in a tokamak magnet with warm reinforcement. The bore of the TF coils is 11 meters, and the on-axis magnetic field is 5 Tesla. Note the wide variation in stress, but also note that the maximum stress is only 6 kpsi. This structure could sustain on-axis fields well over 10 Tesla before reaching magnet stress limits. This magnet was analyzed with a 2-dimensional finite element (2400 elements) elastic stress code.

Results from recent plasma physics experiments, notably ATC at Princeton and ORMAK at ORNL, have been very encouraging. Reactor grade plasma densities have been achieved in a tokamak configuration, and the collisionless regime has been entered with no evidence of enhanced diffusion [16]. These favorable results have helped to accelerate U.S. fusion program goals. It appears that there will be at least one large DT experimental burner operating by 1980 - certainly a tokamak, and very possibly a θ pinch and mirror burner as well. These burners would generate multi-megawatts of fusion power, but would not have tritium breeding blankets.

There is a good chance that scientific feasibility will be demonstrated before these DT burners operate. The Princeton PLT-1 experiment is scheduled to operate in 1976, and it may well achieve the Lawson condition, though it will not burn tritium. The estimated date for a demonstration fusion power reactor has been 2000 AD, but favorable results in the next few years, combined with an accelerated program, could move this date up to the early 1990's.

Fusion researchers have not tended to use superconductivity in plasma experiments except where there was no other choice. There has been good reason for this - generally, the plasma experiment could be done quicker and cheaper if the researcher did not have to add the extra complication of superconductivity. In experiments like Levitron and FM-1, the superconductor was necessary because the floating current ring had to remain suspended for many hours without contact. Normal conductors could not be used for this purpose. In mirror experiments like Baseball II, IMP, and Bumpy Torus, normal conductors would in principle be used, but the extremely high current densities that are required, as well as the high duty factors, make them undesirable. With the exception of some early stability problems in IMP, the superconducting part of these experiments has worked well.

It is not certain that the DT burner experiments scheduled to operate by 1980 will use superconducting magnets. For example, it is surprising to find out that difference in estimated cost between a conventional and a superconducting magnet system for a DT burner tokamak is a relatively small part of the total cost of the experiment, even though the magnet is quite large (typically, 4 meter bore TF coils, with an axis magnetic field of 5 Tesla). This is a result of the low duty factor for such experiments, which will be on the order of a few percent. Since cost and technical factors do not dictate a superconducting magnet, there seems to be a good chance that the conservative conventional magnet approach will be taken. A similar situation exists for the θ pinch and mirror feasibility experiments, though here the case for superconductors is somewhat stronger.

If superconducting magnets are not used for the DT burner experiments, they will be in the following generation of experimental power reactors which should operate in the mid 1980's. Here the operational duty factors will be too high for conventional magnets to be practical.

Because of the concentration on establishing plasma confinement feasibility, work on superconducting technology in the fusion program has been small. This situation should change rapidly in the next few years as confinement feasibility becomes firmly established. Among the relatively few current efforts in superconducting technology for fusion are Henning's [17] development of a cryogenically stabilized Nb_3Sn tape conductor and File's [18] projected test of a 1.0 x 1.3 meter "D" shaped superconducting Nb_3Sn coil. Henning's conductor uses high purity aluminum for the stabilizer, plus high purity nickel as a reinforcement. The combination is very appealing, since aluminum has a much higher electrical conductivity than copper, and the high purity nickel reinforcement allows the direct face cooling of wide conductors by helium. Wide conductors greatly ease problems of magnet winding and assembly.

III. ENERGY GENERATION - MHD

In contrast to fusion, where magnets constitute a major part of the technological development necessary before commercial feasibility, in MHD magnet development is a relatively minor part. Present commercial superconductors are sufficient for MHD power plants though some development work is necessary on the particular magnet shapes that will be required.

The potential capital investment in superconducting MHD magnets is respectable, but small compared to the potential demand for fusion or magnetic storage magnets. To give an idea of scale, if the 10^6 MW(e) installed capacity projected for the U.S. in 1990 were to be entirely coal fired MHD power plants of 1000 MW(e) unit capacity the total capital investment in superconductor would be only ∼1 billion (10^9) dollars at projected superconductor costs. The total magnet investment costs would be several times this value. This situation could dramatically change, however. MHD researchers have tended to build cycles around magnets that are within the current state of the art, i.e., 6 Tesla, and have not investigated cycles using much higher strength magnets, i.e., 20 Tesla. There may be significant advantages in other MHD problem areas if higher field strengths are used. In this event, MHD would be a much larger market for superconducting technology.

Table 5 lists the various MHD concepts now being actively pursued. The open cycle approach claims the most money and attention, while closed cycle runs a distant second, and liquid metal MHD a very poor third. Open cycle is aimed entirely at fossil fuel plants, closed cycle at nuclear reactors, and liquid metal MHD at both.

Open cycle MHD must operate at high temperatures with equilibrium ionization because of the large collision cross sections and energy loss factors for electrons in combustion gases. This further requires that open cycle MHD be used as a topping cycle for a conventional steam plant, if decent thermal efficiencies are to be obtained, and the MHD part of the combined cycle is only about half of the total.

Closed cycle MHD with equilibrium ionization requires nuclear reactors operating at 2000C and above. Such reactors are under development for space propulsion. Some progress has been made, notably the Rover tests, but a high temperature reactor capable of long term reliable operation is still very far away. Gas core and rotating bed reactors seem particularly attractive for MHD cycles, operating at $\sim$ 6000C and $\sim$ 2700C, respectively. Optimists believe that such reactors would be allowed on earth, but this seems unlikely. The hazards of such a power system would be very great since most fission products would be released into the gaseous working fluid. An interesting, far-out alternative avoids this problem by locating the MHD plant in synchronous orbit [19], where the fission products could never return to earth. The output power would be beamed back to earth by microwaves, much as Glaser has proposed [20], with his solar array in synchronous orbit concept. With MHD generation, however, the radiator area would be much smaller than that of the solar cell array. The total weight in orbit would also be reduced.

Non-equilibrium closed cycle MHD has received more attention than equilibrium ionization closed cycles, however, because the reactor temperature can then be kept low enough that fission products are not released from the fuel. HTGR's have thermal efficiencies of $\sim$40%, and future systems employing advanced gas turbines could probably reach 45% thermal efficiency.

A non-equilibrium MHD topping cycle combined with a gas turbine or steam cycle could raise the overall thermal efficiency of nuclear reactors to 50-55%. This could sustantially reduce power generation costs and ease problems of thermal pollution.

Most work on non-equilibrium MHD has concentrated on electric discharge heating of cesium seeded noble gases (argon and helium). The ohmic heating losses in the MHD duct are sufficient to heat the electrons to a temperature much hotter than the kinetic temperature of the gas atoms. Since energy transfer processes between the electrons and gas atoms are quite inefficient (typically one part in 10^4 of the energy difference is transferred in a collision) and some collision cross sections are small,

TABLE 5. MHD Power Generation Cycles

MHD Type	Application	Temperature	Working Fluid Type	Thermal Eff.	Notes
1. Open Cycle Plasma					
a. Equilibrium Ionization	Central Station	>2000°C	Seeded Combustion Gases	50-60%	Combined MHD-Steam Cycle
2. Closed Cycle Plasma					
a. Equilibrium Ionization	Space (Central Station?)	>2000°C	Seeded H_2 or He	30% 50%	MHD; only in space MHD Steam; Terrestrial
b. Non-Equilibrium Ionization	Central Station	>800°C	Seeded A or He	50%	MHD-Gas Turbine or MHD-Steam
3. Liquid Metal					
a. 2-Phase Nozzle-Liquid Flow Separation	Space	>1000°C	Liquid Li	7%	MHD only
b. 2-Phase Expansion in MHD Duct	Central Station	>600°C	Liquid K and He	50%	Combined MHD-Steam Cycle

particularly for argon, only a modest amount of electron heating by the induced electric field in the duct is necessary. Ionization equilibrium is governed by electron temperature rather than gas temperature, with the former typically at 3000K or above, and the latter typically 1500K or lower. The process works very well; unfortunately, in an MHD duct it works too well. Electrothermal instabilities develop in the discharge duct and the $(\vec{v}\mathrm{x}\vec{B})$ field is short circuited in turbulent eddies in the gas, instead of appearing at the electrodes at the sides of the duct. As yet there seems no way to overcome this problem. One possible solution may be to use a low seed concentration, fully ionized everywhere in the duct. If this can be carried out, then instabilities should not grow. However, this has not yet proved feasible.

Because of this problem there has been substantial work on non-electric methods of non-equilibrium ionization in MHD ducts, involving afterglows, radiation induced ionization, metastable nitrogen, electron beams, UV, etc. However, all such concepts have foundered on the very rapid recombination rates occurring at the electron densities needed for practical generators at fields of 5-10 Tesla. This either means there are not enough electrons left before the gas reaches the end of the duct, or that impractically high ionization powers are necessary. With high field generators, however, these alternate methods look much more feasible. This will be discussed in more detail later.

Liquid metal MHD cycles were originally conceived as a source of power for space systems, operating with a nuclear reactor as a heat source. System weight and size were the most important parameters for this application, not efficiency (though this affected radiator weight) or economy. The 2 phase nozzle concept, with separation of the accelerated liquid before decelerating it in a traveling wave MHD generator is a good example of this approach. Thermal cycle efficiency is low, $\sim 7\%$, fluid friction losses high, and generator problems difficult. A better cycle for central station power appears to be the direct expansion of a 2 phase mixture of a liquid metal and a non-condensing gas(e.g., He) directly through a dc MHD generator. By staging generators (in effect MHD topping, bottoming and middle cycles) very respectable overall thermal efficiencies are projected, on the order of 50% at source temperatures of 2000F [21].

Because of the high electrical conductivities of liquid metals, the MHD ducts must be rather small and the magnetic field low. A typical duct would be 1 meter wide, 0.15 meter high, several meters long, operating at ~ 1 Tesla, and generating about 20 MW(e). A large number of these would have to parallel for central station generation. Most likely this would involve running the number of channels required in one or two large air core superconducting magnet for each generator. At the very low fields contemplated, such a magnet would be trivial, both technically and economically.

Returning to open cycle coal fired MHD, Table 6 shows the principal developments required for commercial feasibility. Each of these developments is more, and in some cases much more, difficult than the magnet development. A successful magnet seems virtually certain, but success in these other areas seems much less so.

Electrodes and insulating duct walls are probably the principal problems of open cycle MHD. Figure 8 [22] shows an exploded view of an MHD duct. There are two components of electric field - across the duct (the $\vec{v} \times \vec{B}$ field) and along the duct (the Hall field), with the latter typically 3 to 4 times larger. Electrodes along the duct must be segmented into hundreds of separate units and insulated from each other. The generator can be connected as a Faraday (separate loads to each electrode), diagonal (successive electrodes along the channel at the Hall angle connected in series to a single load), or Hall (separate side electrodes shorted, and end electrodes connected to a single load) machine. It may prove extremely difficult to maintain insulator integrity between electrodes in the presence of high temperatures, electrically conducted slag layers, and high electric fields. Similar problems arise with the duct insulating walls.

Turning now to the more tractable magnet problem, Table 7 gives the important parameters for a 6 Tesla magnet for a 1000 MW(e) commercial MHD plant, as designed by the British CEGB [23]. The size, field, and saddle coil configuration are typical of MHD magnets for central station generation. In his very useful parametric survey of MHD magnets, Stekly [24] describes a variety of winding types, including rectangular, parallel circles, etc. The amount of superconductor required is essentially the same for the various configurations, with the exception of racetrack coils which need twice as much superconductor.

Approximately 1.5×10^6 kA m of superconductor is needed for the magnet. At the superconductor costs projected for a large fusion reactor economy, the superconductor material would cost approximately one million dollars.

Conductors for such MHD magnets will probably be cryogenically stabilized, twisted multifilament NbTi. The overall current density in the windings for the CEGB design is only 2 KA/cm^2, and could be considerably less if desired. The conductor used in the NAL magnet would be satisfactory for such MHD magnets, though more advanced conductors will undoubtedly be developed. Aluminum stabilized conductors could provide even better cryogenic stabilization.

The conductor is twisted at the cross over joints of a saddle shaped coil. However, for such MHD magnets the twist is gentle because of the large scale. The conductor in the Baseball II magnet had a much more severe twist and it performed very satisfactorily.

Field homogeneity is not a problem; variations of several percent in field strength in the duct seem acceptable. In fact, depending on MHD

TABLE 6.   Developments for Coal Fired Open-Cycle MHD

Developmental Area	Problems	Status
1. Combustor	Minimize: slag to duct, seed loss in reject slag	Probably solvable
2. Electrodes	Operate at 1500 C with slag - good electrical emission & conductivity	Difficult Not solved yet
3. Insulating walls in duct	Good insulation strength at 1500 C in presence of slag	Very difficult - Water cooled walls may be necessary
4. Air preheater	Preheater temp. >1100 C exposed to slag, seed, etc.	Difficult Only short term tests
5. Seed recovery	Seed must be leached from slag	Solvable in principal Economics?
6. NO_x in stack gases	1000 ppm NO_x in MHD exhaust 100 ppm emission limit	Probably solvable Catalytic decomposition in preheaters?

TABLE 7.   Dimensions of a Typical Large Magnet*

MHD generator cross-sectional area at exit	6·5 m²
Field on axis	6 T
Spacing between windings	4 m
Winding height	4 m
Winding thickness	0·5 m
Length between cross-overs	12 m
Length of winding overall	16 m
Ampere turns	4×10^7
Stored energy	10^{10} J
Average current density in windings (j)	20 MA m⁻²
Conductor current	5000 A
Mean current density in superconductor (7·2 T mean field)	700 MA m⁻²
Mean resistivity of copper	3×10^{-10} Ω m
Current density in copper	28 MA m⁻²
Mean ratio by vol. copper : niobium titanium	12:1
Rapid shut down voltage	10 kV
Cooling fraction $k = P/\sqrt{A}$	2
Cooling heat flux	3 kW m⁻²
Upper temperature on rapid shut-down	300°K
Coil inductance	800 H
Protection resistor	2 Ω
Time constant for shut-down	400 s
Maximum field inside windings	7·6 T
Maximum field in cross-pieces	7·8 T
Maximum field in corners	~9 T
Maximum field gradients on axis	2·7 T m⁻¹
Maximum field gradient near cross-pieces	4·5 T m⁻¹
Maximum field gradient in windings	10 T m⁻¹
Stress allowed in support structure (aluminium alloy)	190 MN m⁻²
Mass of support structure. Side beams (aluminium alloy)	600 tonnes
Cross ties (stainless steel)	500 tonnes
Mass of winding	270 tonnes
Refrigerator	
Steady-state heat inleak	200 W
Total refrigerator load when energized	500 W
Mass of nitrogen used to cool to 80°K	500 tonnes
Liquid helium required to fill	50 m³

* This magnet was designed for a constant velocity MHD duct.

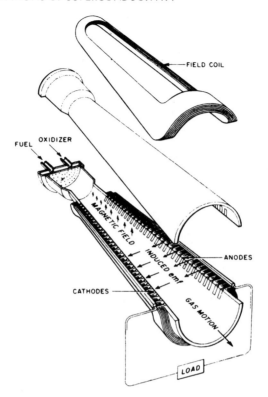

FIG. 8.   Schematic View of MHD Channel [22].

duct design it may be desirable to have a large variation of field along the duct.

A rectangular type of magnet winding seems preferable from a constructional viewpoint than a curved winding, e.g., parallel circles. The principle reason for curved windings is the reduction in amount of superconductor. However, since the magnet fabrication and structure costs are expected to be much larger than the cost of superconductor, it is likely that the overall magnet cost would be less with a rectangular winding. The CEGB estimate for the magnet structure is ~$5 x 10^6.

Relatively few experimental superconducting MHD magnets have been built, principally because the MHD generation experiments have either been of small size and did not need superconducting magnets, or of short enough duration that conventional conductors could do the job. For example, the Mark V generator (Fig. 9) tested by AVCO produced 32 MW(e) for a few seconds, using a self-excited conventional magnet [25].

In 1966 AVCO [22] successfully tested a large superconducting MHD magnet (Fig. 10), intending to demonstrate that such magnets were feasible. The magnet did not have a warm bore. It was 3 meters long, 30 cm bore, with a central field of 4 Tesla, and used NbZr wire.

FIG. 9.   Mark V MHD Generator [22].

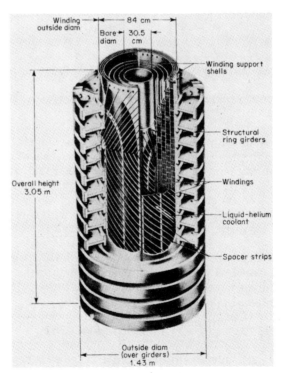

FIG. 10.   View of Large Superconducting MHD Magnet [22].

A 4 Tesla NbTi magnet has been constructed in West Germany at the Institute fur Technische Physik for experiments in connection with the closed cycle plasma loop "Argas" [26].

AVCO is designing a 5 Tesla magnet for a Mark VI test generator [1 MW(e)] which would hold an MHD duct 1 1/2 meters long, ~0.3 meters high at the entrance, 0.5 meters at the exit [27]. Stanford is building a slightly smaller 8 Tesla MHD magnet.

As experiments move from the phase of demonstrating MHD to long term tests of prototype power plants, it is very likely that many more large superconducting magnets will be built.

MHD researchers have concentrated on cycles using superconducting magnets within the present state of the art (~5 Tesla). Table 8 illustrates some of the potential benefits gained if MHD generators operate at much higher fields, e.g., 20 Tesla. The effect on the various operating parameters is considered separately; that is, electrical conductivity cannot be simultaneously reduced by a factor of 16 while operating pressure is increased by a factor of 16.

Some of the possible parametric changes are of no value or even detrimental; for example, a simple decrease in interaction length of a linear generator coupled with the very high field strength would seriously increase end losses. Some changes are of relatively slight value; for example, reducing gas velocity results in less heat loss to the MHD duct walls, but the savings are not very significant for large central station plants. Some changes are of medium value, such as a lower MHD exhaust temperature. This could increase thermal cycle efficiency. Whether the savings in capital and fuel costs would exceed the increased magnet costs requires detailed examination. An increase in thermal efficiency from 52% to 60%, which may be achievable, would justify an increase of $30/KW(e) in capital cost of the MHD magnet. In balance, it would probably pay to do this. More important would be go - no go effects on the feasibility of open cycle MHD. The only way in which this would occur seems to be seed recovery. In coal fired MHD plants it probably will be necessary to leach out seed from slag that blows through the duct. This has been done on a small laboratory scale and will probably be feasible for commercial plants. If it is not feasible, however, or is prohibitively expensive, high field MHD generators would eliminate the requirement for seed recovery.

The Hall parameter, $w\tau$, would be increased by a factor of about four in a high field generator to a value of approximately 12-16. This permits efficient operation in the Hall mode. Electrode construction is simpler and duct walls can sustain much higher electric fields. Linear Hall generators have not been very successful due to shorting and current concentration. Disc Hall generators have operated very successfully,

TABLE 8.   Potential Benefits of High Field Strength (20T)

MHD GENERATORS

Parameter	Change in Parameter (X/X_o) $[X_o$ = value for 5 Tesla Generator$]$	Benefit
Open Cycle		
Interaction Length	$(16)^{-1}$	None - Serious End Effects
Gas Velocity	$(16)^{-1}$	Slight - Less Heat Loss to Duct
Electrical Conductivity	$(16)^{-1}$	
a) Seed Concentration	$(16)^{-2}$	High - Eliminate Seed Recovery for Coal Fueled MHD
b) MHD Exhaust Temperature	Reduced 200°K	Medium - Higher Thermal Efficiency
Hall Parameter ($w\tau$)	$(4)^{1}$	Medium - Operation as Hall Generator
Disc Generator	Simpler Magnet Geometry, Simpler Electrode Geometry, Higher Electric Field Capability	Medium
Operating Pressure	$(16)^{1}$	Medium - Higher Thermal Efficiency
Closed Cycle		
Electrical Conductivity	a) No Non-equilibrium Electrical Ionization (External Ionization)	Very High - Stable Discharge
	b) Eliminate Cs Seed	

however [28]. A serious disadvantage of the disc generator at low fields is that minimum plant sizes tend to be very large, on the order of several thousand MW(e), if air is used for combustion and an efficient cycle is demanded. With high fields the disc generator could be quite compact. The magnet design would also be simpler, i.e., essentially a solenoidal field with the MHD flow radially out from the solenoid's axis.

High field generators would probably be of the greatest benefit to closed cycle non-equilibrium MHD. This could be very important since the overall thermal efficiency of high temperature gas cooled nuclear reactors could be increased from 40% to 55-60%. The much lower electrical conductivity possible with high fields permit one to change from the electric discharge method of non-equilibrium ionization which seems to be inherently too unstable for a practical generator, to an afterglow or radiation induced method of non-equilibrium ionization. Since the ionizing mechanism is independent of electric field, there should be no problems with generator stability. Another advantage is that cesium seeding would not be necessary. This would ease materials compatibility problems, particularly if graphite were part of the high temperature nuclear reactor.

Considering the afterglow mechanism first, Braun [29] has investigated the feasibility of operating generators on He afterglow. The He is pre-ionized just before the inlet and persists for several meters into the generator. Braun's cycle is marginal with a 10 Tesla generator. Only the first 3 meters of the generator are effective ($M = 2$, an upper limit) and the thermal efficiency of the MHD part of the combined cycle is only 10-15% , depending on pre-ionizational level. With 20 Teslas the thermal efficiency of the MHD part of the cycle could be raised to $\sim 30\%$, resulting in an overall plant efficiency of almost 60%. With a 20 Tesla field, the generator could be somewhat shorter, ~ 2 meters, and the pre-ionization level lower. This would yield a more evenly loaded generator.

If the gas in the MHD channel is ionized by radiation (UV, particles, etc.), $\sim 10^{-3}$ watts/cm^3 of ionizing power is required with a 20 Tesla field. The generated electrical power density in the channel is $\sim 10^2$ watts/cm^3 so that the ratio of ionizing to output power is $\sim 10^{-5}$. This ratio scales as B^{-4} for constant channel power density. Thus, from an ionization standpoint, generators using radiation induced ionization look very feasible at 20 Tesla, marginal at 10 Tesla, and not feasible at 5 Tesla.

A disc generator is probably best for such high fields, not only because the disc shaped duct can sustain higher electrical fields, but because magnet construction is simpler. The magnet for a disc generator is essentially a pair of Helmholtz coils.

Twenty Tesla is beyond the capability of Nb$_3$Sn unless the magnet windings operate at ~ 2 K. This temperature has been achieved in Baseball II without much difficulty, and seems practicable for a dc MHD magnet. The refrigerator cost and power input would approximately triple, but this is a minor part of the plant cost. Twisted multifilament NbTi could be used for the low field part of the hybrid winding and Nb3Sn (either multifilament if commercially available or aluminum stabilized tape) for the high field part. If the windings operate at 4.2K, Nb3Sn can only go to $\sim$ 14 Tesla, with the higher field regions requiring a higher field superconductor (V3Ga or NbAlGe) or cryogenic aluminum.

It is difficult to assess the future of MHD. Many difficult problems remain and competing advanced power generation methods may get there first. For example, many researchers believe that combined cycles using advanced high temperature gas turbines for topping, combined with ordinary steam plants should achieve over 50% thermal efficiency at lower cost than MHD.

There is a strong international effort in MHD. The USSR [30] and Japan [31] seem committed to developing commercial plants and other countries are also heavily involved. The U.S. effort, initially strong, has languished for the past few years, but now appears to be on the way back up. My personal feeling is that MHD will have a significant role in future commercial power generation.

IV.   ENERGY STORAGE - PEAKING POWER

Magnetic energy storage to meet peaking power demands is a potential near term very large scale application. Unlike fusion, MHD, and most other applications of superconductivity, magnetic energy storage does not depend on the development of other technologies, and in fact does not even demand any great advance in superconductor or cryogenic technology. However, there are many competing methods of providing peaking power and magnetic energy storage will only displace these if it proves cheaper.

Figure 11 [32] shows the projected load curve for a typical week in 1990 in the U.S. Northeast power region. The installed capacity is 60 GW(e) but the average generation rate is only 35 GW(e). The average load factor for this region is thus only 60%. About half of this inefficiency cannot be helped since some reserve has to be maintained for scheduled shutdowns, accidents, etc. The remaining half results from the daily, weekly, and seasonal swings in the electrical demand. This pattern is typical of any large electrical system.

Utilities reduce the impact of these swings either by using low capital cost generation units (e.g., gas turbines) for the peak power load, or by using an accumulator (e.g., pumped storage) which absorbs energy during low demand periods and feeds it back into the grid during high demand periods. With accumulators a utility can then afford more high capital cost generation units like nuclear reactors because it can run them at high load factors to minimize fixed charges. With accumulators of zero cost and 100% efficiency (all input power returned) the cost of

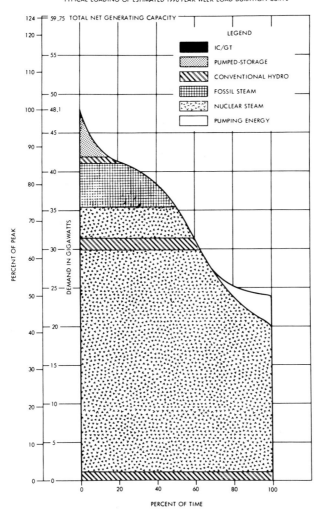

FIG. 11.   Typical Load Curve for Electrical Generation System [32].

electric generation for the system shown in Fig. 11 would be reduced by
~15%. One could never attain the 40% reduction corresponding to the
average load factor since a sizeable reserve capacity is still needed. This
15% reduction would mean that the capital cost of the U.S. electrical generating system in 1990 could be reduced by about 45 billion dollars - a
pleasant prospect for superconductivity, even if it only realized a small
part of this market.

The principal competitors for magnetic energy storage are summarized
in Table 9. It is clear that there are several commercial or near-commer-

TABLE 9.   Methods of Generating Peaking Power

Method	Capital Cost (1985) [$/KW(e)] (33)	Delivered Energy Cost** ($/KWH)	Efficiency (%)	Status
Switching and Sub-Station 20-100 MW(e)				
Gas Turbine*	100	0.018	30	Comm.
Fuel Cells*	180	0.016	80	Devel.
Batteries	200	0.023	70	"
H_2 Electrolyzer-Fuel Cell	250	0.028	60	"
Super Flywheels	?	?	100	Concept
Central Station ≥ 1000 MW(e)				
Surface Pumped Storage (water)	150	0.02	65	Comm.
Underground Pumped Storage (water and air)	150	0.02	65	Devel.

* Requires fuel - other methods are energy accumulators

** 6 hours discharge/day, 15% fixed charges

Fuel Cost - $1.00/$10^6$ BTU

Input Power - 7 mills/KWH

cial alternatives at quite reasonable cost. For purposes of comparing unit costs for energy delivery, a representative 3 hour discharge period twice daily is assumed. Utilities are currently very interested in peak power generators or accumulators at substations, partly because of easier siting, but principally because of transmission limitations. Transmission line capacity (and expense) is set by its maximum load. Current estimates [33] indicate that ~ $60/KW(e) could be saved in capital investment by locating peaking power units at substations. Savings will become even greater as the trend towards underground transmission accelerates.

Unfortunately, magnetic energy storage appears too expensive for sub- and switching-stations, and only appears competitive for very large units linked to central stations. Such a large unit is shown in Fig. 12 (taken from Hassenzahl) [34]. The scale is indicated by the Washington Monument in the background. This unit is conceived of a storing 10^8 MJ (27,000 MWH). The Ludington pumped storage unit, largest in the U.S., has a total storage of 3×10^7 MJ. If magnetic storage were to completely replace pumped storage, the equivalent of approximately 20 of these 10^8 MJ units could be used in the 1990 U.S. electrical system.

Various coil shapes can be used to store magnetic energy, such as the familiar solenoid, a Brooks coil, or a toroid. Of prime concern in a large scale energy storage system will be the minimization of fringe fields. With coil dimensions of the hundreds of meters and fields of ~ 10 Tesla solenoidal fringe fields would extend for 1 mile, requiring very large exclusion areas.

A toroidal magnet, or at least solenoids of alternating magnetic polarity, seems necessary. Some of these geometries are shown in Fig. 13a $\sim$ 13c. One can also use concentric cylindrical or spherical dipoles though these would be more difficult to construct and do not use the superconductor efficiently. The configuration shown in Fig. 13c is particularly attractive; the closed ends eliminate fringe fields yet the bulk of the magnet is a simple solenoid. This has several advantages: efficient use of super-conductor and magnet support structure, relatively easy construction and assembly, and uniform field azimuthally around the coil (except at the ends, which are a small part of the coil volume).

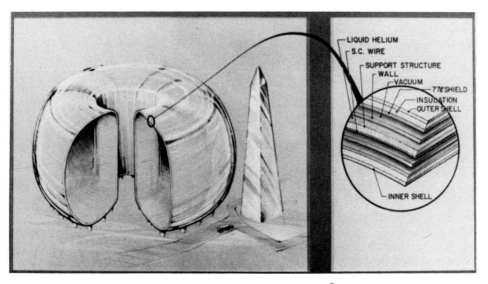

FIG. 12. Magnetic Storage Coil 10^8 MJ [34].

MAGNET GEOMETRIES WITH MINIMUM
FRINGE FIELD

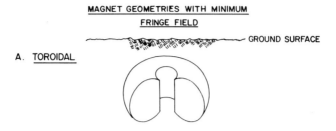

A. TOROIDAL

B. ALTERNATING SOLENOIDS

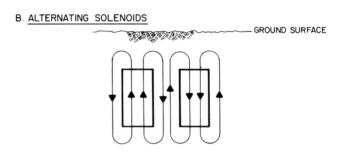

C. SOLENOIDS JOINED AT ENDS

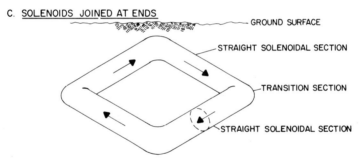

FIG. 13.   Magnetic Storage Coil Geometries.

Very large reductions in the present costs of large superconducting magnets are needed for an economic storage system. For comparison, the NAL magnet stores 400 MJ and costs about $2x10^6$, for a unit cost of $2x10^4$/KWH. The toroidal field coils of the ORNL tokamak fusion reactor design [6] store 40,000 MJ at an estimated cost of $70x10^6$, for a unit cost of $7000/KWH. In order to compete with pumped storage the unit cost will have to be in the range of $30-60/KWH.

Warm reinforcement is essential for economic magnetic storage. Such structures are described earlier in the section on fusion reactors. Either ordinary steel or rock could be used as the warm reinforcement though rock will be much cheaper. Competent rocks (granites, dolomites, sandstones, limestones) can have compressive strengths of 30,000 psi or greater, and tensile strengths of $\sim$1000 psi. They are found in many locations. Tunnels and underground rooms typically cost about $20/m^3 [35] to excavate. A 14T magnet can store about 20 KWH/m^3, so that the excavation costs for a storage magnet will run about $1/KWH stored, negligible compared to other costs. Because of rocks' low tensile strength, tensile cracks will extend radially outward from the superconducting coils. Magnetic forces are then carried in compression by the rock to a point where the tensile stresses are low enough that the rock no longer cracks. For a 12 Tesla solenoidal field, tension cracks will extend radially in the rock to a distance of about 10 times the magnet radius. A large storage coil will have a radius of about 10-20 meters, so that cracks would extend 100-200 meters from the magnet. Such magnets would have to be at least 500-1000 meters below the surface.

The modulus of rock is rather low, $\sim 5 \times 10^6$ psi, so that some sort of adjustable volume fluid chamber may be needed between the magnet winding and the rock to accommodate movement of the rock as it is stressed.

The very large stored magnetic energies, e.g., 10^8 MJ, are not trivial, and correspond to the energy released in respectable earthquakes, (i.e., M = 6 on the Richter Scale). Efficient coupling to produce ground motions would not be possible unless the magnets ruptured, an unlikely though not impossible event. The converse accident, an earthquake triggering the rupture of the magnet is also possible and probably more likely.

A rough but instructive guide to magnetic energy storage costs is shown in Fig. 14 where cold and warm reinforcement (steel and rock) are compared. The geometry of Figure 13c is assumed with costs equivalent to those for a single solenoid (L/D = 20). Dewar costs are taken as $2000/m^2 [34], warm reinforcement barrier costs as $2000/m^2 [15], magnet coil winding and assembly costs as $1000/m^2, cold steel (stainless) as $0.80/lb (including construction), warm steel as $0.15/lb (also including construction) and rock excavation cost as $20/m^3. Working stress in the steel reinforcement is taken as 50,000 psi, which is optimistic. The solenoidal field is 12 Tesla and superconductor costs are taken from a study on the projected costs in a large fusion economy [3], and represent almost an order of magnitude reduction over present prices. Conversion equipment costs are not included; these will be about $20/KW(e) of capacity.

It is clear that magnetic energy storage can compete with other methods only at very large sizes, on the order of 1000-10,000 MWH, and that rock reinforcement must be used. With rock reinforcement, allowing the surface/volume ratio to indefinitely increase, the cost approaches $\sim$$1/KWH for storage and $\sim$ $20/KW(e) for the conversion equipment. The size of such units will be much greater than 10,000 MWH however. The costs shown in Fig. 14 can be lowered by using smaller L/D ratios.

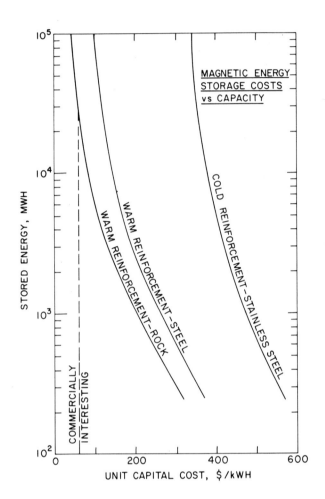

FIG. 14.   Cost of Magnetic Storage vs. Capacity.

Hassenzahl [34] discusses the electrical design of magnetic energy storage devices in detail. Magnetic storage coils can be mated to either ac or dc transmission lines. ac to dc converters will certainly be necessary for the former and they will have to handle the full power flow into and out of the storage coil. With dc transmission lines some type of power handling circuit is still needed to control the power flow to the coil, but its size should be considerably less than with ac lines.

The superconductor requirements seem relatively modest. Twisted multifilament superconductor in a cryogenically stable copper or aluminum matrix would be suitable. Ribbon type conductors would probably not be acceptable (unless ⊥ fields were very small) because of the small hysteresis loss required per charge and discharge cycle. In order to keep the overall storage efficiency at ≥ 96%, the total energy expended in removing heat from the superconducting winding should be ≤ 2% of the total stored energy (1% during charge, 1% during discharge, with the balance lost in ac-dc converters); with a refrigeration factor of 250 watts/ watt(th), at a maximum only 0.004% of the stored energy can actually appear as heat in the superconductor (heat leaks through the warm reinforcement barrier are counted as part of the barrier cost). For the large diameter magnets contemplated this criterion can be easily met with commercially available NbTi. If multifilament Nb_3Sn is commercially developed, storage fields up to 14 Testa are probably achievable: otherwise, storage fields will be limited to the 8 Tesla, available with NbTi. There is some economic penalty if one is limited to 8 Tesla, but it is relatively small since the increased dewar and water barrier costs are partially offset by reduced superconductor costs. The increase in rock excavation costs is minor. Thus it does not appear that Nb_3Sn is essential to an economic magnet storage system.

NbTi can be formed into a hollow superconductor; this type of conductor would probably lead to lower dewar and magnet assembly costs. Because of the brittle nature of Nb3Sn, it may be very difficult to fabricate or handle hollow tube Nb_3Sn superconductor, and this could be a significant argument against its use.

Assuming the geometry of Fig. 13c and an effective L/D ratio of 20/1, a 10,000 MWH storage coil would have a surface area of 80,000 m^2; with a warm reinforcement structure and a 1.0 m thick thermal barrier, 200 KW, 4K refrigerator is $\sim\$7x10^6$, so that with 3 refrigerators (one spare) the capital investment of the 4K refrigeration plant would be only $\sim\$20$/KWH stored. The superconductor hysteresis loss would amount to $\sim$20 KW averaged over a day (2 cycles/day, 1 mil filament diameter) and there would be a much smaller refrigeration load due to eddy current heating (B is typically 0.01 kG/sec). Additional refrigeration costs for the 20K and 77K heat sinks in the thermal barrier would be comparable to those for the 4K sink. Thus warm reinforcement seems quite feasible from the refrigeration standpoint, which would be the principal concern.

The principal U.S. efforts in magnetic energy storage for peaking power are at Los Alamos [34] and the University of Wisconsin [36]. There is a much wider effort, of course, in magnetic energy storage for rapid pulse systems used in fusion research, laser pumping, etc. Some of this has been touched on earlier. To date, the work on peaking power storage units has been principally system studies.

Since the economics of magnetic storage do not appear competitive except at very large sizes, it is worth considering dual purpose concepts, i.e., magnetic storage with some corollary use of the field or low temperature aspect.

One could have the excavated region inside the coil at low temperature, sharing the same dewar. Liquefied natural gas or H_2 could be stored in very large quantities. For example, a single 10,000 MWH coil would store $\sim$600,000 m^3 of liquid H_2, which would have roughly 10^3 times greater total energy than the magnetic field. Besides being a safety problem such storage, even if dewar costs were completely allocated to the storage of liquified gas, would only reduce magnetic storage costs by a few percent. Similar arguments apply to the storage of liquid helium, which at least would not be a safety hazard.

A second dual purpose use would be to vary the magnetic field in a very large low β fusion reactor (e.g., a tokamak) so as to store energy. Here the costs could be fully allocated to the fusion reactor and the storage would be essentially free. Unfortunately, there are serious operational problems. For one thing, the plasma containment would be deteriorating as the field decreased to meet peak power demands, but this is precisely when the reactor should be operating at maximum capacity. Also, unless the reactor was very large, not much energy would be stored. It cannot be ruled out, however, especially since it would make difficult fuel cycles like DD and pB^{11} more attractive.

The third dual purpose use would be to make use of the field for some large scale industrial process. This seems the best hope. For example, very large volumes of water might be purified, food sterilized, etc. This would seem to warrant further exploration to find out what beneficial effects could be obtained from long term exposure to very high magnetic fields. The cyclical field variation could be very helpful in this regard.

It appears that magnetic energy storage will only be competitive for very large systems comparable to pumped storage. Breakthroughs in reducing superconductor and dewar cost could alter this, however. A more immediate applicable use of magnetic energy storage may be as an emergency reserve. With short discharge times on the order of a few minutes, the cost of power from magnetic storage would approach the cost of ac-dc converters, i.e., $\sim$\$20/KW, and would be much cheaper than alternative methods. The stored energy and coil volume are much smaller. A 2000 MW(e) energy reserve discharging in 6 minutes would store 200 MWH in $\sim$15,000 m^3. Full power would be available in a few seconds. During the 6 minute discharge period other generation units could be started and load dropped if need be. Such a reserve could be very attractive to utilities and experience gained by the construction and operation of such storage coils could help lead to the eventual use of magnetic storage for regular peaking power needs.

V.   ENERGY MODIFICATION - AC AND DC GENERATORS AND MOTORS

After contemplating applications of superconductivity for which all sorts of breakthroughs are necessary, it is pleasant to come to rotating electrical machinery applications. Here technical problems seem relatively minor and the economics nicely favorable.

Considering first ac synchronous generators, it appears feasible to reduce the length and weight of generators in large central station power plants by about a factor of 2. This should not only reduce the capital cost of of the ac generator from the present ~$12/KW to an estimated $6-8/KW, but should also increase electrical efficiency by ~ 0.5%. This increased operating efficiency corresponds to an additional saving of $1.5/KW in capitalized operating cost, assuming 7 mills/KWH power costs. At an average saving of $5/KW, several hundred million dollars per year could be saved in the 1980's if all central station generators being installed in the U.S. at that time were superconducting. An additional benefit of superconducting generators is the much larger unit ratings that are possible. Conventional generators are now limited to 1300 MW capacity, while super-conducting generators could operate at capacities far from 2000 MW. With the increase in central station size forecast in the next one-two decades, many such generators will be required.

Figure 15 shows a cross section view of a conceptual superconducting ac generator [37]. Hysteretic superconductor losses at 60Hz are too great to permit using superconducting windings in the generator stator, but they can be used to generate the dc field in the rotor. It is necessary to shield the rotor superconductor with a normal metal can, which may be at cryogenic temperatures, to prevent transient flux changes from causing SC-normal transitions and/or excessive heating at 4K. Besides being lossless, the superconducting rotor winding permits a larger air gap so that stator diameter can be larger than in a conventional machine. This larger stator diameter then permits a shorter length generator, with a net weight savings of approximately 50%. Figure 16 illustrates the savings in size possible with 500 and 1300 Mw ratings [37].

A design of a 500 MW superconducting ac generator is summarized in Table 10, taken from Appleton [37]. Note that the maximum super-conductor field is rather modest, 3.2 T, and that the relatively large rotor permits a low current density in the superconductor, typically a few KA/cm^2. Twisted multifilament NbTi seems perfectly adequate for this purpose. Considerably higher field strengths are achievable with NbTi, let alone Nb_3Sn, but there does not seem to be much point in pushing superconducting technology to its limits, since superconducting ratings will be set by power plant size, not inherent capabilities.

Development of ac superconducting generators will require working out of many new mechanical and electrical details, e.g., cooling and support of stator windings, how to supply liquid helium to a revolving rotor, response of low inertia rotors to electrical transients, but will require relatively little work on the superconducting part. The mechanical development will undoubtedly proceed cautiously through ever bigger units, and with relatively long term testing to find out what the problems are. Thus at this point, the development of superconducting generators for central station seems to be in the hands of manufacturers, which is the best place it could be. Westinghouse [38] and GE [39] are involved in this development. There is extensive non-U.S. work also [40,41].

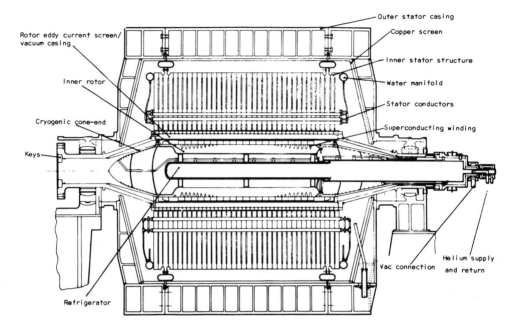

500 MW SUPERCONDUCTING A.C. GENERATOR

FIG. 15.   Cross Section View of Superconducting Generator [37].

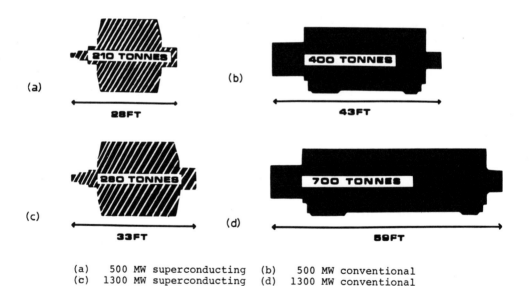

(a)   500 MW superconducting   (b)   500 MW conventional
(c)   1300 MW superconducting (d)   1300 MW conventional

FIG. 16.   Relative Sizes of Conventional and Superconducting
            Generators [37].

TABLE 10.   Parameters of 500 MW
Superconducting ac Generator

Max. rotor magnetic field	3.2T
Max. field at stator	0.8T
Rotor ampere turns	5×10^6
Stored field energy	20 MJ
Overall machine length	8.5 m
Length between bearings	6.4 m
Length of stator winding	3.7 m
Stator nominal dia.	2.1 m
Rotor nominal dia	1.05 m
Screen nominal dia.	4.2 m
Total machine weight	210 tons
Rotor weight	29 tons
Inner stator weight	40 tons
Outer stator weight	120 tons
Max. lift (outer stator bottom)	50 tons
Copper joule loss	1.8 MW
Eddy current loss	0.5 MW
Outer shield loss	1.5 MW
Stray and refrigerator loss	1.0 MW
Windage and friction loss	1.0 MW
	5.8 MW
Overall efficiency	98.8%

Application to other types of rotating electrical machinery
in central station power plants does not appear very significant, with
the possible exception of superconducting acyclic generators supplying dc
for electrolysis purposes (e.g., aluminum or H_2 production). Here the
market for acyclic generators has not developed enough at the present to
tell whether or not this application will be significant. Application of
superconductivity to motors for boiler feed pumps or miscellaneous
power equipment does not seem very important.

For mobile applications (i.e., airplanes and ships) however, superconducting motors and generators are very attractive because of the large
saving in weight. Airborne ac generators have obvious utility, and there
is a vigorous program in this area [42, 43]. Even more interest is
centered on ship propulsion, where both motors and generators are
required to couple the drive power from the ship's turbines to its propellors
[44, 45]. This area is probably where superconducting electrical
machinery will make its first significant commercial impact, most likely
by 1980. See Chapter 4 by Appleton.

VI.   ENERGY TRANSMISSION

There are approximately a dozen projects in the world on superconducting power transmission and these projects receive a large part
of the total superconducting development funds.

If superconducting power transmission is to be widely applied, two
contests have to be won. First, underground transmission has to win
out against overhead. This cannot be done on economic grounds, since
underground transmission is always more expensive than overhead, at
least for single circuit capabilities below about 10,000 MVA. External
factors, like objections to the visual appearance of overhead lines,
restricted right of way, radio noise of overhead lines, etc., in some cases
will force utilities to underground even though it costs more. In the second
contest, superconducting transmission has to win out against alternative
underground transmission methods, such as gas spacer, cryoresistive,
HPOF force cooled, etc. This contest will be decided primarily on costs.
In general, the major expense of superconducting transmission lines is
in the installation of the dewar and refrigeration system, and unit costs
($/MVA mile, for example) tend to drop almost inversely with increasing
capacity. Costs of alternative transmission methods tend to drop more
slowly so there is always a crossover point above which superconducting
lines become cheaper. The future of superconducting transmission depends
on where this crossover point occurs. If it is at 2000 MVA, superconducting cables will probably be used extensively by 1990. If it is at 8000 MVA,
extensive use will probably be delayed till well into the 21st century. A
revealing indication of the U.S. power industry attitude is apparent in
the report to the Electric Research Council by the R&D Goals
Task Force [46]. This report commends only $43x10^6$ cumulative funding

from 1971 to 2000 AD on superconducting ac lines, and nothing on dc lines. This amount is less than 1% of the total R&D on transmission and distribution recommended for the same period. This funding level seems inadequate if superconducting transmission lines are to be a serious possibility.

Returning to the question of underground vs. overhead transmission, there seem to be no estimates of how much and what capacity underground transmission will be required in the next few decades. The many intangible factors and uncertain costs make such predictions very difficult.

Table 11 shows circuit miles of overhead lines for the U.S. for the period 1970-2010. The 1970-1990 values are taken from the 1970 FPC survey [47] while the 2000-2010 values are extrapolations. It is clear that only a relatively small part of the U.S. transmission system is at capacities greater than 2000 MVA where superconducting lines may be viable. A map of the projected 1990 U.S. transmission grid shows that only a small fraction of the high capacity lines traverse urban and suburban areas. An upper limit to the amount of superconducting transmission is on the order of 10% of the circuit miles for capabilities $\geq$ 2000 MVA, since other methods will compete with superconducting transmission. At a cost of 10^6/mile, the cumulative investment in superconducting lines for the U.S. would be at most 3.6×10^9 by 2010 AD. In more densely populated countries, a larger percentage of the transmission system may have to be underground, but the absolute mileage per million of population will probably be of the same order as that in the U.S.

Such estimates are admittedly rough, but they do show that super-conducting transmission will probably not be as economically significant as other applications such as fusion.

The contest between superconducting and other forms of underground transmission is shown in Fig. 17 [48]. The principal competitor to superconducting cables is spacer gas, followed by liquid N_2 cooled cryo-resistive, and then force cooled HPOF or self-contained conventional cable. Spacer gas cable uses a rigid normal metal conductor (aluminum) inside a rigid aluminum pipe at ground potential with SF_6 gas dielectric between. A half dozen or so short lines (1/2 mile or less) are now in commercial operation at power levels up to 2000 MVA [49]. Three separate pipes are used for 3 phase ac transmission. Gas spacer cable manufacturers expect that in the near future a number of very significant improvements will be commercially practiced, i.e., flexible cable and flexible pipe, which should permit much easier cable laying of lengths on the order of a mile; 3 phases in one pipe; and force cooling. With these improvements single circuit capacities up to 10,000 MVA are forecast, with large reductions in transmission cost [49].

It is very likely that in the next 5 years gas spacer cable will be installed in circuit lengths of tens of miles at capacities of 3000 MW at

TABLE 11. Circuit Miles of Electrical Transmission in the U.S.

Circuit Voltage (KV)	Circuit Capacity MVA (1)	Thousands of Circuit Miles in U.S.					Year
		1970 (2)	1980 (2)	1990 (2)	2000 (3)	2010 (3)	
230 & 278 ac	300	42	60	68	75	80	
345 ac	800	15	33	47	55	60	
500 ac	1800	7	20	33	45	55	
765 ac	3800	0.5	4	9	18	36	
1100 ac	7400	0	0	0	?	?	
±400 dc	1440	0.8	2	2	2	2	
Total Circuit Miles with Capacity ≥2000 MVA		1.3	4	9	18	36	
Total Underground Circuit Miles with Capacity ≥2000 MVA		0	0.1 (4)	0.5 (4)	1.8 (4)	3.6 (4)	

(1) Twice SIL Loading

(2) 1970 Power Survey, FPC, p. 1-18-11

(3) Extrapolated from 1970 FPC Survey

(4) 2% of Total Circuit Miles in 1980, 5% in 1990, 10% thereafter

Underground Circuit Miles in 1970: 230 kV - 60
345 kV - 90

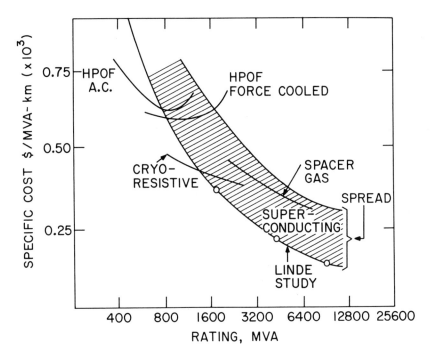

FIG. 17.   Cost of Various Underground Transmission Methods vs.
            Capacity [48].

several locations in the U.S. This headstart could choke off or delay the
development of more advanced transmission systems like superconducting
cables, just as the commercial introduction of light water reactors has
tended to slow-down the development of more advanced nuclear reactors.

The difference in cost between alternative methods of underground
transmission to the average residential consumer will be relatively small.
For example, assuming a $100/MVA mile saving if superconducting
cables are used instead of gas spacer translates into about a 0.5% decrease
in the average residential electric bill for superconducting transmission,
assuming an average underground transmission distance of 20 miles to
the consumer. This marginal benefit, combined with the conservatism
the electric power industry, may seriously retard the development of
superconducting transmission.

A.   DC Transmission

    If one temporarily forgets about the terminals of a superconducting transmission line, there is no doubt that dc transmission wins out over ac. As an example, Table 12 shows some of the advantages of dc over ac if the 3000 MVA BNL ac cable design [48] were converted to dc. The two most important advantages of dc are the much higher power density, which reduces cable cost, and the smaller fault current, which eases the problem of cable reliability.

    The capital cost savings with dc will depend on the particular design and the cable capacity, but one can confidently predict a dc transmission line should be at least 50% cheaper than an equivalent capacity ac transmission line. Now, unfortunately, we remember the temporarily forgotten terminals. At a minimum, two ac-dc conversion terminals will cost $40/kW(e). As an extreme example, suppose that the dc cable costs nothing and the ac cable costs $200/MVA · km. The underground cable run would have to be 200 km before the total cost of the dc line began to be less than that of the ac one. In actuality, the underground run would have to be longer before a crossover occurred, even counting in the lower operating costs of dc transmission and the cost of reactive compensation for ac lines, which will be necessary for lengths much longer than 200 km. In contrast, dc overhead lines crossover at ∼1000 km, and then mostly because of lower line losses.

    Realistically, it will be many decades before underground transmission lines more than 200 km long are necessary, so that at first glance dc superconducting lines do not seem viable, unless none of the advanced ac underground transmission methods (i.e., superconducting, cryoresistive, and spacer gas) prove feasible. There are several caveats to this conclusion:

1.   Very large single circuit capacities (well above 10,000 MW) may become desirable. In a study of power transmission from mine mouth plants located in the U.S. Western coal reserves [50], it was concluded that 50,000 MW dc hydrogen cooled cryoresistive (aluminum conductor) lines would be cheaper than overhead transmission. With dc superconducting lines, underground transmission could probably compete with overhead at single circuit capacities of 10,000 to 20,000 MW.

2.   Asynchronous ties may be desirable for system stability. Here line length may not be important; the New Brunswick tie in Canada, for example, is only 10 meters long (which would hardly use much superconductor). There has been some speculation that a dc superconducting line might serve as a good backbone tie for systems in the U.S. Northeast Corridor. It would run for several hundred miles, from the vicinity of Boston to near Washington, and could be useful in the 1990's.

TABLE 12. Advantages of dc over 3000 MVA BNL ac Cable

Advantages of DC	3000 MW Line Parameters DC	AC
Higher Linear Current Density in Conductor, A/Cm (rms)	2500	320
Fewer Cables in Dewar (same O.D.)	2	15
Smaller Dewar Diameter, cm	20	42
Lower Cable Voltage, Line to Ground (rms)	70	76
Lower Peak Dielectric Stress, KV/cm	105	160
Lower Conductor Hysteretic Losses, watts/km	0	500
Lower Dielectric Losses, watts/km	0	170
Lower Total Heat Leak, watts/km	100	810
Lower Fault Current	X2	X10 (X20 with DC offset)

Disadvantages of DC

Conversion Terminal Costs	$40/KW(e)	0

3.   Cheaper ac-dc converters may be developed. Radically lower costs
        would not be expected since most of the converter station cost is
        associated with filtering, yards, bus-work, etc., and not the valves,
        which could conceivably be greatly reduced in cost.

4.   The possibility of central station dc generation as well as switching
        and sub-station dc consumption cannot be ruled out. Large acyclic
        mechanical generators of 1000-2000 MW(e) capacity are forecast at
        lower costs than for ac generators [51]. The voltage levels are a bit
        low, on the order of 10,000 volts, but could probably be redesigned
        to match the optimum level in a dc line. This would eliminate one
        converter station. MHD generation would also eliminate the con-
        verter at the power station, and the voltage level seems compatible
        with dc transmission. At the far end of the dc superconducting line
        the dc might directly feed accumulators (batteries, H_2 electrolyzer)
        which then discharge for peak loads. These accumulators require
        inverters anyway, so their cost would not be assigned to the dc
        transmission line.

5.   If the superconducting underground line were a link in a long trans-
        mission system, a dc link combined with dc overhead might be
        cheaper than an all ac system. Part of the converter cost would
        be assigned to the overhead and part to the underground lines, making
        the breakeven length for each considerably less. There is difficulty
        matching voltage levels, however. For several thousand MW
        capacity the dc overhead line must run at $\sim \pm$ 1000 kV or employ
        multiple circuits, which would be more expensive. It may be very
        difficult to cary $\pm$ 1000 kV in a dc superconducting transmission
        line.

6.   If the dc line also transmits liquid H_2 as a fuel, the line cost can
        be shared between electrical fuel transmission. Hammel [52] has
        proposed this in connection with liquid H_2 fuel supply for the space
        shuttle program. A dual transmission system supplying both liquid
        H_2 to commercial aircraft at a municipal airport and electric power
        to the city seems more promising. The hydrogen producing and
        liquefier plants would be centralized, and would feed a number of dc
        transmission lines. This would be safer and cheaper than having
        dispersed liquefiers. There appear to be significant advantages to
        liquid H_2 fueled aircraft, and a shift to this fuel could come in
        10-20 years [53]. A typical large airport will require several
        thousand tons/day of liquid H_2. If we allow $0.01/lb for liquid H_2
        transmission (total liquid H_2 cost $\sim$$0.10/lb), the capital investment
        for the transmission line could be $120x10^6 at a delivery rate of
        2500 T/day. This corresponds to a transmission line length of
        $\sim$ 200 km. Thus the electrical transmission cost would be small
        if cables were put in the dewar since the cost of superconductor and
        electrical insulation is small relative to the dewar.

Any of the above factors could radically alter the conclusion that dc cables will not be viable until very long lines become necessary.

Table 13 summarizes the important features of some dc transmission line designs. It is interesting that all designs use helium coolant. LASL [54] has recently begun studies aimed at a reference design for a 1000 km, 5000 MW dc superconducting transmission line. The coolant for such a line has not been selected yet, but there has been strong interest in liquid H_2. With pumped liquid H_2, the inlet coolant temperature must be at or above 14.1K (the freezing temperature of liquid H_2); the outlet temperature will probably be on the order of 15-16K, which indicates that Nb_3Sn would not be suitable for such a line and that some conductor with a transition temperature $\geq 20K$ like Nb_xAl_yGe is necessary. Dahlgren and Kroeger [55] have made $Nb_{12}Al_3Ge$ sputtered conductors with high current densities but transition temperatures are only $\sim16K$. Other researchers have achieved higher transition temperatures, but current densities are disappointingly low. A practical superconductor for a liquid H_2 cooled line may be developed, but it does not yet seem assured.

Another interesting feature of the 3 designs in Table 13 [56, 57, 58] is that two of them use flexible dewars. This would appear to be the key to low cost installation of superconducting transmission lines, where the lines could be laid very rapidly following a trenching machine. The length of open trench would then be minimal. It should be noted that such plowing techniques can only be applied in relatively open country without streets, underground utilities, etc. However, since dc lines in general must be long to offer an advantage over ac, most of the line length will be in such terrain. Flexible dewars would appear to give dc the best chance of success. Flexible dewars can probably be used with ac lines, but the lower power density resulting from the need to keep hysteretic losses tolerable makes such lines larger and much harder to reel.

A third interesting feature of the 3 designs is the different number of circuits in the dewar for each design. The use of two circuits achieves a somewhat better packing fraction than 1 circuit, but the advantage does not seem large. A coaxial conductor circuit, as in the CERL design, appears to have disadvantages in handling and pulling, as compared to two single cables in one dewar. The 1/2 circuit (single lead in one dewar) approach in the AEG-Kabelmetal-Linde design seems disadvantageous at first, as compared to putting a complete circuit inside one dewar, since in the AEG design two dewars are needed, producing a greater heat leak and cost. However, two or more separate lines will certainly be required for transmission reliability. If enough superconductor is put in each of the two leads to carry full line load, then the AEG design would appear to be cheaper overall than building two lines, each with a single circuit.

The principal advantage of Nb_3Sn over one of the ductile Nb alloys is the ability to operate at higher temperature. This benefit seems less important in dc lines than in ac, because of the much smaller total heat

TABLE 13. dc Superconducting Cable Designs

	CGE[1]	AEG-[2] Kabel Metal LINDE	CERL[3]
Cable Voltage, KV	±100	±200	±115
# of Circuits in Dewar	2	1/2	1 (coax)
Dewar	Flexible	Flexible	Rigid
Conductor	Flexible	Flexible	Flexible
Coolant	Helium	Helium	Helium
Conductor Material	Nb-Ti/Al	Nb_3Sn/Cu	Nb-Ti-Zr/Cu
Operating Temperature $^{\circ}$K	4	6	4.2 to 5
Capacity, MW	4000	5000	10,000
Overall Diameter, cm	12	21 (one)	~40
Dielectric	Mylar-He	–	(Nylon on Poly- ethene He impregnated)
Peak Stress at Conductor KV/cm	175	~200	75
Linear Current Density, A/cm	1520	500	1140 (Inner)
Total Heat Leak, watts/km	–	–	87
Cost (#/MVA· km	–	200	52

(1) Dubois, P. et al., p. 173, 1972 Applied Superconductivity
     Conference.

(2) Engelhardt, H., p. 56, 1971 Heidelberg High Voltage
     Direct Current Symposium.

(3) Carter, C. N., Cryogenics, 13, 207 (1973).

leak for dc lines. With NbTi it appears possible to operate in the super-critical flow regime, and there should be no problems of flow oscillations. Under these conditions, the use of a ductile conductor instead of a brittle high temperature one like Nb_3Sn would appear to be advantageous.

The absence of dielectric losses in dc lines allows the use of dielectric with high tan δ's. In general, such material often have higher dielectric strengths than non-polar materials with low tan δ's. Mylar or Nomex tape with He gas impregnations appears to be a good dielectric for dc lines with dielectric strengths above 500 KV/cm. The much higher linear current density possible in the superconductor with dc cables also eases the dielectric problem, since one can readily add more dielectric if needed without seriously increasing the transmission line diameter. For these reasons, dielectrics do not appear to be a problem with dc lines, but could be a serious problem with ac ones.

Finally, it will be much easier to cryogenically stabilize supercon-ducting dc lines with normal metal than ac lines, since the current depth in the normal metal will not be limited by ac skin depth. At 2000 A/cm operating linear current density in the conductor, a x2 fault would only produce 0.2 w/cm^2 heat flux in a 2 mm Al stabilizer (5000 resistance ratio) even if all the current transferred to the stabilizer. Thicker stabilizers could be used if desired.

B.   AC Transmission

Many designs have been proposed for superconducting ac transmission lines. A few of these designs are summarized in Table 14. There are many design choices to be made, and much uncertainty as to the effect of these choices. This has tended to produce a wide variety of designs with no clear cut proof of which are best. The different researchers involved with super-conducting ac transmission lines have differing convictions as to the best route, however, and we shall explore some of these arguments.

One of the earliest designs is that of Meyerhoff et al., at Linde Corporation [59, 60]. This design has many features of spacer gas cable installations. The container enclosure (dewar) is a rigid pipe (Fig. 18) laid in ~ 15 m lengths in a trench, requiring welds at each joint. Three rigid pipe conductors are triangularly spaced inside the dewar with a coaxial shield for each conductor. Each conductor is a separate phase of a 3 phase circuit. The conductor and shields must be joined in the field at 15 meter intervals, along with the dewar. Thus there are 6 joints to be made every 15 meters (3 conductors, 3 shields, and inner and outer dewar walls).

Liquid helium serves both as a coolant and dielectric. It flows inside the 3 pipe conductors between the conductors and their shields.

TABLE 14. ac Superconducting Cable Designs

	Design				
	BNL[1]	LINDE[2]	CERL[3]	CERL-BICC[4]	GRAZ[5]
Cable Voltage, KV	132	230	275	33	20
Dewar	Rigid	Rigid	Rigid	Rigid	Flexible
Conductor	Flexible	Rigid	Flexible	Rigid	Flexible
Conductor Material	Nb_3Sn	Nb	Nb	Nb	Pb on Nb
Operating Temperature °K	6.2-8.2	4.2	5	~5	4.2-5.2
Capacity, MVA	3000	4710	4000	750	1100
Overall Diameter, cm	42	46.7	46.5	20	25
Dielectric	Tape-He	Liquid He	Tape-He	Liquid He	He or plastic
Peak Stress at Conductor, KV/cm	160	34	110	-	~60
Linear Current Density, A/cm (rms)	320	560	250	400	71
Conductor Losses, watts/km	500	50	49	-	30
Total Heat Leak, watts/km	800	180	279	-	210
Cost $/MVA · km (includes capitalized operation costs)	-	190	190	770	440

(1) Forsyth, E., et al., BNL-50325 (1972).

(2) Meyerhoff, R. W., Cryogenics, 11, 91 (1971).

(3) Bayliss, J. A., CERL Report RD/L/N/72 (1972).

(4) Taylor, M. T., Low Temp. & Electric Power, p. 119, Pergamon (1970).

(5) Klaudy, P. A., Bull. Schweiz Electrotech Verein 61, 1179-90 (1970).

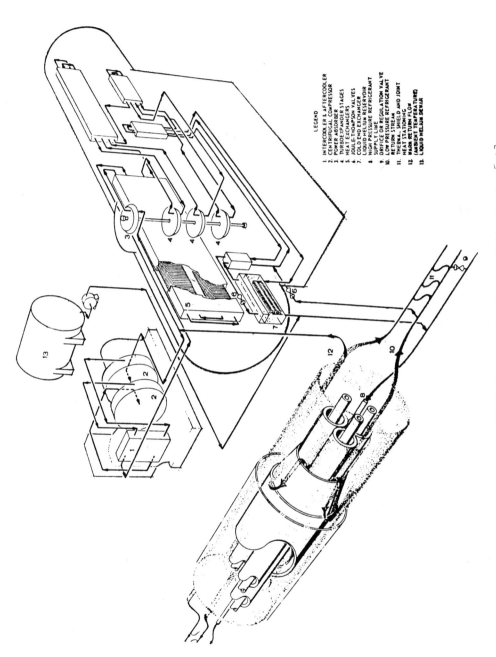

LEGEND

1. INTERCOOLER & AFTERCOOLER
2. CENTRIFUGAL COMPRESSOR
3. POWER ABSORBER
4. TURBOEXPANDER STAGES
5. HEAT EXCHANGERS
6. JOULE-THOMPSON VALVES
7. COLD END EXCHANGER
8. LIQUID HELIUM RESERVOIR
9. HIGH PRESSURE REFRIGERANT
   SUPPLY LINE
10. ORIFICE OR REGULATION VALVE
11. LOW PRESSURE REFRIGERANT
    RETURN STREAM
12. THERMA. SHIELD AND JOINT
    HEAT STATIONING
13. WARM RETURN FLOW
    (AMBIENT TEMPERATURE)
    LIQUID HELIUM DEWAR

FIG. 18. LINDE Superconducting ac Cable Design [59].

This helium flow is termed the "return" helium flow. There is also a "go" helium flow in a separate pipe in the dewar. The "go" and "return" helium flows must be thermally insulated from each other by vacuum insulation. Electric field and magnetic fields only exist between the conductors and their respective shields. The shields are superconducting and at ground potential. The superconducting shields carry currents equal and opposite to the conductor currents, but do not contribute to power transmission. They are necessary, however; otherwise, induced currents in the dewar wall would cause excessive heating.

The superconductor is a thin ($\sim 60\,\mu$) layer of Nb electroplated on copper pipe. It is a type I superconductor with peak surface field limited to ~ 1500 gauss. The design peak magnetic field is 1000 gauss. During faults where the conductor must carry up to 10 times normal current, the Nb goes normal and the current transfers to the copper pipe.

This concept requires that a superconducting joint be made every 15 meters on each of the 3 conductors and 3 shields. Linde proposes to do this by plasma spraying a thin Nb layer on the welded copper pipe. The dewar uses a multi-shield approach with evaporating helium from the inside cooling the dewar shields. Warm helium gas is then returned to the refrigeration station.

This design seems to have been influenced by two beliefs: one, that the hysteretic losses of type II superconductors (e.g., Nb_3Sn) were too great to permit their use for 60 Hz power transmission, and two, that the loss tangent of solid dielectrics was too great to permit their use at 4K and 60 Hz. Subsequent work now indicates that type II superconductors and solid dielectrics can be used.

Leaving aside questions of mechanical practicality and cost of the multitude of joints that are necessary with this design, there are problems with the choice of Nb superconductor and helium dielectric. When an overcurrent fault occurs the Nb will be driven normal. If the current then transfers to a normal metal (aluminum is the best) backing, the conductor temperature will increase well past the transition temperature, and it will require a substantial time, e.g., a few seconds, to recool the conductor to the point where rated current can be carried. This is contrary to ordinary transmission practice where the line is expected to carry rated power immediately after the fault clears. For a typical x10 fault with dc offset and 4 cycle opening time for the breaker the temperature rise with normal metal backing is on the order of 10-20K.

The second problem with Nb is that since its critical temperature is low, the operating refrigeration temperature of the transmission line must be in the range ~ 4 to 5K. Operation in the supercritical pressure range is necessary both to maintain helium dielectric strength and to avoid temperature-pressure oscillations in the helium flow (such oscillations can appear spontaneously in long heated flow lines in certain operating ranges). However, supercritical refrigerators at temperatures of ∼ 5K have substantially lower thermodynamic efficiency than liquid He refrigerators, requiring 2 to 3 times as much input power as a liquid He refrigerator at 4.5K [61]. Efficiency at 8K is much improved, however.

The dielectric strength of helium is relatively low and is very sensitive to small heat inputs (since a small amount of heating can cause large density changes). The surface finish requirements on conductors and joints will be more demanding with helium dielectric than with SF_6 gas dielectric in spacer gas cables. The average corona current must be very low, on the order of 3×10^{-11} A/cm^2, to avoid excessive heating. Impurities carried by the flowing helium will tend to aggravate the problem. Breakdown during overcurrent faults is of concern since the He boundary layer at the conductor can undergo large density changes as the conductor heats up by 15 to 20K. Voltage spikes during the fault could then cause breakdown in an otherwise safe line. The Linde design assured an intrinsic breakdown strength of ∼ 400 to 500 kV/cm. based on the measurements of Goldschwartz and Blaisse for very small gaps (100u) [62]. Subsequent work has shown that breakdown strength for large gaps is considerably less. It appears that the critical condition is not operating stress at the conductor, but the impulse stress during switching surges, faults, etc. It is likely that the operating voltage in the Linde design will have to be somewhat lower if the conductor and shield dimensions remain fixed.

The BICC-CERL design [63] has many of the features of the Linde design, except that the rigid Nb conductor pipes are arranged coaxially instead of triangularly. This seems more mechanically complicated. The 750 MVA capacity is clearly well below the crossover point for superconducting cables. Capacities in the 2000 MVA range and above will certainly demand much higher operating voltages than 33 KV to keep superconductor hysteretic losses in an acceptable range.

The BNL design [48] postdates the Linde and BICC-CERL work by several years. In this interval it became clear that non-polar solid plastic dielectrics had loss tangents of 10^{-5} or less at superconducting temperatures, and that this was sufficiently low to permit the use of helium impregnated tapes for electrical insulation. This has two advantages: higher operating dielectric stress, and flexible cable. Long length (∼ 1 km) of tape wrapped cable can be pulled into an installed dewar, much as cables are now pulled into HPOF pipe. This reduces the number of joints/km by two orders of magnitude, which should increase reliability and cut cost. Figure 19 shows a typical cable. First the inner conductor strip is helically wrapped around a support helix, then tape is wrapped, and then the outer conductor and armor are wrapped. Each of these coaxial cables carries part of the phase current. For the nominal cable dimensions in the BNL design, 3 single cables will carry a total of 600 MVA. Higher capacity lines use more cables, i.e., a 3000 MVA line has 15 cables in a common dewar (Figure 20). It is not clear that this is the optimum design; it may be preferable to use fewer cables in large capacity lines.

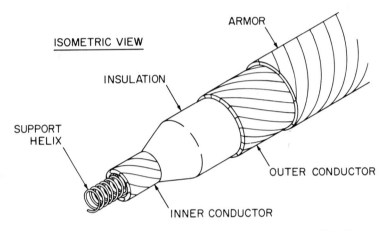

ISOMETRIC VIEW

ARMOR

INSULATION

SUPPORT
HELIX

OUTER CONDUCTOR

INNER CONDUCTOR

NOT TO
SCALE

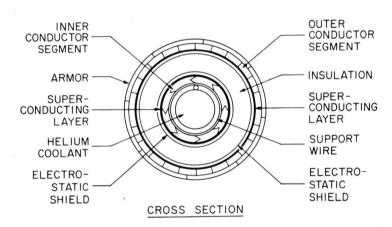

INNER
CONDUCTOR
SEGMENT

ARMOR

SUPER-
CONDUCTING
LAYER

HELIUM
COOLANT

ELECTRO-
STATIC
SHIELD

OUTER
CONDUCTOR
SEGMENT

INSULATION

SUPER-
CONDUCTING
LAYER

SUPPORT
WIRE

ELECTRO-
STATIC
SHIELD

CROSS SECTION

FIG. 19.   BNL Superconducting ac Cable [48].

The use of flexible cable does not appear very radical; gas spacer cables are rapidly moving towards flexibility, both for the conductor and the outer pipe. Such evolution appears natural and destined to produce lower costs. The recent CERL design [64] in Table 14 is part of this trend.

The most significant departure of BNL from previous designs is the use of a type II superconductor, Nb_3Sn. This conductor has very significant advantages in terms of higher operating temperatures and greater fault capability. Measurements at BNL indicate that the hysteretic losses in Nb_3Sn are sufficiently small for practical cable operation. The hysteretic losses in Table 14 are one-half of those in the original BNL design [48]. Recent measurements [65] indicate that losses can be less than one-quarter of the original design values, depending on the method of Nb_3Sn manufacture. At this level other line losses tend to predominate and the difference between Nb and Nb_3Sn losses is not very significant. Measurements further indicate that operation at 8K does not seriously increase losses [66]; with Nb_3Sn conductors several mils thick, it appears practical to carry fault current in the superconductor. The ΔT after a x10 fault is only about 0.5K, so the cable should be able to carry rated current immediately after the fault clears.

CONCEPTUAL 3000 MVA UNDERGROUND SUPERCONDUCTING CABLE

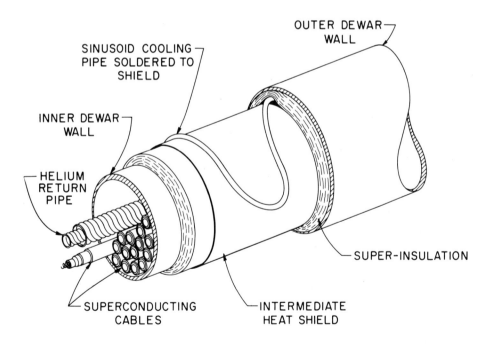

FIG. 20.   BNL Superconducting ac Transmission Line [48].

In summary, ac superconducting transmission seems technically feasible, barring any unexpected pitfalls. Much more work is required before optimized designs can be made, but the most promising direction appears to be towards the use of Nb_3Sn in flexible cable. Undoubtedly superconducting ac transmission will be cheaper than competing methods of ac transmission above some crossover point, but there is no way to realistically determine where that crossover is. If it is at 1000 MVA, then superconducting transmission has a bright future in the next decade; if it is at 10,000 MVA, then application will be many decades away.

VII. MAGNETIC LEVITATION - HIGH SPEED GROUND TRANSPORTATION

Conventional wheeled trains appear to reach the end of the line at speeds much above 150 mph. Track alignment problems become very difficult, very sophisticated springing systems are necessary for safety as well as passenger comfort, and at approximately 200 mph available traction is insufficient to overcome drag.

If trains are to recapture riders from the well developed air and auto transport systems that exist in countries like the U.S., or if they are to avoid going into a long slow decline in other countries where air and auto systems have not yet built up, train operating speeds must increase, costs must become more competitive and service must be more convenient.

Increased operating speed not only attracts customers, but helps to reduce operating cost and permits more convenient scheduling. For this reason, there has been much interest in the past decade on ground transport at speeds of 200-300 mph. These speeds, as we have seen above, are not possible with conventional trains, and one has to go to a train that does not contact the track. Much work has been done since the early 1960's by France, Britain and the U.S. on tracked air cushion trains. In the last few years, however, it has become evident that air suspensions have many serious problems, and development of magnetically suspended trains is now being pushed more vigorously in many countries. It is worth noting these problems so that one can see how magnetic suspensions can overcome them. They include: very low lift/drag ratio; poor payload capability; very small clearance; very stringent tolerances on track smoothness and straightness, making tracks expensive to install and difficult to maintain; safety problems; problems of power collection from track at high speeds; and excessive noise from the suspension blowers. Magnetic suspensions can overcome all these problems if suitably designed. If not suitably designed, however, the same problems will exist and magnetically levitated trains will have no more success than air cushion trains.

Magnetically levitated trains using normal magnets were proposed in the early 1900's, but were never developed because of excessive

weight and power requirements, as well as small clearance. In 1966, Powell
and Danby [67] proposed a new approach involving the use of superconducting magnets on the train and a normal metal track. The original concept
is illustrated in Fig. 21. The moving superconducting train magnets induce
currents in normal metal track, and the interaction of the track currents
and the field of the superconducting magnets generates a magnetic force
that suspend the train. Two sets of track coils are shown in Fig. 21; the
track set in the horizontal plane carries current whenever the train is
overhead and provides the magnetic lift. The track coils in the vertical
plane only carry current when the train moves laterally away from the
equilibrium position (vertical track coil is centered with respect to the
train loop). The train is stable vertically since the magnetic lift force
increases as the vertical clearance decreases; it is stable laterally, since
any displacement laterally generates a force pushing it back to the equilibrium position; it is stable with respect to roll, since two lines of track
are used, one on each side of the train; and it is stable with respect to
pitch and yaw, since discrete loops are used along the track and train. The
train is in effect trapped in a magnetic potential well, with the equilibrium
position at the bottom of the well. It cannot climb out of the well unless a
sufficiently strong external force (wind, track misalignment, etc.) pushes
it out. Like any bound particle, however, it can oscillate in the potential
well with a characteristic decay time set by the resistive properties
of the track. This can cause some problems, and we will examine these
later.

FIG. 21.   First Conceptual Version of Magnetically Levitated Train

The important feature of the Powell-Danby approach is the combination of lossless superconducting train magnets and a normal metal track. One could not afford the expense of superconducting track, while the losses with normal metal train magnets would be unacceptable. Further, current must be induced in the track by the train magnets if the suspension is to be stable and if track losses are to be reasonable. The arrangement of track loops shown in Fig. 21 is not satisfactory since a low speed lateral instability could develop. This could be eliminated with a somewhat different track geometry, however.

Since the 1966 design, several modifications of the original suspension concept have been developed. We shall look at these modified suspensions and examine their relative merits later. They all share the same basic approach, however, that of superconducting train magnets inducing currents in a normal metal track. As interest in superconducting magnetic suspensions began to grow, two companies (Krauss-Maffei and MBB) in West Germany started projects on magnetically suspended trains using iron magnets with normal metal windings. In order to keep power consumption reasonable they were forced to use an attractive force suspension with the train magnets slung under an iron rail. This is inherently unstable, as contrasted to the superconducting suspension with magnets above a normal metal track, which is inherently stable. This inherent instability of the German approach is overcome by continuous servo control of the magnet current. Independent servos are needed for every magnet. The magnet-track clearance is very small, ~ 1 cm, which raises safety problems. Maintaining this clearance requires the servo to handle very large amounts of reactive power. Real power requirements appear reasonable, however. The concept has been successfully tested at modest speeds on short lengths of carefully constructed and maintained track. The real question is how well a full scale system will work, involving track lengths of hundreds of km in which maintenance of track alignment to very close tolerances is required, and where trains must routinely operate at speeds of 200-300 mph. Not only must the system work reliably, but costs must be reasonable. My personal feeling is that such a system can be made to work, but track construction and maintenance costs will be excessively high.

Returning to superconducting magnetic suspensions, Wipf [68] and Coffey [69] in 1968 modified the 1966 Powell-Danby concept by replacing the track loops with a continuous aluminum sheet laid along the track. (Flexible joints or discontinuities are necessary every 20 meters or so to accommodate thermal expansion and contraction). This conducting sheet suspension (Fig. 22) has a somewhat simpler track, but decreases the magnetic lift/drag ratio, which increases the propulsion requirements. This decrease in lift/drag ratio is a result of the condition that in a conducting sheet suspension the total track current must equal the total current in the train magnets. In the 1966 concept, the track currents were made much smaller than is possible with a conducting sheet suspension by series connecting the track loops to inductors. This lowered the track loop current and thus increased the magnetic lift/drag ratio, since the only source of magnetic drag is the I^2R losses in the track. If track currents

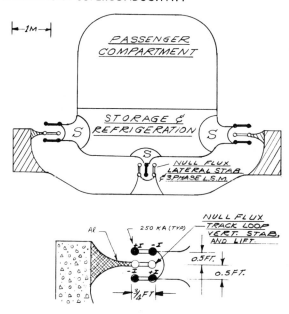

FIG. 22.   Magnetically Levitated Train Using Null Flux Suspension.

are reduced, train magnet strength has to be correspondingly increased to
generate the required lift force. The superconducting train magnets are
lossless, however, so an increase in magnet strength is perfectly
acceptable.

It is not generally appreciated, but because of the equal track-train
current condition in a conducting sheet suspension, the track losses are
comparable to those that would be experienced if the train magnets were
made of normal aluminum conductor. Another feature of the conducting
sheet suspension that decreases magnetic lift/drag ratio is that lateral
stability is achieved by having parts of the conducting sheet in a vertical
plane, usually at the sides of the track. If the train moves laterally towards
the side of the track, a net magnetic force develops to return it to the
center of the track. However, induced currents always flow in these
track side sheets, even when the train is laterally centered. This extra
drag considerably lowers the magnetic lift/drag ratio from that value corresponding to lift only. For adequate lateral stability this decrease
amounts to about a 30-40% decrease in magnetic lift/drag ratio.

When the conducting sheet suspension was first proposed, the magnet
dimensions and track sheet thicknesses in the early designs corresponded
to an overall magnetic lift/drag ratio of about 15. More recent designs
have gone to much longer magnets (typically 5 meters) and thicker track
plates (typically 4 cm), so that overall magnetic lift-drag ratios of about

40 seem possible. Very large weights of aluminum track are required, however, on the order of a million kg/km.

At about the same time in 1968, Powell and Danby [70] proposed a new type of magnetic suspension, the null flux suspension (Fig. 23). The important feature of this concept is that the track and train current loops are arranged so that there exists a position of magnetic symmetry for the train. At this symmetry position there is no net flux through any track loop circuit. If the train moves from this symmetry position, a net flux and current develop producing a magnetic force that tends to return the train to the symmetry position. A train with a constant external force, e.g., gravity, establishes an equilibrium position slightly displaced from the symmetry position, where the steady magnetic force opposes the gravitational force. With this concept the track current can be made arbitrarily small, and the magnetic lift/drag ratio arbitrarily large simply by increasing the train magnet strength. There is some optimum magnet strength of course, since the savings in propulsion cost due to higher magnetic lift/drag ratio at some point became offset by increased magnet cost. Generally, the optimum seems to be at a magnetic lift-drag ratio of 100 to 150 for travel in air at 300 mph. The null flux approach also permits much smaller amounts of aluminum metal in the track loops - typically almost an order of magnitude less than with a conducting sheet suspension. A null flux suspension is inherently much stiffer than a con-

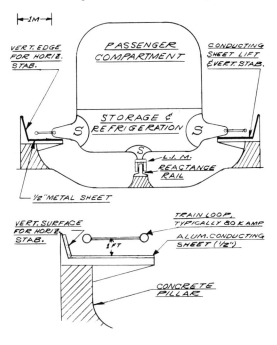

FIG. 23.   Magnetically Levitated Train Using Conducting Sheet
             Suspension.

ducting sheet suspension. The null flux suspension has significant advantages over the conducting sheet suspension: much larger external forces are required to make the train contact the track (e.g., 10 g vs. 1 g equivalent external force); the amplitude of train oscillations due to track irregularities is much smaller (e.g., ~ 1 cm vs. ~10 cm, compared to a nominal track-train clearance of ~ 20 cm); and greater reliability, since a much larger fraction of the train magnets must fail before the train hits the track. It has disadvantages, too: the track structure is more complicated, and a more sophisticated secondary train suspension is required for passenger comfort.

In the last few years both suspension concepts have evolved. Kolm and Thornton [71] have proposed a modification in the conducting sheet suspension that allows self-banking of the train. This involves bending the conducting track sheet into a semi-circular arc. The train can then ride up on the sheet, enabling it to take much tighter curves without passenger discomfort. Kolm has also proposed the elimination of the secondary suspension through active control of the vehicle lift force. In effect, the train flies straight and true regardless of the track irregularities underneath. This will require large clearances, on the order of 30 cm or more, and smooth tracks.

The null flux suspension has been modified [72] by the addition of an iron plate above the train current loops. The iron plate, which is typically 15-20 cm above the train loops provides the magnetic lift force on the train, while the null flux loops only provide stability. The inherent instability of the iron lift is overridden by the null flux loops, with the net stabilizing gradient (e.g., g's of restoring force per unit displacement) being controlled by the number of null flux loops installed. Figure 24 shows an arrangement where the net stability is about 0.2 g/cm. It is important to note that magnetic lift-drag ratio of this suspension is very high, on the order of 500, and that the stability can be set to any desired value without affecting the lift-drag ratio. It is also important to note that this suspension now has essentially no oscillatory component in lift force resulting from discrete track loops (the stabilizing loops are still discrete, but the effects are of second order). This removes one of the possible disadvantages of discrete loop tracks relative to conducting sheet tracks. However, the effect of fluctuating lift components can always be negated by appropriate secondary suspensions, if required.

Magnetically levitated trains cannot be considered in isolation from whatever propels them. At first, the tendency was to try to adapt already existing, at least conceptually, propulsion devices. The most important of these is the linear induction motor (LIM) which was being developed for the tracked air cushion train. Unfortunately, it appears unsuitable for magnetic trains. It has a very small clearance, typically a cm or less, between the reaction track sheet and the stator windings and iron flux path on the train. This negates one of the prime advantages of superconducting magnetic suspensions, that is, the large clearance of 15-30 cm. What is

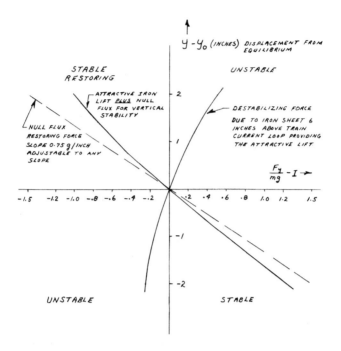

FIG. 24.   Stability of Magnetically Levitated Train with Iron Sheet Lift

worse, magnetic suspensions, even very stiff null flux suspensions, are not stiff enough to prevent external lateral forces from pushing the train onto the reaction track sheet. The conducting sheet suspension is far too weak to permit the use of a LIM, even though this has been proposed. In principle, one could mount the LIM train stator on separate bogeys sprung to the train body and make it independent of the magnetic suspension, but this seems a terrible way to build a high speed train. There are other disadvantages, too, though small by comparison: weight of the LIM stator and power conditioning equipment is very high, and power collection from the rails is very difficult.

In 1969 Powell and Danby proposed [73,74] a new type of propulsion, the linear synchronous motor (LSM). Here the superconducting train magnets interact with a small ac track current to propel the train. An additional set of track loops (Fig. 25) is wound with the same pitch as the train superconducting loops. The train is not only propelled by the ac track current, but is locked in phase with it, so that it moves along the track at a speed controlled by the ac frequency. This is advantageous for traffic control since spacing between trains will remain constant as long as local external forces (wind drag, etc.) do not become great enough to break the synchronous tie. This can be avoided by having adequate reserve thrust capability. This propulsion method would not be feasible with normal conductor train magnets; with superconducting train magnets, however, track current can be low enough that long track energized lengths (e.g.,

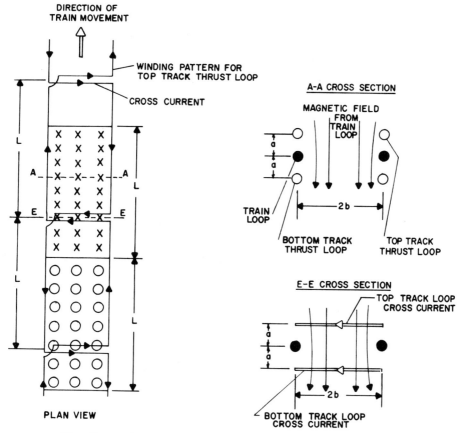

Fig. 25.   Schematic of Linear Synchronous Motor.

5-10 km blocks, switched as train moves along track) can be employed at high motor efficiencies, e.g., 80%. Further, with the null flux suspension the same superconducting magnets that suspend the train can also be used as part of the LSM propulsion.

This description has only skimmed the surface of magnetic levitation. There are vigorous discussions as to the optimum methods of suspension and propulsion, and strong convictions on the subject. However, it will require more analyses and experiments to determine the optimum approach. It is certain, though, that the entire high speed transport system must be optimized rather than just one piece of it. This makes comprehensive systems analyses very necessary.

Table 15 lists some of the factors that I feel will be important in optimizing magnetically levitated train systems, together with desirable goals for each factor. All these goals seem achievable given sufficient development. The safety parameters seem obvious. It should be noted, however, that the conducting sheet suspension will have large oscillations

TABLE 15.   Factors Affecting the Optimum Magnetic
Levitation System

System Factor	Goal
Safety	
Track–Train Minimum Clearance	$\geq$ 10 cm
Oscillation Amplitude (Maximum)	$\leq$ 30% of minimum clearance
Magnet Redundancy	$\geq$ 2/1
Magnet Failure Rate	$\leq 1/10^4$ hours
Magnetic Field in Passenger Compartment	$\leq$ 1 gauss
Passenger Comfort	
Ride Quality	Urban TACV Specs
Track Smoothness	$A \leq 80 \times 10^{-6}$
Suspension Performance	
Lift/Drag Ratio (Magnetic	$\geq$ 100
Lift/Drag Ratio (Air)	$\geq$ 30
Lift Capability (30 meter vehicle)	$\geq$ 50 metric tons
Passenger Capacity (30 meter vehicle)	$\geq$ 150
Propulsion Performance	
Motor Efficiency	$\geq$ 70%
Energized Track Length (LSM)	$\geq$ 5 km
Average/Maximum Thrust	$\leq$ 0.7
Costs	
Track Cost (2 way, not including right of way)	$\leq \$4 \times 10^6$/km
Vehicle Cost	$\leq \$2 \times 10^6$

[75,76] even with much smoother tracks than the A = 80 x 10^{-6} roughness parameter corresponds to, and that statistically there will be a relatively high frequency (e.g., 0.1% of oscillations) for track-train contact. Some sort of auxiliary servo control system seems necessary with a conducting sheet suspension to avoid such contact, which could result in crashes. Such a system would not be necessary with a null flux suspension. One safety parameter that is not considered very much, however, is the strength of the magnetic field in the passenger compartment. Some designs have assumed this can be as high as a hundred gauss or more. However, a number of experiments have demonstrated significant biological effects at a field of a few kilogauss [77,78,79]. In a real system, millions of passengers will be exposed for long cumulative periods to the fields in the passenger compartment. Like radiation, it may be extremely difficult to prove whether or not a threshold level exists for biological damage, and that low level exposures do not cause significant damage to a small fraction of a large population. Because of this uncertainty, it seems wise to design from the start vehicles in which the field in the passenger's compartment is compared to earth's field. If this is not done, a situation could arise similar to that with radiation, i.e., the allowable exposure levels continue to drop with time. It could be very difficult and expensive to modify an already developed magnetically levitated train system to meet lower magnetic exposure levels. It is possible to meet an earth's ambient field standard, though it will be much easier with some magnetic suspensions.

The measures necessary to achieve adequate ride quality are not well defined. The goal is to have a ride quality equal to or better than the urban TACV, which itself is much better than present trains. At 300 mph no track is sufficiently smooth to allow the magnets to be rigidly fixed to the train body unless active servo control of lift force is used, as Kolm [71] has proposed. Without active control, which rouses worries about reliability and safety, a secondary suspension, either active or passive, is required between the train magnets and train body. With such suspensions it appears feasible to meet TACV specs for both the conducting sheet and null flux suspensions.

If magnetic levitation is to be a major mode of transport, rather than a curiosity at a few locations, it must offer more than auto and air transport. One very important factor, now and in the future, will be relative energy consumption. Here, well designed magnetic systems can consume much lower energy/passenger mile than auto or air systems, and further, with magnetic propulsion, use nuclear generated electricity rather than increasingly scarce fossil fuels. The prime factors in energy efficiency will be overall lift/drag ratio and propulsion efficiency. If null flux systems run at 200 mph in air, for example, overall lift/drag ratios of 70 or more can be achieved. In low pressure tunnels, much higher speeds can be achieved with lift-drag ratios of several hundred.

The vehicle cost goal in Table 15 can be met fairly readily. At present prices the superconducting magnets and cryogenic system should cost no more than $250,000 for null flux suspensions and considerably less for conducting sheet suspensions since they require fewer magnets. It is hard to conceive of the total vehicle cost being much greater than 10^6, but even at $2x10^6$ it would be a bargain compared to $8x10^6$ for a 707 aircraft, even though the latter flies at twice the speed. The track cost goal is less certain. One can show that the materials costs are small compared to the $4x10^6$ assumed, but until substantial lengths of track are built for a high speed system and one finds out how expensive it is to achieve desired track quality, costs will remain vague. At $4x10^6$/mile and 200,000 passengers per day, the track capital cost corresponds to 1 cent/passenger mile. The track cost is probably the single factor that will make or break high speed ground transport, and no one really knows what it is.

Table 16 summarizes the significant operational parameters for the various suspension systems that have been proposed. A very important factor exists as to how passenger traffic will be handled in a real system, and it will affect what type of suspension is to be used. At 200,000 passengers per day, which is the present Tokyo-Osaka traffic level [80] if individual 150 passenger vehicles are used, they must operate on about 30 second headings at peak travel periods. I do not believe this to be safe and that in fact, multiple car trains will have to operate in peak periods, with headings of a few minutes between trains. However, it seems very unlikely that vehicles can be safely linked together if the conducting sheet suspension with its weak restoring force is used. With the conducting sheet suspension, the oscillations of one vehicle are a large fraction of the track train clearance; the oscillations of many coupled weakly stable vehicles could be truly frightening.

Table 17 lists some of the important magnet parameters for magnetically levitated trains. The maximum fields contemplated are on the order of 3 Tesla at the conductor surface, well within the capabilities of multifilament NbTi. The ALCOA conductor would appear ideal for magnetic trains because of its light weight and very good cryogenic stability resulting from the high purity aluminum around the NbTi filaments. The conductor could be made additionally safe by winding in sheets of high purity aluminum as heat drains so that any heat due to frictional movement could be removed at the conductor surface. This would make a very rigid strong conductor. The superconducting magnet can be made so that it would be virtually impossible to go normal, and a loss of lift would become practically a miracle. Unfortunately, the most likely accident is one that very possibly could cause a crash, i.e., the collision of a train magnet with an external object on the track. If the object is massive, it might

TABLE 16. Characteristics of Integrated Magnetic
Suspension and Propulsion Systems

System Normalization $W=10^5$ lbs, $\ell=100$ ft $V=300$ mph, D= 12 ft	A	B	C	D	E	F
Lift	Null Flux	Conducting Sheet	Conducting Sheet	Iron	Close Coupled Track Loops	Iron
Vertical stability	" "	"	"	Servo	"	Null Flux
Lateral stability	" "	"	"	Iron	Null Flux	" "
Propulsion	LSM	LIM	LSM	LIM	LSM	LSM
Lift magnet physical clearance (inches)	≥ 3	≥ 3	10	0.5	≥ 3	≥ 3
Minimum clearance for vehicle (inches)	≥ 3	0.5	10	0.5	≥ 3	≥ 3
Vertical restoring force (typ) - g/in.	1.0	0.17	0.07	-	0.17	0.50^+
Lateral restoring force (typ) - g/in.	0.5^+	0.17	0.07	?	0.50^+	0.50^+
Max. random error in track (in.)	±0.5	± 1	± 2	±0.2	± 1	± 1
Aerodynamic drag co- efficient (C_D)	0.2	0.2	0.4	≥ 0.2	0.2	0.2
Aerodynamic drag (MW)	2.6	2.6	5.3	>2.6	2.6	2.6
Magnetic lift/drag ratio	100*	25**	23**	~600	80	~2000
Magnetic drag (MW)	0.60	2.4	2.7	0.10	0.75	0.03
Total drag (MW)	3.2	5.0	8.0	2.7	3.4	2.6
Overall lift/drag ratio for $W=10^5$ lbs	18.5	11.8	7.4	<21.8	17.7	22.2
Total drag(MW) for $W=3\times10^5$ lbs.	4.4	9.8	13.4	>2.9	4.9	2.6
Al. wt/mile of 2-way track(lbs)(inc. propulsion)	5×10^5	1×10^6	1.5×10^6	-	6×10^5	3×10^5
Magnet wt. incl. LIM and servos for systems B&D (lbs)	15,000	30,000	30,000	30,000	10,000	15,000

* Train loop current chosen so as to give this value.
** Includes drag due to lateral stability currents - drag is present even when external lateral force
+ Can be adjusted to any desired value; this value is illustrative   is zero

TABLE 17.   Characteristics of Typical Superconducting
Magnet Systems for Magnetic Levitation

Basis: vehicle length = 30 meters
vehicle gross weight = 45,000 kg
nominal track-vehicle clearance = 15 cm

Parameter	Suspension Type	
	Null Flux	Conducting Sheet
Superconductor Current, KA	300	220
KA m of Superconductor on Vehicle	43,000	5650
Maximum Field in Superconductor, Tesla	3.0	2.1
Cryostat Length, Meters (1 Magnet Loop per Cryostat)	3	1[*]
# of Magnets on Vehicle	20	8[**]
Magnet Weight on Vehicle, kg Superconductor High Purity Aluminum Total, including Structure	 1640 545 6700	 215 75 900
Heak Leak, Watts 4°K Sink 77°K Sink	 40 450	 6 60
Weight of Tankage for 3 days cooling, kg 4°K Sink 77°K Sink	 550 640	 75 85

---

[*]   May be desirable to use longer magnets to increase magnetic life/drag
ratio; this will increase magnet weight and heat leaks.

[**] Two magnets at each corner of vehicle for redundancy.

take out the entire line of train magnets. This type of accident can probably only be minimized and not avoided. Deflection structures can help, together with as redundant a lift system as possible.

Japan and Germany now seem the world leaders in magnetically levitated trains. Japan plans to have a 300 mph, 300 mile system between Tokyo and Osaka by the early 1980's and has mounted a very strong research effort on superconducting suspensions to get there [81, 82]. Germany is comparably committed, but to conventional magnet suspensions [83], though it also has a strong effort in superconducting suspensions. In the U.S. individuals are strongly committed but the government is not. Relatively small experimental efforts on superconducting suspensions are underway at MIT [71], Ford [75], and SRI [76]. The U.S. situation is understandable since a massive air and auto transport system already exists. The official attitude seems to be that if magnetically levitated trains prove successful, the U.S. can always buy them from abroad if it wants to.

In summary, there do not appear to be any serious technical obstructions in the path of magnetically levitated train systems, but there can be serious political ones, and perhaps economic ones, if the track should cost too much. It is extremely important, however, to try to foresee what system requirements will arise and to avoid being locked into an expensive program that could hit a disaster point. I think there are several such points: (1) being committed to a system with fields in the passenger compartment that people later find objectionable; (2) being committed to a system that needs short vehicle headings to handle traffic, and then having new safety rules impose much longer vehicle headings; (3) being committed to any system that requires very carefully constructed track to be safe, and finding out later that such track is impractical in the real world. If any of these happen, magnetically levitated transport might never recover.

VIII. INDUSTRIAL PROCESSING

Industrial processing applications of superconductivity have received relatively little attention, though it is in this area that the earliest large scale commercial impact could be made. This probably results from industrial processes not being very interesting to most workers in superconductivity, and from the conservative nature of process engineers. They prefer small, easily digested improvements, rather than radical ones that seem risky. Also, industrial processes are usually more of an art than a science, and many of the most important details are either not known or not readily available for analysis.

Three properties of a magnetic field are potentially useful for industrial processes: gradient, magnitude, and time variation. Gradient fields are principally useful for separation of constituents from process streams. The magnitude of a field could be useful in accelerating or

inhibiting reactions in process streams. Time variation of a magnetic
field could be useful in a number of ways - for inducing electric fields,
promoting reactions, or generating E-M forces.

Table 18 lists some applications of superconducting magnets for
industrial processes. To date, most interest has centered on separation
processes using gradient fields. This is a natural outgrowth of having well
established processes for the recovery of magnetite from raw iron ore.
Applications based on the magnitude of magnetic fields, as well as time
variation of magnetic fields, may be expected to become important as
more research is done on the effect of magnetic fields.

MIT and MEA working together [84, 85] have made a basic improvement in magnetic separation technology through the use of high gradient
magnetic separation (HGMS). The force on a particle in a magnetic field
is described by

$$F_m = \chi V H \frac{dH}{dx}$$

where $\chi =$ magnetic susceptibility
$V =$ particle volume

$\frac{dH}{dx} =$ magnetic field gradient

Previous magnetic separators used relatively large structures [balls,
grooved plates, etc.] to achieve magnetic field gradients, so that it was
only practical to separate ferromagnetic particles from non-ferromagnetic
ones. The MIT-MEA contribution was to use a structure of fine steel
wool, with much smaller characteristic sizes, and therefore much higher
field gradients. This is illustrated in Fig. 26 where the typical fiber size,
100μ, yields very large gradients. The important separation parameter
[86] in the matrix is the ratio

$$R = \frac{F_m}{F_D + F_G}$$

$F_G =$ gravitational force on the particle

$F_D =$ viscous drag force on the particle

which depends on particle dimensions, magnetic susceptibility, slurry
velocity, magnetic field, and characteristic fiber size. Typical behavior
of this ratio as a function of particle size and slurry velocity is shown
in Fig. 27, for constant magnetic field, fiber size, and particle susceptibility. The maximum separation ratio increases as slurry velocity
decreases, because of decreasing viscous drag forces. At very small
particle diameters, viscous drag predominates, while at very large
particle diameters (diameter > fiber diameter) the effective gradient field
decreases, reducing the magnetic force. Both effects decrease the
separation ratio. Thus there is an optimum particle diameter for which
the separation ratio is maximum.

TABLE 18. Some Industrial Process Applications

of Superconducting Magnets

1. GRADIENT METHODS

    A. CLEANING PROCESSES

        (1) REMOVAL OF IMPURITIES FROM KAOLIN

        (2) CLEANING OF IRON ORE CONCENTRATE

        (3) ASH AND SULFUR REMOVAL FROM COAL

    B. ORE RECOVERY PROCESSES

        (1) RECOVERY OF IRON FROM LOW GRADE TACONITES

        (2) RECOVERY OF WO_3 FROM TAILINGS

    C. WATER FILTRATION

        (1) SEWAGE TREATMENT

        (2) INDUSTRIAL WASTE WATER

2. MAGNITUDE METHODS

    A. CONTROL OF ORGANIC REACTIONS (POLYMERIZATION)?

    B. STERILIZATION?

3. TIME VARYING FIELD METHODS

    A. RECOVERY AND SEPARATION OF RECYCLED SCRAP

SCHEMATIC OF MIT-MEA HIGH
GRADIENT FIELD MAGNETIC SEPARATOR

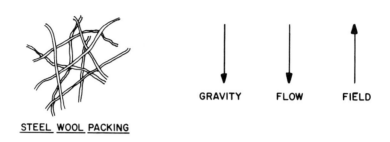

STEEL WOOL PACKING   GRAVITY   FLOW   FIELD

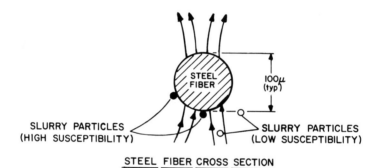

STEEL FIBER CROSS SECTION

FIG. 26.   High Gradient Magnetic Separation Method.

Particles for which the separation ratio exceeds one are trapped on a fiber (though collisions with other slurry particles may temporarily dislodge them). Even if the separation ratio is less than one, however, a magnetic separator could still be very useful, though particles with high susceptibility would not be trapped, but would be slowed in their travel through the matrix. The separator would be more of a magnetic chromatograph than a trapper with characteristic breakthrough curves for materials of different susceptibility.

The virtues of superconducting magnets for separation are that much greater fields can be used than with conventional magnets, and cost/unit volume of field is much smaller. The higher field will obviously increase the magnetic separation ratio; a lower volume cost can also increase the separation ratio, since smaller particles can be used with lower slurry throughput velocities, while still achieving economic operation. With such techniques, it should be economically feasible to separate and classify virtually all paramagnetic substances. Since almost all substances are paramagnetic, magnetic separation with superconductors will become an extremely important unit operation. As yet no superconducting magnet separation systems have been built, principally because the processes considered so far can still use conventional magnets, though less efficiently.

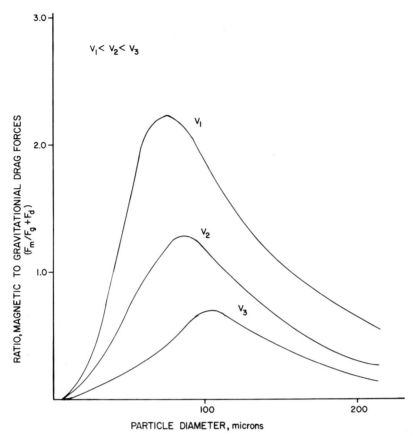

FIG. 27.   Magnetic Separation Parameter vs. Particle Diameter and
Velocity.

There are now a number of commercial HGMS units cleaning
Kaolin of impurities, and commercial units for recovery of iron oxides
from low grade taconites should be in operation in the near future.

Coal cleaning is one of the most exciting applications of
magnetic separation. A major U.S. program is now underway on low
BTU gasification of coal as a clean fuel for power plants. However,
gasification increases fuel costs considerably, i.e., from approximately
$0.40/$10^6$ BTU without gasification to approximately $1.00/$10^6$ BTU with
gasification. The principal virtue of low BTU gasification is the removal of
sulfur and ash from the coal. No satisfactory method now exists for the
removal of SO_2 from stack gasses.

If magnetic separation can clean coal sufficiently that gasification is
not necessary, it might greatly reduce the fuel costs for coal fired power
plants. Approximately half of the sulfur in coal is pyritic and can be
magnetically separated, while half is organic and cannot. Trindade [87]
removed approximately 2/3 of the pyritic sulfur in a coal slurry with one

pass through a non-optimized magnetic separator. Only about 10% of the ash was removed, however. An optimized separator could do much better, but the organic sulfur would still not be removed. If some way to convert the organic sulfur to pyritic sulfur can be developed, then magnetic separation could probably be used in place of gasification. One possible conversion method is to heat ground coal with powdered iron.

An appreciation of the potential economic savings of magnetic coal cleaning is given by the following example. About 300 lbs/sec of coal must be cleaned in a 1000 MW(e) plant. If we assume that the following processes parameters are adequate to clean the coal [30 kG field, 2.5% slurry (vol.) of coal in water, 1 cm/sec slurry velocity, length per pass of 1 meter, and 2 passes], a solenoidal magnet of 10 m bore and 15 meters long is required. A 30 kG magnet of this size if built today would cost on the order of $10x10^6$. This capital cost translates into an additional cost of only 2 cents/10^6 BTU. Additional process equipment (pumps, etc.) plus operating costs would increase the cost somewhat, but not by a large amount. The grinding requirements would not be much different than those now necessary for coal fired plants.

If more exacting process parameters prove necessary (e.g., much lower slurry velocities, many more passes, etc.) and much larger magnet volumes are required, magnetic coal cleaning could still be economic through expected reductions in superconductor prices, use of warm reinforcement, and economies of scale. The cleaning process could be consolidated into a few very large plants with low unit costs which would serve a much larger number of coal fired power plants. A dual purpose magnetic storage coal cleaning magnet would seem a natural combination.

Water purification will probably be the most significant application of magnetic separation, however. Fine particles of iron oxide are added to waste water, either from industrial sources or from primary sewage treatment plants, and then magnetically separated from the water. Suspended solids, BOD, and bacteria are greatly reduced, along with some types of dissolved solids, since they are absorbed on the seeded particles [88]. Marston [89] estimates that magnetic filtration should be much cheaper than tertiary sewage treatment. A $2x10^6$ gal/day magnetic filtration pilot plant is planned for operation on the Charles River in Cambridge, Massachusetts, in the summer of 1974. The unit is not superconducting, but superconducting coils would certainly be economically justified for large scale treatment. For cleaning of very large volumes of waters, e.g., rivers and lakes, a dual purpose magnetic storage water purification magnet seems ideal. Water desalinization does not appear practical with magnetic filtration, but treatment of certain brackish waters may be possible.

If legislated water quality standards are fully implemented, the total investment in U.S. water purification plants will be hundreds of billions of dollars. If magnetic water purification is significantly cheaper than conventional methods, many billions could be saved.

Turning now to applications dependent on magnetic field strength, no practical process has yet been proposed. However, the field strength available to industry has been relatively limited, and the cost/unit volume of field relatively high. With superconducting magnets field strength can be almost an order of magnitude larger than that previously available, and at very low cost. I suspect that in the next few years many processes dependent on magnetic field to control reaction processes will be developed. One possible direction might be polymerization reactions. It would be very interesting to see what effect 15 Tesla magnetic fields have on polymer chain reaction rates and structure. Another example might be sterilization of various materials, e.g., hospital supplies. Effects on metallurgical properties via changes in crystal growths could also be a possibility. The low cost/unit volume of superconducting magnets may well be very impor-tant, since low reaction rates can be compensated by long residence times.

Methods using time varying magnetic fields have not yet been significantly applied. A project at the University of Virginia, with EPA support, has attempted to electro-magnetically separate different types of scrap metal (e.g., aluminum, copper, etc.). Currents in the scrap, induced by rotating magnets, interact with a gradient magnetic field to sort the material according to density and electrical conductivity. As yet, this process has not been commercially applied. The use of superconductors to generate electric heating by transformer action does not appear practical, because of losses in the superconductor. Superconducting RF cavities might be very useful for certain electric-discharge gas phase reactions, because of the much lower losses and higher electric fields achievable with such cavities. Fixation of atmospheric nitrogen and production of hydrazine are examples of such an application.

In summary, superconductors will open up many entirely new industrial processes, as well as displacing many old ones. They will permit economical recovery of minerals from lower grade ores than has been previously possible, thus extending the worlds increasingly scarce resources. They should provide the means to cheaply clean our polluted lakes and rivers. Achievement of the full potentialities of superconductors in industrial processes will require imaginative collaboration between workers in super-conductivity and industrial engineers.

REFERENCES

1.   Powell, J., et al., ANS Trans. 16, 239 (1973).
2.   Hoffman, K., et al., AET-8, Associated Universities, Inc. (1972).
3.   Powell, J., p. 346 Proc. 1972 App. Supercond. Conf., Annapolis.
       Md., IEEE Pub. 72CH0682-5-TABSC (1972). Also BNL 16580 (1972).
4.   Powell, J., et al., to be published.
5.   Golovin, I.N., et al., p. 199 in Proc. Nuc. Fusion Reactor Conf.
       Culham (1969).
6.   Fraas, A.P., ORNL-TM-3096 (1973).
7.   Lubell, M.S., et al., 4th Conference on Plasma Physics and
       Controlled Thermonuclear Research, Madison, June 17-23 (1971).
8.   Gibson, A., Hancox, R., and Bickerton, R.J., 4th Conference on
       Plasma Physics and Controlled Thermo Nuclear Research, Madison,
       June 17-23 (1971).
9.   Burnett, S.C., et al., LA-5121-MS, Los Alamos (Dec. 1972).
10.   Thomassen, K. presented at Texas Symposium on Engineering
       Aspects of Fusion, Nov. 20-22, 1972. Also, LA-5087-MS, Los
       Alamos (1972).
11.   Werner, R., et al., presented at Texas Symposium on Engineering
       Aspects of Fusion, Nov. 20-22, 1972. Also UCRL-74326
       (1972).
12.   Moir, R. and Taylor, C.E., "Magnets for Open-Ended Fusion
       Reactors" presented at Texas Symposium on Engineering Aspects
       of Fusion, Nov. 20-22. Also UCRL-74326, Lawrence Livermore
       Laboratory (1972).
13.   Wood, L., and Weaver, T., COT/Phys. 73-B, Lawrence Livermore
       Laboratory (1973).
14.   File, J., Mills, R.G., and Sheffield, G.V., IEEE Trans. on Nuc.
       Sci. NS-18, 277 (1971).
15.   Powell, J., and Bezler, P., presented at Texas Symposium on
       Engineering Aspects of Fusion, Nov. 20-22, 1972. Also, BNL 17434
       (1972).
16.   Roberts, M. ORNL, personal communication (1973).
17.   Henning, C., et al., p. 521, Proc. 4th Intl. Conf. on Magnet
       Technology, Brookhaven (1972).
18.   File, J., and Sheffield, G.V., Proc. 4th Intl. Conf. on
       Magnet Technology, Brookhaven (1972).
19.   Williams, J. and Clement, J., ed., "Satellite Nuclear Power
       Station: An Engineering Analysis", Georgia Institute of Technology,
       Atlanta, Ga. (1973).
20.   Glaser, P., Proc. 7th IECEC 507 (1972).
21.   Amend, W.E., and Petrick, M., Vol. IV, 125, Proc. 5th Int. Conf.
       on MHD, Munich (1973).
22.   Rosa, R.J., Magnetohydrodynamic Energy Conversion, McGraw
       Hill, New York (1968).
23.   Heywood, J.B., and Womack, G.J., ed., Open Cycle MHD Power
       Generation, Pergamon Press (1969).
24.   Stekly, Z.J.J., Thome, R.J., and Cooper, R.F., Vol.IV, 7,
       Proc. 5th Int. Conf. on MHD, Munich (1971).

25. Brogan, T.R., p. 227, <u>Adv. in Plasma Physics</u>, Academic Press
     (1969).
26. Komarek, P., and Bohn, T., p. 265, Proc. 4th Intl. Conf. on Magnet
     Technology, Brookhaven (1972).
27. Rosa, R.J., AVCO, personal communication (1973).
28. Klepeis, J.K., and Louis, J.F., Vol. I, 649, Proc. 5th Int. Conf. on
     MHD, Munich (1973).
29. Braun, J. Zinko, H., and Palmgren, S., Vol. II, 531, Proc. 5th
     Int. Conf. on MHD, Munich (1973).
30. Kirillin, V.A., et al., Vol. I, 353, Proc. 5th Int. Conf. on MHD,
     Munich (1973).
31. Mori, Y., 569, Proc. 5th Int. Conf. on MHD, Munich (1973).
32. The 1970 National Power Survey, FPC, Part II, p. II-1-15,
     Washington, D.C. (1970).
33. Salzano, F.S., Brookhaven National Laboratory, personal
     communication (1973).
34. Hassenzahl, W., Rogers, J., and McDonald, T., LAUR-73-73,
     Los Alamos (1973).
35. Watson, M.B., EQL Report No. 6, California Institute of Technology
     (1972).
36. Boom, R.W. et al., Paper 0-2, at Cryogenic Eng. Conf., Atlanta,
     Ga. (1973).
37. Appleton, A., and Anderson, A., in Proc. of 1972 App. Supercond.
     Conf., Annapolis, Md., IEEE Pub. 72CH0682-5-TABSC (1972).
38. Mole, C.J., et al, p. 151, in Proc. of 1972 App. Supercond.
     Conf., Annapolis, Md., IEEE Pub.72CH0682-5-TABSC (1972).
39. Fox, G.R. and Hatch, B.D., "Superconductive Ship Propulsion
     Systems", p. 33, in Proc. of 1972 App. Supercond. Conf., Annapolis,
     Md., IEEE Pub.72CH0682-5-JABSC (1972).
40. Eckert, D., et al., "Three-Phase-Synchronoms Alternator with
     Superconducting Field Winding", p. 128, in Proc. of 1972 App.
     Supercond. Conf. Annapolis, Md., IEEE Pub.72CH0682-5-TABSC
     (1972).
41. Appleton, A., p. 16, in Proc. of 1972 App. Supercond. Conf.
     Annapolis, Md., IEEE Pub.72CH0682-5-TABSC (1972).
42. Lowry, L.R., p. 41, in Proc. of 1972 App. Supercond. Conf.,
     Annapolis, Md., IEEE Pub. 72CH0682-5-TABSC (1972).
43. Stekly, Z.J.J., p. 47, in Proc. of 1972 App. Supercond. Conf.,
     Annapolis, Md., IEEE Pub. 72CH0682-5-TABSC (1972).
44. Levedahl, W.J., p. 26, in Proc. of 1972 App. Supercond. Conf.,
     Annapolis, Md., IEEE Pub. 72CH0682-5-TABSC (1972).
45. Liang, S., and Martin, L., in Proc. of 1972 App. Supercond.
     Conf., Annapolis, Md., IEEE Pub. 72CH0682-5-TABSC (1972).
46. Electric Utilities Industry Research and Development Goals
     Through the Year 2000, ERC Pub. No. 1-71 (June 1971).
47. The 1970 National Power Survey, FPC, Part I, p. I-18-11,
     Washington, D.C. (1970).
48. Forsyth, et al., BNL 50325, Brookhaven National Laboratory
     (1972).

49.   Perry, E.R., ITE, personal communication (1973).

50.   Steinberg, M., Powell, J.R., and Manowitz, B., BNL 50187, Brookhaven National Laboratory (1969).

51.   Burnett, J.R., GE, personal communication (1973).

52.   Bartlit, J., Edeskuty, F., and Hammel, E., p. 177 in Proc. 4th Intl. Cryogenic Eng. Conf., Eindhoven, May 24-26, 1972.

53.   Working Symposium on Liquid H_2 Fueled Aircraft, NASA Langley Research Center, May 15-16 (1973).

54.   Bartlett, R., et al., "USAEC-DAT DC Superconducting Power Transmission Line Project at LASL," LA-5271-PR, Los Alamos (Nov. 1 to March 1, 1973).

55.   Dahlgren, S., and Kroeger, D.M., Battelle Northwest, to be published (1973).

56.   Dubois, P., et al, p. 173 in Proc. 1972 App. Supercond. Conf. Annapolis, Md., IEEE, Pub. 72CH0682-5-TABSC (1972).

57.   Engelhardt, H., p. 56 in HGU Colloquim, Heidelberg, Oct. 21-22, 1971.

58.   Carter, C.N., Cryogenics, 13, 207 (1973).

59.   Long, H., et al., EEI Project RP 78-7, Linde Corporation (1969).

60.   Meyerhoff, R.W., Cryogenics, 11, 91 (1971).

61.   Dean, J.W., in Proc. of 4th Intl. Conf. on Magnet Tech., BNL, N.Y. (1972).

62.   Goldschwartz, J.M., and Blaisse, B.S., p. 367 in Low Temperatures and Electric Power, IIR, Paris (1969).

63.   Taylor, M.T. p. 119, in Low Temperatures and Electric Power, IIR, Paris (1969).

64.   Bayliss, J.A., CERL RD/L/N 79/72 (1972).

65.   Forsyth, E., et al., NSF Progress Report for May-June, 1973, Brookhaven National Laboratory, Upton, New York (1973).

66.   Forsyth, E., et al, NSF Semi-annual Report, PTP-11, BNL, Upton, New York (March 28, 1973).

67.   Powell, J., and Danby, G., ASME Paper WA-RR-66-5 (1966). Also in Mech. Eng., 89, 30 (1967).

68.   Guderjahn, C.A., Wipf, S.L., et al, J. Appl. Phys., 40, 2133 (1969).

69.   Coffey, H.T., Chilton, F., and Barbee, T.W., J. Appl. Phys, 40, 2161 (1969).

70.   Powell, J., and Danby, G., presented at Soc. of Eng. Science, Princeton, N.J.(1968). Also in Recent Adv. in Eng. Science, V. 5 159, Gordon & Breach (1970).

71.   Kolm, H., and Thornton, R., in the Proc. of the 1972 App. Supercond. Conf., IEEE Pub.72CH0682-5-TABSC (1972).

72.   Danby, G., and Powell, J., p. 120, in the Proc. of 1972 Appl. Supercond. Conf., IEEE Pub.72CH0682-5-TABSC (1972).

73.   Powell, J., and Danby, G., Trans. 4th IECEC Conf., 931, Washington, D.C. (1969).

74.   Powell, J., and Danby, G., Trans. 6th IECEC Conf., 118, Boston, Mass. (1971).

75.   Coffey, H.T., Chilton, F., and Hoppie, L.O. FRA-RT-72-39,
        Federal Railroad Administration (1972).
76.   Davis, L.C., Reitz, J.R., Wilkie, D.F., Borcherts, R.H.,
        RFA-RT-72-40, Federal Railroad Administration (1972).
77.   Fardon, J., et al., Nature, 211, 433 (1966).
78.   Levengood, W., J. Embryology and Exp. Morphology, 21, 23 (1969).
79.   Barnothy, M.F., Nature, 221, 270 (1969).
80.   Oshima, K. and Kyotani, Y., p. 26 in Proc. 4th Intl. Cryogenic
        Eng. Conf., Eindhoven, May 24-26, (1972). IPC Press (1972).
81.   Takano, N., et al., p. 191, in Proc. 4th Intl. Cryogenic Eng. Conf.
        Eindhoven, IPC Press, May 24-26 (1972).
82.   Oshima, K., and Kyotani, Y., Cryogenic Engineering Conference,
        Aug. 8-10, 1973, Atlanta, Georgia.
83.   Bohn, G., et al., p. 202 in Proc. 4th Intl. Cryogenic Eng. Conf.,
        Eindhoven, IPC Press, May 24-26 (1972).
84.   Kolm, H., p. 17, Proc. of High Gradient Magnetic Separation Sym-
        posium, MIT National Magnet Laboratory, May 22, 1973.
85.   Marston, P., p. 25, in Proc. of High Gradient Magnetic Separation
        Symposium, MIT National Magnet Laboratory, May 22, 1973.
86.   Oberteuffer, J., p. 86, in Proc. of High Gradient Magnetic Separa-
        tion Symposium, MIT National Magnet Laboratory, May 22, 1972.
87.   Trindade, S., p. 102, in Proc. of High Gradient Magnetic
        Separation Symposium, MIT National Magnet Laboratory, May 22,
        1973.
88.   Wechsler, I., p. 38, in Proc. of High Gradient Magnetic Separation
        Symposium, MIT National Magnet Laboratory, May 22, 1973.
89.   Marston, P., Magnetic Engineering Associates, personal
        communication (1973).

Chapter 2

PRACTICAL SUPERCONDUCTING MATERIALS

D.Dew-Hughes

University of Lancaster

Lancaster, England

I. INTRODUCTION

A. Applications of Type II Superconductors

Superconductors are employed in electrical devices because of two advantages: low (ideally zero) power consumption, and the compactness with which superconducting devices can be designed. The first arises from the absence of electrical resistivity in the superconducting state; the second arises partly because the absence of power dissipation removes the need for internal cooling, and partly from the very high current densities which can be achieved in superconductors. A device with a given power rating will have a smaller size if made from superconducting material; this is of great importance when the device must be transported from factory to site. The upper limit on the power of some devices may be set by the maximum size which can be transported. In rotating machinery, the limit on size, and hence on power, is determined by the magnitude of the centrifugal forces that the rotors can withstand. For both these reasons the superior power densities available with superconductors may allow an upper revision of these limits. Superconductors are likely to find employment in electrical engineering applications involving heavy currents and low voltages. The disadvantage of employing superconductors is the necessity for cooling the device down to, and keeping it at, a temperature of a few kelvins. Superconducting devices are economic only if the reduced power dissipation in the device itself is sufficient to compensate for the power requirements of the refrigeration plant. Even then the capital cost of a superconducting device may be so large as to preclude its replacing a conventional device.

At any temperature T below T_c, the application of a magnetic

field greater than a certain value, $H_c(T)$, will destroy superconductivity
and return the material to the normal (resistive) state. The critical
field is, like the critical temperature, characteristic of the material. It
is also a function of temperature, falling from a value $H_c(O)$, usually
10^4 - 10^5 A/m, at OK, to zero at the critical temperature. The current
density carried by a superconductor is limited to a critical value J_c,
which is a function of temperature, and applied magnetic field. Its value
is such that in some materials a dc current density of ~10^9 A/m^2 may be
achieved (cf. the normally allowed 2 x 10^6 A/m^2 for copper conductor at
room temperature). Current densities for alternating resistanceless
current are, however, disappointingly small.

The applications of superconductivity are of three types:

1. Superconductors may be used for the production of large magnetic fields. Large refers not only to the field intensity, but also to the
volume over which it is generated. It is with the materials required for
the generation of large magnetic fields that these lectures will be concerned.

2. Superconductors have been proposed for power transmission
cables. A superconducting cable cannot compete with overhead transmission lines, except over very large distances (several hundred km.).
However, where high power density lines are forced underground for
amenity or other reasons, a superconducting cable may show some economic advantages. Some material requirements for such a cable are essentially the same as for magnet materials, and will not be treated separately.

3. Superconductors can be employed in electronic circuitry.
Devices were originally based on the idea of using the superconducting-
normal transition as a switch. Performance was disappointing compared
to that of thin-film transistor developments, and this application has
been completely dropped. Devices based on a different property of
superconductors, the Josephson effect, are still of great interest. The
critical current which flows between two superconductors weakly coupled
together through a weak link, which may be a very thin (~ 20 Å) oxide
barrier, a layer of nonsuperconducting metal, or a narrow bridge of
superconductor, varies in an oscillatory fashion depending on the local
value of the magnetic field in the weak-link. This effect can be used as
the basis of highly sensitive ammeters, gaussmeters and voltmeters, as
well as logic and memory elements. The materials problems here are
quite different from those associated with (1) and (2) above, and are
basically those of thin film circuitry. Though interesting, they will not
be dealt with here. These are discussed by Matisoo in Chapter 9.

The most important application, in terms of quantity of material
employed, is, and will be for many years to come, the production of
magnetic fields. Firstly, magnetic fields are used in physics laboratories for research purposes, and small superconducting magnets are
now the standard way of producing fields above a few kilo-oersteds.

Highly stable magnetic fields for NMR and for high resolution, high
voltage electron microscopy are provided by superconducting magnets.
Large magnetic fields are also required in high energy nuclear physics,
and superconducting magnets have been built for beam bending and focus-
sing, as well as for large bubble chambers.

Power generation, by MHD or thermonuclear fusion, will require
large fields which can only be provided by superconducting magnets.
Magnetic ore separation, sewage treatment, and water purification are
all areas in which superconducting magnets are expected to make an impact over the next few years. Superconducting motors and generators,
which are now under development, are devices in which the superconductor is used to generate high magnetic fields. Superconducting magnets
have also been proposed for the levitation of high speed tracked vehicles.

B. Properties Required of Commercial Superconductors

The requirements of superconductors for the above applications,
briefly summarized, are:

Critical temperature, T_c. The higher T_c, the higher can be the
operating temperature of the device, thus reducing refrigeration load
and hence both capital and running costs. The other important superconducting properties (to some extent dependent on other factors) in addition scale with the critical temperature. It is, therefore, quite obvious
that a high critical temperature is desirable, and a lot of effort has been
expended in raising T_c to above 20 K.

Critical field, H_c. The superconductor is to be used to generate
high fields, which cannot be higher than some fraction ($\sim 1/2 - 3/4$)
of the critical field of the superconductor. The higher the field to be
generated, the higher must be the critical field of the material.

Critical current density, J_c. One of their main advantages is the
high current densities of which superconductors are capable. The higher
the current density the more compact the device, reducing the amount of
superconducting material, which is usually quite costly, and also reducing the mass which has to be refrigerated. High current densities thus
achieve economies in capital and running costs.

Stability. Superconductors are unstable to sudden changes in
current, field or temperature, and to mechanical shock. If an instability occurs while the superconductor is carrying a large current, the
local energy dissipation can be sufficient to vapourise the conductor.
Superconductors must, therefore, be protected from the effects of instabilities, where the latter cannot be prevented.

Alternating current losses. Electrical engineers are more often
concerned with alternating, or rapidly varying, currents and fields. The
switching-on or -off of a device also constitutes a transient ac. It is unfortunate that under these conditions superconductors do dissipate power.
ac losses must, therefore, be minimized for a superconductor to be used
in any real engineering application.

Fabrication. A superconductor is of no use unless it can be manufactured as a continuous, flexible conductor in ~ 1 km lengths. Ideally
this means the production of wire by conventional metallurgical techniques, though in some cases rather more complicated procedures are
necessary.

Cost. As always, cost is the most important parameter of any
engineering material, and every effort must be made to keep it as low as
possible.

In this paper, each of these properties will be examined in turn.
Methods for producing optimum values will be discussed, and reasonable
estimates of upper limits which can be placed in these values will be
made. An elementary knowledge of the physics of superconductivity and
in particular type II superconductors, is assumed. This latter subject is
covered in detail in the review article by Goodman [1] and the books by
de Gennes [2] and Saint-James, Sarma and Thomas [3]. Several of the
articles contained in the two volumes edited by Parks [4] provide a background relevant to the understanding of superconductors. Superconducting
materials have been previously reviewed by Livingston and Schadler [5],
Catterall [6], and Dew-Hughes [7, 8].

II. THE CRITICAL TEMPERATURE

A. BCS Theory

The Bardeen-Cooper-Schrieffer (BCS) theory [9], now the accepted
theory of superconductivity, predicts that the critical temperature of a
superconductor is given by:

$$T_c \simeq \theta_D \exp[-1/N(0)V]$$

where θ_D is the Debye temperature, $N(0)$ is the electronic density of
states at the Fermi surface, and V, the interaction parameter, is the
matrix element which describes the electron-phonon interaction which is
the cause of superconductivity. $N(0)$ is proportional to the Sommerfeld
constant γ, both γ and θ_D are known for most elements, and for many
alloys. V, however, is unknown, and it is not even known how to predict
V in terms of other atomic or crystalline parameters. Despite its success in many other aspects of superconductivity, the BCS theory has unfortunately given little help in the search for superconductors with higher
values of T_c. This is not a criticism of the theory, but is entirely due to
ignorance of the interaction parameter which is crucial to the theory.

Neglecting V for the moment, metals, alloys or compounds with
high values of γ and θ_D ought to have high critical temperatures. To a
first approximation, values of T_c do seem to follow, at least qualitatively,
variations in γ, but the dependence upon θ_D is not so obvious. Metals

with very low values of θ_D, such as Hg (θ_D = 70 K) and Pb (θ_D = 96 K) have critical temperatures of 4.16 K and 7.22 K respectively. The group Vb elements, V, Nb and Ta, with similar electronic structures, have values of θ_D and T_c respectively of 338 K and 5.3 K, 320 K and 9.2 K, 230 K and 4.4 K. It is possible from a knowledge of T_c, γ and θ_D to extract a value for V. For example, in the TiZr alloy system V shows a variation with composition which is roughly the inverse of that of θ_D [10]. The fact that V should depend upon θ_D is not surprising, since it describes an electron-phonon interaction. It is tempting to assume that the dependence of V on θ_D cancels out the initial θ_D term in the BCS expression, and that T_c governed by the density of states. Thus, the search for high T_c superconductors should be concentrated in the first instance on alloys or compounds in which γ is high.

B. Matthias Rules

From an examination of all the available experimental data, Matthias systematized the occurrence of superconductivity, and formulated empirical rules [11]. The important rules are:

1. Superconductivity occurs only in metallic systems, and never if the system exhibits ferro- or antiferromagnetism.

2. Superconductivity occurs when the electron to atom ratio (e/a) lies between 2 and 8. T_c varies with e/a for metals, and alloys between metals, in the same period of the periodic table. Nontransition metals show T_c increasing as e/a increases from 2 to 6 (beyond 6 the nontransition elements are nonmetallic). The transition metals show a much more complicated behavior, with peaks at e/a = 4.7 and 6.5, and a sharp minimum in between.

3. Certain crystal structures are particularly favorable for superconductivity. The highest critical temperatures are found in the Cr_3Si (A15) and NaCl (B1) structures, with e/a $\simeq$ 4.7. σ phases (D8$_b$), α Mn (A12) and Laves phases are crystal classes which produce compounds with e/a $\simeq$ 6.5, and critical temperatures up to $\sim$ 10 K. The critical temperatures of various alloys and compounds are listed by Roberts [12].

The first rule needs no further comment here. The dependence of T_c upon the electron to atom ratio is related to the effect of the density of states, N(0), and bears out the conclusions of the previous section. Both the Sommerfeld constant γ and T_c for transition metal alloys show maxima at 4.7 and 6.5 electrons/atom, and minima at 4 and 6 electrons/atom. The value of V would appear to be fairly constant between about 4-8 electrons/atom. Beyond 8 electrons/atom the paramagnetic spin susceptibility rises rapidly, which explains the decrease in T_c in this region, despite a third peak in γ at e/a $\simeq$ 10 [13].

The effect of crystal structure is least well understood. The A15 compounds with the highest critical temperatures are those formed between either Nb or V and group IIIa or IVa elements. The stoichiometric

composition is A_3B, with A representing the group Vb elements, and
e/a = 4.5 or 4.75. The critical temperature is a very sensitive function of atomic order and stoichiometry; deviations from perfect order
and the ideal composition cause a sharp decrease in T_c [14, 15, 16].

The high T_c Bl compounds are carbides and nitrides of transition
elements, notably Mo, Nb, Ta and Ti, or mixtures of these, with 4.5 or
5.0 electrons/atom [17]. The highest critical temperatures in a Bl
structure (~ 18 K) are found in mixtures of NbN with either NbC or TiC,
such as to give an e/a ratio very close to the ideal value of 4.7. Thus,
similarly to transition metals, the superconducting behavior of the Bl
compounds appears to be dominated by the electron density of states. T_c
values decline with departure from stoichiometry, though less drastically than for the A15 compounds.

The A12, $D8_b$ and Laves phases with high transition temperatures
all have e/a ratios close to 6.5, the second favorable value. It would appear that the primary role of crystal structure lies in allowing the formation of compounds with a favorable electron to atom ratio, and other effects on superconductivity are secondary.

C. A15 Compounds

There are six binary A15 compounds whose critical temperatures
are sufficiently high to be of real commercial interest. These are:

V_3Ga (15.0 K)	Nb_3Al (18.8 K)	Nb_3Ge (17 K now $\sim$23 K)
V_3Si (17.1 K)	Nb_3Ga (20.3 K)	Nb_3Sn (18.3 K)

The T_c values given here are the best reported values. The compounds
have a wide range of homgeneity, and some do not exist at the stoichiometric A_3B composition. The critical temperature is a sensitive function
of the composition, and can fall by as much as 2-3 K per atom percent
deviation from stoichiometry. For example, the Nb-Ge compound normally exists at a composition corresponding to Nb_4Ge, with $T_c \simeq 6$ K [18].
Rapid 'splat' cooling produces a rather disordered Nb_3Ge whose T_c is
17 K [19] and Nb_3Ga rapidly quenched and annealed has a T_c of 20.3 K [20]
Nb_3Ge films with T_c 23.2 K have been prepared by a high pressure dc sputtering process [119, 120].

T_c is also a function of order, and decreases rapidly if any disorder
between the A and B atoms is present [16, 21]. Most materials are improved by annealing treatments at between 750° C and 900° C for several
hours; this is believed to produce maximum order [15, 20, 22]. It has
been suggested that T_c for fully ordered Nb_3Ge should be ~ 25 K.

The A15 structure is body-centered cubic, with a B atom on each
lattice point. The A atoms lie in orthogonal chains parallel to the principal
axes. The inter-A-atom distance is rather short, approximately 10 % less

than in pure Nb or V, and leads to a rather high density of states at the Fermi surface. Disruption of these chains, by either deviations from stoichiometry or by disorder, is expected to cause a drastic reduction in the density of states. The effects of disorder has been studied by mea- surements of specific heat above T_c. It is possible to separate lattice and electronic contributions to the specific heat, and thus determine γ and θ_D. Disorder is found to have little effect on θ_D, and changes in T_c can be correlated with the effect on γ [23].

It is easy to see that disorder between A and B atoms will disrupt the A chains. It is equally obvious that a deviation from stoichiometry in the direction of excess B atoms will place some B atoms on the A chains and again cause their disruption. Deviations from stoichiometry are, how- ever, usually in the direction of excess A, and on first sight it is expected that the integrity of the A-chains would be preserved. Two explanations have been offered for the observed decrease in T_c with excess A. Neither explanation has yet been proved to be correct. One is that the excess A atoms on the B sites introduce a band structure characteristic of pure A, which interferes with, and leads to the breakdown of, the band structure of the 'perfect' A_3B. The other explanation is that the structure prefers that the excess A atoms do not go into B sites, but that they are balanced by the creation of B site vacancies. These B site vacancies are then pre- sumed to exist only in equilibrium with a lesser number of A site vacan- cies, sufficient to cause disruption of the A-chains and a reduction in T_c.

It would appear that the successful recipe for high critical tempera- tures in these compounds is to ensure that they are prepared with the exact stoichiometric ratio, and subsequently to anneal at the temperature which produces maximum order. This is quite straightforward for V_3Ga, V_3Si and Nb_3Sn, which are all stable at the stoichiometric composition. Nb_3Al, Nb_3Ga and Nb_3Ge, compounds which are capable of higher critical temperatures, are more difficult. None of these compounds is stable at the stoichiometric composition; all are normally produced with excess niobium. This is because of the existence of more stable phases at low- er niobium compositions. These phases, Nb_2Al, Nb_5Ga_3, Nb_5Ge_3, are σ or σ-related phases, and their presence restricts the range of homo- geneity of the A15 phases. No procedure has yet been devised which allows the production of continuous lengths of these binary A15 compounds at the stoichiometric composition, though small samples, as mentioned previously, have been produced by quenching.

In 1967, thirteen years and much effort after the discovery of Nb_3Sn, until then the material with the highest critical temperature, a ternary A15 compound, $Nb_3(AlGe)$ was found to have a critical tempera- ture of 20.05 K [24]. This system has subsequently received attention from many workers, and the critical temperature has been raised to 20.75 K [25]. Its discovery has lead to renewed speculation as to the possibility of substantially improving critical temperatures, and it is important to understand how this improvement in critical temperature has been brought about.

It was reasoned that, though both Nb_3Al and Nb_3Ge are stable only with excess niobium, a solid solution of Nb_3Al and Nb_3Ge might be found in which the stoichiometric composition is just stable [15]. Annealing a composition corresponding to $Nb_3(Al_{0.8}Ge_{0.2})$ for several days below 1000° C resulted in a material with a critical temperature of 20.05 K. T_c for stoichiometric Nb_3Al is expected to be 19.7 K, and for Nb_3Ge 25 K; an interpolation in the ratio 4 to 1 gives $T_c \simeq 20.75$ K. Subsequent work has shown that an optimum annealing temperature of 750° C is required to produce a maximum T_c of 20.7 K. It is suggested that not only is perfect stoichiometry necessary to allow of ordering between the Nb and Al and Ge atoms, but that there might also be ordering between the Al and the Ge atoms on their own (B) sublattice. In a perfectly ordered material T_c may be as high as 23 K.

It would appear that the σ-phases Nb_2Al and Nb_5Ge_3 either do not form a continuous solid solution, or that their range of homogeneity at intermediate compositions is severely limited. The retreat of these phases allows the A15 phase with a 4 to 1 ratio of aluminum to germanium to extend its range of homogeneity up to the stoichiometric composition [26, 27], as depicted schematically in Fig. 1.

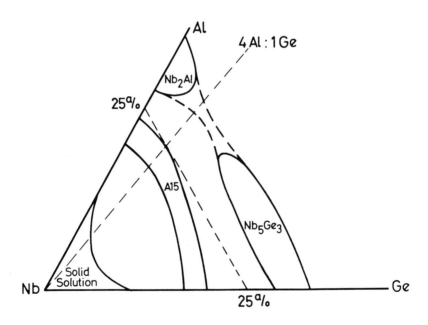

Fig. 1.   NbAlGe Ternary Phase Diagram, Isothermal Section at 750 C (Schematic). (See also Ref. 107.)

The search for higher critical temperature compounds now in-volves the manipulation of phase stability by the addition of third (and possibly fourth) elements. This is an area which does not yet appear to have been exploited by metallurgists. It is, however, unlikely to pro-duce any really startling improvements in T_c.

The intensive effort which has gone into the search for higher critical temperatures has uncovered no other series of compounds with greater promise than the A15s. 25 K seems a reasonable upper limit to place on T_c for A15 compounds based on niobium. It may be possible to improve on this figure for a compound based on vanadium, although theory suggests that really high T_c compounds are mechanically unstable and undergo martensitic transformations to a tetragonal structure with a reduced T_c. Such behavior has been observed in V_3Si. It is too early to say whether 23 K represents an absolute upper limit, but any advance bey-ond that value is likely to be very slow. Superconductivity in the A15 compounds has been reviewed by Hein [108].

III. THE CRITICAL FIELD

A. Type II Superconductivity

A consequence of zero resistivity is that everywhere within a super-conductor, the electric field E is zero. A further effect is also found in superconductors: a magnetic field is completely excluded from the body of the superconducting material, and the magnetic induction B is zero. This is known as the Meissner effect. The magnetic induction does not fall abruptly to zero at the surface of the superconductor, but decreases ex-ponentially over a characteristic distance λ, the penetration depth, from the surface. The theoretical value for λ is given by

$$\lambda_L(0) = \left\{ \frac{m^*}{N(0)e^2} \right\},$$

where m^* is the effective mass of the electron pairs, and e their charge. A typical value of λ for a pure metal superconductor is 5×10^{-8} m. The field is excluded by supercurrents which flow in the penetration layer so as to produce a field within the superconductor which exactly cancels the applied field.

Below T_c the free energy of the superconducting state is less than that of the normal state by an amount Δg_{ns} per unit volume, the super-conducting condensation energy. The exclusion of a field H raises the free energy by an amount $\mu_o H^2/2$ per unit volume, where μ_o is the mag-netic permeability of free space. The critical field is the field at which the two terms balance:

$$1/2 \, \mu_o H_c^2 = \Delta g_{ns}$$

or more correctly, since Δg is a function of the temperature T,

$$\frac{1}{2} \, \mu_0 H_c \, (t)^2 \; = \; \Delta g_{ns} \, (t)$$

where $t = T/T_c$, the reduced temperature. This is only accurate if the dimensions of the superconductor are large compared with λ. The field will penetrate extensively into a thin film or filament of thickness d equal to or less than λ. Superconductivity will then persist to a higher critical field H_d given by

$$H_d \; \simeq \; H_c\!\left(\frac{\lambda}{d}\right) .$$

The condensation energy can be related to the critical temperature and this leads to an expression for the critical field at absolute zero [28]:

$$H_c(0) \; = \; 7 \times 10^{-4} \, \gamma^{1/2} T_c .$$

A high value of the critical field is favored by high critical temperature and high γ. Note the double dependence upon γ; a high γ has already been shown to be a requirement for high T_c. The variation of critical field with temperature is roughly parabolic. Some values of the zero temperature critical field, calculated from the above equation, for A15 compounds with high values of γ are given in Table 1.

Table 1. Calculated values of $H_c(0)$

Compound	$\gamma \, (kJ \, m^{-3} K^{-2})$	T_c (K)	$H_c(0)$ (T)
Nb_3Sn	1.42	18.04	0.53
V_3Si	2.50	16.86	0.61
V_3Ga	3.04	14.83	0.63

Data from Hechler et al [29].

These values are far too low for the material to be of practical interest, but in fact superconductivity is observed at fields 40-50 times greater than these values. If the superconductor could divide itself into lamella or filaments of thickness less than λ, separated by nonsuper-conducting regions into which magnetic field could penetrate, then, as mentioned above, the filaments could remain superconducting up to much higher fields. At this point it is necessary to introduce a second char-acteristic length, the coherence length, ξ. This defines, among other things, the distance over which the electron wave function can change from normal to superconducting behavior. Its effect is to limit to ξ the width of the region which can maintain superconductivity when surrounded by normal conducting material. In pure metals, $\xi \geqslant 10^{-6}$ m, that is 20 x greater than λ. Thus, in a pure superconductor, the smallest filaments which can remain superconducting, in a normal metal matrix, are too large to allow of appreciable field penetration.

However, both λ and ξ, are affected by the purity of the material. For 'dirty' superconductors, in which the normal electron mean free path $\ell < \xi_0$, the coherence length for the pure metal, both the penetration depth and the coherence length are modified to the dirty values [30, 31]:

$$\lambda_d \simeq \lambda_L \left(\frac{\xi_0}{\ell} \right)^{1/2} \quad \text{and} \quad \xi_d \simeq (\xi_0 \ell)^{1/2} \ .$$

The ratio $\lambda/\xi = \kappa$, the Ginsburg-Landau parameter. It can be seen that, as ℓ is reduced by alloying, ξ is also reduced and λ is increased. For very dirty alloys ξ_d may be as small as 10^{-7} m. Superconductors in which $\lambda > \xi$ behave differently from those in which $\lambda < \xi$; the former are called type II superconductors, and the latter type I superconductors. In type II superconductors the postulated division into superconducting and normal lamella or filaments is now possible. These materials show an incomplete Meissner effect; magnetic fields are only partially excluded from the body of the superconductor.

The actual division occurs in a manner predicted by Abrikosov [32]. A reversible type II material behaves like an ideal type I superconductor, with a complete Meissner effect, up to the lower critical field H_{C1}. Beyond H_{C1}, magnetic flux enters the specimen as quantized supercurrent vortices, each carrying one quantum of magnetic flux $\varphi_0 = h/2e \ (= 2 \times 10^{-15} \ Wb)$ where h is Planck's constant. Each vortex or flux line may be thought of as a flexible rod of normal material (the core), with a radius of ξ; the flux within it being maintained by circulating supercurrents flowing in a radius λ.

As the external field is increased the density of flux lines within the material increases, forming a triangular lattice of lines parallel to the external field. At some stage the cores will overlap, the material now becoming completely normal. This occurs at the upper critical field, H_{C2}, given by

$$H_{C2} = \frac{\varphi_0}{2\pi\mu_0 \xi^2} = \frac{\varphi_0}{3.77\mu_0 \xi_0 \ell} \ . \tag{1}$$

Since the normal state resistivity, ρ_N, is also inversely proportional to ℓ, there is a direct relation between upper critical field and normal state resistivity [33] (in SI units):

$$H_{C2}(0) = 3.11 \times 10^3 \ \gamma \ \rho_N \ T_c \ . \tag{2}$$

The importance of resistivity can be seen from the comparison of Mo-Re with alloys of niobium such as Nb-Ti and Nb-Zr. Both Mo-Re and Nb alloys have $T_c \simeq 11 K$; for the former $\rho_N \simeq 8 \times 10^{-8} \Omega m$ and $H_{C2}(0) = 1.4 T$ and for the latter $\rho_N \simeq 8 \times 10^{-7} \Omega m$ with $H_{C2}(0)$ from 8 to 12 T. Resistivity, and hence H_{C2}, can be increased by alloying, deformation, and irradiation. Of these the former is the most effective, and high field materials are either transition metal alloys with high resistivities, or compounds such as the high T_c A15s. These latter have intrinsically small values of ξ_0 leading to upper critical fields in the range 20-25 T. The normal state resistivity is not so important in these compounds, and in fact any of the techniques normally employed to raise ρ_N are likely to have a deleterious effect on T_c, and the resulting H_{C2} will probably be reduced rather than increased.

B. Paramagnetic Limitation of Upper Critical Field

In the foregoing, the magnetic properties of the normal state have been ignored, it being assumed that the free energy of the normal state is unaffected by any magnetic field. This is a reasonable assumption at low fields, as superconductors in the normal state are either diamagnetic or paramagnetic, with very small susceptibilities. These small susceptibilities, however, give rise to a very large effect as the high critical fields, of the order of tens of tesla, found in some type II superconductors, are approached.

If the normal state has a susceptibility, χ, then in a magnetic field, H, the Gibbs function per unit volume is changed by $-\chi H^2/2$. The high temperature, high-field superconductors are based on transition metal alloys or compounds, with a noneven number of electrons per atom. They will, therefore, be paramagnetic rather than diamagnetic, and their susceptibility, χ, will be positive. In a magnetic field the Gibbs function of the normal state will be lowered and there will be a field H_p given by:

$$\chi H_p^2/2 = \Delta g_{ns}$$

above which the normal state always has a lower Gibbs function than the superconducting state in zero field. The Gibbs function of the superconducting state increases as a field is applied. and H_p thus represents the maximum field at which superconductivity will persist; the actual upper critical field is expected to be below this. If the normal state susceptibility is due solely to Pauli paramagnetism then [34, 35]:

$$H_p(0) = 1.84 T_c \text{ (Tesla/per degreee Kelvin)}. \tag{3}$$

Several workers have confirmed that the experimental value of the upper critical field is close to either H_{C2} given by Eq. (1), or H_p given by Eq. (3), whichever is the least, provided that they are widely different

[33, 36, 37]. If they are similar in magnitude, then the effect of the magnetic field on the superconducting state must also be taken into account.

Maki has defined a parameter $\alpha = \sqrt{2}\, H_{C2}(0)/H_p(0)$ where $H_{C2}(0)$
is given by Eq. (1) and $H_p(0)$ by Eq. (3). He showed that in the limit of
small α and no spin orbit scattering the actual paramagnetically limited
critical field, $H_{C2}(0)^*$, is related to $H_{C2}(0)$ by [38].

$$H_{C2}(0)^* = H_{C2}(0)\,(1 + \alpha^2)^{-1/2}. \tag{4}$$

It is found that [39]:

$$\alpha = 2.36 \times 10^3\, \gamma \rho_N. \tag{5}$$

Several workers have confirmed the effect of Pauli paramagnetism
when $H_{C2}(0)$ given by Eq. (1) is greater or the same order of magnitude
as H_p in Eq. 3 ($\alpha \gtrsim 1$) [33, 36, 37]. The situation can be obtained whenever the spin-orbit scattering, which reduces the effect of Pauli paramagnetism is small. Table 2 gives $H_{C2}(0)$ and $H_{C2}(0)$ observed.

The agreement between $H_{C2}(0)$ from Eq. (1) and the observed
critical field is poor. Thus Eq. (1) cannot be used to give quantitative
predictions. For dirty type II superconductors, one can obtain the extrapolated value for the size   of the upper critical field from the slope
of the critical field vs. temperature near the critical temperature. The
equation is simply

$$H_{C2}(T = 0) = -0.69\, \left.\frac{dH_{C2}(T)}{dT}\right|_{T = T_c}. \tag{6}$$

Near T_c, the effects due to Pauli paramagnetism and spin-orbit scattering
is small and can be neglected, thus, the extrapolation $H_{C2}(T=0)$ is the
upper critical field one would have observed at $T = 0$ had Pauli paramagnetism and spin-orbit scattering not been present.

In the case of large spin-orbit scattering, the effects of Pauli paramagnetism are effectively cancelled and the upper critical field is
given by Eq.(6). This is especially noticeable for say $Nb_3(AlGe)$ in which
$H_p \approx 38.5\,T$ whereas the observed critical field is approximately $41\,T$ and
agrees with the value obtained from the high temperature slope [122].

In all the high field type II superconductors in Table 2, the Pauli
paramagnetic effect can be large since the parameter α defined in Eq. (3)
is $\gtrsim 1$. The actual observed $H_{C2}(0)$ shows that the effect of Pauli paramagnetism is moderated by spin-orbit scattering. The qualitative effects
of spin orbit scattering have been discussed by Kim et al [33], and
Neuringer and Shapira [37].

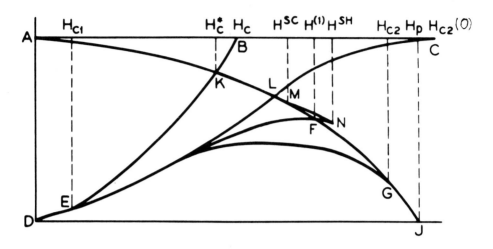

Fig. 2. Schematic Plot of the Free Energy of a Type II Superconductor.

        To understand the effects of Pauli paramagnetism we refer to Fig. 2
which is the plot of free energies of the superconducting and normal
states vs. magnetic field, and of the associated transitions. The horizontal line AC represents the normal state ignoring the Pauli spin susceptibility, while the parabolic curve AJ represents the normal state with
spin paramagnetism. The horizontal line DJ represents the superconducting
state, assuming it to have no response to the magnetic field. The intersec-
tion point J defines the Clogston-Chandrasekhar paramagnetically limited
first-order critical field H_p [34, 35]. A type I superconductor with a
Meissner effect is represented by the parabolic curve DB and the inter-
section point B gives the thermodynamic critical field H_c. Inclusion of
the spin paramagnetism of the normal state lowers the first-order transition to the point K with associated critical field H_c^*. A type II superconductor has field penetration above a point E which defines a critical
field H_{C1}. If spin effects are neglected in the mixed state, the free energy
follows the curve DELC up to a point C with associated second-order
critical field $H_{C2}(0)$, where the curve is tangent to the no-spin normal-
state line AC. Inclusion of spin in the normal state but not in the mixed
state gives a first-order transition at the point L. If the spin susceptibility
of the mixed state is included, the latter may follow the curve DEG to the
point G at which a second order transition occurs to the normal state at
the field $H_{C2}^*(0)$. Alternatively, the mixed-state free energy may follow

Table 2. Upper critical field of type II Superconductors

Material	T_c (K)	γ (mJ m^{-3} K^{-2})	α	$H_{C2}(0)$ (T)	$H_p(0)$	$H_{C2}(0)$ (T) observed
A15 compounds						
Nb$_3$Sn	18.0	1.42	1.27	30	33.5	24.5
Nb$_3$Al	18.7	0.72	1.76	42.5	34.5	32.4
V$_3$Si	16.9	2.50	1.54	34	31.5	23.5
V$_3$Ga	14.8	3.04	1.8	35	27.6	20.8
Nb$_3$(AlGe)	20.7	--	--	-	38.5	>41.0
Nb$_3$Ge	23.2					>36.0
B1 compounds						
NbN	15.7	0.25	0.66	13.5	29.2	15.3
Nb(C$_{0.3}$N$_{0.7}$)	17.4	0.21	--	--	32.5	11
BCC alloys						
Ti$_{0.6}$V$_{0.4}$	7.0	1.1	--	24	13.0	11
Ti$_{0.7}$Nb$_{0.3}$	7.2	1.0	--	17.5	13.4	14
Zr$_{0.25}$Nb$_{0.75}$	10.9	0.79	--	~18	20.3	11

Data from Refs. 29, 36 40, 109, 121 and 122.

the curve DEFNM making a first-order transition to the normal state at the point F corresponding to the field $H^{(1)}$. The point N at the field H^{SH} is the limit of microscopic stability of the superconducting phase ("super-heating"), and the point M at the field H^{SC} is the limit of microscopic stability of the normal phase ("supercooling") Ref. 39.

From Table 2 and Fig. 2 we can see that the highest critical fields can be expected in very dirty type II superconductors with high values of T_c, and large spin-orbit scattering. Thus far the samples of $Nb_3(AlGe)$ have the highest critical fields over 41 T, however, suitable wire has not been produced. The critical field of the highest T_c superconductor to date Nb_3Ge, $T_c \approx 23.2$ has recently been measured as approximately 36 T for a sample with a $T_c \approx 22.5$ K [121].

IV. THE CRITICAL CURRENT

A. Introduction

Silsbee [41] hypothesized that the critical current is that current which just produces the critical field at the surface of the superconductor, and subsequent experiments have verified this. The critical current is reduced by the application of an external field as the field it now has to produce is the difference between the critical field and the applied field. The critical currents of type II superconductors above H_{C1} are much lower than would be expected by inserting H_{C2} in the Silsbee rule, and the variation with an applied field can be quite complicated.

Flux lines in the mixed state of a type II superconductor experience a Lorentz force $\vec{F}_L$ whenever a current flows in the superconductor, given by

$$\vec{F}_{L(1)} = \vec{J} \times \vec{\phi}_o \text{ per unit length of flux line}$$

or   (7)

$$\vec{F}_{L(v)} = \vec{J} \times \vec{B} \quad \text{per unit volume of superconductor,}$$

where $\vec{J}$ is the current density, $\vec{\phi}_o$ is a vector of magnitude $|\vec{\phi}_o|$, directed along the local tangent to the flux lines, and B is the flux density ($= n\vec{\phi}_o$, where n is the number of flux lines per unit area). The force acts in a direction normal both to the flux lines and the current. Unless otherwise prevented, the flux lines will move in the direction of this force, and in doing so induce an electric field:

$$\vec{E} = n\vec{v} \times \vec{\phi}_o = \vec{v} \times \vec{B} \tag{8}$$

where $\vec{v}$ is the velocity of the flux lines. The superconductor shows an induced resistance, as demonstrated by Kim et al. [42].

The critical current is that current which just produces a detectable voltage across the specimen, and is that current which first causes the flux lines to move. If there is no hindrance to the motion of flux lines then the critical current above H_{C1} is zero, and the superconductor is 'reversible'. It is possible to 'pin' flux lines and prevent them from moving by interaction with microstructural features of the material, and a pinning force $\vec{F}_p$ is exerted on the flux lines which opposes the Lorentz force. The critical current density, $\vec{J}_c$, is determined by the magnitude of the pinning force:

$$\vec{J}_c \times \vec{\phi}_0 = -\vec{F}_{p\,(l)} \quad \text{per unit length of flux line}$$

or   (9)

$$\vec{J}_c \times \vec{B} = -\vec{F}_{p(v)} \quad \text{per unit volume of superconductor.}$$

Pinning is due to crystal lattice defects, such as dislocations found in heavily cold-worked materials, impurities, or precipitates of a second phase. Strong pinning materials are metallurgically dirty and by analogy with mechanically strong or magnetically hard materials, irreversible superconductors are called hard superconductors. The critical current is not a property of a particular composition, but of a particular sample of superconductor, and is strongly influenced by the sample's metal-lurgical history, and its value cannot be predicted with any accuracy.

When flux is pinned, it is reluctant to enter the specimen in an increasing field, and the entry of appreciable quantities of flux may not occur until a field several times H_{C1} has been applied. When the flux does penetrate, it does so slowly and penetration is not complete until H_{C2}. On reducing the field the flux is reluctant to leave the specimen, and remanent flux is found when the external field is zero. Reversal of the field causes the specimen to describe a complete magnetic hystere-sis loop. The area of the loop is greater, the stronger the flux pinning, and may be related directly to the critical current of the material. Schematic hysteresis loops and critical current curves for a material in which the pinning has been progressively increased are shown in Fig. 3.

The relation between magnetization and critical current is estab-lished using the concept of the 'critical state' [43, 44]. It is assumed that the superconductor everywhere either carries the critical current, or no current at all. The local critical current density is related to the flux pinning force by Eq. (9), which will be some function of the local value of B, and the critical state is represented by

$$\vec{J}_c \times \vec{B} = \vec{F}_{p(v)} (B).$$ (10)

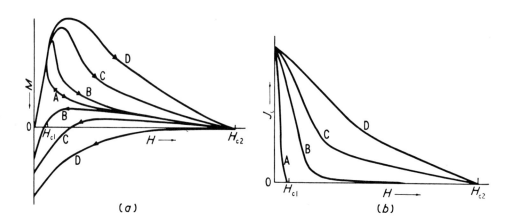

Fig. 3.   (a) Magnetization and (b) Critical Current Plotted Against Mag-
netic Field for Samples of the Same Material with Different
Amounts of Flux Pinning (Schematic): A, Reversible, No Flux
Pinning; B, Very Weak, Mainly Surface, Pinning; C, Irrevers-
ible, Moderate Flux Pinning; D, Very Irreversible, Strong
Flux Pinning.

Integration of this relation, with appropriate boundary conditions, and
knowledge of the dependence of $\vec{J}_c$ and $\vec{F}_p$ upon B, gives the distribution
of flux throughout the specimen from which the magnetization can be calculated. The relation between $\vec{F}_p$ and B is inferred from measured magnetization or current curves, and compared with the theoretical predictions of specific pinning models.

B. Flux Flow

Flux flow is usually investigated by monitoring the longitudinal
voltage developed in a superconductor carrying current in a transverse
magnetic field. The features revealed are:
1. An electric field $\vec{E}$ appears when a certain current density J^1
is exceeded. For current densities smaller than J^1 no field is detected.
2. The field increases with J, ultimately showing a linear dependence upon J (ohmic behavior): $E = \rho_f J$ where ρ_f is the flux-flow resistivity.
3. J^1 depends upon the temperature, the induction B, and is very
sensitive to the sample history. It may be identified with the critical
current J_c.
4. ρ_f is a function of temperature and induction, but is relatively
insensitive to specimen history. As B increases, ρ_f rises to equal the

normal state resistivity, ρ_n. At low temperatures

$$\frac{\rho_f(B)}{\rho_n} = \frac{B}{\mu_0 H_{C2}(0)} \quad . \tag{11}$$

These features are illustrated in Figs. 4 and 5, taken from Ref. [33].

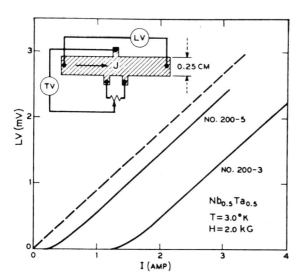

Fig. 4.   Voltage-current Characteristics of Two Specimens of Composition $Nb_{0.50}Ta_{0.50}$ at 3.0 K in a Field of 2000 Oe.

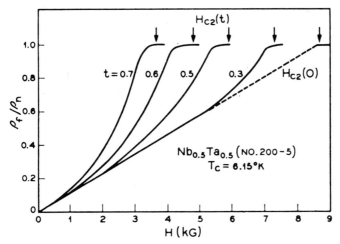

Fig. 5.   Variation of the Flux-flow Resistivity of a NbTa Specimen with Field at Various Constant Reduced Temperatures.

It is possible to account for the above facts by considering the forces which act on a moving flux line. In addition to the Lorentz and pinning forces, there will be a viscous drag force, $\vec{F}_v$, preventing acceleration of the flux line to infinite velocity. This viscous force is assumed to be equal to the velocity $\vec{v}$ with which the flux lines are moving times a viscosity coefficient:

$$\vec{F}_v = -\eta \vec{v}. \tag{12}$$

At constant velocity, these forces must balance:

$$\vec{F}_L + \vec{F}_p + \vec{F}_v = 0 \tag{13}$$

which becomes, on substitution from Eqs. (7), (9) and (12),

$$\varphi_0 (\vec{J} - \vec{J}_c) = \eta \vec{v}. \tag{14}$$

The velocity $\vec{v}$ can be related to the flux-flow field, $\vec{E}$. The number of flux lines per unit area is B/φ_0, and if these lines are moving with a velocity $\vec{v}$, the number that cross a unit boundary, normal to their direction of motion, in unit time, is $B\vec{v}/\varphi_0$. If flux is to be conserved, then whenever the local value of B changes, $\vec{v}$ must also change so as to keep $B|\vec{v}|/\varphi_0$ constant. The value of this constant is from Eq. (8), the flux-flow voltage $\vec{E}/\varphi_0$. Substituting for $\vec{v}$ from Eq. (8) into Eq. (14) gives:

$$\vec{J} - \vec{J}_c = -\frac{\eta \vec{E}}{\varphi_0 B} \tag{15}$$

from which it can be seen that $\vec{J}^1 = \vec{J}_c$, and

$$\rho_f = \frac{\varphi_0 B}{\eta}. \tag{16}$$

Now $\rho_f (H_{C2}(0)) = \varphi_0 \mu_0 H_{C2}(0)/\eta = \rho_n$, which is the same as Eq. (11) if the coefficient of viscosity $\eta = \varphi_0 \mu_0 H_{C2}(0)/\rho_n$. During flux flow the superconductor behaves as if the current density were uniform over the whole superconductor and resistance is developed only in the flux line cores. The origin of viscous drag on the moving flux lines is not yet fully understood, but would appear to be due to some sort of ohmic dissipation in the flux line cores.

The measured value of the critical current depends upon the level of voltage detection (Eq. (15)). If the minimum detectable voltage corresponds to a field $\vec{E}_0$, then $\vec{J}_c = \vec{J}^1 + \eta \vec{E}_0/\varphi_0 B$ where $\vec{J}^1$ is the true critical current, defined by the absence of flux flow. In practice it is found that for high current materials, $\vec{J}_c$ is independent of the level of voltage detection, which means either that $\vec{J}^1 >> \eta \vec{E}_0/\varphi_0 B$, or that, once flux begins to move, when a high current is flowing the resulting dissipation causes the specimen to rapidly revert to the normal state. In nearly reversible materials, the level of detection does become important, and it is necessary to measure the critical current at several values of $\vec{E}_0$, and extrapolate to $\vec{E}_0 = 0$ to get the correct critical current value.

At finite temperatures some flux motion will occur at current densities well below J_c due to thermal activation. This thermally assisted motion is called 'flux creep' [45]; it is negligible in strongly pinned material and will not be considered further here.

C. The Critical State

The critical state, in which it is assumed that the current density is equal to the critical current density $\vec{J}_c$, or is zero, requires that everywhere within the superconductor flux lines are in equilibrium. Any force acting so as to try to move a flux line is just, and only just, opposed by a pinning force. Any imposed disturbance, by changing either a transport current or an external magnetic field, results in a redistribution of flux until the critical state is restored.

The total current density (the sum of magnetization and transport currents) $\vec{J}$ = curl $\vec{H}$ and the driving force on the flux lines is the Lorentz force $-\vec{\varphi}_0 \times \vec{J}$. The critical state Eq. (10) can be written as

$$\vec{J}_c \times \vec{\varphi}_0 = \vec{F}_{p\,(1)}(B) = \text{curl } \vec{H}(B) \times \vec{\varphi}_0$$

or $\hspace{11cm}$ (17)

$$\vec{J} \times \vec{B} = \vec{F}_{p(v)}(B) = \text{curl } \vec{H}(B) \times \vec{B}$$

and the resulting flux distribution is derived from:

$$\text{curl } \vec{H}(B) = \frac{\partial \vec{H}}{\partial B} \text{ curl } \vec{B} = \vec{J}_c(B) = \frac{\vec{F}_{p(1)}(B)}{\varphi_0} = \frac{\vec{F}_{p(v)}(B)}{B} \hspace{1cm} (18)$$

where $\vec{H}(B)$ is the external field that would be in equilibrium with the internal induction $\vec{B}$, and $\partial H/\partial B$ is the slope of the ideal, reversible magnetization curve for the material. The flux distribution within the superconductor can be determined if it is known how either $\vec{J}_c$ or $\vec{F}_p$ varies with B. The critical state had been discussed in detail by Campbell and Evetts [110].

Many empirical expressions have been proposed to describe the relation between $\vec{F}_p$ and $\vec{B}$. These will not be discussed here, as a better procedure is to derive expressions for particular models of the primary interaction between flux lines and the microstructure and compare these with experimental results of J_c versus H. It is found that the theoretical expressions can often fit the general form [46, 47]:

$$\vec{F}_p(B) = \text{const} (H_{C2})^n \text{fcn}(h) \tag{19}$$

in which the temperature dependence of the pinning is contained within the temperature variation of H_{C2}. Plots of $J_c h$ versus h usually fall on one master curve for all temperatures. The shape of this curve is characteristic of the particular pinning process involved.

The next section will summarize the results which indicate the microstructural features which are responsible for strong pinning interactions. The following section will discuss briefly the theories that have been advanced to explain pinning, the various relations between F_p and B predicted by these theories, and the correlation between these and experimental results.

D. Microstructure and Flux Pinning

It is clear from the foregoing that high critical currents result from strong interactions which prevent the motion of flux lines. The problem, familiar to metallurgists, is similar to those of producing mechanically strong, or magnetically hard, materials. In strong materials the movement of crystal dislocations is hindered by regions whose elastic properties differ from those of the matrix. In hard magnetic materials, domain boundaries are obstructed from moving by regions whose magnetic properties are different from those of the matrix. It is reasonable to expect that in high current superconductors, regions whose superconducting properties show a difference from those of the matrix will interfere with the motion of flux lines.

The nature and origin of these regions are varied, but all are 'imperfections' in the regularity of the crystal structure of the matrix. Imperfections whose influence in one direction extends over a distance less than a coherence length, ξ, are not important, as the superconducting properties cannot change appreciably over distances less than ξ. Just as crystal imperfections are classified by the number of dimensions which are large compared to interatomic spacings, so superconducting imperfections may similarly be classified by the number of dimensions which are large compared to the flux line spacing d [48, 49].

Point defects, with no dimension greater than d, are small second-phase particles, small voids and clusters of crystal defects such as those produced by irradiation with high-energy particles. The size of a point defect is such that it can interact directly with only one flux line at a time.

Line defects, with one dimension greater than d, are represented mainly by crystal dislocations. The orientation between the flux line and the defect is important in determining the strength of the interaction.

Surface defects, with two dimensions greater than d, include grain and twin boundaries, stacking faults, martensitic boundaries, certain configurations of crystal dislocations, such as subgrain and polygonized boundaries, and the surface of the superconductor itself. Again the orientation between the flux line and the plane of the surface can be of importance.

Volume defects, with three dimensions greater than d, are large second-phase particles and voids. Each defect is able to interact with several flux lines. At the high fields available in materials of interest, most second-phase particles are likely to be volume defects.

Comparison of measured superconducting properties with observed microstructure have been carried out on three main groups of materials. Volume defects such as precipitates and artificially introduced second-phase particles have been studied in lead alloys; crystal dislocations, twins and the effect of interstitial elements (C, N, O) in body centered cubic metals and alloys; and grain boundaries and interstitial precipitates in A15 compounds. The results of radiation damage have been investigated in all three groups of materials.

The defect may be more strongly superconducting than the matrix (i.e. have a higher T_c and H_c), have a value of $\varkappa$ slightly different from that of the matrix, be less superconducting, be nonsuperconducting or be ferromagnetic. The difference in superconducting properties is more easily investigated and more readily controlled when it is produced either by a precipitation reaction or by powder metallurgy techniques. Voids are a special case of nonsuperconducting precipitates.

It is not possible to cover the vast range of experimental work that has been carried out in this field, the reader who requires further details should consult Ref. (8). From the wealth of the results, the following conclusions may be drawn:

1. Strong flux pinning requires regions whose superconducting properties differ from those of the matrix; the greater the difference, the stronger the pinning.

2. Such regions are most efficiently produced by precipitation, powder metallurgy, dislocation cell structures or neutron irradiation.

3. For a given volume fraction of these regions, the more finely divided the structure, the stronger the pinning. Quantitative metallography indicates that pinning is, in most cases, directly proportional to the area of interface per unit volume.

4. In the niobium alloys which form the basis for much commercial material, pinning is due to the dislocation cell-structure formed as a result of cold work. In A15 compounds pinning is due to grain-boundaries, the martensitic transformation and precipitates.

E. Mechanisms of Flux Pinning

The approach to flux pinning models depends upon whether one be-
lieves that the flux lines within a superconductor can be treated as individual lines or as a fairly rigid lattice. A completely rigid lattice
cannot be pinned, as a displacement through one lattice spacing involves
no net change in energy [50]. If the pinning forces are not strong enough
to disrupt the flux lattice, they may still cause elastic distortion, and a
pinning theory on this basis has been developed by Labusch [51] and investigated experimentally by Feitz and Webb [46]. However, experimental results on high $\varkappa$, strong pinning materials show agreement with
theories which ignore lattice rigidity. Investigation of the flux distribution by the refined Bitter pattern technique has indicated that at high
current densities the flux lattice may break up into an amorphous state
[52]. Herring (private communication) by an extremely elegant combination of this technique and transmission electron microscopy, has
shown that in cold-rolled niobium the flux lattice becomes "poly-
crystalline" in conformity with the underlying dislocation cell-structure,
Fig. 6. It is therefore assumed adequate to calculate flux-pinning forces
in terms of interactions with individual flux lines.

The nature of the interaction between flux lines and the microstructure depends upon the scale of the latter when compared to the
characteristic lengths of the superconductor [47]. When the size of the
microstructural features (dislocation tangles, grain boundaries, precipitates) and the distance between them is of the order of the penetration depth or greater, then the flux density will be able to adjust itself
everywhere so that the induction is in equilibrium with the magnetic
field according to the local superconducting properties. On first sight
this would appear to allow no pinning, since the lattice can readily adjust to the local equilibrium induction as it moves relative to the microstructure. However, a change in induction introduces an irreversible
surface barrier to flux motion [53]. Campbell et al have calculated
this surface barrier pinning and have shown that [48]:

$$J_c \simeq \frac{S_v \Delta M}{\lambda} \left(\frac{\varphi_0}{B} \right)^{1/2} \tag{20}$$

where S_v is the surface area per unit volume perpendicular to the
Lorentz force and ΔM is the difference in equilibrium magnetization
across the boundary.

If the pinning centers are nonsuperconducting precipitates, then
$\Delta M(B)$ is equal to the reversible magnetization of the matrix, and for a
field $H > H_{C1}$ in a material of $\varkappa \gg 1$,

$$\Delta M(B) = \frac{H_{C2} - H}{2.32 \varkappa^2} ,$$

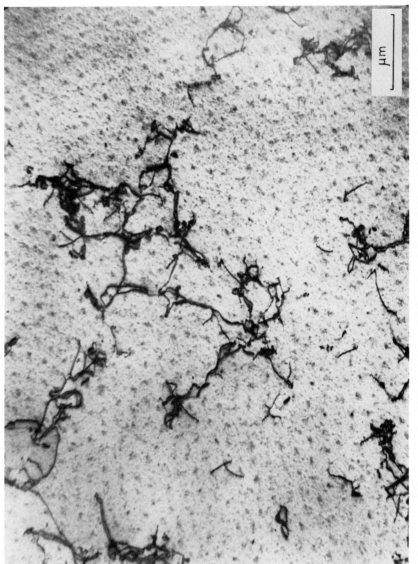

Fig. 6.   Electron Micrograph of Thin Film of Nb. Position of Flux Lines Indicated by Decoration with Iron Particles. The underlying dislocation structure can be seen to have interferred with the flux line lattice. (Courtesy C.P. Herring).

and the Lorentz force is [8, 54]:

$$J_c B = \frac{\mu_0 S_v H_{C2}^2 h^{1/2} (1-h)}{2.48 \, \varkappa^3} . \tag{21}$$

When the pinning is due to a dislocation cell-structure, in which the dislocation cell-walls have a higher $\varkappa$ due to their shorter normal electron mean free path [55], then ΔM (B) is given by [8, 56]:

$$\Delta M (B) = \frac{(H_{C2} - 2H) \Delta \varkappa}{2.32 \, \varkappa^3} ,$$

where $\Delta \varkappa$ is the difference in $\varkappa$ value between cell and cell-wall. The expression for the Lorentz force is now [57]:

$$J_c B = \frac{\mu_0 S_v H_{C2}^2 h^{1/2} (1-2h) \Delta \varkappa}{2.48 \, \varkappa^4} . \tag{22}$$

If the variation in microstructure is on a scale finer than λ, the flux density cannot vary with sufficient rapidity to conform to the value in local equilibrium with the microstructure. Pinning arises because of the variation in free energy of the flux line core with change in $\varkappa$ across the microstructure. Hampshire and Taylor [47] have shown that, in this case, the Lorentz force is given by:

$$J_c B = \frac{\mu_0 S_v H_{C2}^2 h(1-h) \Delta \varkappa}{2.67 \, \varkappa^3} . \tag{23}$$

When plotted as $\vec{J_c} \times \vec{h}$ (the difference between H and B in high $\varkappa$ superconductors at fields $\gg H_{C1}$ is negligible) versus h, all of these functions show at least one maximum at some value of h between 0 and 1. This maximum is at h = 0.33 for Eq. (21) and at h = 0.5 for Eq. (23). When pinning is due to a coarse dislocation structure and Eq. (22) is appropriate, maxima should occur at h = 0.17 and at a value of H = H_{C2} for the cell interiors (regions of minimum $\varkappa$). As the measured H_{C2} corresponds to that for cell walls, this second maximum is expected at h just < 1 [56]. In practice, there is a range of $\varkappa$ values between cells and cell-walls, $\Delta \varkappa$ does not have a single value, and the maxima are very broad and exist at values below those predicted.

Equation (21) has been verified for pinning by normal bismuth particles in a Pb-Bi eutectic [48], and for various specimens produced by powder metallurgy [58]. Eq. (22) has been used to explain the pinning by dislocation cell-structures in cold-rolled Mo-Re [56], and

Eq. (23) gives very good agreement with experimental results on Nb-Ti alloys in which pinning is due to a fine dislocation cell structure [47]. All three expressions have been invoked to explain results in heat-treated Nb-Ti alloys in which pinning is due to precipitates of titanium oxides and sub-oxides [57]. It is concluded that the mechanism of flux pinning in high $\varkappa$ superconductors is reasonably well understood, once the micro-structural features responsible for pinning have been identified.

F. Maximum Critical Currents

Strong flux pinning, and hence high critical currents, are given by fine dislocation cell-structures, precipitates, or both. Extreme cold work, $\sim 99.995\%$ reduction in area, of ductile niobium-based alloys can result in dislocation cell structures with cells down to ~ 0.1 μm diameter. S_v and hence J_c is proportional to (cell diameter)$^{-1}$ [59]. Subsequent annealing at $\sim 400^\circ$C allows dislocation rearrangement which enhances the $\varkappa$ difference between cell and cell-wall, increasing pinning strength. This treatment has an additional advantage, as the conductor is usually incorporated in a copper matrix, in that the effects of cold-work in the copper are reduced, thus decreasing its resistivity.

It would appear that even greater effects can be obtained by "doping" the alloy with an interstitial solute, usually oxygen. During the annealing treatment this segregates to the cell-walls, and may precipitate as an oxide or sub-oxide [57], giving greatly enhanced pinning. Further cold work will subdivide the precipitate particles, and an even finer cell structure will form around the particles as nodes of the structure [7]. Unfortunately this prospective improvement is illusory, as the presence of interstitials makes deformation of the alloy considerably more difficult, and the very large reductions necessary to produce the stable multi-filamentary conductors, cannot be achieved.

The compound superconductors are too brittle to be fabricated by cold work. Strong pinning is produced by a fine grain size [60, 61] and by precipitates [62, 63]. It is difficult to distinguish the two effects, as the presence of precipitates can cause refinement of the grain size. Precipitates have been introduced into Nb$_3$Sn via the substrate [61, 62] or by gases [63, 88, 111] during the formation from the vapor phase. Oxides [63] and carbides [112] form a fine substructure within the Nb$_3$Sn grains. Critical currents increase with increasing quantity and dispersion of the precipitates; maximum values being given by particles ~ 50 Å diameter at a spacing of 200 Å.

The low temperature martensitic transformation may contribute to flux-pinning and critical currents in A15 compounds [7] and has been demonstrated in In-Tl alloys [113]. Brand and Webb [114] showed that pressure, which inhibits the martensite transformation in V$_3$Si, also reduced the magnetic hysteresis. Livingston [115] found that in In-Tl maximum flux-pinning was associated with an incomplete martensite transformation. A partial transformation provides regions of differing superconducting properties required for high critical currents.

If the mechanisms of pinning by surface interaction given in the previous section are accepted, then the critical current should be proportional to the surface area of inhomogeneity present. The surface area is maximized by a dispersion of very fine particles, but their efficiency falls off as their diameter decreases below a few times the flux lattice spacing, due to the greater difficulty of the flux lattice adjusting itself to the particle interface. Efficiency also falls off as the diameter increases, due to a decreasing proportion of the surface being correctly oriented parallel to the flux lines. Increasing the volume fraction of precipitate, f, reduces the volume of matrix able to carry current; the reduction is greater than f, as parts of the matrix are "shadowed" by the precipitates. Thus a law of diminishing returns operates. Bibby [58] has discussed these geometrical effects in detail.

Maximum values of critical current density may be estimated in one of two ways. One is to optimize the various parameters which appear in any one of the expressions, Eqs. (21) - (23). Consider for example, a Nb-Ti alloy, in which H_{C2} is $\simeq$ 10 T, $\varkappa \simeq$ 35, $\Delta\varkappa$ is ten percent of this, 3.5, and with a cell diameter of $\sim 0.1\ \mu$m, S_v is $10^7\ \text{m}^{-1}$. Substituting these values into Eq. (23) gives, at a reduced field h = 0.5 (i.e. H = 5 T) a critical current density of $\sim 1.5 \times 10^9$ amps m^{-2}. This is of the order of the best values so far achieved in commercial Nb-Ti conductors.

Equation (23) can be rewritten, using $H_{C2} = \sqrt{2}\ \varkappa\ H_C$ and the relation between $H_C(0)$ and T_c:

$$J_c = (3.7 \times 10^{-4} S_v\ \gamma^{1/2} T_c\ (1-h) \Delta\varkappa)\ \varkappa^{-2}. \qquad (24)$$

Adjustment of the composition of a material to increase H_{C2} by an increase in $\varkappa$, is therefore likely to produce a reduction in J_c, unless accompanied by a corresponding increase in γ and T_c. Methods for improving J_c should concentrate upon increasing S_v. In the alloys this means refining the dislocation cell-structure without impairing ductility, a process for which there exist very few guidelines. For the A15 materials this implies formation of the compound under conditions which give minimum grain size; usually achieved by keeping the reaction temperature as low as possible.

The alternative approach is to realize that when flux lines are pinned, the Lorentz force is transferred to the conductor as a mechanical stress, and that this stress cannot exceed the yield strength of the material. A simple analysis shows that that the hoop stress on a conductor, bent to a radius of R m, carrying a current density J A m^{-2} in a field of B tesla, is equal to JRB. If the conductor is unsupported, the maximum current density will be that which produces a hoop stress equal to the tensile yield stress σ_{yt}, and :

$$J_{max} = \frac{\sigma_{yt}}{RB}. \qquad (25)$$

The tensile yield stress of Nb-Ti at $4.2\,K$ is $\sim 1\,GN\,m^{-2}$. A Nb-Ti conductor bent to a radius of $0.1\,m$, in a field of $5\,T$, could just carry a current of $\sim 2 \times 10^9\,a\,m^{-2}$ ($2 \times 10^5\,a\,cm^{-2}$), which is only slightly better than that achieved in practice in existing materials. As these materials are incorporated in a matrix of much weaker copper, the allowable overall current density is less than this.

In a solenoid the windings give some mutual support and the ana-lysis is slightly different. The solenoid is considered to be a thick walled cylinder, of outer radius $= \alpha R$ and half-length βR, where R is the inner radius. The magnetic field exerts a pressure $p = 1/2\,HB$, subjecting the windings to a radial compressive stress σ_{rr} and a tangen-tial, tensile stress $\sigma_{\theta\theta}$. These stresses vary with position; $\sigma_{rr} = p$ at the inner surface and 0 at the outer surface, $\sigma_{\theta\theta} = p(\alpha^2+1)\,(\alpha^2-1)^{-1}$ at the inner surface and $2p(\alpha-1)^{-1}$ at the outer surface.

The solenoid central field is given by $H = RJF(\alpha, \beta)$ where $F(\alpha, \beta)$ is the Fabry factor. If the coil fails when the tensile stress at the outer surface $= \sigma_{yt}$, the maximum current density is given by:

$$J_{max} = \frac{(\alpha-1)\,\sigma_{yt}}{RBF\,(\alpha,\,\beta)}\,. \tag{26a}$$

$F(\alpha, \beta) \simeq 2$ when $\alpha = \beta = 3$, and Eq. (26a) becomes equal to Eq. (25). Failure can be prevented by an external jacket of some strong material. The stress at the inner surface now becomes important; if it can be assumed that degradation of superconducting properties occurs when this exceeds some critical value σ_c, then:

$$J_{max} = \frac{2\,\sigma_c}{RBF\,(\alpha,\,\beta)}\,\frac{\alpha^2-1}{\alpha^2+1}\,. \tag{26b}$$

The microstructures which promote flux pinning also increase the mechanical strength, though at a slower rate; e.g., $J_c \propto d^{-1}$, whereas $\sigma_y \propto d^{-1/2}$ (d is the dislocation cell diameter). The ultimate current density will always be limited by the strength of the material. Small improvements in J_c are to be looked for, but spectacular increases are unlikely.

V. STABILITY

A. Introduction

Instabilities in type II superconductors are a natural consequence of the critical state. Everywhere within the superconductor flux is in a metastable condition, the pinning force exactly balancing the Lorentz force. Local flux motion can be generated spontaneously by thermal activation (flux creep), or by an external stimulus such as a sudden change in transport current, ΔI or mechanical shock. The flux motion, if large enough to be detected in, for example, a magnetization experiment, is called a "flux jump" [44, 64]. The movement of flux is dissipative, and causes a local rise in temperature. The pinning force usually decreases as the temperature increases, (dJ_c/dT is negative), more flux is able to move and a small flux jump can develop into a "flux avalanche". If the avalanche is not halted, the temperature rise will ultimately be large enough to cause the superconductor to exceed its critical temperature and become normal. The normal state can also be directly induced by an external temperature pulse, ΔT, which may be generated by friction if the superconductor is able to move under the action of electromagnetic forces, or by the release of elastic energy consequent upon the cracking of brittle components (such as "potting" compounds) under differential thermal contraction when subjected to cryogenic temperatures.

Protection against instability can be provided by switching off the current the moment that any flux motion, or the emf that results therefrom, is detected. This may be acceptable for small laboratory magnets, but it is undesirable to interrupt the operation of many devices in this way. It is possible to operate at currents below which flux jumps occur spontaneously, but it is then also necessary to shield the device from any external stimulus. A more satisfactory approach is to ensure that a local flux jump cannot propagate into a flux avalanche. There are effectively three ways of preventing flux propagation:
   a. Minimizing the temperature rise, by using a material with a high specific heat [65]. This has been referred to as "enthalpy stabilization".
   b. By arranging for dJ_c/dT to be positive such that a local temperature rise causes an increase in pinning [66]. This is called "intrinsic stability".
   c. By reducing the cross section of the conductor, so that the reduction in pinning strength is counteracted by a greater reduction in the Lorentz force [65]. This is called "filamentary stabilization".

None of these approaches gives protection from a temperature rise produced externally from the conductor. A superconductor is fully stabilized if the energy dissipated ohmically as the current passes through the conductor in the normal state can be removed faster than it can be generated. "Cryostatic stabilization" [67] is arranged to do just this, and allow the conductor to recover to the superconducting state. An outline of the stability problem is presented in schematic form in Fig. 7.

B. Criteria for Propagation of Flux Jumps

The motion of flux into a superconductor will cause a reduction in

It must be noted that R is the overall radius of the conductor, and that ρ_N is the average resistivity of the conductor in the normal state. Provided that the current density is kept below J_R, the superconductor will always recover from any disturbance.

The crucial quantity is seen here to be $\dot{Q}$. This depends on the nature of the interface, the nature of the coolant, and the temperature difference across the interface. The usual coolant for superconductors is liquid helium. Chester [65] has pointed out that the heat flux $\dot{Q}_f$ at which the evaporation of helium changes from nucleate boiling to film boiling is critical. Once film boiling has begun, the surface of the conductor is covered with an insulating blanket of gaseous helium, and the rate of heat transfer drops abruptly. $\dot{Q}_f$ depends on the nature and geometry of the surface, but lies in the range $0.1 - 1\,W\,cm^{-2}$.

Seshan [70] has shown that the maximum recovery current of Pb-Bi specimens of differing radii fits Eq. (31) with $\dot{Q} = \dot{Q}_f$. He has also shown that the results of Kobayashi, Yasukochi and Ogasawara [71] on Nb-Zr wire, also fit this equation. Much greater heat fluxes can be achieved by cooling with supercritical helium gas at about $5\,K$. Large devices are now being planned with this type of cooling.

D. Techniques of Stabilization

The three methods of stabilization can be understood by reference to Eq. (30). The maximum stable current, J_p, is increased if the specific heat of the material is increased. Specific heats do rise rapidly with temperature at the low temperatures at which superconductors operate, and an instability may be arrested as the temperature and specific heat rise. This is known as partial stabilization [72].

It can be seen that if dJ_c/dT is positive the conductor is stable to any current [66]. No practical material has yet been developed which utilizes the concept of intrinsic stability.

The final, and most important method of stabilization, is to make d as small as possible. This is done by fabricating superconductors as fine filaments, of which a number, from 10-1000, are embedded in a non-superconducting matrix. The usual matrix material is high conductivity copper, which has several advantages. D_{th} in copper is $10^7 - 10^8\,m^2\,s^{-1}$ and $D_M \sim 10^4\,m^2\,s^{-1}$, thus the copper assists in removing heat from the superconductor, while slowing down the propagation of any magnetic disturbance. The stability conferred by the fine filaments is assisted, and flux jumps in one filament are not transferred to neighboring filaments. In addition, the electrical conductivity of copper is $\sim 10^3 - 10^4$ times greater than that of the superconductor in the normal state. The value of ρ_N for the composite conductor, appropriate for Eq. (31), is therefore much lower, and the recovery current much higher, than for a non-composite material.

By reference to Eqs. (30) and (31), it is possible to design a composite which is completely stable to flux motion, and which is able to recover if driven normal by other disturbances. J_p is the current density in each filament, and ideally the filament diameter is chosen so that J_p is equal to the value of J_c at the highest field at which the conductor is to operate. The overall diameter of the composite, and the ratio of superconductor to normal material, and hence the number of filaments, are chosen such that the overall current density of the composite conductor is equal to J_R.

This concept of complete stability was first utilized by Kantrowitz and Stekly [73]. The insertion of suitable values of material parameters into Eq. (31) indicates that some thirty times as much copper as superconductor must be used to give complete recovery, and the resulting overall current densities are low. If, however, partial stabilization is allowed for, and it is assumed that only a few of the filaments fail at any one time, the quantity of copper required can be considerably reduced. Successful composites are now made with the copper/superconductor ratio very close to one.

Filamentary stabilization is only effective provided that the filaments are independent of one another. Independence is achieved under conditions of steady current and field, there being no voltage to drive current across the normal matrix separating the superconducting filaments. Every conductor will at some time be subjected to a changing field, either externally imposed or due to a change in the current flowing through it, if only when the device is switched on or off. The changing field induces an emf across the normal matrix, and current flows between neighboring filaments as a result of this emf. The filaments are completely coupled, when the rate of change of field, $\dot{H}$ exceeds a critical value [74]:

$$\dot{H}_C = \frac{2\rho\bar{\lambda}J_c\,d\omega}{\ell^2(\omega+d)} \, , \qquad\qquad (32)$$

where ρ is the resistivity of the matrix, $\bar{\lambda}$ is a space factor for the composite, d is the diameter of the filament, ω the interfilamentary spacing, and ℓ is the length of the conductor. Stability at high rates of change of field can be achieved by using a high resistivity matrix, though this is undesirable from the recovery point of view, or by twisting the filaments, in which case the value of ℓ to be inserted in Eq. (32) is the twist pitch. Twisted composites are now common in large magnets. Transposing the filaments by cabling gives even greater improvement in performance.

With the introduction of filamentary composites, instabilities no longer limit the performance of high-current superconductors. The correct choice of such factors as filament diameter, number of filaments, packing fraction, resistivity of matrix, pitch of twist, transposition and cabling, make it possible to attain any reasonable level of desired stabil- ity. Rise time of large coils can still, however, be a problem.

VI. AC LOSSES

Consideration of the electrodynamics of the superconducting state [75] indicates that type I superconductors, and type II superconductors below H_{C1}, should show no electrical resistance at frequencies below $\sim 10^{12}$ Hz. This is found, in practice, to be so, and losses occur at all frequencies. This is thought to be due to surface imperfections [76], and certainly losses at radio frequencies in Nb are reduced by several orders of magnitude after annealing close to the melting point in a high vacuum [77]. A possible source of rf losses is phonon genera- tion [78] assisted by surface fissures. Low frequency losses are be- lieved to be due to surface roughness [79] which in some way makes the surface a multiply connected "sponge" with an ability to trap flux. Seebold [80] has shown that surface flux trapping can account for ac losses in Pb, In, Sn and dilute Pb alloys.

If a type II superconductor is subjected to an alternating field or current, such that H_{C1} is exceeded at the surface, then flux penetrates into the superconductor and regions of the material will be cycled round minor hysteresis loops. If the maximum field amplitude exceeds H_{C2}, the whole specimen will experience a major hysteresis loop. The energy loss per cycle should be independent of frequency and be given by:

$$W = \int_V \int_{H_1}^{H_2} (\vec{H} \cdot d\vec{M})\, dV, \tag{33}$$

where H_1 and H_2 represent the minimum and maximum field amplitudes, the second integration being carried out over the volume of the sample. There have been several solutions given to this equation in terms of the various models of the critical state. Dunn and Hlawiczka [81] suggest that the energy dissipated per unit area of a flat strip should be:

$$W_S = \frac{4.22 \times 10^{-9}}{J_c} (H_m')^2 \left\{ 1.5 H_{C1} + 3\Delta H + H_m' \right\} \text{ J cm}^{-2} \tag{34}$$

where $H_m' = H_m - \Delta H - H_{C1}$, H_m is the maximum field amplitude, and ΔH is the field difference between the exterior and interior of the sample, due to surface screening currents. According to this model, no bulk losses should be found for field amplitudes below $H_{C1} + \Delta H$. The above expression has been verified for Nb by Easson and Hlawiczka [82] and by Linford and Rhodes [83]. ΔH is found to depend on surface condition, increasing as the surface finish is improved.

If the maximum amplitude of the applied field does not exceed the field, H_p, which may be many times H_{C1}, at which the static magnetization curves shows a peak, then the ac losses are lower the stronger the pinning. This can be seen in Eq. (34) where $W_S \propto 1/J_c$. If H_m exceeds H_p, then losses begin to rise rapidly, and become larger the stronger the pinning, and hence the greater the irreversibility of the material. Results of ac loss measurements in superconductors have been the subject of a detailed review by Wipf [84]. In general losses in superconductors appear to be less than in normal metals subjected to the same cyclic fields or currents, at least at fields below H_p.

As losses rise rapidly with H_m', approximately as H_m^3 if $H_m \gg H_{C1} + \Delta H$, superconductors exposed to ac conditions are used in configurations in which the surface area is maximized, and surface current densities kept as small as possible. Filamentary composites would appear to be ideal, but unless the matrix is insulating, the outer filaments effectively shield the current from the inner ones. With the largest practical degree of twist, and of transposition by cabling, it is only possible to ensure that all filaments are carrying current provided a frequency ~ 1 Hz is not exceeded. There appears to have been little work done on developing a filamentary material with a completely insulating matrix. One method involves the hydrostatic extrusion of lead filaments in a plastic matrix. Some preliminary work carried out by the author in collaboration with the U.K.A.E.A. Springfields laboratory indicated that this could be a feasible proposition.

An alternative approach would be to reduce H_m' in Eq. (34) by increasing ΔH. This latter quantity is a measure of the reluctance of flux to penetrate the superconductor because of the surface barrier predicted by Bean and Livingston [53]. The theoretical value of this barrier is applicable only to the case where the surface of the conductor is absolutely flat and smooth; surface roughness causes it to be considerably reduced. Quantitative relations between roughness and ac losses are rare, due to the highly sophisticated techniques necessary to properly characterize surfaces. The very high quality of surface perfection required to give substantial reductions in ac losses is unlikely to be produced economically in an industrial environment.

VII. FABRICATION OF SUPERCONDUCTING MATERIALS

A. Ductile Alloys

Ductile Alloy superconductors based on Nb are produced by conventional metallurgical techniqucs. Compositions of alloys which have been, or are now being, manufactured commercially include

Nb+ 25-33 at% Zr, Nb+ 55-80 at%; Nb-Ti-Zr and Nb-Ta-Ti alloys
have also been reported. Details of the manufacturing methods and
processing routes have not been disclosed, but an educated guess can be
made as to their nature.

Ingots of the alloys, probably ~10 cm in dia. and 1 m long are
produced from the pure constituents by consumable electrode arc melting
in a vacuum. The ingots are then broken down by hot extrusion, prior
to wire drawing. The latter process is interrupted at appropriate stages
for brief recrystallization anneals. These will be carried out in vacuum
at about 800°C for 30 min. The final reduction (in area) from the ultimate recrystallization anneal is about 99.99% to ensure a dislocation
cell-size of approximately 0.1 μm. A final treatment at ~400°C may
be given to enhance the definition of the cell structure by recovery processess, and thus optimize the critical current density. The original
monofilamentary conductors were then given an electroplated copper
finish, to assist in making contacts and to confer some stability.

It is believed that considerable difficulty was experienced with
lubrication of the Nb-Zr alloys during drawing, and serious galling
(welding of the material to the die) took place. The Nb-Ti alloys were
found not to be quite so difficult to process (the presence of Zr hydride
may also have rendered the Nb-Zr alloys prone to brittle failure). Cladding of the ingot with copper prior to reduction not only eases the subsequent processing, but also produces the composite material required
for stability.

Multifilamentary composites are produced by inserting several
Nb-Ti ingots (probably of a lesser dia. than for monofilament material)
into holes previously drilled in a much larger copper, cupro-Ni, or
Al ingot. The composite ingot is, as before, broken down by hot extrusion, followed by wire drawing. The final passes are through hexagonal shaped dies. The resulting hexagonal wires are then more easily
stacked into Cu tubes prior to further reductions. Annealing treatments are carried out as before (below 660°C, the melting point of Al,
for filaments in an Al matrix). In this way up to ~10000 filaments of
dia. less than 10 μm may be incorporated in Cu, Cu plus cupronickel,
or Al matrices. The conductor may then be subjected to a twisting
operation to produce a pitch of one twist per 25 mm. Flat strip can be
formed by simply rolling the wire product. Other shapes, including
hollow conductors, are also available.

The multifilamentary composites were first developed by Imperial
Metal Industries in cooperation with the Rutherford High Energy
Laboratory, and represent a considerable technical achievement. All
other manufacturers of Nb-Ti materials have since produced multifilamentary conductors of various configurations. The reluctance to
reveal details of the processes involved arises from the fact that the
techniques are in the main, conventional, and patent protection is dubious.

B. A15 Compounds

The A15 compound superconductors are brittle, and it is therefore not possible to process ingots of the compounds by the techniques
just described for the ductile alloys. Either the constitutent elements
must be processed to final configuration and then reacted to form the
compound, or vapor deposition techniques must be used. Nb_3Sn and
the methods for its production has been the subject of a recent review
[85].

Wires of Nb_3Sn were first produced by the Kunzler technique [86].
Mixed Nb and tin powders were sealed in a Nb tube which was then
drawn down to final wire size. After winding the Nb_3Sn was formed by
reaction at $\sim 1000^\circ$ C. The limitation placed by this high temperature on
the materials for other components of the coil, plus the lack of precise
metallurgical control of the microstructure of the resulting compound,
have led to the abandonment of this method.

The vapor deposition method was developed next by Hanak at
RCA [87]. Gaseous chlorides of Nb and tin are separately produced by
passing chlorine over the metallic elements held at $800 - 900^\circ$ C. The
chloride vapors are together introduced into the reaction chambers
where they meet a counter current of hydrogen gas. The chamber is
maintained at 700° C which ensures that the chlorides do not condense.
A substrate ribbon, which may be Nb, SS or Hastelloy, passes through
the chamber and is resistively heated to $\sim 1000^\circ$ C. At this temperature
the chlorides are reduced by the hydrogen and Nb_3Sn is deposted on the
tape. HCl vapor is introduced with the hydrogen to suppress the formation of $NbCl_3$ which is solid at 700° C and therefore would block the
chamber. The thickness of the compound layer, controlled by the rate
at which the substrate is passed through the chamber, can be up to ~ 1
µm. Stabilization can be conferred by electroplating with copper or
silver, or by soldering a layer of copper tape to either side of the conductor. Cladding in SS will give the required mechanical strength. Current densities can be increased by additions to the reaction gases [63, 88,
111].

The substrate material is chosen such as to have a higher coefficient of thermal expansion than the compound. This ensures that,
on cooling down, the Nb_3Sn layer is in compression and therefore has
less tendency to crack. It was at one time thought necessary to first
plate the substrate with platinum, in order to produce a thin layer of
Nb_3Pt to promote adhesion between compound and substrate. This expensive practice is now found to be not necessary. Variations on the
RCA process are described by Echarri and Spadoni [85].

The most straightforward method of producing compounds is the
diffusion process. For example, tin is deposited on the surface of a
Nb wire or tape, and diffusion is carried out at a temperature, usually
$900 - 1000^\circ$ C, which allows of the formation of Nb_3Sn. The tin may be

introduced to the Nb by exposing the latter to tin vapor, by electro-
phoretic deposition of tin powder into the Nb, or by simply dipping the
Nb into molten tin. The latter is now the process by which most of the
commercial Nb_3Sn is produced. V_3Ga is also produced by this method.

Nb tape or ribbon is fed into a bath of molten tin maintained at
a fairly low temperature ($\sim 300^\circ$ C). As the tape leaves the tin bath,
excess tin is removed by rollers. It is important that the compound not
be formed in the presence of excess tin, which can penetrate along
the compound grain boundaries, thus interrupting the superconducting
current paths and reducing J_c. The tape then passes into the reaction
chamber, held at a temperature of 900-1000° C at which the tin diffuses into the Nb and forms Nb_3Sn. The kinetics of this process are
discussed in Ref. 85. Copper tapes are subsequently soldered to
either side of the reacted ribbons, and as with the vapor deposited
material, cladding in stainless steel provides any necessary mechanical
strength. The production of V_3Ga tape is similar to the above in
principle [89], though it is essential to clad the tape with Cu before reacting, or add Cu to the Ga bath, to prevent the formation of V_3Ga_2, VGa_2
[116]. Superconductors produced in this way are reasonably stable, as
the layer of compound is quite thin. They are not as satisfactory as
would be multifilamentary composites of the same material, especially
at the outer ends of coils, where the field has a component normal to the
tape plane. The value of d in Eq. (30) is now the width of the tape.

C. New Processess

A method for producing multifilamentary Nb_3Sn has recently been
developed [90]. Rods of Nb are placed in previously drilled holes in
an ingot of a copper-tin alloy (bronze). This material is co-processed
exactly as for the niobium-titanium multifilamentary materials. When
the final size is reached, with the Nb filaments ~ 5 μm dia., the conductor is annealed for several hours at 650-750° C. The tin in the matrix
adjacent to the Nb reacts and coats the Nb filaments with a thin layer of
Nb_3Sn. The composite then consists of Nb filaments which provide
mechanical strength, a layer of the high field superconductor and a matrix
of high resistivity bronze which helps to decouple the filaments. Stability
can be improved by the introductions by various means of strands of
pure copper.

This "bronze route" has also been used to produce V_3Ga [91, 92]
and V_3Si [92, 93] by co-processing vanadium rods in copper-gallium
or copper-silicon matrices. A particular advantage of these processes,
in addition to producing a composite with the desirable characteristics
listed above, is that the presence of the copper somehow stimulates the
formation of the A15 compound at a temperature lower than that required
for direct combination of the elements [116]. This results in a much finer
grain size in the compound [117], and hence a higher critical current density
in the superconductor. The process unfortunately does not seem to
work for Nb_3Al or Nb_3Ga [94]. This may be because these compounds

will form only at a temperature much higher than the melting point of the matrix ($\sim 1000^\circ$ C). It has been reported that, for example, Nb_3Al does not form below 1850° C [95].

A disadvantage of this process is that the matrix, being an alloy, work-hardens rapidly during the reduction process, and frequent annealing is necessary. An external diffusion process with a pure copper matrix removes the need for these anneals. The resulting Nb filaments - copper matrix wire is coated with tin, which is then diffused through the copper to react with the Nb at temperatures similar to those used in the bronze process [96]. This method is also applicable to V_3Ga (D. Dew-Hughes, unpublished) though there are complications due to the high concentrations of Ga in both Cu and V in equilibrium with V_3Ga.

Both of these methods will receive extensive development over the next few years, and it is expected that fully stabilized filamentary compounds made by one or other of these techniques will replace the tape products currently in commercial production. Efforts will be made to discover similar routes for producing multifilamentary Nb_3Al, Nb_3Ga, and in particular $Nb_3(AlGe)$.

VIII. COMMERCIAL SUPERCONDUCTING MATERIALS

The commercial situation with regard to superconducting materials is complex and rapidly changing. The potential demand for material consequent upon the adoption of rotating superconducting machines is sufficient to have attracted many companies into the business of development and manufacture of superconducting materials. This level of activity, and the high cost of development,is not really justified by the current demand for superconductors. Consequently manufacturers are continually improving specifications, and adjusting prices, in order to gain a slight advantage over their rivals. Some of the larger companies with an original interest in superconducting materials have since opted out; some of the smaller companies set up specifically to exploit a particular material or process have now gone out of business. To attempt to review the commercial scene in detail is a difficult and risky business. This has, in fact, been recently done [97, 98] and will therefore not be repeated here; just a brief summary of the current position will be given.

Only three types of material are readily available commercially. Niobium-titanium is produced in a variety of single and multifilamentary configurations, with up to ~ 1000 filaments, in copper, aluminum or cupro-nickel matrices. The ratio of conductor to matrix varies from 1:1 to 1:10. The filaments may be twisted, and the cross-section be round, rectangular or even hollow for internal cooling. This material can be used to build magnets capable of generating fields of 8.5 T at

4.2 K, and up to 10 T if the operating temperature is reduced to the
λ-point of liquid helium, 2.17 K [99]. Current densities of
$\sim 1.5 \times 10^9$ A m^{-2} at 5 T are attained within the superconductor. Overall current densities depend upon the relative amount of normal metal
matrix present, but are typically one half to one third of this value.

Niobium-tin tape is available, produced either by vapor-deposi-
tion or by diffusion. The tape, complete with copper stabilization and
stainless steel cladding, is typically ~ 0.2 mm thick and up to 12.7 mm
(1/2") wide. The ratio of copper to conductor is $\sim 5{:}1$. Critical current density in the superconductor at 5 T is $\sim 5 \times 10^9$ A m^{-2}, but the
overall current density in the composite conductor is reduced by a factor of 10, and is thus about the same as that for the fully stabilized
Nb-Ti. As the latter is much cheaper, Nb3Sn is only used for fields in
excess of 10 T, at which Nb-Ti cannot operate. There appears to be
little significant difference between vapor-deposited and diffused
material.

Vanadium-gallium tape, also produced by the diffusion process,
is now available from two Japanese sources. Despite a lower critical
temperature (15 K vs. 18 K), V$_3$Ga can have a higher critical
current density at fields above 12-16 T. At 5 T the current density is
$\sim 3 \times 10^8$ A m^{-2} (cf. 5×10^8 A m^{-2} for Nb3Sn), at 12 T both can carry
$\sim 2 \times 10^8$ A m^{-2}, and at 16 T J_c is $\sim 1.5 \times 10^8$ A m^{-2} for V$_3$Ga and
$\sim 1 \times 10^8$ A m^{-2} for Nb$_3$Sn ([97] and manufacturers' literature). The
reason for this is not known. V$_3$Ga is formed at $\sim 700^\circ$ C, as opposed to
900-1000° C for Nb3Sn resulting in a finer grained material, and a less
complete martensite transformation [118]. Lorentz force ($\vec{J}_c \times \vec{h}$) versus
h curves for V$_3$Ga generally have a maximum at h $\simeq 0.5$, corresponding
to ΔK pinning (Eq. 23), due to an incomplete martensite transformation,
as suggested in IV F. In Nb3Sn the maximum occurs at h $\leqslant 0.3$, in-
dicative of pinning by nonsuperconducting precipitates (Eq. 21).

The relative costs, expressed as cost in pounds per unit current
density, per unit length of conductor, as wound in a typical coil, of
examples of all three types of material, are shown in Fig. 8, (kindly
provided by P.E. Hanley of the Oxford Instrument Co.). The higher
current density of V$_3$Ga at high fields is reflected in its lower cost. It
is quite clear from this figure that each material is likely to be employed
in a specific field range: Nb-Ti up to 8.5 - 10 T, Nb3Sn from 10 - 14 T,
and V$_3$Ga from 14 T up to, possibly, 18 T. High field magnets are likely
to be built with two or three concentric windings, of Nb-Ti for the low-
field outer portion, of Nb3Sn in the intermediate field region, and V$_3$Ga
for the center, high field, component.

An interesting comparison between Nb-Ti and Nb3Sn magnets has
been prepared by the Oxford Instrument Co., and is given in Table 3.
Both magnets are capable of producing 10 T in a 25 mm bore. The Nb-Ti
magnet is operated at 2.17 K, the Nb3Sn at 4.2 K. Capital and operating
costs are seen to favor the Nb-Ti magnet; the lower operating costs

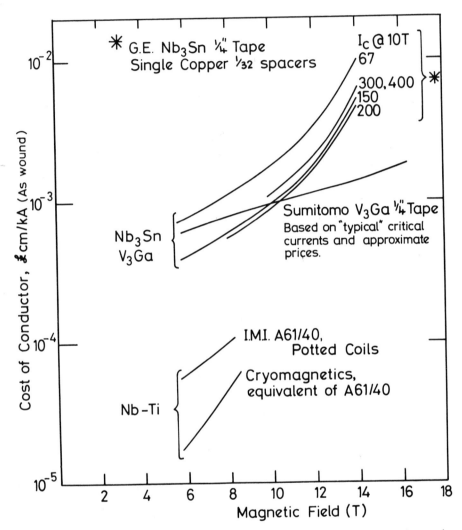

Based on manufacturers guaranteed short sample critical currents, and likely turns density using standard winding techniques.

Fig. 8. Relative Costs of Superconducting Material (Courtesy P.E. Hanlon and the Oxford Instrument Co., Ltd.).

Table 3. Comparison between Nb-Ti and Nb_3Sn magnets

Material	IMI Multifilamentary Nb-Ti	G.E. Nb_3Sn Tape
Maximum field	10 T	10 T
Bore	25 mm	25 mm
O.D.	150 mm	120 mm
Length	160 mm	200 mm
Operating temperature	2.17 K	4.2 K
Operating current	47 amps	110 amps
Length of superconductor	7 km	1.7 km
Capital cost	£ 2600	£ 6000
Helium consumption	~5 ℓ to cool down then 0.25 ℓ/hr.	0.8 ℓ/hr.

Table 4. Manufacturers of superconducting materials

	Nb-Ti	Nb$_3$Sn	V$_3$Ga
Canada		Canada Super-conductor and Cryogenics Co. Ltd. (ex RCA vapor deposition process)	
France	Thomson-Houston HB CGE Marcoussis (also with Al matrix)	Thomson-Houston HB	
Germany	Vacuumschmeltze		
Japan	Hitachi Ltd. (Nb-Ti-Zr) Mitsubishi Electric Co. Ltd. (Nb-Ti and Nb-Ta-Ti) Vacuum Metal-lurgical Co. Ltd.		Sumitomo Electrical Industries Ltd. Vacuum Metal-lurgical Co. Ltd.
Switzerland	Brown-Boveri & Co. Ltd. (Oerlikon)		
U.S.A.	Airco/Magnetics Corp. of America Supercon Supertechnology	Intermagnetics General Co. (ex G.E.) Union Carbide (Linde)	
U.K.	Imperial Metal Industries		

The above is believed to be accurate as of June 1973.

arise out of the lower rate of helium boil-off, resulting from the lower current required to energize the Nb-Ti magnet. The manufacturers currently supplying superconducting material together with the type of material produced, are listed in Table 4.

IX. FUTURE PROSPECTS

The most important problem in superconducting materials is that of raising the critical temperature. This will reduce costs by allowing of operation at higher temperatures, as well as contributing to possible increases in H_{C2} (Eqs. (3) and (6)) and J_c (Eq. (24)). Current densities have been shown to be limited by mechanical, rather than by super-conducting considerations (Eqs. (25) and (26)). A similar conclusion applies to the upper field at which a magnet can operate; as the magnetic pressure $1/2 \mu_0 H^2$ on the solenoid windings cannot be allowed to ex-ceed some critical stress. An increase in H_{C2} is only of value in that it will permit operation at a higher temperature for a given field.

Stability is no longer a problem; the design of fully stabilized materials is thoroughly understood. ac losses depend upon surface perfection and upon H_{C1} (Eq. (34)). The latter is also proportional to critical temperature. Thus the greatest bonus will accrue from an increase in T_c.

The first step is not to seek an improvement in T_c as such, but to learn how to utilize, in a practical material, the increases already achieved over Nb_3Sn in such compounds as Nb_3Al, Nb_3Ga, and $Nb_3(Al, Ge)$. A start has already been made with Nb_3Al [100] and $Nb_3(Al, Ge)$ [101]. The latter material was prepared by rolling together foils of Nb, Al and Ge to form a laminated tape, which was then heat-treated for 1 min. at $1600^{\circ}C$, followed by a 2 hr. anneal at $750^{\circ}C$. The critical current density of this material was disappointingly low at only $4 \times 10^6 A\ m^{-2}$ in 2.4 T at 4.2 K. In fact, the best reported value of current density in the ternary compound is only one tenth of that for Nb_3Sn [102]. Current densities in Nb_3Al are believed also to be low. (Recent improvements in J_c have been reported for Nb_3Al [123], Nb_3Ga [124], and $Nb_3(Al, Ge)$ [125].) A possible explanation for this is that precipitates, such as oxides, normally assist in limiting the grain size, giving a high J_c. However, in compounds involving aluminum the excess of aluminum which must be present during their formation will, because of the great avidity of aluminum for oxygen, take up all of the available oxygen and prevent other oxides forming. The solution may be to alloy with the niobium an element such as zirconium which forms an even more stable oxide than aluminum. This technique has already produced improvements in Nb_3Sn [62].

Once the presently known high T_c materials have been commer-cially exploited, the incentives to find even higher T_c materials will increase. It is obvious that a part of this search must concentrate on new ternary, and perhaps quaternary, A15 compounds. An understanding of the factors that affect phase stability, in particular that of the σ-phases, is, as pointed out in Section II-C, necessary to prevent the search being purely empirical. This is a problem which is concerning metallurgists in another context. The appearance of σ-phases in many high tempera- ture, high strength alloys is, due to their brittleness, extremely deleterious. The theoretical approaches developed for this application, together with a sound knowledge of the relevant binary phase diagrams, will assist in a rapid solution of this problem. It is, however, un- likely that T_c will be raised above $\sim 25\,K$ in an A15 compound.

High critical temperatures are found in only one other class of compounds, the carbides and nitrides of the transition metals which crystallize in the B1 structure. The critical fields are in all cases lower than those for the high T_c A15 compounds, and the B1 compounds are consequently unlikely to be developed. Of rathter great interest is the compound V_4HfZr, which has the cubic C15 structure. Although its critical temperature is only $10\,K$, it has an upper critical field at $4.2\,K$ of $23\,T$ [103]. The electron to atom ratio for this compound is 4.66, very close to one of Matthias' favorable values (Section II-B). It may well be that other C15 compounds will be found with even higher critical temperatures.

A new class of superconductors was discovered when it was found that the intercalation of weakly superconducting transition metal di- chalcogenides by various materials, such as alkali metals or organic complexes, in many cases caused spectacular increases in T_c [104]. For example, TaS_2 has a critical temperature of $0.8\,K$; this is raised to $3.5\,K$ after intercalation with pyridine. Unfortunately the dichalcogenides with a high critical temperature suffer degradation upon intercalation; e.g., $NbSe_2$ has $T_c \simeq 7\,K$ which is reduced to $\sim 3.5\,K$ upon intercalation. The intercalating substances are all able to donate electrons to the host material, and it now appears that the change in critical temperature can be correlated with the change in density of states as the Fermi level is altered by the addition of electrons. The critical temperature is found to be insensitive to the spacing of the host layers [104]; theoretical cal- culations indicate a similar insensitivity of the band structure to layer separation (P.M. Lee: private communication). High critical tem- peratures are thought to be unlikely in these intercalated compounds. These materials are highly anisotropic.

The possibility of an organic superconductor, first suggested by Little [105], is again in the news [106]. Fluctuations in conductivity have been observed at $58\,K$ in tetrathiofulvalene-tetracyanoquinodimethane (TTF-TCNQ) and may indicate a possible lower T_c superconducting state. The organic superconductor has not yet been found, and though this dis- covery brings it closer, there is still a long way to go! Even should organic superconductors becomes a reality, it is difficult to conceive of them carrying the high current densities required for electrical engineering applications. Other suggested mechanisms for superconduc- tivity have yet to receive experimental confirmation.

The average rate at which the critical temperature of known super-conductors has been raised since the original discovery of superconduc-tivity is ~ 1 K every four years [97]. This suggests that T_c should be raised to ~ 22 K in the near future and that 25 K could be reached in 1985. A more conservative projection assumes a logarithmic rate of improvement, with 25 K not being obtained until nearer the end of the century [8]. Any advance beyond 25 K will almost certainly require a new mechanism of superconductivity, and further extrapolation is quite unsound. Nb_3Ge with a $T_c = 22.3$ K has just been reported [119] and has been extended to a critical temperature of 23.2 K [120].

ACKNOWLEDGMENTS

I am indebted to P. McDonald of the Oxford Instrument Co. for providing much of the information in Section VIII, and to C.P. Herring of the Metallurgy Department, Oxford University, for providing Fig. 6.

REFERENCES

1. B.B. Goodman, Rep. Prog. Phys., 29 445 (1966).
2. P.G. de Gennes, "Superconductivity of Metals and Alloys", Benjamin, New York (1966).
3. D. Saint-James, G. Sarma and E.J. Thomas, "Type II Super-conductivity", Pergamon, Oxford (1969).
4. R.D. Parks (Ed), "Superconductivity", Dekker, New York (1969).
5. J.D. Livingston and H.W. Schadler, Prog. Mater. Sci., 12, 183 (1964).
6. J.A. Catterall, Metall. Rev., 11, 25 (1966).
7. D. Dew-Hughes, Mater. Sci. Eng., 1, 2 (1966).
8. D. Dew-Hughes, Rep. Prog. Phys., 34, 821 (1971).
9. J. Bardeen, L.N. Cooper and J.R. Schreiffer, Phys. Rev., 108, 1175 (1957).
10. E. Bücher, F. Heiniger, J. Müheim and J. Müller, Rev. Mod. Phys., 36, 146 (1964).
11. B.T. Matthias, Prog. Low Temp. Phys. Vol. 2, Ed. C.J. Gorter, North Holland, Amsterdam (1957) p.138.
12. B.W. Roberts, Nat. Bur. of Stand. Tech. Notes 482 (1968), 724 (1972).
13. G. Gladstone, M.A. Jensen and J.R. Schreiffer, in Ref. 4, p.665.
14. J.J. Hanak, G.D. Cody, J.L. Cooper and M. Rayl, Proc. 8th Int. Conf. Low Temp. Phys. Ed., R.O. Davies, Butterworth, London (1962) p.353.
15. G. Arrhenius, E. Corenzwit, R. Fitzgerald, G.W. Hull, Jr., H.L. Luo, B.T. Matthias and W.H. Zachariasen, Proc. Nat. Acad. Sci., 61, 621 (1968).
16. R.D. Blaugher, R.E. Hein, J.E. Cox and R.M. Waterstrat, J. Low Temp. Phys., 1, 539 (1969).

17. L.E. Toth, "Transition Metal Carbides and Nitrides", Acad. Press, New York (1971).

18. B.T. Matthias, T.H. Geballe and V.B. Compton, Rev. Mod. Phys., 35, 1 (1963).

19. B.T. Matthias, T.H. Geballe, R.H. Willens, E. Corenzwit and G.W. Hull, Jr., Phys. Rev., 139, A1501 (1965).

20. G.W. Webb, L.J. Vieland, R.E. Miller and A. Wicklund, Solid State Comm. 9, 1769 (1971).

21. R. Flückiger, P. Spitzli, F. Heiniger and J. Muller, Phys. Letters, 29A, 407 (1969).

22. R.H. Willens, T.H. Geballe, A.C. Gossard, J.P. Maita, A. Menth, G.W. Hull, Jr., and R.R. Soden, Solid State Comm. 7, 837 (1969).

23. F. Heiniger, R. Flückiger, A. Junod, J. Muller, P. Spitzli and J.L. Staudenmann, Proc. 12th Int. Conf. on Low Temp. Phys.

24. B.T. Matthias, T.H. Geballe, L.D. Longinotti, E. Corenzwit, G.W. Hull, Jr., R.H. Willens and J.P. Haita, Science 156, 645 (1967).

25. J. Ruzicka, Zeit. Physik. 237, 432 (1970).

26. A. Cave, Ph.D. Thesis, UMIST (1972).

27. A. Birch, Ph.D. Thesis, UMIST (1972).

28. B. Mühlschlegel, Zeit. Physik, 155, 313 (1959).

29. K. Hechler, G. Horn, G. Otto and E. Saur, J. Low Temp. Phys., 1, 29 (1969).

30. L.P. Gor'kov, Sov. Phys. JETP 9 (1960) 1364; 10 998 (1960).

31. C. Caroli, P.G. de Gennes, and J. Matricon, Phys. Kondens. Mater., 1, 176 (1963).

32. A.A. Abrikosov, Sov. Phys. JETP, 5 1174 (1957).

33. Y.B. Kim, C.F. Hempstead and A.R. Strnad, Phys. Rev., 139, 1163 (1965).

34. A.M. Clogston, Phys. Rev. Lett., 9, 266 (1962).

35. B.S. Chandrasekhar, Appl. Phys. Lett., 1, 7 (1962).

36. T.G. Berlincourt and R.R. Hake, Phys. Rev. 131, 140 (1963).

37. L.J. Neuringer and Y. Shapira, Phys. Rev. 140, A1638 (1965).

38. K. Maki, Physics 1, 21, 127, 201 (1964).

39. N.R. Werthamer, E. Helfand and P.C. Hohenberg, Phys. Rev., 147, 295 (1966). A.L. Fetter and P.C. Hohenberg, Superconductivity Vol. 2, Ed., R.D. Parks, Marcel Decker, Inc., N.Y. (1969).

40. R.R. Hake, Appl. Phys. Lett. 10, 189 (1967).

41. F.B. Silsbee, J. Wash. Acad. Sci., 6, 597 (1916).

42. Y.B. Kim, C.F. Hempstead and A.R. Strnad, Phys. Rev., 131, 2846 (1963).

43. C.P. Bean, Phys. Rev. Lett., 8, 250 (1962).

44. Y.B. Kim, C.F. Hempstead and A.R. Strnad, Phys. Rev., 129, 528 (1963).

45. P.W. Anderson, Phys. Rev. Lett., 9, 309 (1962).

46. W.A. Fietz and W.W. Webb, Phys. Rev., 657 (1969).

47. R.G. Hampshire and M.T. Taylor, J. Phys. F (Met. Physics), 2, 89 (1972).

48. A.M. Campbell, J.E. Evetts and D. Dew-Hughes, Phil. Mag., 18, 313 (1968).

49. J.D. Livingston, Proc. Summer Study on Superconducting Devices and Accelerators, Brookhaven National Laboratory (1968) p.377.
50. J.I. Gittleman and B. Rosenblum, J. Appl. Phys., 39, 2617 (1968).
51. R. Labusch, Crystal Lattice Defects, 1, 1 (1969).
52. H. Trauble and U. Essmann, J. Appl. Phys. 39, 4052 (1968).
53. C.P. Bean and J.D. Livingston, Phys. Rev. Lett., 12, 14 (1964).
54. R.I. Coote, J.E. Evetts and A.M. Bampbell, Can. J. Phys., 50, 421 (1972).
55. A.V. Narlikar and D. Dew-Hughes, Phys. Stat. Solidi, 6, 383 (1964); J. Mater. Sci., 1, 317 (1966).
56. D.Dew-Hughes and M.J. Witcomb, Phil. Mag. 26, 73 (1972).
57. M.J. Witcomb and D. Dew-Hughes, J.Mater.Sci. 8, 1383 (1973).
58. G.W. Bibby, Ph.D. Thesis, Cambridge (1970).
59. D.F. Neal, A.C. Barber, A. Woodcock and J.A.F. Gidley, Acta Met., 19, 143 (1971).
60. J.J. Hanak and R.E. Enstrom, Proc. 10th Int. Conf. Low Temp. Phys. Ed., M.P. Malkov, Viniti, Moscow (1966) p.10.
61. E. Nembach, Zeit, Metall., 61, 734 (1970).
62. M.G. Benz, Trans. Metall. Soc. AIME 242, 1067 (1968).
63. P.B. Hart, C. Hill, R. Ogden and C.W. Wilkins, J. Phys. D. (Appl. Phys.), 2, 521 (1969).
64. J.E. Evetts, A.M. Campbell and D. Dew-Hughes, Phil. Mag., 10, 339 (1964).
65. P.F. Chester, Rep. Prog. Phys., 30, 561 (1967).
66. J.D. Livingston, Appl. Phys. Lett., 8, 319 (1966); J. Metals, 18, 698 (1966).
67. Z.J.J. Stekly, J. Appl. Phys., 37, 324 (1966).
68. I.D. McFarlane and D. Dew-Hughes, J. Phys. D. (Appl. Phys.), 3, 1423 (1970).
69. H. Brechna, Proc. Summer Study on Superconducting Devices and Accelerators, Brookhaven National Laboratory (1968) p. 478.
70. K. Seshan, M. Sc. Thesis, Lancaster (1971).
71. H. Kobayashi, K. Yasukochi and T. Ogasawara, J. Appl. Phys. 9, 889 (1970).
72. R. Hancox, Proc. I.E.E., 133, 1221 (1966).
73. A.R. Kantrowitz and Z.J.J. Stekly, Appl. Phys. Lett., 6, 56 (1965).
74. M.N. Wilson, C.R. Walters, J.D. Lewin and P.F. Smith, J. Phys. D. (Appl. Phys.), 3, 1518 (1970).
75. F. London, "Superfluids" Vol. 1, John Wiley, New York (1950).
76. T.A. Buchold, Cryogenics 3, 141 (1963).
77. J.P. Turneaure and I. Weissman, J. Appl. Phys. 39, 4417 (1968).
78. J. Halbritter, A. Appl. Phys., 42, 82 (1971).
79. R.J.A. Seebold, R.M.F. Linford and R.G. Rhodes, J. Phys. D. (Appl. Phys.), 2, 1463 (1969).
80. R.J.A. Seebold, Ph.D. Thesis, Warwick (1969).
81. W.I. Dunn and P. Hlawiczka, J. Phys. D. (Appl. Phys.), 1, 1469 (1968).
82. R.M. Easson and P. Hlawiczka, J. Phys. D. (Appl. Phys.), 1, 1477 (1968).

83. R.M.F. Linford and R.G. Rhodes, J. Appl. Phys. 42, 10 (1971).
84. S.L. Wipf, Proc. Summer Study on Superconducting Devices and Accelerators, Brookhaven National Laboratory (1968). p.511.
85. A. Echarri and M. Spadoni, Cryogenics 11, 274 (1971).
86. J.E. Kunzler, E. Buehler, F.S.L. Hsu and J.H. Wernick, Phys. Rev. Lett., 6, 89 (1961).
87. J.J. Hanak, Metallurgy of Advanced Electronic Materials, Ed., G.E. Brock, Interscience, New York (1963) p.161.
88. R.E. Enstrom, J.J. Hanak and G.W. Cullen, R.C.A. Review, 31, 702 (1970).
89. S. Fukuda, K. Tachikawa and Y. Iwasa, Cryogenics 13, 153 (1973).
90. A.R. Kaufmann and J.J. Pickett, J. Appl. Phys., 42, 58 (1971).
91. M. Suenaga and W.B. Sampson, Appl. Phys. Lett., 18, 584 (1971).
92. K. Tachikawa, Y. Yoshida, and L. Rinderer, J. Mater. Sci., 7, 1154 (1972).
93. M. Suenaga and W.B. Sampson, Applied Superconductivity Conference, Annapolis, Maryland, May 1972.
94. W.B. Sampson, IVth Int. Conf. on Magnet Technology, Brookhaven, Long Island, September 1972.
95. J.G. Kohr, T.W. Eagar and R.M. Rose, Metallurgical Trans. AIME, 3, 1177 (1972).
96. M. Suenaga and W.B. Sampson, Appl. Phys. Lett. 20, 443 (1972).
97. M.S. Lubell, Cryogenics 12, 340 (1972).
98. P.A. Battams, Cryogenics 12, 356 (1972).
99. M.N. Biltcliffe, P.E. Hanley, J.B. McKinnon and P. Roubeau, Cryogenics 12, 44 (1972).
100. J.G. Kohr, B.P. Strauss and R.M. Rose, I.E.E.E. Trans. Nucl. Sci., NS-18, 716 (1971).
101. E. Tanaka, T. Fukuda, S. Kuma, T. Yamashita and Y. Onodera, Appl. Phys. Lett., 14, 389 (1969).
102. B.W. Howlett, Proc. 11th Int. Conf. Low Temp. Phys. Ed., J.F. Allen, St. Andrews (1968) p.869.
103. K. Inoue, K. Tachikawa and Y. Iwasa, Appl. Phys. Lett., 18, 235 (1971).
104. F.R. Gamble, J.H. Osieki, M. Cais, R. Pisharody, F.J. DiSalvo and T.H. Geballe, Science 174, 493 (1971).
105. W.A. Little, Phys. Rev., 134, A1416 (1964).
106. New Scientist, 58, 198 (1973).
107. A. Muller, Zeit. Naturforschung 25a, 1659 (1970).
108. R.A. Hein, Science and Technology of Superconductivity, Ed. W.D. Gregory, W.N. Mathews, Jr. and E.R. Edelsack, Plenum Press, New York (1973) p. 333.
109. S. Foner, E.J. McNiff, Jr., G.W. Webb, L.J. Vieland, R.E. Miller and A. Wicklund, Phys. Letters 38A, 323 (1972).
110. A.M. Campbell and J.E. Evetts, Advances in Physics, 21, 199 (1972).
111. G. Ziegler, B. Blos, H. Diepers and K. Wohlleben, Zeit. Angewandte Physik 31, 184 (1971).

112. Y. Uzel and H. Diepers, Zeit. Physik 258, 126 (1973).
113. G.J. van Gurp, Phys. Stat. Sol. 17, K135, (1966).
114. R. Brand and W.W. Webb, Solid State Comm. 7, 19 (1969).
115. J.D. Livingston, J. of Metals 18, 698 (1966).
116. Y. Tanaka, K. Tachikawa and K. Sumiyama, J. Jap. Inst. Metals 34, 835 (1970).
117. E. Nembach and K. Tachikawa, J. Less Common Metals 19, 359 (1969).
118. E. Nembach, K. Tachikawa and S. Takano, Phil. Mag. 21, 869 (1970).
119. J.R. Gavaler, Appl. Phys. Letters 23, 480 (1973).
120. Physics Today, 26, 17 (1973).
121. S. Foner and J.R. Gavaler, private communication.
122. S. Foner, E.J. McNiff, Jr., B.T. Matthias, T.H. Geballe, R.H. Willens and E. Corenzwit, Phys. Letters 31A, 349 (1970).
123. J.C. Kohr, T.W. Eagar and R.M. Rose, Metall. Trans. 3, 1177 (1972).
124. S. Foner, E.J. McNiff, Jr., L.J. Vieland, A. Wicklund, R.E. Miller and G.W. Webb, Proceedings of the 1972 Applied Superconductivity Conference, IEEE Pub. No. 72CHO682-5-TABSC, p. 404.
125. R. Löhberg, T.W. Eagar, I.M. Puffer and R.M. Rose, Appl. Phys. Letters 22, 69 (1972), and further improvements have been made (private communications).

Chapter 3

SUPERCONDUCTING MAGNETS

H. Brechna

Department of Electrical Engineering

Federal Institute of Technology, Zurich, Switzerland

I.   INTRODUCTION

Electromagnets were used since the turn of the century in physics laboratories in limited numbers. However, very extensive work on magnet development started around 1950 with the establishment of magnet laboratories and high energy physics establishments throughout the world. The literature in the form of theoretical and experimental papers, reports published in journals, research and progress-reports and engineering notes on various methods of magnetic field generation is quite extensive. Some books are devoted entirely to the subject of magnet design such as those by De Klerk [1], Parkinson and Mulhall [2], Montgomery [3] and Brechna [4]. More specialized books on beam optics by Banford [5], on magnetostatic theory by Durant [6], on applied magnetism by Zijlstra [7] and on field problems by Binns and Lawrenson [8] may serve as a basic references.

The application of magnets for high energy accelerators are shown by Livingstone and Blewett [9] and by Neal [10] and review articles on the generation of high magnetic fields are written by Montgomery [11], Herlach [12], Brechna [13] and Brown [14]. Superconducting magnets and materials are discussed by Chester [15], Wipf [16] and in the RCA [17] review as well as by other authors. Proceedings of conferences devoted to magnet engineering and applications of superconductivity are important sources of information. To keep up with the rapid developments, we refer to the advances of cryogenic engineering [18], the proceedings of magnet engineering conferences [19] and proceedings of applied superconductivity conferences [20]. A survey of literature on superconducting devices was published first in 1967 by Goree and Edelsack [21] and is continuously being updated.

Superconducting magnets for high magnetic fields were introduced around 1960, about 50 years after the discovery of superconductivity. At first superconducting magnets were used exclusively for dc fields applications, however in 1967 the first pulsed superconducting magnets were tested. To date there are several dc type superconducting magnets operating with field energies in the range of 100-800 MJ and dipole type ac magnets in the energy range of 40-200 kJ.

The importance of laboratories devoted entirely to magnet research was foreseen by F. Bitter, the founder of the National Magnet Laboratory at MIT. The success of this laboratory spurred the founding of several new national laboratories such as the Magnet Laboratory in Braunschweig, Germany, the laboratories in Leiden, Netherlands, Canberra Australia, the Lebedev Institute in Moscow and others. Laboratories for particular research in high intensity magnetic fields are those in Grenoble France, McGill University in Canada, and the Max Planck Institute in Garching, Germany.

In high energy physics research laboratories, individual magnet groups were established, which developed, designed and build or subcontracted magnets for their own particular needs. Some of these worldwide establishments include ANL, BNL, NAL and SLAC in the U.S. RHEL in England, CEN in France, Frascati in Italy, CERN and SIN in Switzerland, DESY and IEKP in Germany, and Serpukov in USSR.

Practically all the above laboratories and many more are also engaged in building superconducting magnets. In the last two decades high field technology has developed rapidly such that there is hardly any area in physics, chemistry, biology, medicine, electrical machinery, transportation, ecology, energy generation and conversion in which these new magnets of some sort are not being used.

Let us summarize a few major achievements in superconductivity and magnet technology since the discovery of type II superconductors in 1957:

1957 Nb_3Sn ribbon was manufactured in long lengths and small size solenoids producing about 6 T were built.
1961 NbZr was drawn in form of wires and small size solenoids generating up to 5 T central fields were developed.
1963 NbTi monofilament conductors. These strands were twisted in the form of cables and wound into solenoids. A new generation of magnets emerged yielding about 6 T and field energies of about 0.3 MJ.
1965 The first MJ high field magnets became operational. Mainly in form of solenoids and split coils. At the same time the Nb_3Sn tape magnets for beam transport magnets were successfully tested. Superconducting magnets for high energy physics experiments were utilized. Multifilament conductors were produced after the cryostatic stability theory was developed.

1968   The first multifilament conductor with about 50 filaments were available commercially and the era of ac superconducting magnets began. The technology for coextrusing two and three component multifilamentary conductors developed fast. In 1972 conductors with 14,000 filaments ($\sim 8\mu$m filament diameter) and three components became available.

1968   Ternary alloys such as $Nb_3(GeAl)$ with critical temperatures above 20K were discovered. Since then other superconductors on the same basis were obtained having a T_c close to 21 K. In 1972 these alloys had been improved to an extent that they had already a current density of $\sim 10^4 A/cm^2$ at 14K exposed to a field of 12 T. Such conductors are a great asset to magnet technology.

1968   The first large iron bound superconducting bubble chamber magnet with a bore diameter of 4.8m and a field energy of 80MJ became operational at ANL.

1970   Multifilamentary cables became commercially available carrying over $10^3 A$ at a field of 4 T.

1972   Direct cooled composite conductors utilizing supercritical helium as a coolant were used in a large 3 m diameter iron bound split coil, 1.8 T magnet at CERN.

1972   The largest superconducting dc magnet with a field energy of about 800 MJ, with a central field of 3.6 T and a bore diameter of 4.72 m was tested successfully at CERN

1973   A binary alloy based on NbGe was discovered having a critical temperature of 23.2K.

The list can be extended by including magnets successfully applied in many other areas, specifically in neurosurgery, biology, fusion reactors, propulsion, beam optics and others.

Parallel to these achievements ac accelerator and beam transport magnets were developed with fields ranging from 4 to 6 T and lengths of about 1.4 m. Theories of field aberation, ac losses, dynamic stability could be checked experimentally and the superconductors started to advance towards power engineering and other areas.

Using implosive methods, solenoids have been used in high energy physics experiments up to 200 T and in solid state physics and magneto-optical experiments up to 400 T. Water cooled dc magnets with a bore diameter of about three cm have produced 25 T fields. Pulsed magnets with central fields of 100 T, energized from fast acting capacitor banks, have been developed by several laboratories.

II.   APPLICATIONS OF SUPERCONDUCTING MAGNETS

Many large scale industrial applications of superconducting magnets have been proposed. Some alternative methods not using superconducting magnets now exist, but if superconductivity is to be applied in these areas, significantly better and economical performance is expected. In some areas the use of superconducting magnets made the idea technically and economically feasible. Since superconducting magnets introduce the new dimension of low temperature physics and technology in the process, their success is intimately connected to the reliability of low temperature apparatus such as refrigerators and liquefiers, control units and availability of cryogens. Questions on safety and reliability of superconducting systems become more critical and the failure rate must be very low to protect both lives and property. Although the advantages of the new concepts are fully realized and appreciated for large scale applications, where field energies may exceed several hundreds of Megajoules, there is still hesitation whether to apply this new technique for fear of failure. In this regard the first major commercial application of superconductivity magnets must operate as reliably as their conventional watercooled counterparts. Continuous close supervision by professional crews of the system operation must be kept to a minimum.

The effective translation of new inventions and new devices to industrial practices, however, is vital to economic growth and the realization of the new magnet technology in industrial application will benefit the society. Only when need for new technologies becomes acute, the funding and support for new applications will become available. This is the reason, that major interest has developed for superconducting motors for ship propulsion, for energy conversion methods and electromagnetic storage of energy.

It is beyond the scope of this paper to treat all types of superconducting magnets used or proposed for different purposes. Only a few applications are mentioned:

Biological Applications:
Magnetic fields have been employed to study the growth of plants and behavior of insects, birds and animals.

Chemical Applications:
Magnetic fields may change chemical reactions and could be used as a catalyst.

Medical Applications:
Magnetic fields have been applied to repair arteries and dry up tumors and heal aneurisms without surgery. The influence of magnetic fields on vital functions of the human body is being studied.

Material Science, Solid States Physics and Electron Microscopy:

Levitation:
A major application is economical transportation.

Energy Generation:
Use of superconducting magnets for the generation of fusion power. In
MHD power plants the superconducting magnet has already been employed
successfully.

Magnetic Separation:
Magnetic separation is applied commercially to separate ferromagnetic
and paramagnetic materials, in the Kaolin industry to remove magnetic
discolorants from clay; selective magnetic cleaning of coal, i.e., removal
of mineral matters from the organic substance.

Sewage Treatment:

Cleaning Polluted Water:

Inductive Storage Systems:
Energy storage coils are proposed for C.T.R. systems and for power
peak shaving schemes.

Shielding and Shaping Magnetic Fields:
Shielding and shaping magnetic fields by means of superconducting sheets
and screens have been employed in high energy systems. Methods to
"freeze" the magnetic field within superconducting tubes are being
investigated.

High Energy Accelerators:
Accelerator dipole and quadrupole magnets capable of pulsing magnetic
fields up to 8 T/s have been developed. Direct current experimental magnets having field energies up to 800 MJ are already in operation.

The list can be extended to include many more applications. In all
the above and other applications the field geometry, particular field
distribution and the field strength or field gradient will be different.
Reliability, ease in operating the magnets, and economical considerations
will be important in the design of the magnet systems being used. For
most technical applications the magnet operation must be fully automatic
which requires sophisticated control and feedback systems. Closed
circuit refrigeration systems interconnected with the energy supply
system will be a necessity in most cases.

A possible arrangement of a superconducting magnet, the refrigeration unit and the energy supply is illustrated in Fig. 1. The superconducting magnet is cooled by means of liquid helium, which is supplied either
from a storage dewar or from the cold box. For large magnets the cooling
scheme may be uneconomic and forced single phase pressurized helium
(or supercritical) or two phase slush helium is passed through coolant
passages incorporated in the conductor or the coil. For small size, high
field or high gradient magnets, the method of pool boiling is still quite
popular, since it is simple and the control of the liquid helium level
is not difficult.

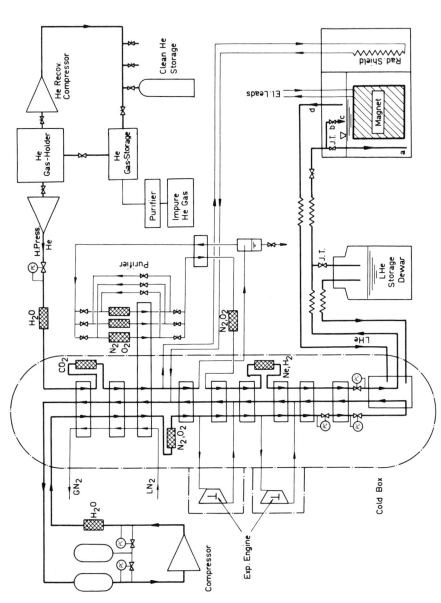

Fig. 1. Superconducting Magnet and Refrigerator Unit. During filling helium is pressurized through the bottom (a) of the cryostat. When the level is reached (a) is closed and (b) opens. Cold gas is passed through (d) over heat exchangers to the cold box. The gas passing through electrical counterflow leads is warmed up and accumulated in the warm helium storage facility.

III.   EXAMPLES OF OPERATING SUPERCONDUCTING MAGNETS

Among the number of large superconducting magnets in operation we mention five magnets. Four magnets are used as bubble-chamber magnets and the fifth is a detector magnet.   Chronologically the 4.78 m ANL split coil magnet is already a veteran among the large magnets. It has been in intermittent operation from December 1968 to date. The coil has been cooled down to liquid helium 8 times, and until March 1972 the number of days this magnet had been filled with liquid helium was 520 days. For about 210 days the magnet had been energized to produce fields exceeding 1 T. High energy experiments have been carried out at 1 T, 1.5 T and 1.8 T. The split coil arrangement with a heavy iron yoke surrounds a 3.7 m bubble-chamber with vertical axis.

Table 1 illustrates some of the pertinent data for this magnet. It is impressive to note that the magnet was completed in less than a 3 year period. The construction of the magnet started in 1966. The magnet contains a field energy of 80 MJ. Thus, the designer of this magnet had to proceed quite cautiously and use a modest current density of 1700 A/cm^2 in the composite conductor with only 6 superconducting wires. It was realized that the magnet could be built with a conventional copper coil requiring about 10 MW of power to produce the field of 1.8 T. But the opportunity to open a new era for large superconducting magnets was present and fortunately realized. The success of this magnet spured other laboratories such as CERN and NAL to build their bubble-chambers. The design of this magnet started in 1967 and the first operation of the magnet was carried out successfully in 1972. The bubble-chamber is intended for experiments using particle beams from the intersecting storage ring and from the European 300 GeV synchrotron. The inner coil diameter is 4.72m, about the same as the ANL magnet, but its central field is 3.5 T resulting in a stored field energy of 800 MJ. The magnet design had to be carried out very carefully, simply because of the enormous Lorentz forces on the conductor, which made the reinforcement of the conductor with stainless steel tapes necessary. The coil axial compressive force is about 90×10^6 N. The conductor constraining stress is 10^8N/m^2 and the constraining stress in the reinforcement $2 \times 10^8 \text{N/m}^2$. Pertinent magnet data are given in Table 1. The overall size of the coil is shown in Fig. 2. It is easily seen, that in case of the CERN bubble-chamber magnet the use of superconductors was essential. If the magnet would be wound of OFHC copper, about 60 MW power would be required to energize the coil. Even if such a power source were available, at an operating time of 2000 hours per year, the power bill would exceed 1.2×10^6 assuming 10^{-2} per kWh electrical energy. By using a superconducting coil, the refrigerator providing 900 W of refrigeration at 4.5K has an installed power of 360 kW.

One may observe in these magnets the progress in the construction of composite conductors. The ANL coil conductor has a cross-section of $5 \times 0.25 \text{ cm}^2$ with 6 NbTi wires embedded in the copper matrix,

Table 1. Superconducting dc experimental magnets

	BEBC CERN	NAL	ANL	BNL	Omega CERN
Central field (T)	3.5	3.0	1.8	3	1.8
Operating current (A)	5,700	5,000	1,800/2,200	6,000	5,000
Bore diameter (m)	4.72	4.5	4.78	3	3.5
Coil/Pol separation (m)	0.9	0.99	0.3	0.2	1.5/2
Stored field energy (MJ)	800	400	80	72	50
Overall conductor current-density (A/cm^2)	1,350	3,700	1,440/1,700	2,600	1,550
Length of conductor (m)	60,000	40,800	40,000	12,000	10,000
Conductor cross-section (cm^2)	6 x 0.3	3.8 x 0.381	5 x 0.25	5 x 0.2	1.8 x 1.8
Weight of coil (kg)	166×10^3	73×10^3	45×10^3	19×10^3	24×10^3
Weight of iron yoke (kg)	2×10^6	None	1.6×10^6	None	1.4×10^6
Magnet charging time (h)	$\leqq 6$	3	$\leqq 1$	<1	<1
Refrigeration output at (4.5 - 5 K). (W)	900	?	400	-	800
Test to full field (year)	1972 (Oct.)	1972 (Sept.)	1968 (Dec.)	1970 (Nov.)	1972 (Sept.)
Cost (U.S. Dollars) of the conductor	4×10^6	2×10^6	2.5×10^6	1×10^6	2.1×10^6

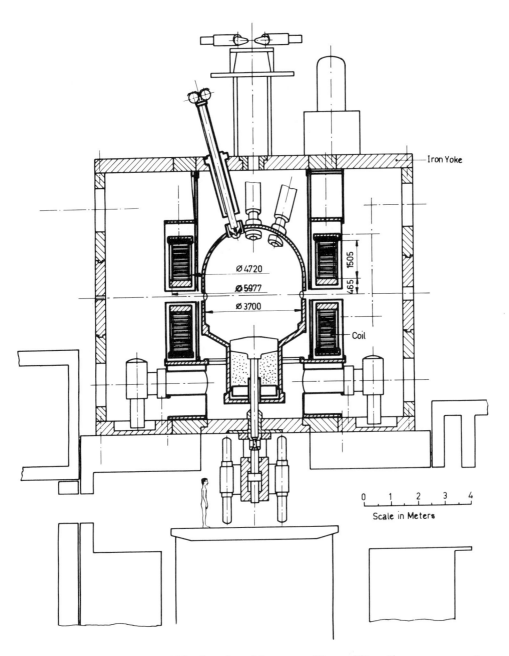

Fig. 2. CERN 3.7 Bubblechamber Magnet. The split coil arrangement energized to 5,700 A accounts for the operating field of 3.5 T. The thin iron yoke serves for field shielding purposes only.

(Cu: SC ratio 26:1). The ANL magnet can be charged to full field in about 1h to the operating current of 1800 A, corresponding to a central field of 1.8 T. The CERN magnet can be charged in less than 6 hours to 5700 A corresponding to 3.5 T.

Among the giant superconducting magnets we note the NAL super-conducting magnet for the 4.6 m diameter spherical bubble-chamber. The magnet central field is 3 T and the total stored field energy 400 MJ. In the incredibly fast time of three years this magnet became operational. The builders of this magnet attributed their success to their past experience with the other magnets. In modesty they indicate, that the properties of the conductor, the predictability of the performance and methods of restraining Lorentz forces were already known.

It is true, that there are a number of common features in all the magnets, but the building of a system, which does cost several million U.S. dollars is certainly a venture. The success of these magnets have opened new vistas for superconducting magnets to be used in other areas of physics.

The idea of using supercritical helium as a coolant was developed at MIT and utilized by pressurizing it through a hollow composite conductor at SLAC around 1965. The first large scale application of the idea utilizing supercritical helium and pumping it through hollow conductors is the CERN-Omega magnet, which has a central field of 1.8 T and a bore diameter of 3 m. The conductor used has a cross-section of 1.8×1.8 cm^2 with a coolant passage of 0.9×0.9 cm^2. It consists of a central copper tube. Around the copper tube 30 multifilament strands, each enclosing four NbTi wires (0.025 cm) are wound in parallel, forming the inner layer, while a second copper layer is wrapped over the first layer as reinforcement and protection. Silver solder is used to impregnate the copper and composite wires to attach them to the copper tube, thus yielding a high efficiency cooling. The magnet consists of six double pancakes, which are electrically connected in series, but for helium cir-culation, are hydraulically connected in parallel. The conductor is wrapped in a glass tape and each double pancake is impregnated under vacuum in an epoxy resin. The refrigeration capacity to cool the magnet down from ambient to 4.5K and keep it operational is 800 W. The magnet was first operated in 1972.

The magnet refrigeration circuit is shown in Fig. 3. The super-critical helium stream from the refrigerator passes through a heat exchanger, immersed in liquid helium containined in an auxilliary cryo-stat. It then circulates through the pancakes and finally expands to atmospheric pressure through an expansion valve. The cold helium vapor is returned to the low pressure side of the refrigerator, except a small fraction, which is used to cool down the current leads. The pres-sure drop through the coil is limited by the pressure available at the refrigerator output, typically about 10^6 N/m^2, and by the maximum avail-able increase in the helium temperature due to its isenthalpic expansion through the coolant passages of the double pancakes.

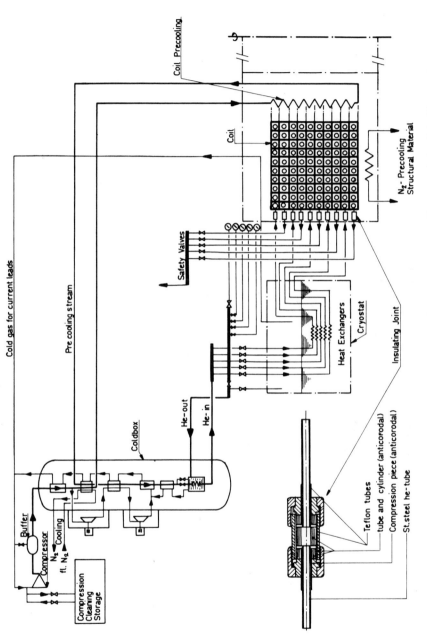

Fig. 3. The CERN-Omega Magnet and Refrigerator Arrangement. Supercritical helium at a pressure of 10 MN/m² is passed through hollow composite conductors. The helium passes through heat exchangers. Each hydraulic passage is provided with an insulating joint (shown enlarged).

Among the five presently largest magnets, shown in Table 1, we note, that two of the magnets are not equipped with an iron yoke. The iron yoke does not contribute much to the central field, but has the major advantage of shielding the electromagnetic fields from the surroundings. This is important in laboratories, where experimental equipment and tools may be around. The other major advantage of the iron shell is the reduction of the stored energy.

Parallel to these magnets a generation of medium size magnets have emerged for plasma and fusion experiments. Research devices were built at Culham (Levitron), at Livermore Laboratory (Baseball and Levitron), at Princeton (Levitron) and are currently being tested. The magnetic fields at the conductors are quite high and exceed in some cases 6 T. Most important is the confinement of the electromagnetic forces.

The new generation of fusion reactor magnets, already in a planning stage for fusion reactors, envision 10 - 15 T fields in working volumes of several m^3. With the development of new superconductors (V_3Ga) and $Nb_3(Al, Ge)$ the achievement of such goals seems to be in the near future (See Powell, Chapter 1).

There are many areas where small volume high field solenoid type magnets are applied. Magnets producing fields up to 15 T in bores of 3 cm, with or without a radial access to the median plane, are commercially available in a variety of sizes and shapes, with persistent switches, with fluxpumps, retractable current leads, with and without iron shields etc. For particular application one may best consult the coil manufacturers.

In most laboratories bulk liquid helium is available in storage dewars. Small dc magnets are energized from available power supplies. The experimental cryostat may also be available. The magnet itself and the conductor are the major cost of the system. The optimization of high field coils is accomplished in two ways:

1.   Current optimization: The solenoid is subdivided into a number of modules. According to the field amplitude at the particular module the appropriate current density in this module can be selected.

2.   Combination of superconducting and normal conductor coils: This type of coils generally known as "hybrid" systems, consists of a large bore superconducting magnet enclosing one or more small bore high current density "resistive" magnets. Hybrid systems are an economical way to generate magnetic fields above 20 T.

Presently there are three hybrid systems in a testing stage: The hybrid system at Oxford University with a central field of 16 T at a power level of 2 MW, the system at MIT with a central field of 22T at a power level of 5 MW, and the system at McGill University with a central field of 25 T. This magnet system has a cryogenic Al-coil insert, while the other magnets will use Bitter type copper or copper alloy discs. The combination of superconductors with conventional conductors can circumvent the limitations on each component if used separately. By restricting the use of superconductors to regions in which the ambient field does not drastically limit its current density, one needs to employ a modest amount of superconductor to replace the resistive conductor. We note from Fig. 4, that NbTi can carry about one-fifth the current density at 10 T as at 5 T and it does carry very little current at fields above 12 T. Nb_3Sn is practically useless at fields above 22 T. In fact, at fields above 15 T its current carrying capacity will be so low, that the coil wound with this material would be quite expensive. V_3Ga may prove to be a more economical conductor up to fields of 20 T, depending on its cost. $Nb_3(GeAl)$ conductors are still not available commercially in useful lengths. If the superconductor in a hybrid magnet is replaced by a resistive wire, the required power to generate even a modest field in a large bore will be exorbitant. A hybrid magnet can be used to generate a field of 20 T with about 2.25 MW power, which is a factor of three less than for a purely resistive wire. The total wire required would cost about \$30,000, a saving of a factor of four compared to a purely superconducting magnet. To review the existing hybrid system, we describe the three magnets currently being tested.

A.   Oxford University Hybrid Magnet

The system is composed of two sections: The inner section has a working bore of 5.0 cm diameter, an outer diameter of 24 cm and a length of 10 cm. It generates 9.4 T at a power level of 2 MW. Due to the moderate field of 16 T (total field) a copper helix is used, which is adequate in strength. The coil is water cooled and has a heat flux of 600 W/cm^2. The outer section consists of a composite conductor of rectangular shape. The outer cryostat diameter has a diameter of 91 cm. It generates a 6.6 T field. The overall current density in the coil is 6,250 $A.cm^{-2}$. About 144 filaments with 12u m diameter are embedded in a copper conductor of 0.6 x 0.1 cm^2 with a Cu-SC ratio of 6:1.

B.   The MIT Hybrid Magnet

The system includes two resistive coils and an outer superconducting coil. The innermost coil has a 3.81 cm i.d., 11.25 cm o.d. and 10.16 cm length. It requires 1.4 MW to generate 7 T. The coil is watercooled and has a heat flux of 1200 W/cm^2. Because of its high central field, 0.5% Be-Cu with 7.85 x 10^8 N/m^2 yield strength, has been selected. The intermediate resistive coil consists of Bitter type copper discs with an i.d. of 12 cm, an o.d. of 33.34 cm and a length of 21.6 cm. It generates with 3 MW power a field of 9 T. The heat flux of this water cooled coil is 550 W/cm^2. The outer coil is superconducting. It generates 5.8 T in a 40 cm i.d. The conductor is NbTi-copper composite. The heat flux of this magnet is 0.25 W/cm^2.

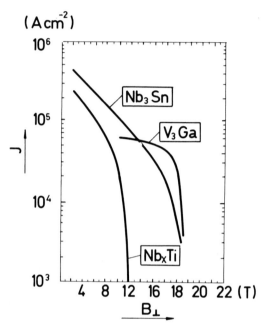

Fig. 4. Critical Current Densities of Commercially Available Type II
Superconductors vs. Transverse Magnetic Field. Nb_xTi is used
in the field ranges up to $\sim 8\,T$, Nb_3Sn up to $16\,T$, and V_3Ga in the
range of $13 - 18\,T$.

C.   The McGill University Hybrid Magnet

The coil consists of three concentric sections. The innermost coil employs ultrapure Al tape cooled by supercritical helium initially at 8 K but which is warmed up to 18 K. The gas pressure rising from $10^6 N/m^2$ to $4 \times 10^6 N/m^2$. The helium circulates through the system. The builders claim, that the resistivity of the aluminum tape is reduced 5000 times below the room temperature value at these high fields and at high stresses. Since pure Al tape is mechanically very weak, it has to be supported by means of stainless steel bands and radial clamps. The coil generates 10 T. The superconducting coils has two sections. Each section generates 7.5 T. The inner superconducting coil consists of Nb_3Sn tape, the outer coil uses a NbTi copper composite.

Only a few years ago, the region of high dc fields above 20 T was exclusively the domain of water cooled resistive coils. The trend is steadily towards using superconductors when possible. It is not surprising that there is great interest in developing higher field hybrid magnets by using newer type superconductors, such that only the innermost coil section has to produce fields above 20 T.

From the time the first type II superconductor became available to date hardly 15 years have past. Still the technological development in these years without great economical insentive is unparalleled in modern history. The development of superconducting magnets would not have been realized without the need for high field magnets at a moderate cost in many areas of physics. Many research activities would not have been possible without superconductors, which enable the generation of high fields in small coil volumes economically. There are still many applications for high field superconducting magnets, but the emphasis is changing from high energy and solid state physics to technological applications.

IV.   MAGNET SYSTEMS

A magnet system comprises essentially of three major parts:

1.   Magnet, including the cryostat, the current leads, the helium transfer connection and safety and control units.

2.   The refrigerator including cold box, compressors, gas purification systems, storage for gas, compressors, helium transfer lines and control and safety parts.

3.   The power supply, the energy dump, switchgears, current and voltage control units.

Since the refrigerator and power supply with all auxilliary parts account for about 60% of the total cost of the installation, system optimization based on the magnet alone is not correct. A close cooperation between the magnet designer, the cryogenic engineer and the power supply manufacturer is essential.

In the optimization of the system one should not forget the importance of the control and safety circuits. The goal in the design of the system is utmost reliability. Even in a fully stabilized magnet (see Section VII) there is always the possibility of a failure. Power failures, faults in compressor units (still a sore area), discontinuity in the helium supply or internal magnet failures may lead to a fault and thus to a transition from the superconducting to a "normal" state or simply to a catastrophic "quench".

The major advantage of a superconducting magnet is the low maintenance cost of its energy density. Type II superconductors permit the choice of magnetic fields in a useful aperture or working field area, which are a factor of two or even three higher than in conventional magnets. In conventional coils the cost of the power necessary to produce these fields is significant.

In the largest superconducting experimental magnets, the overall current density in the coil is a factor of two or three higher than in the counterpart copper or aluminum coil. Thus, the forces in geometrically identical coils will be a factor of four to ten higher in the superconducting winding. Even if we consider the fact, that the yield strength of conductor materials are about 30% higher at operating cryogenic temperatures, and that the yield strength of Nb_xTi exceeds the strength of copper by a factor of about four, the high Lorentz forces necessitate special precautions, such as reinforcement of the conductor itself, encapsulating the coils in such a form, as to avoid wire motion. In multimegajoule magnets one must encase the coils in order to assure full coil rigidity.

Every superconducting magnet is thermomechanically and magnetomechanically cycled. In many cases the coil is warmed up to room temperature or some intermittent value where the effective thermal capacity of the coil is low. The coil is a complex structure consisting of different metals for stabilizations and reinforcement, superconductors, insulations, potting agents, insulating spacers. All these materials have essentially different thermal contractions. If these materials are not matched, strain energy is stored in different components leading to fatigue. It is appropriate, of course, to assume, that the coil designer will use available data on material properties [22, 4] and does limit the magnetomechanical properties to values well below the yield strength of the conductor. By selecting appropriate potting and encapsulating materials, the strain energy in these materials can also be minimized. It may be pointed out, however, that fatigue data on thermosettings, composite materials, even reinforcements such as stainless-steels at liquid helium temperatures are available only in scattered form and more experimental data are urgently needed.

From the three major parts of the magnet system (power supply, refrigerator, magnet) we discuss only the optimization of the magnet and deal with magnet safety. To indicate the important parameters of the magnets, which affect primarily the cost of the magnet, we discuss a general coil arrangement and treat a few specific coil shapes. The cost of the magnet scales with the coil volume, since the conductor does account for the major part of the magnet cost. The size of the refrigerator and the cryostat depend on the magnet cooling surface. One attempts to minimize the coil volume without jeopardizing the reliability of the magnet and make arrangements within the coil to utilize the enthalpy of the helium.

The general form of the magnetic field in the median plane of a magnet $B(x, 0, z)$ is given in terms of the axial field $B(0, 0, z)$ and super-imposed multipole components:

$$B(x, 0, z) = B(0, 0, z) \left[1 + n(\frac{r}{r_o}) + p (\frac{r}{r_o})^2 + q (\frac{r}{r_o})^3 + \ldots \right]$$

$$= B(0, 0, z) \, r_o \, (k_o + k_1 r + k_2 r^2 + k_3 r^3 + \ldots) \qquad (1)$$

$$= B(0, 0, z) \, r_o \sum k_m(z) r^m.$$

The significance of Eq. (1) is, that it shows the multipole content of any field configuration. $k_m(z)$ is the multipole strength defined by

$$k_m = (\frac{1}{Br_o}) \frac{1}{m!} \frac{\partial^m B}{\partial r^m} \Bigg|_{r=0}. \qquad (2)$$

$m = 0$ corresponds to the dipole component, $m = 1$ corresponds to the quadrupole component, $m = 2$ corresponds to the sextupole component, etc., and Br_0 = field rigidity. From Eq. (2) we obtain:

$$k_o = \frac{1}{r_o} \, ; \qquad k_1 = \frac{B_o}{r} \frac{1}{(Br_o)} \, ; \qquad k_2 = \frac{B_o}{r^2} \frac{1}{(Br_o)} \, \ldots$$

where B_o is the flux density at $r = 0$.

For any particular application a specific field shape containing either pure elements (dipole, quadrupole...) or a combination will be required. By an appropriate coil design, the other terms can be eliminated or reduced to values, where their effect is negligible. If the reduction of multipole components prove to be costly and cumbersome, field correction coils [4] are used, which are placed around the aperture or around the coils.

A.   Axially Symmetric Magnets

In this section we consider the special cases of axially symmetric magnets as shown in Fig. 5. The two field components (axial and radial) at any point in space are given by:

$$B_z(R, \theta, z) = \sum_{n=0}^{\infty} \frac{1}{(2n)!} B_z^{(2n)}(0, 0) R^{2n} P_{2n}(\cos\theta) \tag{3}$$

$$B_r(R, \theta, z) = \sum_{n=0}^{\infty} \frac{1}{(2n)!} B_z^{(2n)}(0, 0) R^{2n} P_{2n}'(\cos\theta) \tag{4}$$

with the central field expressed by:

$$B(0, 0, 0) = (\lambda J)_0 \, a_1 \mu_0 \beta \ln\left[\frac{\alpha + (\alpha^2 + \beta^2)^{1/2}}{1 + (1 + \beta^2)^{1/2}} \right] \tag{5}$$

where we introduced dimensionless parameters α as the ratio of the inner diameter ($2a_1$) to the outer diameter ($2a_2$) and β the ratio of the coil length ($2b$) to the inner diameter.

The axial field at any point of the coil axis is accordingly:

$$B_z(0, 0, z) = \mu_0 \frac{\lambda J}{2} a_1 F(\xi) \tag{6}$$

with:

$$F(\xi) = (\beta + \xi) \ln\left[\frac{\alpha + (\alpha^2 + (\beta + \xi)^2)^{\frac{1}{2}}}{1 + (1 + (\beta + \xi)^2)^{\frac{1}{2}}} \right] + (\beta - \xi) \ln\left[\frac{\alpha + (\alpha^2 + (\beta - \xi)^2)^{\frac{1}{2}}}{1 + (1 + (\beta - \xi)^2)^{\frac{1}{2}}} \right] \tag{7}$$

and $\xi = z/a_1$. The coil volume is simply

$$V = 2 \pi a_1^3 \beta (\alpha^2 - 1) . \tag{8}$$

Equations (6) and (8) can be expanded to obtain the fields from separate solenoids forming split coil or multicoil arrangements, or coaxial solenoids. The values of α and β have to be obtained from the coil shape and arrangement. For a split coil arrangement consisting of two coils of identical bore diameters and the same ratio of the outer to inner radii, but differing lengths, we may write,

$$B_z(0, 0) = \mu_0 a_1 \left[\beta_1 (\lambda J_1)^2 + \beta_2 (\lambda J_2)^2 \right]^{\frac{1}{2}} F(\alpha, \beta_1, \beta_2) \tag{9}$$

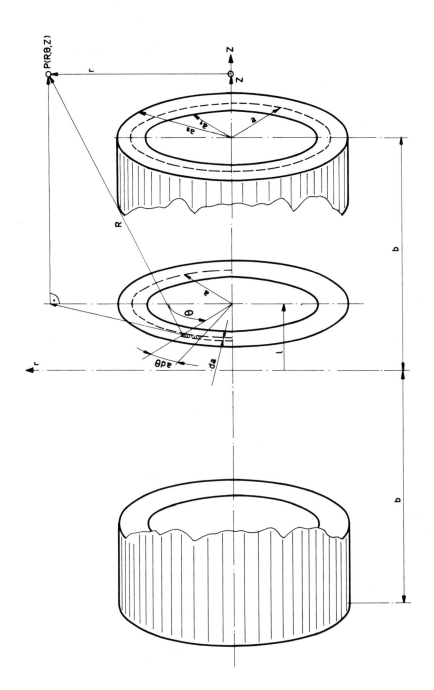

Fig. 5. Axially Symmetric Coil Arrangement to Calculate the Magnetic Field at any Point in Space.

with the geometry factor

$$F(\alpha, \beta_1, \beta_2) = \beta_1 \ln \left[\frac{\alpha + (\alpha^2 + \beta_1^2)^{\frac{1}{2}}}{1 + (1 + \beta_1^2)^{\frac{1}{2}}} \right] - \beta_2 \ln \left[\frac{\alpha + (\alpha^2 + \beta_2^2)^{\frac{1}{2}}}{1 + (1 + \beta_2^2)^{\frac{1}{2}}} \right]$$

and the coil volume,

$$V = a_1^3 \, 2\pi (\alpha^2 - 1)(\beta_1 - \beta_2). \tag{10}$$

Let us now study two practical cases.

1.   Short solenoids

Short solenoids with $\alpha \gg \beta$ and $\beta < 1$ are common in large bore, split coil arrangements, such as in bubble-chamber magnets. From Eq. (5) we obtain:

$$B(0, 0) \cong \mu_0 (\lambda J_0) a_1 \beta \ln \alpha. \tag{11}$$

For all values of $1 < \alpha < 2$ we can simplify Eq. (11) to:

$$B(0, 0) \cong \mu_0 (\lambda J_0) a_1 \beta (\alpha - 1). \tag{12}$$

Inserting the value of α from Eq. (12) into Eq. (8) we obtain,

$$V = \frac{2\pi a_1}{\beta} \frac{B(0, 0)}{\mu_0 \lambda J_0} \left(2a_1 \beta + \frac{B(0, 0)}{\mu_0 \lambda J_0} \right). \tag{13}$$

We calculate the coil surface area in terms of the parameter $B(0, 0)/(\mu_0 \lambda J)$ by assuming the coil is divided into sections or pancakes with $2n$ radial cooling surfaces including coil end surfaces. The circumferential cooling surfaces are added to form the external (at a_2) and internal (at a_1) cooling surfaces. If only the fraction f of each cooling surface is exposed to the coolant, we may write for the entire cooling surface:

$$S = \frac{2\pi f}{\beta^2} \left(2a_1 \beta + \frac{B(0, 0)}{\mu_0 \lambda J} \right) \left(a_1 \beta^2 + \frac{nB(0, 0)}{\mu_0 \lambda J} \right). \tag{14}$$

Since the coil volume is proportional to $B(0, 0)/(\mu_0 \lambda J)$ the increase of (λJ) will yield a smaller coil. The situation is not obvious from Eq. (14), as $a_1 \beta^2$ is the same order as $B(0, 0)/(\mu_0 \lambda J)$.

2.   Long solenoids

If $a_1 \beta \gg a_2 > a_1$, i.e., if the coil length is large compared to the radial dimensions, we obtain from Eq. (5)

$$B(0, 0) = \mu_o \lambda J \, a_1 (\alpha - 1) \tag{15}$$

which yields the expression for the coil volume and cooling surface:

$$V = 2\pi a_1 \beta \frac{B(0, 0)}{\mu_o \lambda J} \left(2a_1 + \frac{B(0, 0)}{\mu_o \lambda J} \right) \tag{16}$$

$$S = 2\pi f \left(2a_1 + \frac{B(0, 0)}{\mu_o \lambda J} \right)\left(2a_1 \beta + n \frac{B(0, 0)}{\mu_o \lambda J} \right). \tag{17}$$

Again we find the same parameter $B(0, 0)/(\mu_o \lambda J)$, which affects the coil volume.

3.   Multipole magnets

Superconducting dipoles and quadrupoles have been developed for dc and pulsed operation. Multipole magnets with cylindrical apertures are employed for various applications. We study the effect of the parameter $B/(\mu_o \lambda J)$ on the coil volume.

Dipoles:
Over the median plane the magnetic field generated by a dipole coil surrounded by an iron shell, shown in Fig. 6 is given by:

$$B_y = \frac{\mu_o}{\pi} (\lambda J_o) \, a_1 \left[(\alpha - 1) + \frac{a_1^2}{3b_1^2} (\alpha^3 - 1) \right]. \tag{18}$$

If the radius of the iron shell is placed at a distance where the iron shell is not saturated, then with $a_1 < b_1$ we can approximate Eq. (18) to:

$$B_y = \frac{\mu_o}{\pi} (\lambda J_o) a_1 (\alpha - 1) \cos\theta . \tag{19}$$

The distribution of the current density over the aperture-circumference is sinusoidal. The coil volume (not counting end sections) is simply,

$$V = 2\pi a_1^3 (\alpha^2 - 1) \tag{20}$$

and thus,

$$V = 2\pi^2 a_1 \beta \frac{B_y}{\mu_o \lambda J_o \cos\theta} \left(2a_1 + \frac{B_y \pi}{\mu_o \lambda J_o \cos\theta} \right). \tag{21}$$

Various dipole designs are given in Figs 7 and 8.

Quadrupole:
Ignoring the effect of the iron shell, we can write for the field gradient:

$$G = \frac{\mu_o}{\pi} (\lambda J_o) \cos(2\theta) \, \ln\alpha \tag{22}$$

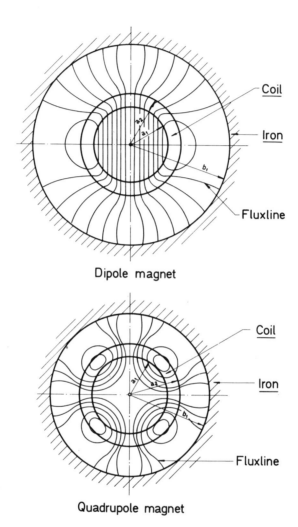

Dipole magnet

Quadrupole magnet

Fig. 6. Schematic Representation of a Dipole and Quadrupole Field
Distribution. In the calculation we assumed, that the relative
permeability in the iron shell is infinitely. Flux lines penetrate
the iron at right angles.

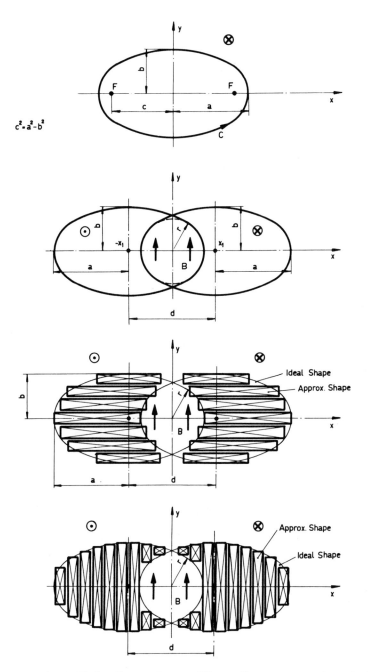

Fig. 7. Dipole Coil Configurations. The coil cross-section, indicated by the ellipses are approximated by rectangular current blocks.

or

$$G \cong \frac{2\mu_o}{\pi} (\lambda J_o)\cos(2\theta) \frac{\alpha - 1}{\alpha + 1} . \tag{23}$$

After inserting the value of α from Eq. (23) in the Eq. (20) for the coil volume we obtain

$$V = 4\pi^2 a_1^3 \frac{G}{\mu_o(\lambda J_o)\cos(2\theta)} \left(1 - \frac{\pi G}{2\mu_o(\lambda J_o)\cos(2\theta)}\right)^{-2} . \tag{24}$$

In all four cases shown, the minimization of $B/(\mu_o \lambda J)$ is indicated. A high overall current density means high Lorentz forces which generate mechanical stresses in the conductor, which may exceed the yield strength of the material. Thus, one parameter, which limits the over-all current density in the coil is the Lorentz force and the material strength. Designs of quadrupole magnets are shown in Fig. 9.

For solenoid type coils we may use the hoop-stress equation derived by Appleton et al. [23] (see Figs. 10 and 11).

$$\sigma_{c,max} \cong a_1(\lambda J_o) \frac{Ba_1}{6} \frac{\alpha^2 + \alpha - 2}{\ln\alpha} . \tag{25}$$

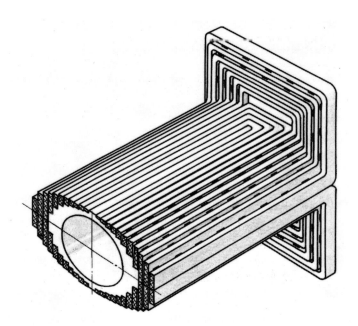

FIG. 8. Coil Ends Bend 90° to the Coil Axis to prevent Excessive Field Enhancement at the coil End Sections. See Fig. 7.

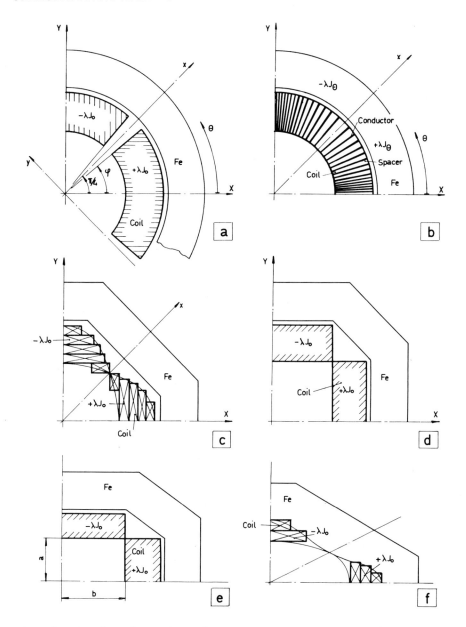

Fig. 9. Quadrupole Coil Configurations: a) Current blocks with constant current density. The angle ψ is selected $\pi/3$. b) Current density variation according to cos θ distribution. c) Coils approximating intersecting ellipses. d) Square quadrupole. e) Panofsky type quadrupole. f) Morpurgo type quadrupole. In all these cases the iron shell is also cooled to liquid helium temperature and is used for the coil support and for preventing coil sagging.

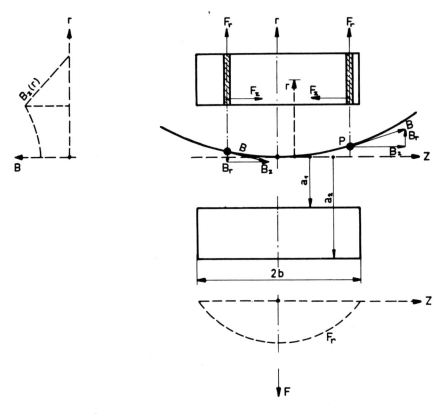

FIG. 10. Stress Distribution in an Axially Symmetrical Coil.

If we assume, that $Ba_2 = 0$, and insert the value of $(\alpha-1)$ from Eq. (12) in Eq. (25) we obtain the hoop-stress for short solenoids

$$\sigma_{c,\,max} = \sigma_c(a_1) = \frac{k}{6\beta}\,(\lambda J)B(0,0)\left[3a_1\beta + \frac{B(0,0)}{\mu_o\lambda J_o}\right] \tag{26}$$

with $k = Ba_1/B(0,0)$. For long solenoids the hoop stress is given by:

$$\sigma_{c,\,max} = \sigma(a_1) = \frac{k}{6}\,(\lambda J_o)B(0,0)\left[3a_1 + \frac{B(0,0)}{\mu_o\lambda J_o}\right] . \tag{27}$$

It can be noted, that $B(0,0)/(\mu_o\lambda J_o)$ is the same order of magnitude as $3a_1$ only for small ($<10^6$J) coils with high central fields. As an example a 10 T coil can easily be built with an overall current density of 10^8A/m^2, giving a value of $B(0,0)/(\mu_o\lambda J_o) = 0.08$ m. The inner

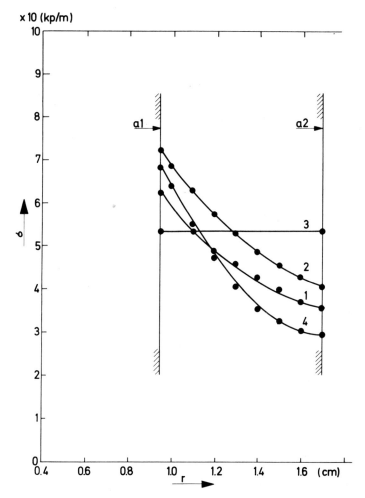

Fig. 11. Field Distribution Over a Coil According to Exact and Approximate Stress Equations. 1) Stresses according to:

$$\sigma_c = a_1 \ (\lambda J_o) \left[\frac{Ba_1(\alpha^2 + \alpha - 2) + Ba_2(2\alpha^2 - \alpha - 1)}{6\left(\dfrac{a}{a_1}\right) \ln \alpha} \right] ;$$

2) Stresses according to

$$\sigma_c = \frac{\lambda J_o \ a_1^2}{4 a \ln \alpha} \ (Ba_1 + Ba_2)(\alpha^2 - 1);$$ 3) Stresses according to:

$$\sigma_c = \frac{\lambda J_o a_1}{4} \ (Ba_1 + Ba_2)(\alpha + 1);$$ 4) Stresses calculated from body
forces integrated over the coil cross-section (exact equation).
The dimensions of the coil were: $a_1 = 0.95$ m; $a_2 = 1.7$ m;
$2g = 0.15$ m; $N \ I = 1.02 \times 10^7$ A/coil, $2(2b+g) =$ total length $= 2.3$ m.

diameter of such a coil will be also about 0.4m. Thus, the increase in stress in a long solenoid will be about 30%.

In large bore ($2a_1 \geq 1m$) but short solenoids, the current density with a central field of 5T will be about $5 \times 10^7 A/m^2$, resulting in $B(0,0)/(\mu_0 \lambda J) = 0.08m$. If $\beta \leq 0.2$, the stress enhancement will be about 30%. With increasing values of a_1 and decreasing β, Eq. (27) shows that stress may reach values, which could exceed the yield strength of the conductor material. The conductor must be reinforced. A high value of λJ_0 yields a reduction in the characteristic number $B(\mu_0 \lambda J_0)$, but this value becomes insignificant in the overall stress number. A high value of λJ_0 is, however, desirable from economic point of view and various possibilities to limit or restrain stresses have been explored, which are summarized briefly:

i)   The axially symmetric coil is subdivided into a number of modules. As the maximum field value in each module is known, the current density according to the B-J properties of the superconductor and the stability criteria can be selected. Each module can be reinforced separately. This subdivision of the coil is called current optimization.

Since the price of the composite conductor is not a trivial item for purposes of cost estimation the quantity of magnet current times, conductor length (Am), is an important parameter and must be optimized. This parameter is defined by $Q = (\lambda J)V$. Thus, for a current optimized coil we may write:

$$Q = \sum_{i=1}^{n} (\lambda J_0)_i V_i.$$

In smaller solenoids generating central fields of about 16 T, overall current densities in the order of $15 \times 10^3 A/cm^2$ have been achieved using the current optimization method.

ii)   The conductor is sandwiched between stainless steel reinforcement tapes. Using high strength materials, the coil can be built very rigidly.

iii) Using either heat drains or internally cooled conductors the coil can be impregnated in high tensile strength thermosettings. Heavy iron shells or restraining clamps, or even rocks may be placed around the coil, preventing any coil motion. This solution assumes a completely rigid coil, where internal conductor motion is entirely eliminated.

V.   SUPERCONDUCTOR STABILIZATION

For applications, where large field volumes are required, (high energy experimental magnets, magnets for ore separation, water and

sewage treatment, energy storage), the construction of reliable
magnets with a predictable performance is of prime concern. Another
important factor besides stress which must be considered in selecting
the appropriate current density in magnet systems is based on the cryostatic stability effect. The superconducting filaments are embedded in
a matrix of normal metal, such as copper or aluminum. In case of a
disturbance, portions of the superconductor may be driven normal.
Part of the current will flow through the matrix material. If the disturbance is of short duration, the conductor can cool down again below
the critical temperature of the superconductor and the current will
return gradually into the superconductor, until a complete superconducting
condition is restored. The amount of the substrate in combination with
the superconductor is calculated from the cryostatic stability criteria [4]
for the critical current I_c,

$$I_c^2 \rho_n = 2h(T_c - T_b) A_n S_n \tag{28}$$

where T_c is the critical temperature of the superconductor, T_b the bath
temperature and h the coefficient of heat transfer. The design heat
flux $h(T_c - T_b) = h\Delta T$ may be obtained from Figs. 12 and 13.

Assuming, the magnet is immersed in liquid helium and has an open
structure, we may use a value of $h\Delta T = 3 \sim 4 \times 10^3 W/m^2 K$. We can
relate the conductor perimeter per unit length S_n to the conductor cross
section A_n by, $S_n = f \cdot (A_n)^{1/2}$, where the factor f accounts for the
conductor shape, e.g., for a circular cross section, $f = 2(\pi)^{1/2}$, for a
square cross section, $f = 4$; etc. If we consider the case, that all the
current flows through the substrate and the superconductor is in a
normal state, we may write for the current density:

$$J = (2h\Delta T f/\rho_n)^{2/3} I^{-1/3}. \tag{29}$$

No matter how stable the conductor, there is always a chance, that
some kind of failure may occur, which could lead to a quench. The
current, which then flows through the substrate will heat up the substrate.
The heat will either propagate along the conductor or will persist at the
spot of disturbance. If all the conductor is heated up above T_c and the
coil is driven normal, the field energy is dissipated into the conductor
and the coolant. Due to the field decay, a voltage appears over the
conductor and the conductor temperature will rise rapidly to values
indicated by the heat capacity of the conductor and the enthalpy of the
coolant.

To protect the coil against damage, overheating, insulation
breakdown or simply prevent excessive helium boil off, we set an upper
current limit, which can flow through the conductor from the following
criteria: 1) The maximum temperature at any point within the coil should
not exceed a certain value, indicated by the characteristics of the superconductor and the insulation.

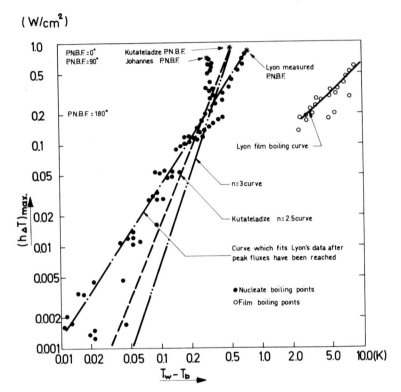

Fig. 12. Peak Nucleate Boiling Flux Data vs. Temperature Rise. The
upper limit of p.n.b.f. of heated horizontal surfaces is about
0.9 W/cm², the lower film boiling heat flux is 0.2 W/cm². The
p.n.b.f. in a coil with coolant passages varies between 0.3 –
0.4 W/cm².

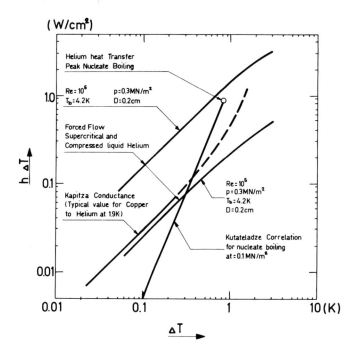

FIG. 13. Heat Transfer vs. Temperature Rise Data in a Helium Bath, and Pressurized Helium Passed through Channels at Low (Re = 10^5) and Higher (Re = 10^6) Velocities (NBS).

2) The maximum voltage at the coil terminals should not rise above some preset value and the magnet should be disconnected from the power supply, whence this value is reached.

The magnet is connected to a protective circuit shown schematically in Fig. 14. The field energy is dissipated by an external ohmic shunt, and auxillary parts, such as the iron yoke, the cryostat, etc. The resistance of the shunt is determined by the desired field decay time constant.

In case of a catastrophic quench, the field energy dissipated into the coolant is generally sufficient high, that all the helium will evaporate. It can be noted, that the helium evaporation succeeds a rapid mass transfer within the bath between gaseous and liquid helium and is quite turbulent. The heat exchange effect in helium is efficient and only in the case, where the magnet current flows in the conductor for some time (fault in the safety circuit). All the helium in the vessel will eventually evaporate resulting in an internal pressure rise in the cryostat which will open safety and vent valves or crack open rupture discs.

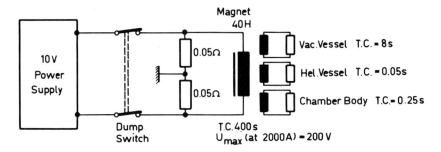

FIG. 14. Coil Protective Circuits. The superconducting coil is connected to a low value resistance for fast energy extraction
(ANL). If the coil consists of two or more sections, the
voltage difference between sections in case of a quench is used
to open the switch connecting the magnet to the power supply.

For our study we assume the worst case, where all the helium has
evaporated and only the heat capacity of the conductor limits the maximum
temperature. The current density is determined from the field decay time-
constant and the energy balance

$$\left[J(f) \right]^2 \rho dt = c_p \delta dT, \tag{30}$$

Since ρ and $c_p \delta$ are temperature dependent, we can rewrite the Eq. (30) after
introducing a time constant $\tau = L/R$ for the coil,

$$\frac{1}{2} J^2 \tau = \int_{T_b}^{T_{max}} \left(\frac{c_p \delta}{\rho} \right)_n dT = g(T_m) . \tag{31}$$

Tables and curves of $g(T_m)$ are given in the literature [4], or may be obtained from c_p and ρ data, compiled over the temperature ranges of interest. Inserting for $L = 2W/I^2$ and $R = U_{max}/I$ in Eq. (31) we get for the
current density,

$$J \leq \left[g(T_m) U_m I/W \right]^{\frac{1}{2}} . \tag{32}$$

This equation accounts for the thermal properties of the substrate, the
maximum allowable voltage across the coil U_m and the total field energy
of the magnet W.

Eliminating "I" from Eqs. (29) and (32) as an independent variable, we derive the optimum current density,

$$J_{opt} \leq \left[g(T_m) U_m /W \right]^{0.2} \left[2h\Delta Tf/\rho_n \right]^{0.4}. \tag{33}$$

To give an example, we select $U_m = 10^3 V$, which is safe value for most insulations, $g(T/m) = 10^{17} A^2 s/m^4$ for copper in the temperature range of 100 - 300 K. Assuming $h\Delta T = 3 \times 10^3 W/m^2$ and using a rectangular shaped conductor with $f = 2(1 + \frac{b}{a})/(b/a)^{1/2}$

we obtain for an aspect ratio of $b/a = 0.4$ the value of $f = 4.427$. The resistivity of copper is about $\rho_n = 4 \times 10^{-10} \Omega m$, at 4.5K, including strain and magnetoresistance effects. With these values we get $J_{opt} = 3.3 \times 10^9 W^{-0.2} (A/m^2)$. For a 100 MJ coil we obtain $J_{opt} = 8 \times 10^7 A/m^2$.

Figure 15 illustrates the maximum field energy vs. current density for a number of magnets either in operation or in a design stage. The value calculated is above the dotted line in the Fig. 15 showing the design trend, but still within the stability limit. It is interesting to note, that the nominal current density decreases by about two orders of magnitude when the field energy is increased to 1 GJ.

VI.   INDUCTIVE ENERGY STORAGE MAGNETS

As an example of large scale application of magnets, we discuss the inductive storage system. These systems could serve many important purposes. Depending on the magnetic field, inductive storage systems have the advantage of very high energy densities compared to capacitive or other types of energy storage. Depending on the energy discharge time these magnets are used mainly in fusion reactors and inductive energy storage for network systems.

Energy discharges in the (ms) range may be used to start a fusion reaction of DT or DHe_3. Although the procurement of magnets storing about $10^{10} J$ seems feasible with present magnet technologies, the switch, which enables the fast energy discharge from the magnet into the load, is not. At present, about 1 MJ energy may be discharged in about 100 ms using superconducting switches. Considerable research activities are devoted to making such switches operational.

Inductive storage systems for energy storage in network systems, especially for energy peak shaving are of considerable interest. New design philosophies, material improvements, and the simplicity of storing energies in the range of $10^{13} J$ have accelerated research activities. Although the capital cost of these magnets are about a factor of two to four higher than pumping stations used for peak shaving operations, the hope of reducing the cost of conductor and developing new methods of coil

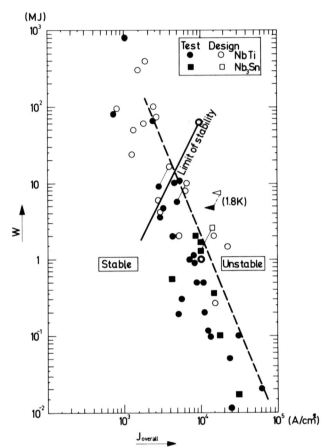

FIG. 15. Maximum Coil Overall Current Density Versus Field Energy
Plotted for Magnets in Operation or in a Design Stage. The
current density in coils with low energies ($\leqq 1$ MJ) is dictated
by the stability criteria and at high energies by the stresses
in the conductors (ORNL).

reinforcement and structure makes the superconducting energy storage
system quite attractive. From a number of possible coil designs, the
magnet shapes selected for closer scrutiny have been the classical Brooks
coil, the spherical coil and the toroidal coil. The first two magnet types
require an iron or superconducting shield to screen the magnetic field
from the environment. The toroidal coil has no appreciable fringing
field. It may consist of a number double pancakes arranged in a manner
shown in Fig. 16. In the design discussed, we have selected direct
cooling of the composite by pumping supercritical helium through hollow
conductors, similar to the method used in the CERN-Omega magnet in Fig. 3.

    Hollow superconducting composite conductors are already available
commercially in lengths of 200m carrying about 1000A. The coolant is
pressurized through the coolant passages at pressures well above the

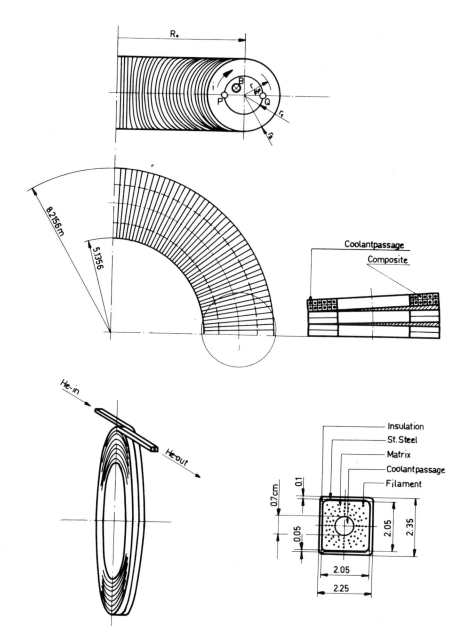

Fig. 16. Toroidal Coil for Energy Storage. The coil consists of double pancakes. Supercritical helium is pressurized through coolant passages of hollow composite conductors. The pancakes are electrically connected in series, hydraulically in parallel. The toroidal coil is capable of storing 10^{10} J at a maximum field of 15 T.

critical value to avoid the generation of a two phase flow and with it flow instabilities. In normal operation the helium flow can be reduced to that required to keep the superconductor at an operational temperature below its critical temperature and to compensate for the heat inleak. Only when energy is being pumped into or out of the magnet need the helium flow be increased to values reaching a Reynolds number of 10^6 or higher (Fig. 13) to prevent coil quenching. The hydraulic passages are connected in parallel to prevent internal joints and to limit the length of individual coolant passages, while electrically all turns may be connected in series. Each hydraulic passage is insulated from the rest of the coil by a vacuum insulated vacuum joint shown enlarged in Fig. 3. Each turn can be mechanically reinforced by means of stainless steel bands, or the superconducting filaments may be embedded in a high tensile strength normal metal, such as Be-copper or non magnetic stainless steel. Studies to place the toroidal coil underground and use the reinforcing effects of rocks or concrete structures to prevent the coil from moving are possible and should be compared to designs, which make the coil completely self-supporting. This requires, that the Lorentz forces acting on the coil should be limited to values, where the tensile and hoop stress do not lead to conductor motion.

To give some data on the magnitude of such a project, we calculate a toroidal magnet used for the energy storage of 10^{10}J. We refer to Fig. 16 which illustrates a toroidal coil. With the current flowing in the circumferential direction, the flux density will be parallel to the toroidal coil axis. The magnetic field can be expressed by

$$B(r, \varphi) = \frac{\mu_o}{2\pi} \frac{NI}{R_o + r \cos\varphi} \cdot \tag{34}$$

The relation between Ampturns and overall current density is given by

$$NI = 2\pi (\lambda J) R_o (r_2 - r_1) \text{ for } r_1 < r < r_2; \tag{35}$$

$$NI = 2\pi (\lambda J) R_o (r_2 - r) \text{ for } r < r_1. \tag{36}$$

Introducing the dimensionless parameters

$$\alpha = \frac{r_1}{R_o} \; ; \; \beta = \frac{r_2}{R_o}$$

we obtain the maximum flux density at the point P:

$$B_{max} = B(r_1, \pi) = \mu_o (\lambda J) R_o \frac{\beta - \alpha}{1 - \alpha}$$

which is only one boundary condition for the selection of the superconductor and its current density. For illustrative purposes we do not attempt to optimize the current in the coil.

Although the field distribution within the bore of a toroidal coil is not homogeneous, as we have seen, we use for simplicity the value of B_{max} to give approximate equations for the hoop and radial stresses:

$$\sigma_{\theta,\,max} = r_1 B_m (\lambda J_0)\,\frac{\beta}{\beta - \alpha}\,\frac{2 - \beta}{2(1 - \beta)}\quad \text{and}$$

$$\sigma_{r,\,max} = r_1 B_m (\lambda J_0)\,\frac{\beta}{2(\beta - \alpha)}\,.$$

(38)

(39)

The stored field energy in the coil is given by

$$W_{tot} = 2\mu_0 \pi^2 R_0^{\,5} (\lambda J_0)^2\, f(\alpha,\,\beta)$$

(40)

with

$$f(\alpha,\,\beta) = (\beta - \alpha)^2 \left[1 - (1 - \alpha^2)^{\frac{1}{2}} \right] + \beta\,(\beta - \alpha)(1 - \alpha^2)^{\frac{1}{2}}$$

$$+ \left[\sin^{-1}(\alpha) - \sin^{-1}(\beta) \right] + (1 - \alpha^2)^{\frac{1}{2}} - (1 - \beta^2)^{\frac{1}{2}}$$

$$+ \frac{1}{3}\left[(1 - \beta^2)^{\frac{3}{2}} - (1 - \alpha^2)^{\frac{3}{2}} \right].$$

Curves of constant $f(\alpha,\,\beta)$ are given in Fig. 17, which may be used to optimize the coil. The overall volume of the coil is given by

$$V = 2\pi^2 (\beta^2 - \alpha^2) R_0^{\,3}\,,$$

(41)

and the required amount of superconductor,

$$Q_{sc} = 2\pi^2 (\beta^2 - \alpha^2)\, R_0^{\,3} (\lambda J_0)$$

$$= 2\pi^2\,\frac{B_{max}}{\mu_0}\,(1 - \alpha)\,(\beta + \alpha)\, R_0^{\,2}\,.$$

(42)

We select the two parameters $\alpha = 0.2$, $\beta = 0.3$ and obtain $f(\alpha,\,\beta) = 2.468 \times 10^{-4}$. For the energy of $W = 10^{10}$J, after selecting a field of $B_m = 15$ T(V_3Ga) we obtain from the energy equation

$$W = 2\pi^2\,R_0^{\,3}\,\frac{B_m^{\,2}}{\mu_0}\,\frac{(1 - \alpha)^2}{(\beta - \alpha)^2}\,f(\alpha,\,\beta)\quad \text{where}$$

the radial distance, $R_0 = 5.1356$ m; and with it $r_1 = 1.027$ m and $r_2 = 1.54068$ m. The overall current density is given by

$$(\lambda J)_{min} = \frac{B_{max}(1 - \alpha)}{\mu_0 R_0 (\beta - \alpha)}\,,$$

which gives, $(\lambda J)_{min} = 1.86 \times 10^7$ Am^{-2}. The amount of superconductor is obtained from Eq. (42), $Q = 7.9 \times 10^8$ Am. Choosing a current of 10^4A, we need 8×10^4 m of composite conductor with an overall effective cross-section of 3.8 cm^2. A hollow composite conductor with the dimensions 2.05 x 2.05 cm^2, and a coolant passage of 0.70 cm diameter with rounded corners could be selected.

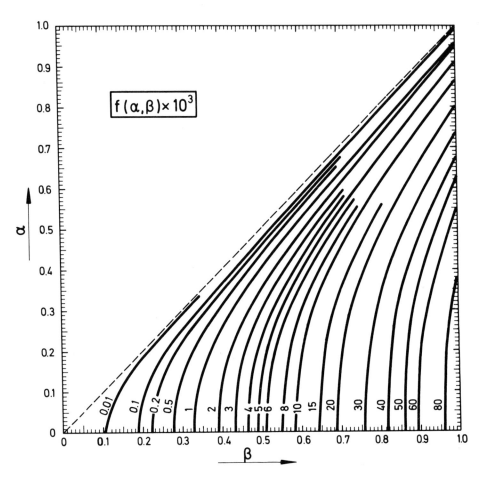

Fig. 17.   Geometry Factors for Optimization Toroidal Coils. The parameters $\alpha = r_1/R_0$ and $\beta = r_2/R_0$ and the mean radius R_0 determine the coil volume.

According to Eqs. (38) and (39), the stresses on the conductor will be

$$\sigma_{\theta, max} = 10.4 \times 10^8 N/m^2; \quad \sigma_{r, max} = 4.3 \times 10^8 N/m^2.$$

At a field of 15T, the cross-sectional area of the superconductor (Fig. 4) will be: $A_{sc} = 0.2$ cm^2, which gives a ratio of the matrix to superconductor of 19.

In the estimate of the overall coil size we choose a space factor of 0.7 for reinforcement and insulation. The insulation thickness of each side of the conductor is 0.1 cm and stainless-steel reinforcement strips of 0.1 cm thickness give an overall conductor size of 2.35 x 2.25 cm^2. The reinforcement strip is necessary due to the relative high circumferential stresses.

The toroidal coil is composed of double pancakes. Each double pancake has 36 turns with a continuous length of 288 m. Selecting 50μm diameter filaments, we require 10^4 filaments to be codrawn with the substrate. A hollow conductor of 290m length, with a copper matrix is not available. Recently aluminum stabilized superconductors have become commercially available, but can not be used due to stress limitations. Stainless-steel stabilized conductors are presently not commercially available and may pose problems due to coextrusion of V$_3$Ga and steel.

Thus, two hollow composite conductors with a hard copper matrix, each having a length of 150m can be wound, between stainless-steel sandwiches, in parallel and connected electrically in series. The heat transfer coefficient of pressurized liquid helium at 4 x 10^{5}N/m^2, passing with a velocity of 1 ms^{-1} through the hollow conductor, is obtained from:

$$\frac{hd}{k} = 0.0259 \times (Re)^{0.8} (Pr)^{0.4} \left(\frac{T_w}{T_b}\right)^{-0.716}$$

with the parameters

Reynold number factor: δ/η	4 x 10^7(s/m^2)
Reynold number, Re	2.8 x 10^5
Prandtl number, Pr	0.28
Thermal conductivity k	3.5 x 10^{-2}W/mK

and assuming: $T_w/T_b = 1.1$ we obtain h = 1.67 x 10^3 W/m^{2}K.

We can calculate the optimum current density, using the stability criterion of Eq. (33):

$$J_{opt} = 3.4 \times 10^7 A/m^2, \text{ assuming a voltage of } 10^3 V$$

when the energy is dissipated during switching, or

$$J_{opt} = 6.8 \times 10^7 A/m^3; \text{ with } U_m = 10^4 V.$$

In both cases J_{opt} is higher than $J_{opt} = 2632 A/cm^2$. In the above calculation we used for $\rho_n = 3.5 \times 10^{-10}$ Ohm.m, which is due to the high mechanical strain (0.5%) and the magnetoresistance effect (average of 8T). It is further of interest to calculate the pressure drop along the hollow conductor ($10^5 N/m^2$); the coil weight (83×10^3 kg), the total weight of the coil including reinforcement and supports (150×10^3 kg); the amount of liquid nitrogen to cool down the coil from room temperature to 78K(160×10^3 liters); and pressurized liquid helium to cool down the magnet from 78K to 4.3K (83×10^3 liters). In both cases the specific heat of the liquids can be utilized.

A new study at LASL (W. Keller private communication) for an energy storage coil of 4×10^{13}J uses a current density of $10^3 A/cm^2$ at a maximum flux density of 8 T. The coil (a Brooks coil without screening) has a diameter of about 100 m. The capital cost of the coil is based on two structural reinforcement systems. The coil reinforcement is at liquid helium temperature, as discussed above, or the coil is placed in a rocksystem and the reinforcement is then at room temperature.

The capital cost of the two systems with cold and warm reinforcements are given as follows:

	Cold-reinforcement structure (millions of U.S. $)	Warm-reinforcement structure (millions of U.S. $)
Superconductor	100	100
Cryostat	80	150
Structural reinforcement	1000	250-500
ac, dc convertor	50	50
Refrigerator	2	15
Total	$1232	$565-815

An equivalent pump storage station would cost about $450 million. Even if the cost of the system seems optimistic, it illustrates the importance of the structural support of the magnet. At a possible cost saving of $500 million, research activity in this area is certainly justified.

VII.   MAGNETS FOR PULSED FIELD APPLICATIONS

Superconducting pulsed magnets were proposed around 1967, but are still in a prototype stage. Several laboratories in the U.S. (BNL, LRL) and Europe GESSS (the GESSS-laboratories in Europe are CEN-Saclay France, IEKP-W, Germany and RHEL, England) have produced and tested ac magnets for low frequency operation ($\leqq 1$ Hz), but the lack of a long history of experience has prevented their large scale utilization in synchrotrons or other applications.

When a type II superconductor is exposed to a changing magnetic flux, or an ac transport current is passed through the superconductor, the bulk superconductor is subject to a varying flux, which penetrates its surface and generates hysteretic and self-field losses. In both cases, the distribution of the magnetic flux is controlled by the pinning forces in the bulk superconductor and by a thin vortex free surface layer, which can support a surface screening current.

A type II superconductor used in a coil winding carries a transport current, caused by the terminal voltages of the power supply. It is also exposed to an external field, which is either generated by all the other coil windings, or simply by the coil itself. If the superconductor is subjected to a field less than its lower critical field H_{C1}, it behaves in the same manner as a type I superconductor. At higher fields (Schubnikov Phase) up to its upper critical field H_{C2}, the conductor shows a hysteretic behavior as illustrated in Fig. 18 according to Wipf [24]. If a current or a magnetic field varies in a superconductor, an electric field is generated and the energy flowing into and out of the superconductor can be obtained by integrating the Poynting's vector $\vec{S} = \vec{E}_s x \vec{H}$, over a complete current or field cycle. In type II superconductors, the integral is not zero. In type I superconductors, the field penetrates only a thin surface layer and is prevented from penetrating the interior of the superconductor by screening currents. The field energy associated with the surface of the type II superconductor is stored partially as magnetic energy and partially as kinetic energy by the electrons that provides the screening current. If the external field is removed, the energy is absorbed by the external circuit and thus completely recovered.

The hysteretic effects and with it the ac losses are quite substantial in the superconducting materials with high current densities, such as those being used in magnets. Hysteretic loss calculations have been performed by a number of investigators, such as Bean [25], London [26], Hancox [27], Hart [28], and Wipf [24]. In any metallic conductor, eddy current losses are induced, if $\dot{B} \neq 0$ in a composite conductor. The induced currents are modified by the superconducting filaments, which do not admit a resistive longitudinal component of the electric field. The induced currents generate losses in the conductor. A third category of losses in the conductor are losses due to field inhomogeneities over the conductor. Self-field effects are observed in most ac machines and are reduced by subdivision of the conductor into strands and transposition of these individual strands,

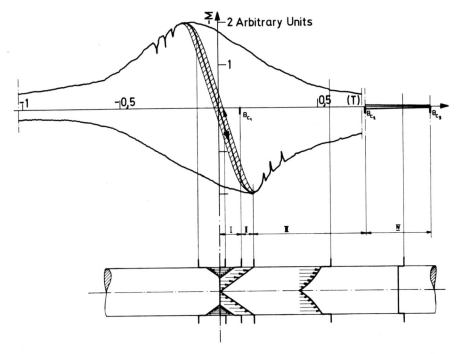

Fig. 18.   Magnetization Loop and Flux Penetration Profiles in a Type II
Superconductor. The flux profiles are determined for external
fields only. If a transport current is present in addition, the
electric center will shift from the center towards either conductor surface   (Wipf).

(Roebel = and Punga-conductors). Essentially, self-field effects, due to transport currents in the wire, have the same origin as the skin effect in the normal conductor and tend to exclude current density changes in the interior of the wire. The simple twisting of individual strands of a conductor does not entirely remove self-field losses. The strands must be transposed and since filaments within a strand can not be transposed, they should be twisted. Self-field losses are of electromagnetic nature and may be obtained by integrating Poynting's vector over the conductor surface. The irreversible part of this energy integrated over a cycle is dissipated as heat.

Losses are generated by a changing magnetic field in all metallic parts of the magnet in addition to the coil. Losses are generated in the iron shell and the heat must be removed by the helium. In the iron shell, the dissipative losses are hysteretic and eddy current type losses. In laminated silicon steel yokes and shells self-field losses can be neglected. A major source of heat is the metallic parts of the cryostat. The cylindrical inner or outer shell of the cryostat is either of nonmetallic materials (Glass-reinforced thermosettings) or the stainless steel tubes are corrugated to permit the use of a thin wall and to increase the current path. Reinforcing rings, bolts, nuts etc. may also be sources of heat in an ac magnet. Reinforcing rings may also be made of unidirectional glass filament epoxy - tapes. Losses, due to bolts, etc., are small and not important.

In ac magnets we have to deal with phenomena summarized below:

1.   Joule heating due to hysteretic, eddy current and self-field losses.

2.   Flux jump instabilities due to the occurrence of perturbances (field, current), or due to motion of flux or current in the super-conductor.

3.   Variable and time dependent Lorentz forces, which cause the conductor to move, generating fractional losses.

4.   Conductor degradation and training.

In addition to these fundamental phenomena in pulsed magnets we have to deal with the following problems:

1.   Generation of undesired field multipoles due to the coil configuration and conductor and coil manufacturing tolerances (deviations from the ideal shape).

2.   The influence of the finite conductor size on the coil geometry and with it the field aberrations.

3.   Influence of thermal contraction on the coil shape.

4.   Heat transport due to heat conduction, thermal diffusivity and to heat transfer.

5. Distortion of the coil shape due to Lorentz forces.

6. Long range material behavior, when subjected to fluctuating fields and currents (fatigue).

7. Influence of various thermal contractions and strain energies of materials (insulations, metals,) on the magnet performance.

8. Insulation breakdown due to surge voltages occurring from quenches.

9. Environmental influences, mainly irradiation, and impurity in the cryogens.

Many of the problems listed are common in ac and dc magnets. Others are specific items encountered only in ac magnets. In what follows we summarize the various types of losses without giving their derivation.

A.   Field Profiles in a Superconductor

The pinning force per unit length of a flux line is given by:

$\vec{\Phi}_o \times \vec{J}_c = \vec{F}_p$ with $\vec{J}_c$ the critical current density and $\vec{\Phi}$ the unit flux quantization (h/2e). Using Amperes law

$$\frac{1}{\mu_o} \vec{\Phi}_o \times \text{curl } \vec{H} = \vec{F}_p , \tag{43}$$

we can modify this expression for a semi-infinite slab with its surface in the yz plane as:

$$\frac{\vec{\Phi}_o}{\mu_o} \frac{d|H|}{dx} = \vec{F}_p . \tag{44}$$

Green and Hlawiczka [29] have defined the pinning forces to have the form: $F_p = \pm k_n |H|^n$ where k_n and n are constants. The sign is determined by the direction of the Lorentz force. Eliminating $\vec{F}_p$ we obtain: $\vec{\Phi}_o \times \vec{J}_c = \pm k_n |H|^n$, which gives for our simplified one dimensional case the relation:

$$\frac{\Phi_o}{\mu_o} \frac{dH}{dx} = \pm k_n |H|^n . \tag{45}$$

The field is obtained after integration

$$|H| = \left[\pm k_n (1-n) \frac{\mu_o}{\Phi_o} (x-x_p) \right]^{1/(1-n)} \quad \text{where} \tag{46}$$

k_n is a constant whose dimensions depend on n. For n = 0.

$$H = k_o \frac{\mu_o}{\Phi_o} (x - x_p) \quad \text{and}$$

$$J = \frac{dH}{dx} = k_o \frac{\mu_o}{\Phi_o} = \text{const.} \tag{47}$$

This leads to the well known "Bean Model", where $J = J_c = \text{const.}$. The field lines are straight, as shown in Fig. 19. For $n = -1$,

$$H = \left[2k_1 \left(\frac{\mu_o}{\Phi_o} \right) (x - x_p) \right]^{1/2}. \tag{48}$$

From Eq. (48) we see,

$$J_c H = k \cdot \frac{\mu_o}{\Phi_o} = \text{const.}$$

which is "Kim's Model". Experimentally Kim and co-workers [30] obtained:

$$J_c = \frac{\alpha}{H + H_o} \quad \text{or,} \quad J_c = \frac{J_o H_o}{H + H_o} \tag{49}$$

where α, J_o and H_o are constants. Field profiles for Kim-type penetration are shown in Fig. 20.

Using Maxwell's equation, $J_c = dH/dx$ and integrating Eq. (49) we obtain the field profile in an imperfect type II superconductor:

$$H(x) = H_o \left\{ \left[\left(1 + \frac{H_{ext}}{H_o}\right)^2 - \frac{2J_o x}{H_o} \right]^{1/2} - 1 \right\}. \tag{50}$$

The distance x_p from the surface of the superconductor, where the field has dropped to zero is readily given by

$$x_p = \frac{H_{ext}}{J_o H_o} \left[1/2 \, H_{ext} + H_o \right]. \tag{51}$$

For $x_p = a$ (half the width of the superconductor) we obtain the maximum field, indicated by $H_m = H(a)$.

The flux per unit length through the superconducting sample is obtained from

$$\Phi = \mu_o \int_{x_p}^{x} H dx \tag{52}$$

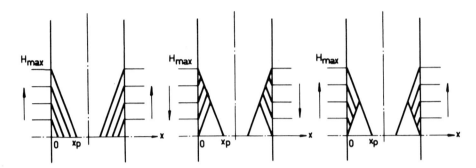

FIG. 19. Profiles of a Penetrated Field into a Type II Superconductor due
to Increasing and Decreasing External Fields, according to
Bean's Critical State Model.

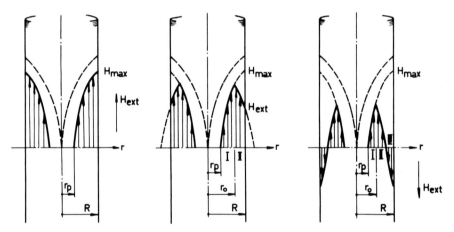

FIG. 20. Profiles of a Penetrated Field into a Type II Superconductor due
to Increasing and Decreasing External Fields, According to the
Kim Model.

which after inserting the H value from Eq. (50) in Eq. (52) and integration yields

$$\Phi = \mu_0 H_0 \left\{ \frac{H_0}{3J_0} \left[\left(1 + \frac{H_{ex}}{H_0} - \frac{2J_0}{H_0} x \right)^{\frac{3}{2}} - \left(1 + \frac{H_{ex}}{H_0} - \frac{2J_0}{H_0} x_p \right)^{\frac{3}{2}} \right] - (x - x_p) \right\}.$$

(53)

Using the Eq. (46) we get the penetrated flux into the sample:

$$\Phi = \mu_0 \left(\frac{k_n \mu_0}{\Phi_0} \right)^{1/(1-n)} \frac{\left[(1-n)(x - x_p) \right]^{2-n/(1-n)}}{2-n}$$

(54)

which gives for the particular case of n = -1,

$$\Phi = \mu_0 \left(\frac{k_1 \mu_0}{\Phi_0} \right)^{1/2} \frac{[2(x - x_p)]}{3}.$$

(55)

We assume now that the magnetic field changes sinusoidally according to $H = H_{ext} \sin(\omega t)$. Inside the slab, the field is time dependent, but not necessarily sinusoidal. Inserting the time dependent H in Eq. (46), we obtain the penetration depth,

$$x_m = \frac{(H_{ext})^{1-n}}{(1-n) kn \mu_0 / \Phi_0}.$$

(56)

With reference to Fig. 20 the corresponding values are, at x = 0 the field $H = H_{ext}$; at $x_p = x_m$ the field H = 0. With the changing flux inside the superconductor, one has to calculate the flux $\Phi(t)$ for each case as illustrated in Fig. 20.

B.   Hysteretic Losses

The hysteretic losses per cycle per unit surface area are proportional to the area of the hysteresis loop. Denoting the loss by P_{hl} we may write:

$$P_{hl} \sim \oint H_{ext} \sin(\omega t) d\Phi.$$

Equations for H and $\Phi(t)$ have been calculated as indicated above and the hysteretic losses associated with differently shaped conductors are summarized below.

For a rectangular conductor, with the field swept between $-H_{max}$ and $+H_{max}$, the loss is given by

$$W_{hl} = \mu_o d^2 b 1 J_c H_{max} \quad (Ws) \ . \tag{57}$$

For a cylindrical conductor with a diameter d and length 1, the hysteretic loss per cycle is given by

$$P_{hl} = \frac{8}{3\pi} \mu_o V J_c \frac{d}{4} \dot{H}_{ext} \quad (Ws/cycle) \tag{58}$$

or in explicit form, $P_{hl} = \frac{1}{6} \mu_o d^3 1 J_c \dot{H}_{ext} \quad (Ws/cycle) \ . \tag{59}$

For a superconducting coil, the total power dissipation in a coil of volume V, subject to a field change between H_{min} and H_{max} is obtained by using Kim's relation,

$$W_{hl} = \frac{\mu_o J_o H_o V d}{2} \ln \left[\frac{H_{max} + H_o}{H_{min} + H_o} \right] \quad (Ws) \ . \tag{60}$$

The term $\mu_o H_o$ is approximately 1T.

For a superconductor carrying a transport current less than the critical current, with the field varying between $-H_{max}$ and $+H_{max}$ the loss is given by

$$W_{hl} = \mu_o V d J_c H_{max} \left[\frac{2}{f} - 1 - \frac{2(1-f)}{f} \ln \frac{1}{1-f} \right] \tag{61}$$

with $f = I_T / I_c$ the ratio of transport to critical current. Usually $f < 1$. For $f = 1$, where the transport current is equal to the critical, the equation for a rectangular conductor given in Eq. (57) is obtained.

C.   Eddy Current Losses

Eddy current losses can be calculated, if the matrix material, the current distribution in the matrix, and the rate of change of the field is known. For simplicity we assume uniform distribution of filaments throughout the composite conductor and the external field perpendicular to the composite axis. The eddy current losses can be expressed in the form (Fig. 21):

$$P_e = \frac{1}{\bar{\rho}_c} \left(\frac{\dot{B} 1_p}{2\pi} \right)^2 V \tag{62}$$

where $\bar{\rho}_c$ indicates the effective transverse resistivity of the composite. For a one component matrix, e.g., copper or aluminum

$$\frac{1}{\bar{\rho}_c} \simeq \frac{1}{\rho \, matrix} \left(\frac{w-d}{w} + \frac{\bar{C}}{R} \right) \text{ where}$$

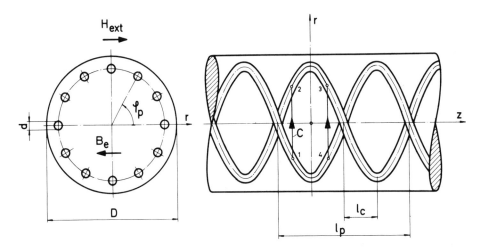

FIG. 21.   Schematic Representation of a Two Component (Superconductor
and "Normal" Metal) Composite Conductor. The filaments
are distributed uniformly over a circle of radius r and twisted
with a twist pitch if l_p.

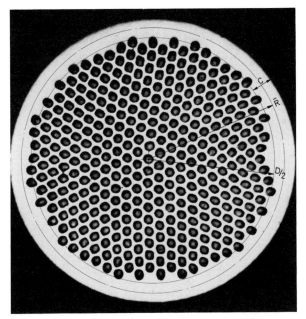

FIG. 22.   Two Component Composite Conductor. Eddy currents will
flow between filaments located in opposing sides of the
center. Eddy currents will flow not only in radial direction but also through the high conductivity metal in the
conductor circumference.

$\overline{C}$ and R are given in Fig. 22, w is the distance between two filaments and d is the filament diameter. If the filaments are not twisted enough, the diamagnetic current can flow through the matrix (paths $\overline{12}$ and $\overline{43}$) material to complete the circuit with filaments on the other side of the wire. The higher $\dot{B}$ becomes, the more the wire must be twisted to prevent coupling between filaments. How much the wire must be twisted is given by the critical twist length given by

$$\ell_c^2 = 2J_c \, \rho_{matrix} \, d \, \lambda_s^{1/2} ((w-d)/d + \overline{C}/R)/\dot{B} \, (m^2) \tag{63}$$

where λ_s is the fraction of the cross section occupied by the superconductor.

D.   Self-field Losses

Changes in the flux pattern over the superconductor induced by self-fields lead to the movement of fluxoids within the superconductor and thus to dissipative losses. These losses are of electromagnetic nature and can be obtained by integrating the Poynting vectors $\vec{S} = \vec{E} \times \vec{H}$ over the conductor surface. The irreversible part of the energy, integrated over a cycle is dissipated as heat. For the conductor sizes contemplated in pulsed magnets, the external field generated in the magnet is much higher than the self-field of the conductor itself. Thus J_c is the same across the conductor and the Bean-Model modified for cylindrical geometries can be applied. Referring to Fig. 23, the dissipative part of $\vec{S}$ is obtained from the difference of Φ, corresponding to maximum and minimum field values if the transport current I_T is varied.

The self-field losses are obtained from the expression

$$W_{sf} = \pi \mu_o J_c^2 R^4 \ell \left\{ \left(1 - \frac{H_{max} - H_{min}}{J_c R}\right) \left[\ln\left(1 - \frac{H_{max} - H_{min}}{J_c R}\right) \right. \right.$$

$$\left. \left. + \frac{H_{max} - H_{min}}{J_c R} \right] + \frac{1}{2} \left(\frac{H_{max} - H_{min}}{J_c R}\right)^2 \right\}. \qquad \text{(Ws/cycle)} \tag{64}$$

Self-field losses for a slab ($R \to \infty$) are accordingly:

$$W_{sf} = \frac{\mu_o}{6} \frac{\pi R \, l}{J_c} (H_{max} - H_{min})^3 . \qquad \text{(Ws/cycle)} \tag{65}$$

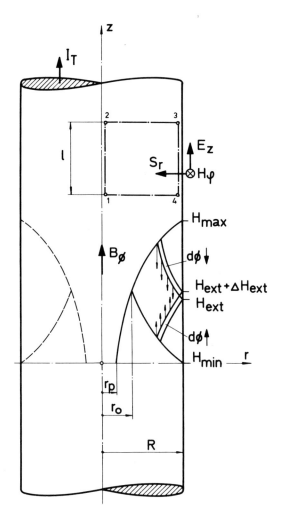

Fig. 23. Field Profiles in a Superconductor Due to External Changing
Magnetic Fields. If the magnitude of the external fields on
both sides of the conductor are different, additional losses
(self-field losses) are generated.

E.   Magnetic and Thermal Instabilities

Nonideal type II superconductors become resistive, when the macroscopic current density, either impressed by the transport current, or induced,   exceeds the critical current I_c. The critical current density J_c forces the flux to move causing energy dissipation proportional to the flux flow resistivity $\rho_f = \rho_n(H/H_{c2})$, where ρ_n is the normal state resistivity of the superconductor. The current density must exceed the depinning value of J_c before flux lines can move. The resulting electric field (Fig. 22) is given by:

$$E = \rho_f(J - J_c) \, .\tag{66}$$

Perturbations occurring in the voltage-current characteristic, forces the superconductor to revert to the resistive state even before the true critical current is reached. We call this case "degradation". A disturbance changes both the Lorentz force

$$\vec{F}_L = \int_V \left[\vec{B}(r) \cdot \vec{J}_T(r) \right] d^3 r \, ,\tag{67}$$

and the pinning force

$$\vec{F}_p = \mu_o \left[\vec{J}_c \cdot \vec{H}_{ext} \right] . \tag{68}$$

The disturbance usually does not spread immediately over large areas of the coil, but may be initially localized. If the decrease of $\vec{F}_L$ is smaller than that of $\vec{F}_p$, the equilibrium is called "unstable" and the disturbance may spread. If the reduction of $\vec{F}_L$ is larger than that of $\vec{F}_p$, the condition is stable and the area of disturbance may diminish. In the design and operation of a magnet it is important that the magnet performance is stable and self-healing. In recent years the criteria for stability have been treated quite extensively theoretically, verified by experiments, and we summarize the results.

In section V we treated the cryostatic stability criterion, where the superconductor was embedded in a high thermal and electrical conductivity matrix and the heat generated by the flux movement was absorbed by the substrate. We found that a substantial amount of normal metal compared to the superconductor was required and therefore overall current densities were small. In ac type magnets with relatively small stored field energies, it is of interest to increase the overall current densities and thus two other types of stability criteria have been investigated.

The first criterion is the enthalpy stabilization. The thermal capacity of the conductor must be sufficiently high to prevent an

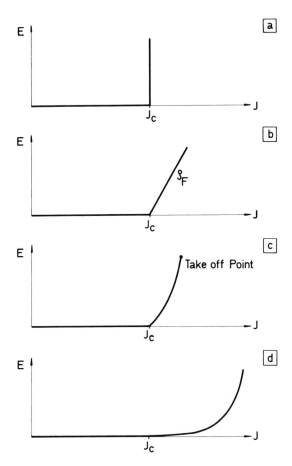

Fig. 24.   Relation between Electric Field and Current Density in a Superconductor (a,b) and a Composite Conductor (c,d). ρ_F represents the effective flux flow resistivity of a superconductor. J_c
is clearly determined in (a,b,c) whereas the gradual increase of
E, (d) in a composite leads to errors in determining the critical
current density. It is now generally agreed, that J_c is that
value at which an effective resistivity of 10^{-12} Ωcm can be measured over the sample.

abiabatic flux jump from driving the superconductor normal. The condition of enthalpy stabilization requires that

$$(J_c d)^2 \leq 3 \; \frac{(c_p \delta)}{\mu_o} \; \left(-\frac{1}{J_c} \frac{dJ_c}{dT} \right)^{-1} \tag{69}$$

where d is the diameter of the superconductor, $c_p \delta$ is the heat capacity and $\frac{-1}{J_c} \frac{dJ_c}{dT}$ refers to the subtangent at the J-T characteristic of the superconductor. For a NbTi conductor with $J_c = 10^5 A/cm^2$, and $\left(\frac{-1}{J_c} \frac{dJ_c}{dT} \right)^{-1} \cong 3K; c_p \delta = 1.1 \times 10^{-3} Ws/cm^3 K$ we obtain a wire diameter of d = 0.9 x 10^{-2} cm. For diameters smaller than this value, the conductor is stable. Figure 25 illustrates magnetic stability effects.

The second criterion is the dynamic stabilization. A sufficient amount of normal metal is added to the superconductor to magnetically dampen any flux motion. The energy released by a flux jump is removed by the thermal diffusivity of the matrix. The growth of possible instability regions is prevented. The amount of normal metal, used for dynamic stability, is obtained from the equation giving the diameters of the superconducting filaments.

$$d^2 \leq \frac{138}{\pi^2 \mu_c} \frac{(D_{th})_s}{(D_m)_n} (c_p \delta)_s \frac{1-\lambda_s}{\lambda_s} \frac{1}{J_c^2} -\frac{1}{J_c} \frac{dJ_c}{dT}^{-1} \quad \text{where} \tag{70}$$

D_m D_{th} are the magnetic and thermal diffusivities of the normal conductor and the superconductor, λ_s is the space factor of the superconductor within the composite. Assuming the filament is embedded concentric within the matrix, then the diameter of the matrix is simply $D = d/\sqrt{\lambda_s}$.

In type II superconductors the thermal diffusivity is quite small ($< 1 cm^2/s$) compared to copper (280 cm^2/s). The magnetic diffusivity in superconductors is substantially higher ($> 200 cm^2/s$) depending on the flux flow resistivity or H, than that of pure metals ($< 1 cm^2/s$). In the superconductor, the magnetic field can penetrate much faster than the heat can be removed, leading to a temperature rise in the area of disturbance. The temperature rise may spread along the conductor. In the composite conductor this effect is clearly remedied. As a special case of the enthalpy-stabilization we note that intrinsically stabilized superconductor has a diameter which is much smaller than indicated by the above two requirements. Such a superconducting filament would not necessitate the use of any normal metal for stabilization. Some normal metal is still used for mechanical rigidity, as a shunt material, and for added safety.

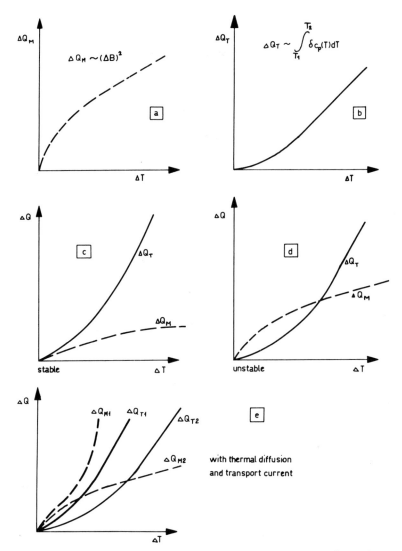

Fig. 25. The Occurrence of Magnetic Instabilities (Hart): a) The heat generated is proportional to $(\Delta B)^2$; b) The heat absorbed by the superconductor is proportional to the enthalpy of the material; c) The superconductor is stable against small disturbances. More heat is absorbed than generated; d) The superconductor is unstable against large disturbances; e) The absorbed heat includes thermal diffusivity. In the case of ΔQ_{M1} and ΔQ_{T1} the condition is clearly unstable. In the case of ΔQ_{M2} and ΔQ_{T2} the heat dissipated in the conductor is carried away by the thermal diffusity and at the crossover point a stable condition is restored.

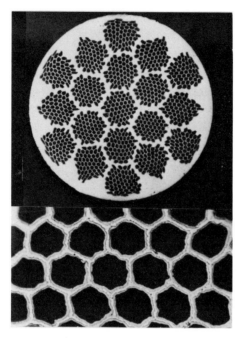

FIG. 26.   Three Component Composite Conductor. In each bundle the filaments are twisted. The filament bundles are also twisted inside
the composite conductor. Each superconductor is surrounded by
a cupronickel layer to reduce the flow of eddy currents and by a
copper layer for dynamic stability purposes. In most recent conductors cupronickel barriers are foreseen between filament bundle
and radially at the outer conductor circumference (IMI).

Dynamically stabilized composite conductors can have very high
current densities. In such a composite conductor, as shown in Fig. 26,
the ratio of superconductor to matrix material have even reached about
1:1. Recently, the trend has been somewhat reversed and ratios of
normal metal to superconductor of about 1.5:1 are preferred in the composite to improve stability.

F.   Removal of Heat

The energy dissipated in a pulsed coil must be removed as quickly
and efficiently as possible. Since the thermal capacity of the coil at
liquid helium temperature is quite low, very small energy densities
(10^{-2}J/cm^3) will be sufficient to drive the superconductor normal. The
thermal diffusivity of the insulating material is also too low ($2-4 \times 10^{-3}$W/cm
to transport the heat efficiently across the coil cross-section to the cooling
surface.

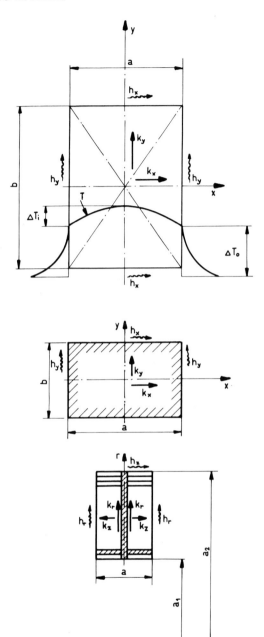

Fig. 27. Coil and Cooling Arrangements. Symmetric cooling arrangements in axially symmetric coils. ΔT_i is the temperature
drop from the coil hot-spot to the coil surface (conduction).
ΔT_0 is the temperature drop from the coil surface to the bulk
helium (heat transfer).

One major step in the direction of better cooling of superconductors is the transposition of individual strands. Although not intended primarily for this purpose, the transposition brings each strand at some point along the cable to the cable surface. If the coil is potted in a thermosetting, the heat has to travel essentially through an insulation barrier only, which seldom exceeds a thickness of about 0.2 cm.

In multilayer coils with no internal coolant passage, the heat must travel through several layers of insulation and the removal of heat may pose problems. Three solutions have been employed successfully:

1.   Subdivision of the coil into modules, where the helium can reach two or more surfaces easily. The width of these passages should be such that any helium gas bubble formation does not block the free passage of helium. Designs of this type of coil are illustrated in Fig. 27. Assuming symmetrical conditions as shown, the temperature rise in a module of thickness "a" will be approximately [4] given by

$$\Delta T = w_v \left[\frac{1}{2k_x} \left(\frac{a}{2}\right)^2 + \frac{a}{2h_x} \right] \text{ where} \tag{71}$$

k_x is   the effective thermal conductivity of the coil and h_x is the heat transfer coefficient of the coil surface to the bath and w_v is the power density and "a" the module thickness between two coolant passages. The effective thermal conductivity is obtained from the thermal conductivities of individual components and from the arrangement of individual cables and strands. If k_x and h_x are known, "a" can be determined. For a power density of $10^{-2} W/cm^3$, limiting the temperature rise to 0.2K, one may obtain as a lower limit of "a" = 1cm.

Subdivision of the coil into modules has advantages but also some drawbacks. The advantage of coil subdivision into appropriate modules leads to current optimization. Each module must also be mechanically supported which may result in an expensive solution. In coil designs with concentric layers Fig. 28, coil shells are fitted on each other very tightly and the whole coil is supported externally and by the iron yoke. Small missmatches in the thermal contraction will result in relative motion of individual shells. Coil training is one result of this .

2.   The second design scheme is the employment of heat drains between layers. In impregnated coils this scheme has given excellent results leading to very compact coils, Fig. 29. As a heat drain, copper or aluminum mesh is used, which become after impregnation an integral part of the coil. The heat drain is either extended over the coil width into the bath or soldered to cooling tubes. The heat removal is not as efficient as in the first case utilizing coolant passages, and there is always the danger that voids between the screen and the thermosetting (cracks) may develop when the magnet is pulsed. The heat drains may be grounded. If not, when they are at a certain potential and are brought

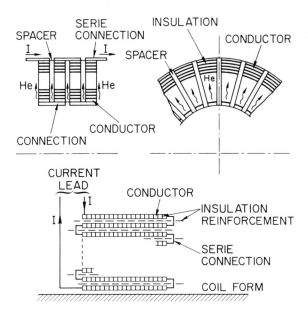

FIG. 28. Coil Design with Radial and Axial Coolant Passages.

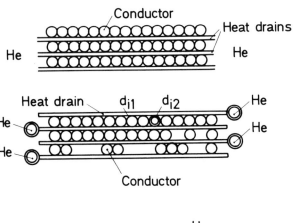

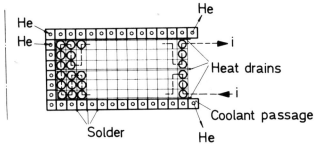

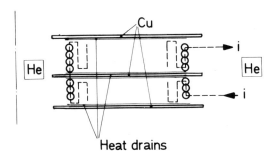

Fig. 29. Coil Design with Heat Drains. The thermal conduction
         shields are placed in the coil, surrounding conductor layers
         or facing pancakes. To avoid eddy currents in the metallic
         shield litz wire or interwoven flat tapes may be used. The
         heat drains are either connected to a coolant channel or are
         protruding into the bath.

up to some intermediate electrical potential, voltage breakdown may occur through the capacitance effect between the coil and the screen. Experience with heat drains to date has shown, however, that these difficulties can be overcome.

The third possibility is to use hollow cables and pressurize super-critical helium through internal passages as discussed in Section IV. Internally cooled multistrand cables are quite common in the power industry. A flat thin stainless-steel ribbon is wound spirally inter-spersed with an insulating tape to form the coolant passage. Individual insulated strands are wound around this spirally formed channel. These cables are flexible and reliable. Coolant passages with diameters on the order of a few mm (2-5), including the wall thickness of the stainless-steel -insulation tape, will be sufficient for the passage of helium. If the temperature rise in the helium due to insenthalpic expansion becomes a problem, the solution given by Morpurgo for the CERN-Omega magnet [31] can be used successfully as shown in Fig. 30. The cable is wrapped with a glassfiber tape. Wound into coils it can be impregnated to form a rigid coil.

Limiting the temperature rise during a pulse is necessary not only from the stability point of view, but also because of the J_c-T character-istic of the superconductor. With increasing temperature the critical current density in the superconductor is reduced. Referring to Fig. 31 an increase in temperature from 4.2K to 5K would result in a reduction in the critical current density by 26% if the conductor is exposed to a transverse magnetic field of 6T. Generally, the operational temperature should not exceed about 5K. At lower flux densities the reduction in the critical current density due to a temperature rise is less critical.

VIII. COIL STRUCTURE AND PULSED MAGNET DESIGN

There are several ways to produce a field of a given amplitude and distribution within the area used for beam transport or for experimental purposes (useful aperture). The field at a point Z_o within the useful aperature of a magnet is given by

$$B^*(Z) = j\frac{I}{2} \mu_o \sum_{n=1}^{n} \frac{1}{Z-Z_o} = j\frac{I}{2\pi} \sum_{1}^{N} \frac{Z_o^{n-1}}{Z^n} ,$$

or   (72)

$$B^*(Z) = \sum_{n=1}^{N} c_n (Z_o^{n-1})$$

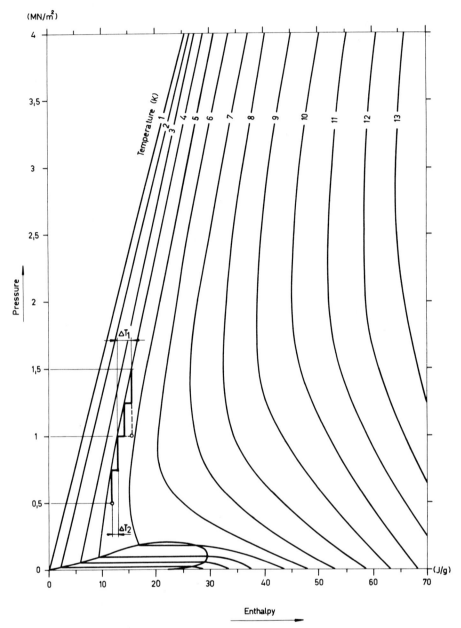

Fig. 30. Pressure-Enthalpy Diagram of Helium. Supercritical helium
          passing through hollow conductors is   subject to pressure
          reduction resulting in a temperature rise. To reduce the
          temperature ΔT_1 (isenthalpic expansion due to a pressure
          drop of $0.5\,\mathrm{N/m^2}$) to ΔT_2 intermediate cooling stages can be
          incorporated (Morpurgo).

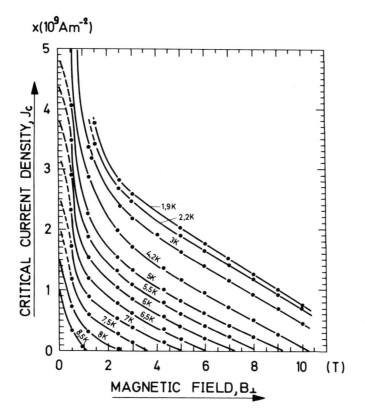

FIG. 31. Critical Current Density of Nb$_x$Ti Versus Transverse Magnetic
Field in the Temperature Range of 1.9 K to 8.5 K.

with

$$c_n = j \frac{I}{2\pi} \mu_o \sum_{n=1}^{2n} \frac{(-1)^{n-1}}{Z^n} \, .$$

Generally, $N = 2n$. As usual: $B^*(Z) = B_X - j B_Y$ represents the complex conjugate of the two dimensional field at $Z = X + jY$, $n=1$ represents a dipole coil, $n=2$ a quadrupole, etc., where n is a multipole number.

To construct a coil which generates a certain multipole field, the idea of intersecting elipses or circles was employed. In coils with concentric boundaries, a current density distribution according to $\cos(n\theta)$ was utilized over the coil blocks of constant current density, which makes the magnet construction much simpler and more susceptible to manufacturing procedures. Two possible designs were shown in Fig. 9a and 9b with a detailed design shown in Fig. 32.

In Fig. 33 the modified $\cos(n\theta)$ current distribution is illustrated. In this case the coil is divided into a number of blocks and surrounded by an iron shell. We can write for the two field components

$$B_r = \frac{\mu_o (\lambda J_o)}{\pi} \left(\sum_{i=1}^{m} (-1)^{i+1} \cos\alpha_i \right) r^{n-1} \left[\frac{1}{2-n} (a_2^{2-n} - a_1^{2-n}) \right.$$

$$\left. + \frac{1}{(2+n)b_1^{2n}} (a_2^{2+n} - a_1^{2+n}) \right] \sin(n\theta),$$

$$B_o = \frac{\mu_o (\lambda J_o)}{\pi} \left(\sum_{i=1}^{m} (-1)^{i+1} \cos\alpha_i \right) r^{n-1} \left[\frac{1}{2-n} (a_2^{2-n} - a_1^{2-n}) \right.$$

$$\left. + \frac{1}{(2+n)b_1^{2n}} (a_2^{2+n} - a_1^{2+n}) \right] \cos(n\theta).$$

In these equations $n=1$ represents a dipole, $n=2$ a quadrupole coil, etc.

As was mentioned, the conductors to be used in a coil may be a cable or braid, of rectangular, square or flat shape. In the flat conductor (cable or braid) case, a one layer coil configuration may be adopted. To omit field errors, the placement of the conductors in this case is of great importance, since the iron contribution to the aperture field is rather small ($<10\%$) and can not be utilized for field shaping. In the case where several conductor shells or double layers are placed radially over

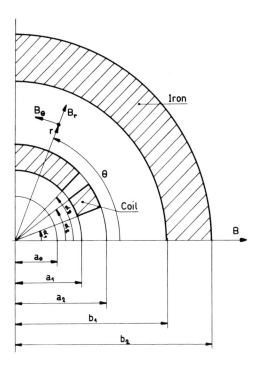

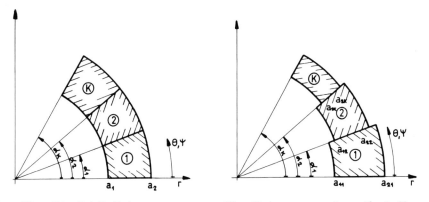

Fig. 32. Dipole Coil Arrangement. The five concentric coil shells are
subdivided into blocks of uniform current density. With the
angular distribution of individual coil blocks the field homogeneity of 10^{-3} over the useful aperture was obtained. The
central field is $4.5\,T$. The iron shell is cooled to liquid helium
temperature and placed at a certain distance from the coil
such that the iron is not saturated. The laminated iron yoke is
welded at the outer surface to restraining the steel plates and the
two yoke halves are bolted together (IEKP).

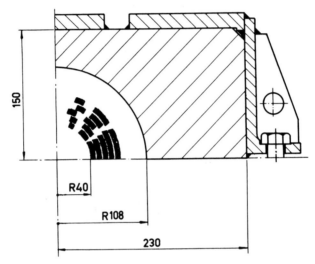

FIG. 33. Schematic Arrangement of a Coil and Iron Shield for Field
Calculation.

each other, the placement error of one layer may be corrected by the
other adjoining shells. Placement tolerances to obtain a field or
gradient homogeneity of $\Delta B/B = 10^{-3}$ within the useful aperture ($\sim 70\%$
of the inner coil diameter) will be $0.1 \sim 0.3$ mm on a diameter of 6 cm.
Therefore, the shells have to be placed on carefully machined formers
with grooves and accurately placed wedges of nonmetallic material,
such as glass reinforced epoxies. Each shell is then impregnated in a
suitable thermosetting. Usually half-shells are assembled on a cylinder
and accurately placed over each other to form concentric cylinders.
Between coil shells, spacers are located to permit free flow of helium
on both the inner and outer cylindrical surfaces after the coil is
assembled and reinforced. The iron shell (usually laminated silicon
steel sheets) is placed around the coil. The impregnation of coil shells
with thermosettings is imperative as single layers cannot be reinforced individually by means of strong bandages in order to prevent conductor motion.

In Fig. 34 we illustrate two possible dipole designs. In the top
figure the coil consists of four shells with coolant passages between
cylinders surrounded by an iron shell. In the middle figure the current
density distribution is approximated by current blocks. To correct for
sextupole and decapole components, correction coils are placed around
the aperture as shown in the bottom picture. The conductors producing
a certain multipole are parallel to the coil axis and interconnected. When
the field is raised or reduced, currents are induced in these correcting
coils, cancelling the undesired multipole field produced by the main
coil. The positioning of the iron shell depends on the magnetic field
around the coil. If the core is in close proximity to the coil ($b_1 \cong a_2$),
portions of the iron yoke are saturated and may also produce field
errors (Fig. 35). The placement tolerance of the iron shell around the
coil should be within 0.15 mm, assuming a coil aperture of 6 cm diameter.

Coil end effects in short beam transport magnets with the magnet
length of about $1 = 10a_1$ influence   the effective field length of the magnet
if the transport current is changed from zero to a peak value. Since
the effective length can vary over the aperture due to end effects,
additional field aberrations are generated. The field distribution at
the coil ends are three dimensional and exact calculation is difficult. A
possible coil-end design is shown in Fig. 36 for a saddle type magnet.
The coil-ends are placed on a cylindrical surface and arranged such
that the field enhancement at the ends is small and their effect on
$\int Bdl$ is not serious. Simplified calculation methods for coil-ends, approximating three dimensional coilends by two dimensional geometries are
given by Mills [32], Green [33] and others.

IX.   CONDUCTOR CONFIGURATIONS IN PULSED MAGNETS

Superconductors used in pulsed magnets must be free from flux
jumping instabilities and their energy dissipation in form of heat must be

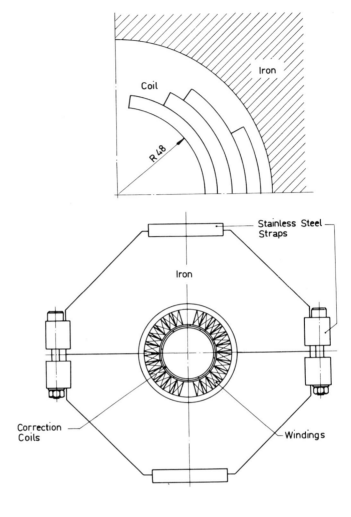

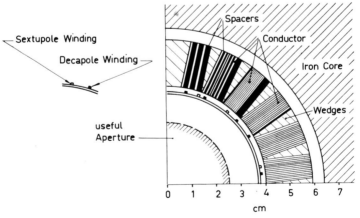

Fig. 34. Dipole Coil Configurations. The variation of the current density as a function of the angle θ was possible by changing the
arc length of shells (top) or by placing insulating spacers
between coil layers (bottom). Correction coils placed around
the aperture do compensate multipole field components (BNL).

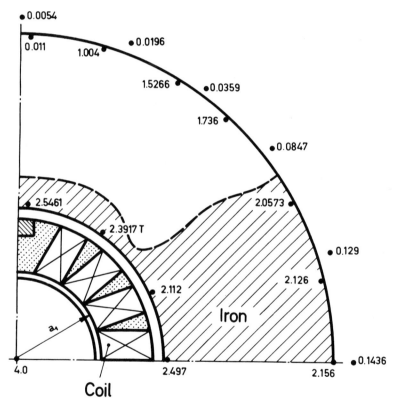

FIG. 35. Fluxdensity Distribution Over the Iron Shell. The yoke is in
close proximity of a dipole coil generating 4T central field.
The nonuniform fluxdensity in the iron does produce additional
field aberrations.

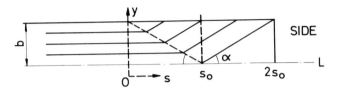

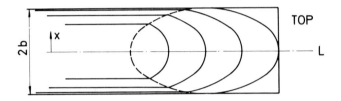

FIG. 36. Coil-end Configuration for Saddle Shaped Race Track Coils.
The model presented is used to simulate three dimensional
problems to two dimensional geometries.

low. Attempts to comply with these requirements are:

1.   Composite conductors (strands) with superconducting filaments
      embedded and twisted in a matrix of normal metal.

2.   Composite conductors (strands) according to (1) transposed into
      a cable or braid (Fig. 37).

Flux jump instabilities can be eliminated by using multifilament
conductors ($d \cong 5 \times 10^{-3}$ cm diameter). In the new composite conductors,
nearly 50% of the cross-section of the composite is "filled" with superconductors, resulting in high overall current densities. At fields of
5T, current densities on the order of $2 \times 10^4 A/cm^2$ or higher are common.
The composite conductors Fig. 38 are twisted or transposed into a
cable or braid shown in Fig. 37. A 5μ m diameter filament carries a
maximum current of 20 mA at 5T. Typical magnet currents are between
2000 A to 5000 A, so that 10^5 filaments must be connected in parallel.
Commercially available conductors have 10^3 to 10^4 filaments. Thus,
10 to 100 strands are used in a cable.

Cables with many strands have a poor packing factor, and com-
pacting the cable to required tolerances may result in the damage of in
dividual strands at crossover points. To eliminate coupling between
strands when the field is varied, the strands may be individually insulated.

At crossover points, the insulation may be damaged and the cable
may behave as a single wire and thus self-field losses, which are proportional to the fourth power of the conductor diameter (not filament

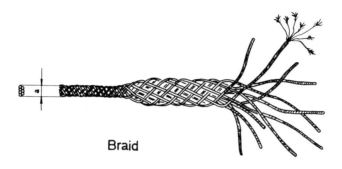

Braid

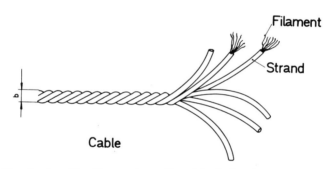

Filament

Strand

Cable

Fig. 37.   Conductor Configurations Used in Pulsed Magnets. Strands
are transposed into a braid and into a cable.

Principal Scheme of transposition

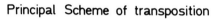

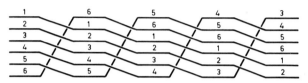

transposed Cable

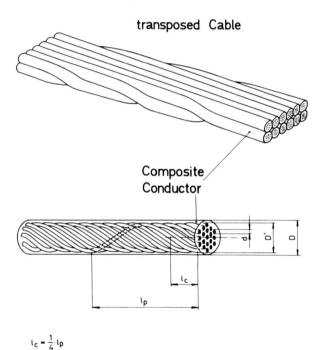

Composite
Conductor

$l_c = \frac{1}{4} \cdot l_p$

Fig. 38. Fully Transposed Cable. A single composite conductor
(strand) with twisted filaments is shown at the bottom.

diameter), may become excessive. Only a fully transposed cable or braid exhibits no self-field losses.

There have been attempts to increase the number of filaments in a single wire to 10^5 or more. But it can be shown, that this is only possible without excessive coupling losses, if a high resistivity material is present between the filaments and filament bundles. The high re-sistivity barrier (usually cupronickel) around each filament prevents transverse currents from flowing from one filament to another. Other-wise a composite conductor with many filaments would behave like a single filament and the subdivision of filaments would not help. Trans-position of filaments in a strand has not been possible. Thus, some additional self-field and coupling losses will always be encountered. But in designs according to Fig. 38, these losses are small com-pared to hysteretic losses.

X.   EXPERIENCE WITH PULSED MAGNETS

With presently available composite conductors, pulsed magnets for 0.1 to 1 Hz operational frequency have been built. This range satisfied the present need for superconducting accelerator magnets. For operation at frequency of 0.1 Hz, the refrigerator will see a load of 20-30 W/m of magnet, which are detailed as follows, hysteretic losses, (10 μ m filaments) ≈7 to 10 W/m; eddy current losses ∼ 1 W/m; self-field losses (0.04 cm strand with no shorts) 0.05 W/m; iron core losses ∼ 2 to 4 W/m. Thus, the total refrigeration losses, due to the magnet only, is 10 to 15 W/m magnet. In addition we have to consider (5 ∼ 10 W/m) losses due to heat leaks through the cryostat and the helium transfer lines. The refrigeration and energizing circuits for a pair of pulsed magnets are illustrated in Fig. 39. In this application, two-phase helium, supercritical or simple bulk helium may be used to remove the heat. To improve heat conduction through the cable or braid in various applications they are impregnated in In or InSn, AgSn and diluted with Th to increase the resistivity of the solder. For such cables, additional losses between strands called coupling losses are encountered similar to the magnetization losses of individual strands [34]. These coupling losses may be quite considerable, since a metallic bond between indiv-idual insulated strands exists. The calculation of coupling losses is according to the equation

$$P_{av} = \frac{D^2 l_p \dot{B}^2}{16\pi \bar{\rho}_c} \left(1 - \frac{8}{3\pi} \frac{d}{D} \frac{\dot{B}^2}{\dot{B}_c^2}\right). \tag{60}$$

In this equation d could mean either the filament diameter (single strand) or the strand diameter(multistrand cable), D the strand diameter (single strand), or the cable diameter (multistrand cable), and $\bar{\rho}_c$ will be either

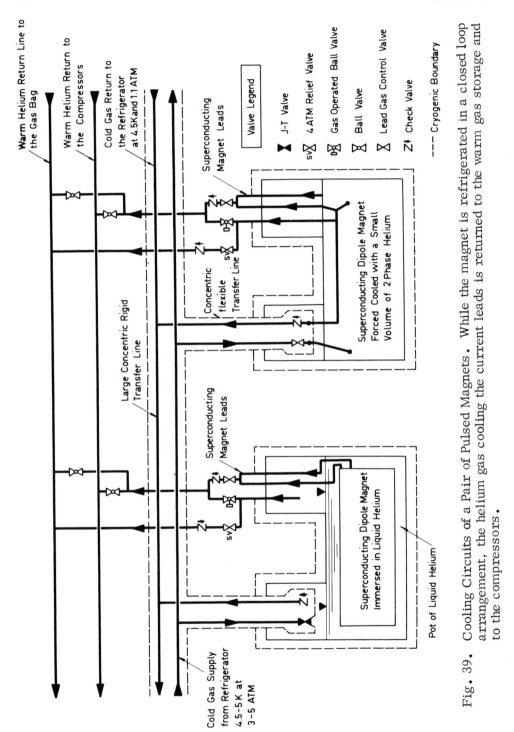

Fig. 39. Cooling Circuits of a Pair of Pulsed Magnets. While the magnet is refrigerated in a closed loop arrangement, the helium gas cooling the current leads is returned to the warm gas storage and to the compressors.

the effective resistivity of the strand or the effective resistivity of the cable. Depending on the field rise time the coupling losses may reach values which are a factor of two (slow pulses), or 5 (pulses with about 1.5 s rise time higher than calculated by Eq.(60)). At faster field rise times, the coupling losses may reach an order of magnitude higher value than the mere magnetization losses of a strand.

These loss values are certainly not tolerable, even if no degradation or training are observed, and even if the heat generated during pulsing can be removed fast enough, a soldered cable should not be used. Fully insulated strands transposed in cables and impregnated in suitable thermosettings after coil winding is recommended.

A far better solution is, however, the single strand multifilament conductors shown in Fig. 26, with enough filaments embedded in the two component matrix to carry the full operational current.

In the above compilation of the losses, we did not account for losses due to electrical leads. Counterflow leads for currents up to 10,000 A are available commercially. Optimized current leads have a steady state loss value, which corresponds to about 1 W per 10^3A and lead. However, the cold gas passing through the lead will not be available to the refrigerator and additional coolant capacity must be provided. The additional capacity will be 12 ~18W for a 2000 A lead pair. Thus, conservatively one may add another 15 ~20 W to the refrigeration load for a pair of leads.

For a superconducting magnet system consisting of several coils with the same current, one seldom uses a pair of leads per coil, which would be current optimized and counterflow cooled, but rather connects a number of magnets electrically by means of superconducting leads. Also, the transfer of helium may be carried by one transfer line to several cryostats or one may place several magnets in the same cryostat. This way we may essentially use the above limit of cooling requirement of 20 ~30 W per meter magnet, which would include helium transfer and current lead losses.

The drawback of this scheme would be the behavior of the magnets. Since each magnet has a certain stored energy, if one magnet in the cryostat quenches the helium in the cryostat may evaporate partially and can effect the performance of other magnets through the sudden pressure rise in the cryostat, and due to inductive coupling between coils. The control system of such an arrangement has to be studied rather carefully and each magnet shielded separately.

Of particular interest is also the instantaneous value of the hysteretic losses in a cylindrical conductor which can be expressed by

$$W_{hl} = \frac{8}{3\pi} \, \mu_o \, VJ_c \frac{d}{4} \dot{H}_{ext} \left[1 + \left(\frac{I_T}{I_c} \right)^2 \right].$$

For a constant value of H_{ext}, only $J_c(H)$ depends on the field level. The critical current density in individual superconducting filaments is particularly high at low flux densities, resulting in high hysteretic losses in the filament fundle at the outside circle of a composite conductor. At higher fields current sharing between filament bundles will occur, as $J_c(H)$ is lower and the inner filament circles have to take over.

The heat transfer due to the average losses over the time length of a pulse presents no major difficulties. The heat generated due to the instantaneous loss values, mainly at low field levels, may pose a problem due to the thermal diffusivity of the system.

When magnets are operated, individual strands or turns may move under the influence of Lorentz forces. As stated, one effective way to prevent wire motion was to impregnate the coil in a thermosetting. Considerable work has been done in selecting and testing suitable epoxy-filler systems, which have a high thermal conductivity at 4.2 K ($\sim 3 \times 10^{-3}$ W cm^{-1} K^{-1}) and practically the same thermal contraction coefficient as the composite conductor. Pockets of pure thermosettings must be avoided between turns and pancakes. Since it is difficult to impregnate coils with cable or braid conductors with filled thermosettings, the potting of the coil can be performed in two steps: a) impregnation with low viscosity pure thermosetting under vacuum; b) potting the coil with the same but filled thermosetting under vacuum, while the first epoxy is still liquid ($\eta < 1000$ cP). By applying pressures ($\Delta p > 10$ kp/cm^2) after the coil is potted, the filled epoxy will penetrate the coil and replace the pure epoxy in the larger ducts and spaces. The thermosetting can be cured while the pressure is applied to the coil.

The coil structure must be rigid to withstand the radially outward magnetomechanical forces, which are on the order of $6 \sim 8 \times 10^4$ kp/m in a 4.5 T coil with an inner diameter of 8 cm and an outer diameter of 16 cm. These forces are restrained by means of strong supports (clamps, bandages of glass reinforced epoxies) around the circumference of the coil.

In all pulsed magnets with potted or unpotted coils, training has been observed after the coil was cooled down and operated for the first time. In a few coils the design current, which may be 90% of the short sample critical current value, and the design field could be reached after many quenches (in some coils after up to 100 quenches). The coil does train much less after a second or further cool down. It is interesting to note that when the coil is energized for the first time after the first cool down, a quench on the order of $0.3\ I_{max}$ has been measured for many magnets. Early studies explained this premature quenching to local overheating of the conductor. Due to the poor thermal diffusivity of the insulation the heat could not be conducted and dissipated fast enough to the bath. The possibility that the thermosetting may crack while exposed to time variable Lorentz forces and that the released strain energy may account for local heating leading to a premature quench is currently accepted for potted coils. Training has also been observed in unpotted coils where the braid was exposed to helium. In this case we may explain training by wire movement. Due to frictional losses heat is generated driving the coil normal. When the coil has established its position physically, wire motion is reduced to values which do not lead to extended training. While the effect of strain energy in thermosettings on training can be greatly reduced by utilizing new types of epoxies and the use of fillers, the macroscopic wire motion is difficult to eliminate in unpotted coils without the introduction of elaborate and costly structural designs and reinforcements.

It is important to study the effect of a two componnet matrix in a composite conductor on the stability criteria. Generally, individual superconducting filaments embedded in a two component normal matrix have a layer of cupronickel and are then surrounded in a copper layer. Since the thermal diffusivity of cupronickel ($\sim 36\,\%$ Ni) is poor compared to copper, the idea, that first copper should be placed around individual filaments for dynamic stability effects surrounded by supronickel to reduce coupling is certainly correct. The procurement of such a conductor is possible, and will be incorporated in the future.

Two more problems should be mentioned, although practical solutions have not been developed:

1. Insulation breakdown between turns, coils and shields, due to surge voltages. If heat drains are used, faulty insulation on both sides of the heat shields or around the conductor, which may lead to corona discharges and eventual voltage breakdown, must be avoided.

2. Irradiation problems in accelerator or experimental magnets being placed close to targets, collimators or dumps. Irradiation damage may occur also from mis-steering the beam of charged particles. Experimental results indicate that the heating of the conductor through deposited irradiation energy (beam loss) is the most severe problem. A dose rate of only 10^{-2} J deposited into one cm^3 conductor is sufficient to drive the conductor normal, regardless how well the conductor is cooled. Irradiation affects the superconductor (production of defects) and can change the critical temperature and the critical current. The organic insulation is also damaged if about 10^{11} rads are deposited into the coil and conductor insulation. This may lead to interturn and interlayer shorts. Irradiation also changes the resistivity of the matrix material and may lead to a more flux-jump sensitive conductor and thus impair the original dynamic and cryostatic stability of the conductor. The best way to protect the coils against irradiation in areas, where the rate of beam loss is high is by shielding the magnet faces by means of thick iron shields which can absorb the loss of primary and secondary particles of the beam.

REFERENCES

1. DeKlerk, D., The Construction of High Field Electromagnets, Newport Instr. Ltd., U.K. (1965).

2. Parkinson, D.H. and Mullhall, B.E., The Generation of High Magnetic Fields, Plenum Press New York (1967).

3. Montgomery, D.B., Solenoid Magnet Design, Wiley-Interscience, New York, (1969).

4. Brechna, H., Superconducting Magnet Systems, Springer, Berlin, New York (1973).

5. Banford, A.P., The Transport of Charged Particle Beams, E.&F.N. SPON Ltd. London (1966)

6. Durant, E., Electrostatique et Electrodynamique, Masson & Cie., Parks (1953).

7. Zijlstra, H., Experimental Methods in Magnetism, North Holland Publishing Co., Amsterdam (1967).

8. Binns, K.J.and Lowrenson,P.J., Electric and Magnetic Field Problems, Pergamon Press, MacMillan, New York (1963).

9. Blewett, J.P. and Livingston, M.St., Particle Accelerators, McGraw Hill, New York (1962).

10. Neal, R.B., The Stanford Two Mile Accelerator, Benjamin Inc., New York (1968).

11. Montgomery, D.B., Rep. Prog. Phys. $\underline{26}$, 69 (1963).

12. Herlach, F., Rep. Progr. Phys. 31, I, (1968).

13. Brechna, H., Methods of Experimental Physics, Vol. 8. Academic Press New York (1969).

14. Brown, K.L. and Howry, S.K., SLAC 91 (1970).

15. Chester, P.F., Rep. Progr. Phys. $\underline{30}$, Part. II

16. Wipf, S.L., Phys. Rev. $\underline{161}$, 404 (1967).

17. RCA Review, Special Issue on Nb_3Sn, Vol. XXV, 3, (1964).

18. Advances in Cryogenic Engineering, Plenum Press. New York, (1954-1973...).

19. Proceeding of International Conference on Magnet Technology, Stanford (1965), Oxford (1967), Hamburg (1970), Brookhaven (1972).

20. Jour. of Appl. Phys. $\underline{39}$, 6 (1968(, Jour. of Appl. Phys. 42, 1, (1971), Jour. of Appl. Phys. $\underline{43}$,

21. Goree, W.S. and Edelsack, E.A., Stanford Research Inst. (Publication 1967-   ...)

22. Johnson, V.J., Properties of Materials at Low Temperatures, A compendium, New York, Pergamon Press (1961).

23. Appleton, A.D., et al., Proc. Second Internation Conference on Magnet Technology, Oxford, 553 (1967).

24. Wipf, S.L., Proc. 1968 BNL Summer Study Brookhaven III. 1042 (1969).

25. Bean, C.P., Rev. Mod. Phys. 36, 31 (1964).

26. London, H., Phys. Letters, 6, 162 (1963).

27. Hancox, R., Proc. IEE, $\underline{113}$, 1221 (1966).

28.   Hart. H.R., Proc. 1968 BNL Summer Study Brookhaven II (1969).
29.   Green, I.M. and Hlawiczke, P., Proc. IEE, 114, 1321 (1967).
30.   Kim, Y.B. et al., Phys. Rev. 131, 2486 (1963).
31.   Morpurgo, M., Proc. of International Conference on Magnet
       Technology, Hamburg, 908 (1970).
32.   Mills, F.E. and Morgan, G.H., BNL note 17814 (1973).
33.   Green, M.A., LBL Engineering note UCID 3492 (1970).
34.   Rutherford, S.Cond. Magnet Division, Exp. and Theoretical studies
       of Filamentary uperconducting omposits. R.PP/A73 (1969), and
       J. Phys. D3, 1518 (1970).

Chapter 4

SUPERCONDUCTING DC MACHINES

A. D. Appleton

International Research and Development Co. Ltd.

Newcastle Upon Tyne, England

I. BACKGROUND SITUATION

A rotating electrical machine may be manufactured in a few minutes using some copper wire, a battery and a few odds and ends. However, when the power required from a machine is increased, more sophisticated methods are necessary and ultimately the available technology is pushed to its limits. For large direct current motors and generators, the subject of this paper, the limit of power using the conventional design approach is around 10 MW or so and, of course, the use of iron magnetic circuits is essential (we shall see later how essential is the iron if normal temperature materials are employed). The magnetic flux density in the iron may be between 1 and 2 Tesla depending upon the selected machine geometry, and certainly flux densities much below this level are not much use for rotating electrical machines. It is for this reason that the early superconductors (type 1 superconductors) were of only academic interest to machine designers; the best of them is niobium with a practical working level of less than 0.2 Tesla. It is this fact which explains the absence of superconducting machine development from that historic day in 1911 when superconductivity was discovered to the mid 1960's when the first machine (in the power range) was manufactured. Superconducting machines became possible with the development of type 2 superconductors, initially with niobium-zirconium, now with niobium-titanium and perhaps in the future niobium-tin. The potential of superconductors to electrical engineers is graphically illustrated in Fig. 1 which compares the current carrying capacities of copper and niobium-titanium.

Let us consider some of the problems which faced a machine designer when he first came to grips with superconductors. The first necessary condition is that a very low temperature is required.

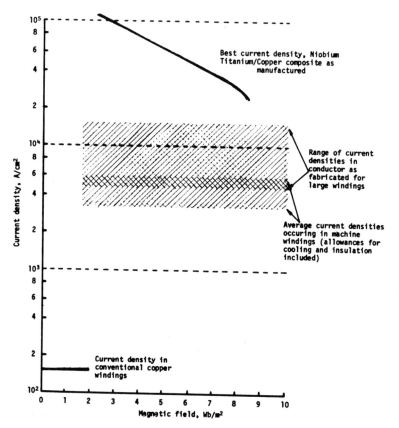

FIG. 1.   The Current Carrying Capacity of Niobium-Titanium Superconductor.

A. Cooling to the Limit

        Improvements in the techniques of cooling electrical machines have
resulted in rather dramatic design and economic advances (particularly
in respect to ac generators). Modern conventional electrical machines
may well employ water cooled copper circuits, and by improved cooling
methods the specific rating of machines is increased. Indeed, it is only
by improving the cooling of conductors that the cost per kVA of machines
has been kept within reasonable bounds. How far, then, can improved
cooling methods be pushed? When the temperature of the cooling medium
is reduced below ambient a penalty is incurred, namely the provision of
refrigeration power. It will be shown later that, in general, it is uneconomic to cool normal conductors, and that benefit is derived only by taking
advantage of the transition to the zero-resistivity characteristic of superconductors. At the very low temperatures required for superconductors
(for niobium-titanium about 4.5 K) a new type of engineering is required,
and electrical designers must understand the subtleties of cryogenics.
This subject will frequently reappear for all the large scale applications.

FIG. 2. An Early Superconducting Coil Using Niobium-Zirconium Wire.

B. Early High Field Superconductors

When niobium-zirconium wire (typically about 0.010 in. diameter)
became available in the early 1960's many laboratories began to experiment with small superconducting solenoids; Fig. 2 shows a coil typical
of the period. The coils were novel and provided the means for numerous
scientific experiments in flux densities of 4 or 5 Tesla or so, but when the
performance of the coils was examined with the critical eye of the industrial machine or magnet designer, the results were not particularly
attractive. The performances achieved were often not consistent, and
there appeared to be almost a black art in making coils perform satisfactorily. The poor performance was due to two main causes: basic
instabilities in the superconductors, and poor cooling conditions. The
resistivity of superconductors when not superconducting is quite high
$(30 \times 10^{-6}$ ohm cm) and the transition from the superconducting to the
normal state was all too readily induced. In this situation it was impractical to consider investing large amounts of money in very large magnetic
systems or in the construction of large electrical machines.

C. Stabilized Superconductor

The development of a solution to the problem of unstable superconductors began by the realization that the thermal capacity of the superconducting coil must be increased and also that an alternative path must be
provided for the current when the superconductor (over small regions)
reverts to the highly resistive normal state. Numerous stability criteria
were involved but the most widely used method was to clad the superconductor with copper with a good metallic bond between the copper and the
superconductor. If, in addition, the copper was provided with a good

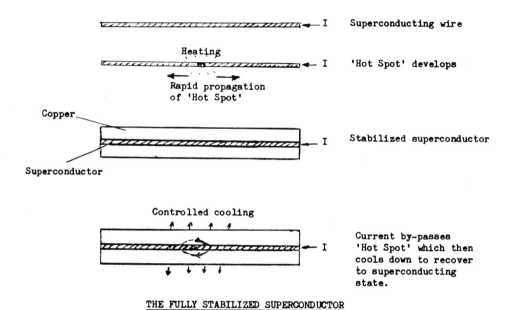

THE FULLY STABILIZED SUPERCONDUCTOR

FIG. 3. The Use of Copper to Stabilize the Performance of Superconductors.

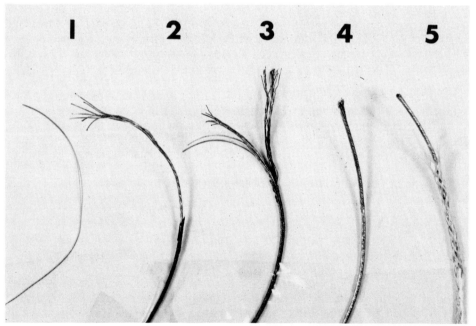

FIG. 4. An Early Attempt to Stabilize a Seven Strand Niobium-Zirconium Cable for the IRD 50 HP Motor.

helium cooling system, a greatly improved performance was achieved. The principle involved is quite simple; if, due to an instability, a small normal region appears in the superconducting wire, this region is short circuited by the much less resistive copper and the current therefore flows in the copper. Now, provided that the heat $J^2 \rho$ generated in the copper can be removed by heat transfer from the copper to the helium coolant, a stable condition exists and the normal region in the superconductor will cool down and once again carry the current. This situation is illustrated in Fig. 3. An early attempt to achieve this condition in a seven strand niobium-zirconium cable is shown in Fig. 4, while Fig. 5 shows a properly engineered fully stabilized niobium-titanium superconductor. The characteristic of the latter is shown in Fig. 6, and it is seen that up to about 1000A there is no voltage across a sample in the field of 3.7 Tesla; when this current is increased a small voltage is measured and a voltage current characteristic may be obtained. An important feature is that the curve is reversible, and if the current is reduced the voltage reduces. What is happening is that the superconductor is on the border line of stability, and the current is continually moving in and out of the superconductor. If the current is increased too much, it will spend more and more of its time in the copper, and eventually the cooling system will be saturated and a breakdown of stability will occur. A similar situation will occur if the rate of change of the current is too high, a most important limitation of the fully stabilized superconductor; the current cannot be increased too rapidly, probably with a time constant of not less than a few minutes. However, the fact that stability could be reasonably guaranteed enabled a number of large magnets to be constructed and, as we shall see later, machine development was initiated.

D. Intrinsically Stable Superconductor

The limitation on the rate at which current may be changed in the fully stabilized superconductor was a nuisance, particularly for dc generators where the most convenient method of changing the output voltage is by control of the field current. It was therefore of considerable benefit when this problem was overcome by the development of the intrinsically stable superconductor. It was shown that if the diameter of the superconductor is made sufficiently small, typically about 20 microns, then the problem of stability is largely eliminated. The new superconductor is made by drawing down a large number of niobium-titanium strands in a copper matrix. This development took place in about 1968 (although it had been predicted earlier) and some of the early materials had 50 or 60 filaments of about 50 microns diameter (Fig. 7); it is now possible to produce conductors with many thousands of filaments of just a few microns diameter. The multi-filament superconductor enables magnet windings to be produced in which the current may be changed from zero to its maximum value within a second or two without incurring unstable behavior and without excessive hysteresis losses. We are, however, still a very long way from having a high field superconductor which will operate satisfactorily at power fre-

FIG. 5. A Fully Stabilized Niobium-Titanium Superconductor (IMI LTD).

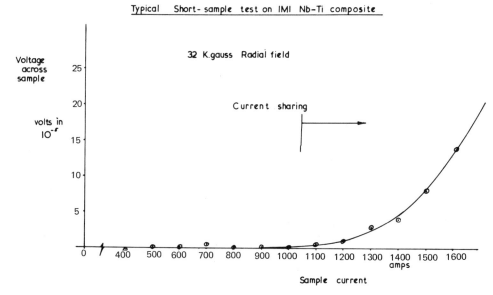

FIG. 6. The Performance of the Fully Stabilized Superconductor Shown in
Fig. 5.

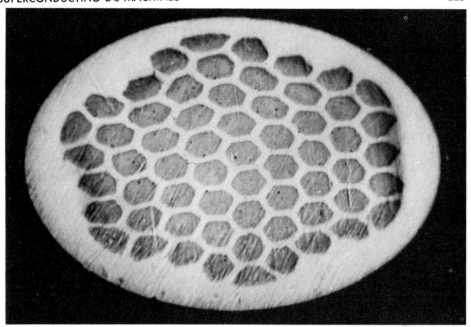

FIG. 7. A Multi-Filament Fully Stabilized Niobium-Titanium Superconductor.

quencies; however, it will be shown later that this is not very significant
for electrical machine designers. The availability of the multi-filament
superconductor means that windings may be encapsulated in an epoxy resin,
rather than having to be in intimate contact with helium, and this greatly
simplifies the design and manufacturing problems. The most important
fact to note, however, is that the superconducting materials now available
commercially are capable of achieving the performance required for electrical machines, and there is nothing to prevent the latter from being
developed for a wide range of industrial applications. We shall touch upon
the design of superconducting windings as we proceed with our studies of
dc machines.

E. Limitations of Conventional DC Machines

The first dc machine was that produced by Michael Faraday in 1831 -
the Faraday disc or homopolar (also known as acyclic) machine. The
simple principle illustrated in Fig. 8 consists of a copper disc rotating
between the poles of a permanent magnet so that it cuts the magnetic field
at right angles. Unlike the now more commonly used heteropolar machines,
the rotating conductor always cuts the magnetic field in the same direction;
thus the generated emf (which is the rate at which the magnetic flux is cut)
is always in the same direction, i.e., either radially outwards or radially
inwards, according to direction of rotation and the direction of the magnetic
field. To produce a machine it is necessary to provide electrical brushes

on the outside perimeter and also at the center of the disc.

Let us consider a simple example; if we take a disc of 1 meter diameter in a uniform magnetic flux density of 1 Tesla, rotating at 3000 rev/min, the generated voltage between the axis and the perimeter is about 39 volts. Now, if we wish to increase the voltage of the machine, it is necessary to provide a number of discs, each with a set of electrical brushes, and connect them in series. This is quite different from a heteropolar machine where, because the rotor conductor (armature) passes through a magnetic field first in one direction and then the other (Fig. 9), it is a simple matter to connect a large number of conductors in series. It was the difficulty of achieving a reasonable terminal voltage that prevented the wider exploitation of the homopolar machine, but it has been used as a generator where a large current is required at a relatively low voltage.

Consider a heteropolar dc machine with an armature of diameter D meters, length L meters, maximum air gap flux density B_g Tesla, armature conductor current I_a, total armature conductors Z_a, and ratio of average flux density over a pole pitch to maximum air gap flux density K_f.

In one revolution the total flux cut by a conductor is given by
$\Phi = \pi DLK_fB_g$ (Webers); the work done in one revolution is ΦZ_aI_a

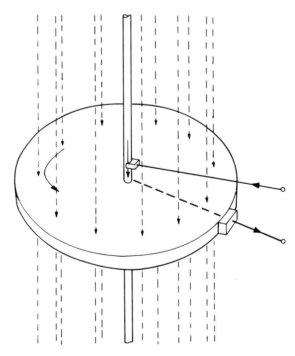

FIG. 8.   Schematic of a Simple Homopolar Machine. Disc Rotates in a Magnetic Field.

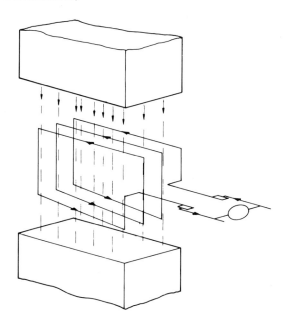

FIG. 9. Schematic of a Heteropolar Machine. Armature Passes Through
         the Magnetic Field in One Direction, then in the Other.

Defining specific electric loading as q ampere conductors per meter
of circumference we have $I_a Z_a = \pi D q$ and the work done per revolution
becomes $D^2 L(\pi^2 q K_f B_g)$.

If the machine power is P kW at N rev/min

$$\frac{P}{N} = D^2 L(0.164 q K_f B_g \times 10^{-3})$$

The term in brackets is called the Output Coefficient (ESSON) and
its numerical value does not vary widely for a given class of machine.
The value of B_g is typically about 1 Tesla; K_f is about 0.7 and a high value
of q for large machines is about 50,000. These values yield

$$\frac{P}{N} = 6D^2 L.$$

The limiting values of D and L are determined by, _inter alia_, periph-
eral speed and voltage between commutator segments.

The output coefficient of 6 applies to large diameter machines, e.g.
D = 4 meters and falls with smaller diameters. Figure 10 shows the maximum outputs which are possible with conventional machines as a function
of speed. The outputs are derived from published data which is a few years

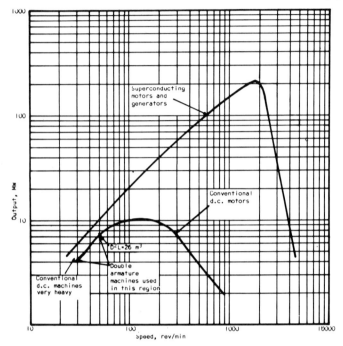

FIG. 10.   Maximum Outputs of Conventional and Superconducting dc
             Machines.

out of date, but it is understood that present day capabilities are not much
different from those indicated. The outputs may be increased perhaps by
60% if duplex lap windings are employed, but some manufacturers regard
these as troublesome. The maximum power is in the region of 10 MW at
speeds of a few hundred rev/min. At lower speeds the machines become
very heavy, i.e. hundreds of tons, and double armature machines are
employed for certain ratings.

In practice, most dc machines are motors because dc power is
achieved more readily from solid state rectifiers; the large motors are
used where a variable speed drive is required, such as in a steel rolling
mill. There are a number of cases where, although a variable speed
drive is required, a constant speed (or perhaps a two speed) ac motor
is used because dc motors are not available at the required ratings.

F. Limitations of Superconducting Homopolar Machines

It is not possible to produce an output coefficient for a superconducting machine in the relatively simple form of the conventional machine.

The armature conductors of either a disc or drum-type homopolar
machine will cut a total flux Φ in one revolution (corresponding to $\pi D K_f B_g$
for conventional machines). However, there is no simple relationship

between Φ and armature diameter in a superconducting machine. The specific electric loading in a conventional machine refers to ampere conductors in the armature, but for superconducting machines it is more useful to define q as the total slipring current per meter of circumference.

The work done per revolution becomes

$$\frac{P}{N} = \frac{\Phi\pi Dq \times 10^{-3}}{60}$$

Since Φ is defined as the total flux between the sliprings of either a disc or a drum-type machine, this expression applies only to a machine with one pair of sliprings.

This expression may be simplified to

$$\frac{P}{N} = Cq\Phi D$$

where the value of C is defined in Fig. 11. For a simple disc machine with 2 pairs of sliprings (Fig. 11b)

$$C = \frac{2\pi\, 10^{-3}}{60} \approx 10^{-4}.$$

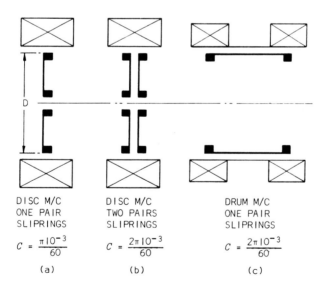

DISC M/C ONE PAIR SLIPRINGS	DISC M/C TWO PAIRS SLIPRINGS	DRUM M/C ONE PAIR SLIPRINGS
$C = \dfrac{\pi\, 10^{-3}}{60}$	$C = \dfrac{2\pi\, 10^{-3}}{60}$	$C = \dfrac{2\pi\, 10^{-3}}{60}$
(a)	(b)	(c)

FIG. 11. Effect of Machine Geometry on Factor C.

Therefore $\qquad$ $\dfrac{P}{N} = q\Phi D \times 10^{-4}$

Let us take as an example a homopolar machine with a major slip-ring diameter of 2 meters and an average flux density in the bore of 3 Tesla; if we assume that the diameter of the minor slipring is 0.5 meters, the value of Φ is $\frac{\pi}{4}(2^2 - 0.5^2)3 = 8.8$ Wb. Taking a speed appropriate to a diameter of 2 meters, say 800 rev/min, we have $P = 0.7q$ kW. Thus, if we require a power of 30 MW, the value of q required is about 43,000 A/m or 430 A/cm; if the brush current density is limited to that of a copper graphite brush, say 200 A/in^2 (31 A/cm^2), the axial length of the slipring becomes about 14 cm. In practice, because it is not possible to cover the entire circumference of the slipring with the brushes, this length would be more like 20 cm, which is becoming rather large. This simple example illustrates the power that may be obtained from a superconducting machine, but shows the desirability of an improved method of current collection. The probable design limits for superconducting dc machines are also shown on Fig. 10. The problems of current collection and a further discussion on machines and ratings are dealt with in later lectures.

II. BASIC DESIGN OF SUPERCONDUCTING DC MACHINES

In the first section we examined the limitations of conventional dc motors and indicated how machine ratings could be increased by the use of superconductors with the homopolar type of geometry. In this section we will consider the design of the superconducting machine in more detail and build up the picture towards the realization of practical hardware. For more details see references [1] to [15].

A. Elimination of Iron Magnetic Circuits

The ability of superconductors to carry very high current densities means that a field winding for a machine may be produced with a very large number of ampere-turns. The fact that the superconductors can maintain this current density in the presence of high magnetic fields means that electrical machines may be produced without the use of iron magnetic circuits and, therefore, without the limitations that are imposed by the latter.

Let us examine this with a simple example; consider a straight conductor carrying a current i and producing a flux density of B Tesla at radius r from the conductor.

We have $\qquad$ $i = \int \vec{H} \cdot \vec{dl}.$

$\qquad\qquad\qquad\qquad$ $\vec{B} = \mu_0 \vec{H}$

then
$$i = \frac{2\pi r B}{\mu_0}$$

$$= rB \times 10^7 \text{ A (or A-turns)}$$

If $r = 0.5$, $B = 2$ Tesla, then $i = 10^7$ A-t; if the air magnetic circuit is replaced by iron, the value of i drops to about 10^4 A-t. Of course, if B is
required to have a much higher value of about 4 Tesla, the value of i is
2×10^7 A-t whether or not iron is employed and, therefore, of course,
there is no point in using iron.

Table 1 shows how impractical it is to consider using an air magnetic
circuit with normal conductors; it relates to the field condition for the Fawley motor which is to be described later. If the superconductor is removed
from the composite material to leave only copper operating at room temperature, the situation is illustrated in Table 1.

Table 1: Characteristics of Fawley motor assuming normal copper conductor only.

Current Density A/cm^2	Weight of Copper Tonnes	Useful Magnetic Flux Wb	Power Consumption kW
4000*	5.1	6.46	14 000
1500	19.7	5.45	5 000
620	36.7	4.40	2 400
155	38.3	2.00	713

*the condition for the Fawley motor is the first row.

It is seen that an attempt to reduce the power consumption by reducing the current density rapidly increases the weight of copper and brings
down the useful flux of the machine. The position is not improved if an
attempt is made to operate the normal conductor at cryogenic temperatures of, say, 77 K (liquid nitrogen) or 20 K (liquid hydrogen) because although the joule losses are lower, the refrigeration power must be supplied.

At 77 K power consumption = 30 MW.
At 20 K power consumption = 20 MW.

We reach the conclusion that, for large machines at least, an air magnetic
circuit is possible only by the use of superconductors.

What limitations do we have with superconductors as far as electrical
machines are concerned? One of these is quite certain - they will not operate at power frequencies; it will emerge in later sections that for dc
machines, full benefits may be achieved if only the excitation winding is made
superconducting.

We have, therefore, a means for providing the excitation of electrical machines without the use of magnetic iron, and we will now consider the form of the design.

B. Superconducting Heteropolar DC Machines

We may ask why are superconductors employed on homopolar machines and not on the more widely used heteropolar machines? Considering the two-pole machine shown in Fig. 9, the armature current produces a flux which links with the excitation field and, of course, the unbalance of magnetic flux density on each side of the armature conductor produces the force which develops the motor torque (or the load of the prime mover if the machine is a generator). Since the armature current is time varying the net result is that the superconducting field winding will be subjected to a time varying magnetic flux; this was a good reason for avoiding the heteropolar machine in the early days of unstable superconductors, and is certainly not attractive at present. The next fact to note is that the superconducting winding is subjected to the full torque reaction of the machine, and must therefore be supported through to the machine foundations; the effect of this is increased refrigerator load. To some extent the two problems may be alleviated by providing a third winding, as shown in Fig. 12. If the third winding carries the armature current, it will relieve the field winding of the torque reaction and reduce the armature reaction effects. However, even if a satisfactory arrangement could be conceived it would be complex and inefficient, and, furthermore, there are other problems. To achieve an uprated armature with higher flux densities and currents it is necessary to avoid the use of magnetic iron, or commutation will be an extremely serious problem; it is therefore necessary to produce non-magnetic laminations for the rotor, and even then it is unlikely that the eddy current problems will be overcome. In the homopolar machine these problems do not arise, but there is substituted the difficult problem of current collection which is discussed in section III.

C. Superconducting Homopolar Machines

The basic features of a disc-type superconducting homopolar machine are shown in Fig. 13. It utilizes a single field winding enclosed within a cryostat to minimize the heat leak to the winding by conduction and radiation. The magnetic field produced by the superconducting winding is a maximum at the inside surface of the winding and, in the bore, falls to a minimum on the axis. The armature consists of a Faraday disc with sliprings near the shaft and at the outside of the disc. Means for collecting the current at the sliprings are provided, and a stationary disc takes the current radially between the outer slipring and the machine terminals. The useful magnetic field of the machine is that which cuts the Faraday disc between the sliprings, and Fig. 14 shows why it is important for the disc to occupy as much of the bore as possible. A small increase in the diameter

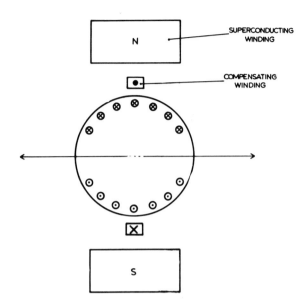

FIG. 12. Superconducting Heteropolar Machine with Compensating Winding.

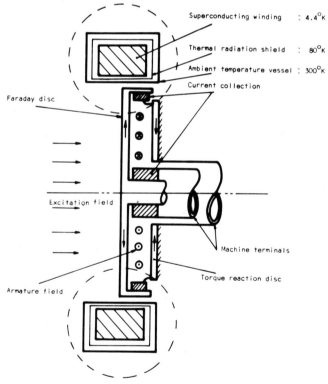

FIG. 13. Basic Features of Superconducting Homopolar Machine.

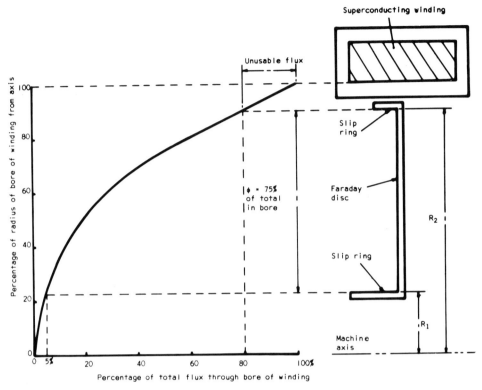

FIG. 14. Curve Showing Flux which cuts Faraday Disc in a Superconducting
Machine.

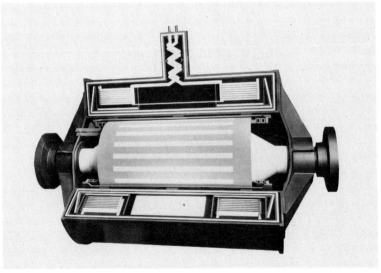

FIG. 15. Drum Type Superconducting Machine.

of the Faraday disc produces an increasing increment of useful flux as the diameter is increased. In the space between the field coil and the Faraday disc there must be accommodated the coil former, a vacuum space, a radiation shield, a further vacuum space, the thickness of the ambient temperature casing of the cryostat and a clearance gap for the rotating armature. Using conventional type graphite brushes for current collection, there would be a substantial loss of useful flux if these were arranged in the conventional manner. To avoid this loss the slipring is designed with an over-hung lip and the slipring is at the underside, i.e. the brush pressure is radially outwards (Fig. 13).

The stationary disc (Fig. 13) carries the armature current through the same magnetic field as the Faraday disc and therefore experiences the same torque as the latter. In other words, the stationary disc takes the torque reaction of the machine and none of this force appears on the field winding. It may also be noted from Fig. 13 that the magnetic field produced by the armature current is in the azimuthal direction between the Faraday disc and the torque reaction disc. This flux is in quadrature with the excitation field and does not give rise to any armature reaction.

Thus, the superconducting winding is not subjected to any mechanical or magnetic influences, precisely the situation required.

The basic features of a drum-type superconducting homopolar machine are shown in Fig. 15. The working conductor is now a drum and the armature current flows in the axial direction. The sliprings are now at the same diameter and located near the mid-plane of each coil. The torque reaction is taken on a stationary cylinder co-axial with the armature. Means must be provided to accommodate the large force between the windings, and this is most readily achieved if they are contained within a common cryostat.

The choice of the homopolar machine remains the most attractive for superconducting dc motors and generators and it is most unlikely that the heteropolar type will be able to displace it. There are two major problems associated with homopolar machines; efficient current collection and low generated voltage. It will be shown later that the performance of the current collection system has a profound effect upon the optimization of a design for a given rating. The problem of low voltage has been overcome in a manner which will be described later in section III.

D. Calculation of the Magnetic Field

The calculation of the magnetic field of the coil of a homopolar machine is straightforward but tedious and is therefore best performed by a computer. The requirements are to determine the total flux which links between the sliprings of the Faraday disc and to calculate the maximum magnetic flux density in the winding, since it has a profound effect upon the cost of the winding. There are two methods available for the calculation.

At distances greater than about two or three times the stator radius it is convenient to use the following method to find the magnetic field. In polar coordinates the magnetic potential at any point r, θ, due to a current i flowing in a coil of radius A is given by

$$V = \frac{i}{2}\left[\frac{1}{2}\left(\frac{A}{r}\right)^2 P_1(\cos\theta) - \frac{3}{8}\left(\frac{A}{r}\right)^4 P_3(\cos\theta) + \frac{15}{48}\left(\frac{A}{r}\right)^6 P_5(\cos\theta) - \ldots\right],$$

the origin being taken at the center of the circle and the axis of the circle being the line $\theta = 0$, and $P_n(\cos\theta)$ is the Legendre polynomial of nth degree.

The expression is valid for r > A and a similar expression is available for r < A but the latter is not convenient for present purposes because the series does not converge very rapidly.

The magnetizing force in the directions r and θ is obtained as

$$H_r = \frac{-\partial V}{\partial r}$$

$$\text{and} \quad H_\theta = \frac{1}{r}\frac{\partial V}{\partial \theta}.$$

Thus

$$H_r = \frac{i}{2A}\left[\left(\frac{A}{r}\right)^3 P_1(\cos\theta) - \frac{3}{2}\left(\frac{A}{r}\right)^5 P_3(\cos\theta) + \frac{15}{8}\left(\frac{A}{r}\right)^7 P_5(\cos\theta) - \ldots\right]$$

$$H_\theta = \frac{i}{2r}\left[\frac{1}{2}\left(\frac{A}{r}\right)^2 P_1{}'(\cos\theta) - \frac{3}{8}\left(\frac{A}{r}\right)^4 P_3{}'(\cos\theta) + \frac{15}{48}\left(\frac{A}{r}\right)^6 P_5{}'(\cos\theta) - \ldots\right]$$

$$\text{where } P_n{}'(\cos\theta) = \frac{dP_n(\cos\theta)}{d\theta}.$$

The radii of the Legendre polynomials are given in published tables, e.g., Tables of Functions, Jahnke and Emde.

$$P_1(\cos\theta) = \cos\theta, \qquad P_1{}'(\cos\theta) = -\sin\theta,$$

$$P_3(\cos\theta) = \frac{1}{8}(5\cos3\theta + 3\cos\theta), \quad P_3{}'(\cos\theta) = -\frac{3}{8}(5\sin3\theta + \sin\theta).$$

For points close to the winding, for example, if the value of $\frac{A}{r}$ (or $\frac{r}{A}$) is greater than about 0.5, the expressions involving the Legendre polynomials do not converge very rapidly and a more convenient method is required to calculate the magnetic field strength.

Using Cartesian coordinates, let the coil be in the plane z = 0; the coil radius is at y = A (see Fig. 16).

$$H_z = \frac{i}{2\pi[(A+y)^2+z^2]^{\frac{1}{2}}} \left[K + \frac{A^2-y^2-z^2}{(A-y)^2+z^2} E\right] \quad \text{and}$$

$$H_y = \frac{i\,z}{2\pi[(A+y)^2+z^2]^{\frac{1}{2}}} \left[-K + \frac{A^2+y^2+z^2}{(A-y)^2+z^2} E\right]$$

where K and E are elliptic integrals of the first and second kind and of modulus k,

where $\quad k^2 = \dfrac{4\,A\,y}{(A+y)^2+z^2}$. The values of the elliptic integrals

are given in Tables of Functions, Jahnke and Emde.

Where the magnetic field strength is required for calculating the machine performance, only the values of H in the plane z = 0 are required. Thus $H_y = 0$.

Writing $\Delta = \dfrac{y}{A}$,

$$H_z = \frac{i}{2\pi A}\left[\frac{K}{(1+\Delta)} + \frac{E}{(1-\Delta)}\right].$$

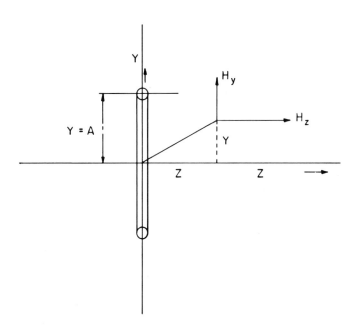

FIG. 16.   Definition of Cartesian Coordinators for the Coil.

The choice of method depends upon the distance of the point at which the flux is being calculated from the winding. The computer may be programmed to calculate the contribution of an element of current to the magnetic field at any point. The size of current selected depends upon how close to the winding the field is required. The program also integrates the flux between the sliprings; there is no real difficulty with these calculations. It is interesting to note, however, that a "back-of-the-envelope" calculation will produce the answer to within 10% or so.

E. Geometry of Armature

It will be apparent that one of the sliprings must be in the high field region as close as possible to the field winding. However, the other slipring may be at any convenient position as indicated in Fig. 17. Consideration of Fig. 17 will show that there is an advantage in moving the low field slipring away from the center line of the machine; this is because the magnetic flux lines are moving away from the axis and therefore the further away the low field slipring is moved the larger the radius it may assume without losing any useful flux. In fact, it if is moved a distance equal to the bore of the winding, it will have the same radius as a high field slipring. It is important to note that the profile of the slipring must follow the direction of the flux lines otherwise a voltage will be generated across the surface of the slipring and this will drive a circulating current through the brushes.

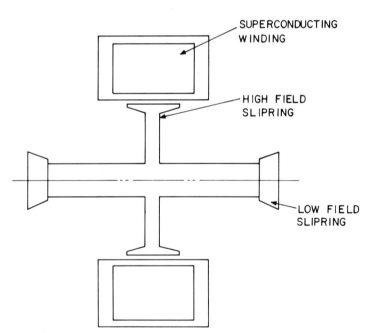

FIG. 17. Schematic Showing Position Low Field Slipring.

F.   The First Superconducting DC Machine

The first embodiment of the above ideas was in the 50 hp motor
built by IRD for MOD(N) in 1966, often referred to in IRD publications as
the model motor. It was required to demonstrate that the principles discussed could be achieved in a practical manner, and the design of the model
motor was commenced on 1st January 1965 and the machine was commissioned on 1st June 1966. The principle employed is as indicated in Fig. 13
and the principal features will now be discussed.

When the superconductor and coil former were designed, niobium-
zirconium was the most suitable superconductor, and Nb-25%-Zr in the
form of wire of 0.010 in. diameter was chosen for this machine. Stabilization of superconductors was not then formalized and it was decided to
reduce possible degradation of the coil performance by laying up the wires
in the form of a seven strand cable. When it later became apparent that
copper would increase the stability of the coil and help to protect the superconductor in the event of a quench, an outer wrapping of copper wires was
wound onto the cable and the whole bonded together by indium. The cable
was then insulated by braided glass fiber with small spaces to permit
some helium access. The resulting conductor was only partially stabilized
and the coil quenched at 240 A in a maximum field of 2.7 Wb/m^2 which is
somewhat below the short sample limit. Thereafter, in experiments on the
motor the current was limited to 230 A to avoid quenching.

The entire cryostat was of stainless steel with the annular coil vessel
suspended from a 12 in. long neck tube of 2 in. diameter and 0.010 in wall
thickness inside the chimney. Figure 18 shows the coil box inside the cryostat which has the end plates removed. A liquid nitrogen cooled radiation
shield surrounded the coil box in the space between the latter and the cryostat outer vessel.

The diameter of the cryostat bore was fixed by cost considerations
at 15 in. and, with the flux available in it (the expected flux was obtained
in operation), it was possible to generate a total of about 10 V at 2000
rev/min. This meant that to attain 50 hp something approaching 4000 A
armature current was necessary, and this current would have to pass
through four sets of sliding contacts in the armature circuit.

The development of current collection techniques is discussed in
the next section, but as a result of the tests it was apparent that the sliprings of the model motor would require water cooling and that 1% chromium-
copper would be the most suitable slipring material because of its good
thermal conductivity, although it was not the best material to minimize
brush losses. Cupro-nickel had proved much better for losses, but its
inferior thermal conductivity would not permit its use in the cooling configuration of the model.

All of the brush materials tested were basically mixtures of copper

FIG. 18. Superconducting Winding Inside Cryostat of 50 HP Superconducting Model Motor.

and graphite. Morganite grade CM1S proved to be most suitable for inner
brushes and CM2 for the outer brushes (because of the difference in slipring speeds). However, because of supply difficulties CM1S was used
throughout the machine. This departure of both brush and slipring materials from the ideal caused considerably increased friction at the outer
brushes with a reduction in the motor efficiency.

The highest recorded efficiency of the model motor was 69% and with
the ideal brush slipring combination this would have approached 80%. The
greatest source of losses in the motor is the brushgear and the importance
of minimizing brush losses in large machines is very clear. The relatively low efficiency of the model motor is due largely to the small scale of
this machine. Large motors can, despite their brush losses, achieve
efficiencies well above 95%.

Tests were conducted to try to measure an armature reaction effect,
but none was detectable as would be expected from the arrangement of the
armature conductors. It was also quite apparent that there was no force
on the superconducting coil due to the armature currents as the coil was
satisfactorily supported from above only by the very thin walled tube.

Some of the experiments on the motor were conducted with iron in
the magnetic circuit (placed in or near the cryostat bore and arranged
symmetrically about the coil). An increase of about 20% was obtained in

the maximum flux using about 400 lb. of iron, and a maximum power output was, of course, obtained with the iron in use.

The performance figures of the motor under typical running conditions are shown in Table 2 which also shows the figures obtained at maximum power output.

Table 2: Major parameters of the model motor.

Typical running conditions	
Armature current I_a	3500 A
Field current	200 A
Speed	1900 rev/min
Voltage drop in busbars	0.75 V
Voltage drop across rotor	0.99 V
Voltage drop in armature conductors	0.20 V
Total of brush voltage drops	0.79 V
Generated voltage (E_b)	8.36 V
Terminal voltage	9.35 V
Total brush electrical loss	2.76 kW
Total armature electrical loss (incl. brushes)	3.50 kW
Developed power ($E_b \cdot I_a$)	29.3 kW
Maximum load conditions	
Terminal voltage	10.74 V
Generated voltage	9.45 V
Armature current	3800 A
Developed power	35.9 kW

The output power was measured on calibrated generators used as loads for the motor.

G. Preliminary Optimization Methods

We will continue now our discussion of the expression developed in section I,

$$\frac{P}{N} = q \, \Phi \, D \times 10^{-4},$$

which relates to a homopolar machine with two Faraday discs as in the 50 hp model motor described above. When consideration is given to the values of the product ϕ D it is found that there are technical and economic limits according to the value of D (which is closely related to the bore of the field winding). Figure 19 shows an approximate range of values for ϕ D as a function of D. Consider the case of a machine of 40,000 kW at 3000 rev/min; there are limits to the maximum practical values of D and selecting a value of D = 0.8 we see that ϕ D is about 1.4. For a value of C = 10^{-4}, curve I, Fig. 19, shows that the value of q is about 10^5 A/m. The tabulation to the right of Fig. 19 shows the importance of achieving a high brush current density. For q = 10^5 A/m, a brush current density of only 310 kA/m² (200 A/in²) would mean that the slipring must have an axial length of at least 0.32 m. This is impractical and therefore it is seen that brush current density is a limiting factor for the rating of super-conducting dc machines. Curve II on Fig. 19 shows that a greater choice of design parameters is available if the speed of the 40,000 kW machine is reduced to 1500 rev/min. Curve III is for a slow speed motor and if we

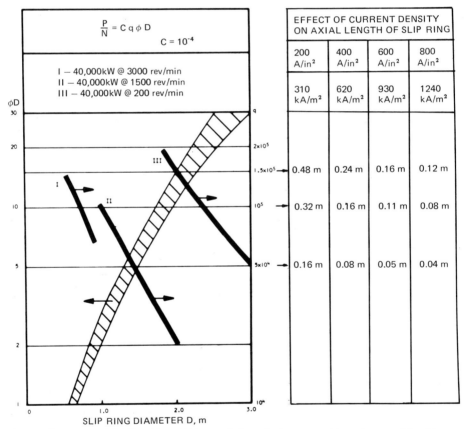

FIG. 19. Curves Showing Effect of Brush Current Density on Machine Design.

select D = 2, the value of Φ D is about 14 thus Φ = 7.0 Wb. The stage
voltage V = ΦN/60 and for these parameters V = 23 V; here we see the
advantage of the IRD segmented slipring approach. With a series/parallel
segmented system a terminal voltage of about 1 kV is feasible. These
simple examples show the need for a high brush current density and the
advantage of the segmented slipring system. The importance of a low
contact voltage drop is also evident; in the case of the low speed motor
above, each 0.23 V drop will account for 1% of machine efficiency.

With this brief insight into the design we will consider some of the
more important design problems in section III.

III. PROBLEM AREAS

The commercial exploitation of the superconducting homopolar
machine cannot be achieved without providing solutions to a number of problems which are the subject of this section. Fortunately, with a determined effort and faith that these machines have an important role to
play on the industrial scene, the solutions are being found. The problems
which have been, or are being, tackled are as follows: 1) low voltage of
the armature, 2) current collection and 3) magnetic screening. There are
many other problems which have presented themselves and been overcome,
but in this section we will confine ourselves to the above.

A. Low Voltage of Homopolar Machines

The voltage in a Faraday disc is generated between the sliprings and
for the present discussion we may assume that these are in the plane of
the disc with diameters D and d. The problem to be solved is the achievement of a higher voltage without employing a large number of Faraday discs.
A fairly obvious step is to divide the disc (Fig. 20) into a number of segments which are insulated from each other, but how is it possible to connect them in series in a practical manner? The solution is shown in
Figs. 21 and 22 (the disc is opened out for clarity). In Fig. 21 a brush is
placed upon every other segment and the brushes are connected as shown.

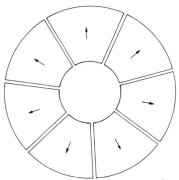

FIG. 20. Schematic Drawing of a Faraday Disc.

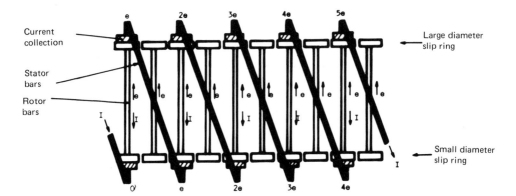

FIG. 21. Principle of Segmented Slipring (Position with Brushes on the
Segment).

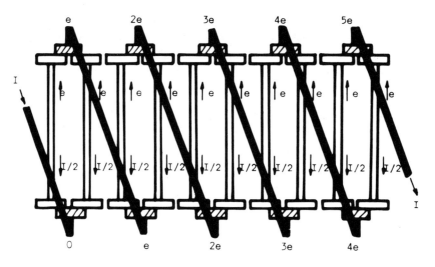

FIG. 22. Principle of Segmented Slipring (Position with Brushes on the
Two Segments).

If the voltage generated in the Faraday disc is e it can be seen that the voltage of adjacent brushes is 0, e, 2e, 3e, etc., and that a series addition of voltage is obtained. The segments in Fig. 20 which are not in contact with the brushes are floating electrically and, of course, no current flows in them. In Fig. 21 the rotor has moved such that the brushes are now bridging the gaps between adjacent segments and the current previously flowing in one segment (and rotor conductor) is now shared between two segments. When the rotor moves further the condition of Fig. 20 is achieved again. This system provides, therefore, a means of achieving the series addition of voltage which will operate at whatever position the rotor happens to stop, and provides insulation between the different voltages of the segments. This arrangement is, however, not completely satisfactory because consideration of Fig. 20 will show that there must be a position on the slipring where the voltage between adjacent segments is ne, where n is the number of voltage stages. It is common practice, and a good practice, that the voltage between segments of commutators should not exceed about 30V and clearly this is a serious limitation to the design thus far presented. The solution to this problem is shown in Fig. 22 where the brushes are arranged in two parallel circuits around the slipring. Consideration of Fig. 22 will show that nowhere does the voltage between adjacent segments exceed the voltage generated in one stage. Figure 23 shows the positions of the connections to the sliprings when the arrangement of Fig. 22 is employed. Since it is necessary for the power supply to the motor to be in the form of a co-axial circuit it is necessary to provide a connection from the point where the current leaves the low field sliprings back to the supply to the high field sliprings. This is achieved as shown in Fig. 23. Consideration of Fig. 20 shows that the segmented slipring concept provides the equivalent of a single turn of armature current which, of course, introduces an element of armature reactance. The arrangement of Fig. 22 changes this position to two loops of current in opposite directions between the terminal points, and consideration of Fig. 23 shows that the connection between the terminal points can completely compensate for the current loops. Thus a solution to the armature voltage problem is found which is satisfactory in all respects.

A further feature of this system, however, requires some discussion, namely the switching of the current between the segments. It will be noted that the system requires a switching action and not commutation as in the heteropolar machine. This feature of switching was very carefully studied because it was feared that it might give rise to sparking at the brushes; however, it was shown theoretically and later demonstrated practically, that provided the rotor conductors are not embedded in magnetic iron their reactance is too low for switching problems to be encountered.

B. Current Collection

1. Solid brushes

The importance of current collection has already been mentioned in the two previous sections and was realized at the very start of the IRD

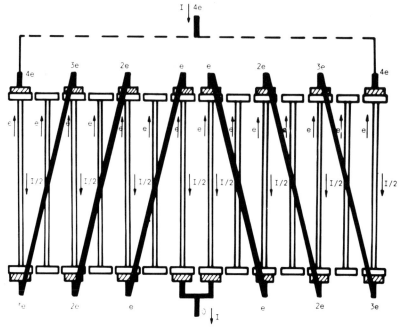

FIG. 23. Principle of Segmented Slipring (Series Parallel Arrangements).

program on the development of superconducting machines. At the time
that the 50 hp model motor was being developed, work was initiated to
determine the limits of performance for solid brushes, particularly those
with a high copper content. Several brush grades and slipring materials
were tested under conditions of homopolar machines, that is, with a large
number of brushes on a single slipring. Measurements were made of contact voltage drops, friction losses, wear rates, current sharing, and
cooling requirements. Figure 24 shows one of the test rigs employed for
this purpose and it may be seen that a current shunt was used in series
with each of the brushes. It is necessary for satisfactory operation of
conventional brushes that the slipring track forms a stable film of oxide
and it was thought that a large number of brushes running on the track
might inhibit this film formation.

It was found that the brushes could be run with no deterioration of
performance with a 50% brush cover of the slipring track. There were
indications that significantly more cover of the slipring with brushes led
to increased slipring wear. As a result of the tests it was decided to
employ a 1% chromium copper for the sliprings of the 50 hp model motor,
and also that the sliprings would have to be water cooled. Later tests
showed that the preferred material, cupro-nickel, was acceptable, and this
was used on the Fawley motor to be discussed in the next section.

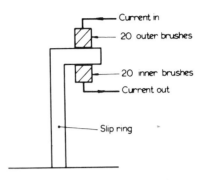

TEST RIG WITH SLIP RING REMOVED   ARRANGEMENT OF SLIP RING AND BRUSHES

(a)   (b)

FIG. 24. Test Rig for Examining the Performance of Solid Brushes on
Homopolar Machines (Slipring Removed to Show Brushes).

The results of intensive work on the performance of conventional
grades of brush led to the conclusion that the limit of performance was
a current density of about 300 kA/m^2 (200 A/in^2) and a slipring speed of
about 40 m/sec (8000 ft/min). It has been shown in the previous section
that if the current density is limited to this value, a severe limitation on
the possible ratings of superconducting homopolar machines will result.

2. Liquid metal current collection systems

Many brush systems employing liquid metals have been used or
proposed in efforts to overcome the drawbacks of solid brushes for homopolar machines. Materials used include: mercury, sodium, sodium-
potassium alloys, gallium, gallium-indium, mercury-indium, and other
low-melting-point alloys. The ability of liquid metals to wet suitable
solid metals produces extremely low potential drops at the brush/slipring
interface and, in principle, liquid metals appear to be the ideal current
transfer medium for homopolar machines. However, in practice, several
problems arise which make them less attractive than would appear at first
sight.

For example, mercury vapor is toxic, so that a machine which uses
this metal must be carefully enclosed and/or ventilated. Sodium and
sodium-potassium alloys are highly reactive and, to ensure safe operation,
enclosure in a pure inert atmosphere is essential. Gallium, mercury-
indium and other low-melting-point alloys are less toxic than mercury and

less reactive than the alkali metals, but experience has shown that they degrade rapidly in use. Despite these difficulties, various types of homopolar machine employing liquid-metal brushes have been built by several organizations. Nevertheless, the relatively limited voltage and current output ranges attainable with solid brushes, together with the practical difficulties relating to liquid metals, clearly illustrate the requirement for an alternative current-transfer system which, for maximum versatility, should be capable of operating on the segmented sliprings of multistage machines.

3. Carbon fiber brushes

The suggestion that a brush made from carbon fibers could have a performance potentially superior to that of conventional solid brushes was first made by the Scientific Adviser, Director General Ships, MOD(Navy). The essence of the idea was that a brush made from a bundle of flexible carbon fibers had the advantage of a very large number of potential contact points, compared with the small number available with rigid conventional brushes. Subsequently work commenced on carbon-fiber brushes at IRD in 1966, and has continued to the present date. Plain carbon or graphite fiber (i.e. as supplied by the manufacturers) was employed in these brushes. Experiments were undertaken on a variety of slipring surfaces at speeds of up to 30 m/s and usually in air. Three main categories of slipring materials were evaluated: metal, carbon or graphite, and silver-impregnated graphite; and several externally applied lubricants were also investigated. The total brush losses (electrical plus friction) were, broadly speaking, similar to those of conventional solid brushes, with the exception that very high friction coefficients (up to nine) were observed at positive brushes on metal surfaces under certain circumstances. This condition, together with the rather high total (positive plus negative) voltage drop, which was frequently more than 4V at current densities in the region of 300 kA/m^2, made the use of such brushes for homopolar generators singularly unattractive.

However, during one test in which brush and slipring wear was being measured, it was noticed that the brush performance had improved considerably with substantial reductions in the voltage drops and friction coefficients. Careful examination of the brushes revealed that this had been caused by the accumulation of metal debris from the slipring in the brush. Experiments with a plain carbon-fiber brush preimpregnated with small solid-metal particles confirmed this improvement in electrical performance and, as a result, it was decided during 1968 to investigate the practicability of plating carbon or graphite fibers with a thin layer of metal.

The first plated brushes were produced on a small scale by a batch process and using short lengths of fiber. Even though the plating was highly nonuniform on these initial brushes, short-duration experiments on metal sliprings gave promising results: typical voltage drops at 300-900 kA/m^2 and 20 m/s being about 0.2V for the negative brush and 0.8V for the positive brush.

Having shown that satisfactory performance could be achieved during short-term tests with metal-plated fiber brushes, a more comprehensive program was inaugurated in September 1968. The objectives were to establish production-plating and brush-fabrication techniques, together with performance and wear characteristics for fiber brushes under conditions appropriate to future superconducting homopolar machines.

As a result of this program, pilot-scale production-plating techniques have been established which enable tows containing 10,000 carbon fibers to be continuously plated. The greatest problem in the plating process is to obtain a good and even penetration of plating solution and current throughout the fiber bundle so as to ensure that every fiber is uniformly and individually plated. This has been made possible by the use of a novel fiber-spreading technique developed at IRD, and Fig. 25 shows a cross-section of a fiber plated by this method.

In recent months further improvements have been achieved at IRD with metal plated fiber brushes. Figure 26 shows brush wear rate against slipring velocity in air for positive and negative brushes for a range of current densities of from $465kA/m^2$ (310 A/in^2) to $930kA/m^2$ (620 A/in^2). It is seen that the wear rate of the negative brush is approximately an order of magnitude lower than the positive brush. One test has been conducted for over 4000 hours at a slipring velocity of 40 m/s (8000 ft/min) with the current density increased gradually up to $930kA/m^2$, the wear on the negative brush being about 1 mm and on the positive brush about 10 mm.

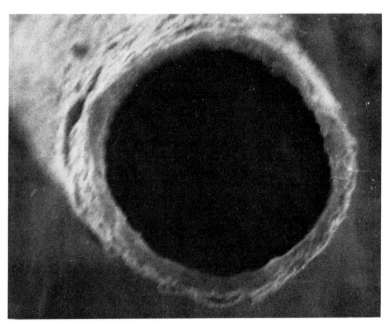

FIG. 25. Metal Plated Carbon Fibre.

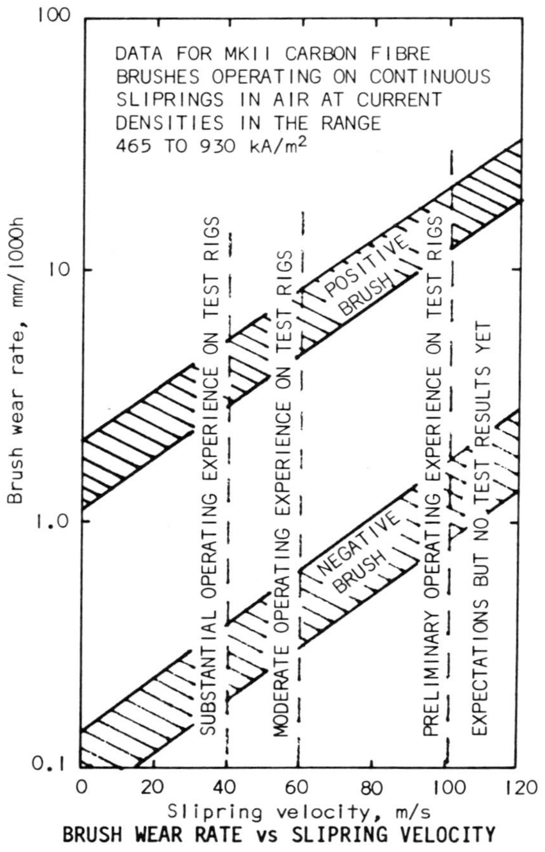

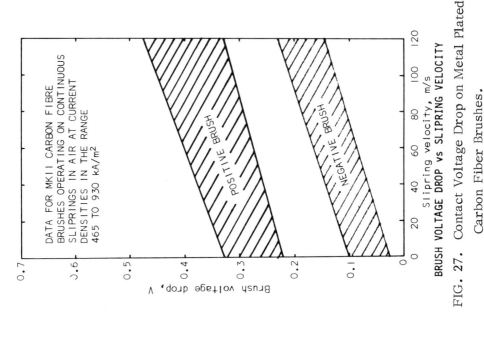

FIG. 27. Contact Voltage Drop on Metal Plated
Carbon Fiber Brushes.

FIG. 26. Brush Wear Rate of Metal Plated
Carbon Fiber Brushes.

Figure 27 shows the contact voltage drop for the positive and negative brushes, and for the 4000 hour test mentioned above the combined voltage drop was less than 0.5 volts. The voltage drop is not too sensitive to slipring velocity and the coefficient of friction for the tests is about 0.5; this relatively high figure is not serious because of the very low brush pressures required, typically 1 lb/in^2. Figure 26 indicates the operating experience at the different slipring velocities, and it is seen that up to 60 m/s a considerable amount of data are available though for the higher values only preliminary information is available. However, we are now confident that high current densities are achievable with the new IRD brushes, with acceptable contact voltage drops and friction coefficients. The wear rates are low and proven up to about 60 m/s, with a reasonable expectancy of operation up to about 100 m/s.

These results are adequate for the superconducting homopolar machine applications which have been considered.

C. Magnetic Screening

It is apparent that a machine with a simple field winding with no means for controlling the position or magnitude of the external magnetic field is not acceptable for many applications. This problem was one of the first to be tackled at IRD as early as 1964, and very precise design procedures are now available. There are two methods of screening the magnetic fields of the machines, by using iron or by screening coils. It was decided in 1965 to concentrate upon the use of screening coils because it seemed a retrograde step to reinstate the iron which the superconductor had made redundant. The simple principle of screening is shown in Fig. 28, a coil of larger diameter than the field winding carries current in the opposite direction to the latter. By careful design the degree of degrada- tion of the useful flux of the machine may be kept to an acceptably low level and a satisfactory design produced. Depending upon the application, it may be advantageous to employ a multiplicity of screening coils and the problem may be eased by employing a drum-type machine rather than a disc machine.

There are, of course, large machanical forces between the field winding and screening coils and an adequate support structure must be provided. The whole of this structure is at low temperature and one severe problem which presents itself is eddy current heating in this structure when the current is changed rapidly, for example in changing the output of a generator; this problem can be overcome by careful consider- ation of flux movement.

For some applications, where weight is not particularly important, there may be some advantage in employing some iron in the magnetic circuit. The advantage arises by having to employ less superconductor for a given machine flux; it was shown on the 50 hp model motor discussed

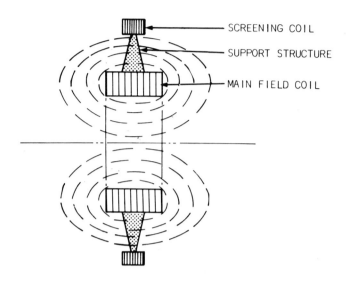

SCREENING COIL

SUPPORT STRUCTURE

MAIN FIELD COIL

FIG. 28, Screening of a Superconducting Homopolar Machine.

in the last lecture that about 20% less superconductor is required if 2 inch thick iron plates are provided around the machine.

IV.   FAWLEY MOTOR

A. Introduction

The development of the 50 hp homopolar model motor on behalf of MOD(N), discussed in section II provided a convincing demonstration that superconductors could be harnessed to produce electrical machines. It is a fact of life that no matter how convincingly one may talk or how elegant may be one's calculations, the probability of gaining credibility for a new device is immeasurably enhanced by a practical demonstration. In 1966, after the successful operation of this model, the world's first superconducting motor, IRD decided to seek support, through the National Research Development Corporation, for the construction of a much larger machine. The NRDC, set up by the British Government in 1947 to assist in the development and exploitation of new ideas, became interested and set up a working party of eminent scientists and engineers to advise them on the merits, both technical and commercial, of the new machine. After a very thorough investigation the working party was unanimous that the project should proceed. The NRDC, after giving consideration to the rating of a machine which could break even on cost with a conventional design, decided to support the construction and test of a motor of about 3000 hp at a speed of about 200 rev/min.

The machine to be constructed was required, of course, to be tested on full load and, if possible, in an industrial environment. The location of a suitable load for the motor was not as easy as might at first appear; it was difficult enough to find a convenient load, but when linked with the fact that the new machine was a prototype and therefore liable to have defects, the problem was much more severe. No industrial manufacturer likes to be a "guinea pig" and would certainly not risk a loss of production. The problem was solved by courtesy of the Central Electricity Generating Board. A new power station was under construction at Fawley, near South-ampton, which had four 500 MW turbogenerators with oil-fired boilers (oil supplied from the nearby Esso refinery); the cooling water for the con-densers of the power station was provided from four 3250 hp, 200 rev/min pumps, and the station design is such that any of the generators can oper-ate with any one of the cooling water pumps. The phasing of the construc-tion program required that the cooling water pumps would be installed long before all of the 500 MW turbogenerators were commissioned, and a solu-tion to our problem thereby presented itself. The cooling water pumps were designed to be driven by ac induction motors which were 6-pole machines (with a speed, therefore, of about 1000 rev/min) and drove the pumps through a gearbox to achieve the speed of 200 rev/min. The CEGB agreed that a 3250 hp, 200 rev/min superconducting motor could be sub-stituted for the ac motor and trials carried out for a limited period.

B. Some Design Considerations

Since the superconducting motor was not intended to be a permanent installation it was necessary to design the system such that the tests could be carried out with the least possible disturbance to the pit which housed each pump and its drive motor. It was very quickly established that the optimum design of the superconducting motor would have a larger diameter than the higher speed ac motor and, therefore, one of the first design decisions was that, since the foundations of the ac motor could not be dis-turbed, it was necessary to provide a gearbox; this was because the shaft of the pump could not be aligned to the shaft of the superconducting motor (the speed ratio was, of course, 1:1).

A number of other design problems peculiar to the installation pre-sented themselves before the system could be finalized, two of the most important were as follows:

a) To minimize the cost of the prototype machine it was desirable to have no screening winding and it was therefore necessary to estimate the effects of any magnetic iron in the cooling water pit; the problem with the iron is that it would exert a force on the superconducting winding. There was the massive structure of the iron pump itself, the large main water pipes and the reinforcing iron in the walls of the pit and in the motor foundations.

b) It was highly desirable to site the low temperature parts of the helium refrigerator as close as possible to the motor, since an attempt to transfer liquid helium over large distances would have greatly increased the refrigeration load. The magnitude of the problem may be appreciated when it is realized that the base of the motor was 42.5 feet below mean sea level.

A number of other problems arose because the cooling water pit had not been designed to accommodate a superconducting motor. However, the problems were solved, and the design of the motor and its associated plant was able to proceed. The design of the Fawley motor started on May 1, 1967, less than 12 months after the first demonstration of the 50 hp model motor for the Ministry of Defence, and it was first commissioned at IRD on October 31, 1969.

C. The Design of the Fawley Motor

It is not possible to give all of the design details of the Fawley motor here, and neither is this permissible, because, although its design has now been overtaken by more advanced work at IRD, it remains one of the very few machines of its class in the world, and contains many proprietary features. The selection of the major parameters was achieved by a careful optimization procedure based upon a reasonable compromise between the achievement of a minimum cost with minimum design risk. Comments on the selection of parameters will be made as we proceed with the description.

1. The superconducting winding

The superconductor available in 1967 was the fully stabilized Nb-Ti/ copper strip, and it was possible to design the winding with a reasonable assurance that a satisfactory performance would be obtained. It is a fact that, for a given quantity of superconductor, more useful machine flux will be obtained as the diameter is increased; of course, there comes a point where other factors such as increasing cost of the cryostat and rotor with increasing diameter begin to overtake the saving in superconductor cost. It was decided that the coil current should be less than 1000 A to permit an easy design of current lead with low losses, and that for the application in view, the value of the inductance was of no particular significance.

The coil takes the form of a rectangular toroid of 2.4 m bore and 2.8 m outside diameter with an axial length of 0.53 m; a total of $5\frac{1}{4}$ tons of copper-stabilized niobium-titanium superconductor is used to produce a total flux of 7.1 Wb. The stabilized superconductor is of rectangular section, 10 x 1.8 mm, and is wound in the form of 18 double discs to obviate making joints at the inner bore of the coil during winding. The current carrying capability of this composite superconductor has already been illustrated in Section I.

Explosively welded joints were made at the manufacturer's works because it was not possible to produce the superconductor in longer lengths than sufficient for single discs; these joints were located at the inner bore of the winding. The discs are separated axially by rings of radially aligned spacers covering approximately 20% of the faces of each disc.

The coil is cooled by liquid helium in natural convection over the edges of each disc, the design of the spacers and the laths upon which the coil is wound being such that gas bubbles evolved within the coil are ejected to its outside faces thus avoiding parts of the coil being poorly cooled. Figure 29 shows the coil being wound and the spacers which separate the discs.

The stresses within the coil due to the electromagnetic forces on the conductors are mainly the tensile hoop stresses tending to burst the coil and the compressive axial stresses tending to compact the coil; as the hoop stresses alone were of the same order as the yield stress of annealed copper, it was obvious that a rigorous approach was required.

The field plot within the coil was obtained by a computer program and, in conjunction with the appropriate current density, the body force on the conductors and the mechanical stresses within the coil were found directly using three-dimensional elastic theory; the analysis developed by Middleton and Trowbridge of the Rutherford Laboratory proved particularly helpful. Up to this point no consideration has been taken of the fact that the axial loads are not transmitted through the coil uniformly, but are concentrated on the radial spacers. The axial stresses within the coil are governed predominantly by the area of each disc which is covered by spacers, but this in turn affects the stabilization of the conductor on heat transfer grounds.

By choosing a certain degree of cold-work in the copper and knowing the mechanical properties of the copper at 4.2 K, it was possible to optimize the covered area of the discs so that the required heat flux from the conductor was minimized throughout the coil without exceeding the design stress. The condition chosen was a 3% cold worked copper with a yield stress at 4.2 K of 17,000 lb/in^2.

As already mentioned, the coil is fully immersed in liquid helium at a bath temperature of 4.4 K, and the conductor is designed for edge cooling at a maximum heat flux of 0.325 W/cm^2; experimental work showed that a nucleate boiling flux of 0.55 W/cm^2 corresponded to a temperature difference of 0.45 K. This data was obtained under conditions corresponding closely to design conditions -- a bath temperature of 4.4 K and with a Bicelex varnish covering 0.0005 in. thick, the peak nucleate boiling flux under these conditions being 0.61 W/cm^2. The proximity of the adjacent disc in the actual coil was of small importance, as work by M. H. Wilson of the Rutherford Laboratory had shown that the cooling channels in the coil afforded virtually open-bath conditions.

FIG. 29. Superconducting Winding for Fawley Motor Being Wound.

The figure referred to above of 0.325 W/cm^2 relates to only a small portion of the coil; the variation of required heat flux value over the complete coil is considerable, the minimum value being 0.204 W/cm^2.

The conductivity of the explosively welded joints was found to be very similar to that of the parent material; it was thought advisable, however, to ensure that the joint was wound in the inner turn of each disc where the enhanced cooling could be advantageous.

The transition temperature of the niobium-titanium filaments at 3.5 Wb/m^2 is 7.9 K; this compares with the expected conductor temperature of 4.65 K when a heat flux of 0.325 W/cm^2 is being dissipated from the conductor edges.

The production lengths of superconductor were tested at IRD to establish the short sample performance, the resistance ratio of the copper and the mechanical strength. The protection system for the winding consisted of a bridge system of voltage tappings on the winding whereby a normal region would cause unbalance and a trip. Further protection was afforded by the measurement of helium pressure and the rate of change of helium pressure. The completed winding is shown in Fig. 30.

2. The cryostat

The cryostat is a fabricated stainless steel vessel which contains the

Table 3. Basic coil parameters.

Dimensions

Inner bore	2.40 m
Outer diameter	2.84 m
Length	0.535 m
Total weight	$5\frac{1}{4}$ tons
No. of discs	36
No. of turns per disc	110

Electrical parameters

Conductor filament current density	72,000 A/cm^2
Conductor current density	4,000 A/cm^2
Overall coil current density	2,550 A/cm^2
Inductance	55 H
Ampere-turns	2.86 x 10^6
Stored energy	1.46 x 10^7
Coil current	725 A

Mechanical parameters

Maximum hoop stress in conductor	6940 lb/in^2 tensile
Maximum mean axial stress in conductor	1310 lb/in^2 compressive
Maximum (axial) stress in spacers	6840 lb/in^2 compressive
Minimum proof stress of copper in conductor at 4.2 K	17,000 lb/in^2
Degree of cold work in conductor	3%

Materials

Superconductor	Composite conductor of NbTi in H.C. copper; 10 mm x 1.8 mm with 5 filaments of 0.020 in. diameter
Length: Piece length	1825 m
No. of pieces	18
Rated field	3.5 Wb/m^2 axial
	2.7 Wb/m^2 radial
Current	727 A
Laths	Cold worked EN58B stainless steel with phenolic resin insulation strip
Coil former	Stainless steel AISI 304L
Spacers	Glass reinforced epoxy-resin mouldings
Varnish insultation	Bicelex 0.0005 in. film thickness
Interturn insulation	Glass fiber/isophthalate-polyester tape 0.005 in. thick

---

FIG. 30. The Superconducting Winding for the Fawley Motor.

superconducting winding and the liquid helium. The thermal losses are minimized by surrounding the helium vessel with a radiation shield which is held at a temperature of about 80 K. Of course, conduction losses are minimized by the use of a vacuum and careful design of the coil supports. The weight of the coil and the containment vessel is taken by four stainless steel cables, but because of the effects of iron in the vicinity of the machine it was necessary to provide support in the other two planes. A schematic view is shown in Fig. 31 and Fig. 32 shows the inner (low temperature) vessel after completion of the fabrication. The design of the latter presented some difficulties. In order to reduce to a minimum the distance from the coil bore to the cryostat bore (Fig. 31) the former on which the coil is wound was designed to be the inner cylinder of the vessel, and various methods of construction of the remainder of the vessel were examined.

The criteria to be met were:

(1) During cool-down the vessel is subjected to a high pressure for a period of a few days;

(2) When at low temperature, the differential contraction of coil and former would result in high compressive hoop stresses in the latter;

(3) The vessel must remain helium leak-tight after cycling to 4.4 K;

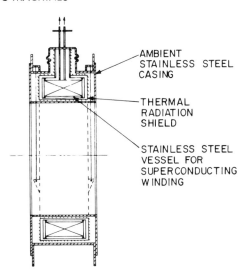

FIG. 31. Schematic View of Superconducting Motor.

FIG. 32. Superconducting Winding of Fawley Motor After Fabrication of Helium Containment Vessel.

(4) In the manufacture of the vessel the coil temperature must not exceed 250 C;

(5) The material must have adequate strength at room temperature and at 4.4 K. Following BS 1515, the maximum design stress used was 9.6 ton/in^2. In addition, the resistivity must be high so as to minimize circulating currents caused on coil discharge, and the material should be non-magnetic.

The high pressure requirement results from the need to keep the cool-down time reasonably short, the limiting factor being the heat flux obtaining in the coil, which varies with coolant pressure. It was decided, therefore, to circulate pressurized helium gas through the refrigerator and cryostat to cool both together. The maximum pressure in the refrigerator cycle (a turbine expansion Claude cycle) was 9 atmospheres and, allowing for pressure drops, a convenient operating pressure for the coil vessel was chosen to be 7 atmospheres. The design pressure was increased to 7.7 atm. to allow for the functioning of safety devices.

The most suitable material was stainless steel, AISI 304L being chosen because of its strength and resistivity and availability. This steel is not stabilized, but the low carbon content ensures good weldability. Weld specimens were prepared and tested, the most critical test in this case being a Charpy impact test carried out at 77 K.

A welded construction was considered necessary to meet the requirement for helium tightness over a long period of thermal cycling, but the welding procedure was complicated by the fact that the superconducting coil would be in position inside the vessel during the welding operations. Four circumferential welds were required with full depth penetration so that there was a serious risk of overheating the superconducting material with consequent permanent damage. A compromise had therefore to be reached between the small number of heavy current weld runs required for good impact strength and the larger number of light runs required to reduce the heat input.

The temperature problem was further eased by the device adopted to reduce the differential contraction stresses on cool-down. A coil wound directly onto the inner cylinder/former would produce high local stresses at the ends of the winding in the former adjacent to the welds, and these stresses would be in the same sense as the stresses resulting from the internal pressure. A number of longitudinal spring laths were therefore introduced between the former and the winding which would distribute the load more evenly on the former and, also, introduce an additional thermal resistance between the weld area and the winding.

With a correctly designed system of laths, the resulting shrinkage stresses could be treated as a uniform pressure on the inner cylinder so that the effective internal pressure on the inner cylinder was 135 lb/in^2 abs. and on the outer cylinder, 115 lb/in^2 abs. (7.7 atm). One further load on the inner cylinder was the weight of the winding, resulting in a maximum of 7 lb/in^2 additional pressure at the top of the cylinder.

These pressure conditions roughly defined the thickness of plate required for the cylinders, and the main problem thus became the end closures of the toroidal vessel.

After examining various methods ranging from a thick, flat plate closure to a costly dished end (which proved to be unobtainable) a design was evolved which made use of flat end plates suitably profiled to provide elastic hinges which would relieve the stresses at the welded corners, potentially the critical areas.

The outer casing of the cryostat was to be basically a simple vacuum vessel, but, in addition, it was required to transmit the full load torque reaction of the motor to the foundations; this torque appears on the stator frames and it was decided that these should be mounted on the cryostat casing, but in such a way that no forces appeared across the demountable main cover plates of the vacuum vessel.

This meant, therefore, that the cryostat casing was to function also as the main location between the bearing housings at each end of the motor shaft. Problems of deflection and machining were significant in the design, and the final design incorporated location devices which were not affected by the dimensional changes occurring between the machining operations and the finished assembly. Figure 33 shows the outer casing of the cryostat.

FIG. 33. Outer Casting of Cryostat.

Calculations showed that the optimum thermal insulation was achieved
by the use of a single refrigerated radiation shield at between 60 and 100 K
with multi-layer superinsulation on the warmer side and none on the
helium vessel side, the whole inter-vessel space being evacuated to better
than 10^{-5} torr.

To simplify the refrigerator system and its supplies, helium gas at
80 K was chosen as the cooling medium for the shield. This also avoided
the two-phase flow problems which would have occurred in a shield of this
size with the more common liquid nitrogen coolant, and the temperature
fitted in very well with that of the first expansion stage of the refrigerator.

Theoretically, a copper shield would have been marginally superior
in thermal performance, but stainless steel was again chosen, primarily
because its high resistivity eliminated the danger of large circulating
currents and collapsing forces in the shield on sudden coil discharge. It
is also much easier to maintain the required polish on stainless steel than
on copper during manufacture.

3. The rotor

It has been stated above that the total flux of the winding is 7.1 Wb,
but of course not all of this is available for the generation of voltage.
This has been discussed in Section II and, for a reasonable design, the
useful flux is reduced to about 6.45 Wb; therefore, at the maximum speed
of 200 rev/min, the generated voltage is about 21.5 V. If the principle of
the model motor described in Section II had been employed, i.e. with two
Faraday discs, one on each side of a stainless steel strength disc, the
power supply requirement would have been 43 V, 58,000 A. It was because
this was considered to be unsatisfactory that the segmented slipring design
discussed in Section III was developed at IRD. The number of voltage
stages on the armature is then increased from 2 to 40 and the generated
voltage (back emf) at full field and full speed is designed to be 428 V.
Figure 34 shows the stainless steel strength disc of the rotor being machined
and Fig. 35 shows the finished rotor being assembled into the cryostat.

4. The stator

The stator on the motor consists of the stationary armature conductors
which carry the current between the brushes and the brush-gear. The
stator conductors are subjected to the full torque reaction and must therefore be well supported; this is achieved by means of an aluminum alloy
disc which may be seen on Fig. 35. The torque reaction from the aluminum alloy disc is transmitted to the cryostat structure and thence to the
foundations by a stainless steel cylinder which may be seen in Fig. 36.

The brushes employed were of a conventional type with a high copper
content and were arranged in multiple brush boxes one on every other
segment of the sliprings.

FIG. 34. Strength Disc of Fawley Motor Being Machined.

FIG. 35. Rotor of Fawley Motor Being Assembled into Cryostat.

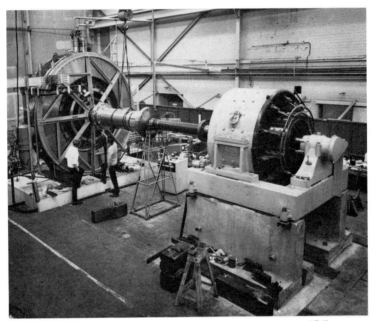

FIG. 36. The Fawley Motor Under Test at IRD.

5. The refrigerator

For a motor which was to replace an existing conventional motor as a demonstration prototype, it was essential that all the ancillary equipment should be as simple and reliable as possible. With normal power station operation in mind, it was decided that the refrigerator should be fully automatic in operation. The design chosen is in fact capable of fully automatic cool-down and filling of the cryostat as well as automatic steady-state operation.

The refrigerator cycle, designed by British Oxygen Cryoproducts Ltd., in close collaboration with IRD, uses a low pressure turbine expansion system which is able to supply gas at maximum cycle pressure to maintain the cryostat radiation shield at 80 K and to provide, in steady-state conditions, a continuous supply of liquid helium to the coil vessel. The major part of the boil-off gas from the coil vessel is returned to the refrigerator at or near its saturation temperature, but a portion of it is used to cool the coil current leads and is then returned at ambient temperature.

The specified refrigeration loads are 27 watts at 4.4 K + 100 watts at 80 K + 220 watts lead cooling between 4.4 and 300 K.

As mentioned above, a high pressure supply of refrigerant was required during cool-down of the coil and this was very neatly provided by a diverter valve immediately upstream of the Joule-Thomson (J-T) valve at the cold end of the refrigerator.

Control of this diverter valve is entirely automatic, a temperature sensor causing the valve to switch from cryostat to J-T stream when the coil temperature is down to 5 K. In this condition the high pressure gas is throttled to its saturation pressure and temperature and the wet vapor passed to a separator in a liquid helium reservoir feeding the cryostat.

The operating specification of the motor called for maintenance-free running of ancillaries, including the refrigerator, of up to 5000 hours. If the refrigerator did require attention within this period, the supply of liquid helium to the coil was to be maintained while repairs were carried out.

A liquid reservoir was therefore provided using a standard 1000 liter storage dewar normally maintained half full to allow 12 hour shut-down of the refrigerator. The reason for the extra capacity was to allow for decanting of the liquid in the coil vessel into the storage dewar should any work on the coil become necessary.

Because of the layout of the prototype, involving low temperature transfer lines, and because of various safety margins in the design, the refrigerator is rather larger than, theoretically, it need be. The use of vacuum insulated transfer lines with valves which had to be provided between refrigerator, storage dewar, and cryostat increased the refrigerator rating by 50%.

To reduce as far as possible the length of transfer line exposed to ambient radiation, the liquid feed and gas return lines entered the cryostat at the point closest to the refrigerator terminals, continuing to the bottom and top of the coil vessel inside the liquid helium space (Fig. 37).

The liquid line from the storage dewar to the cryostat was fitted with a proportional control valve actuated by a level sensing differential pressure cell on the cryostat. This system was designed to control the level to within one inch and was backed up by low and high level alarms actuated by carbon resistors in the cryostat chimney.

A similar level control was fitted to the storage dewar to regulate the output of the refrigerator by controlling the position of the J-T valve.

The gas emerging from the two coil leads requires to be maintained at ambient temperature and is therefore passed through a temperature controlled flow valve on each lead outlet before joining a common return to the warm end of the refrigerator.

The cool-down of the whole system would have been best achieved by cooling cryostat and refrigerator together, but a slight variation was called for in order to fill the cryostat with liquid in a reasonable time

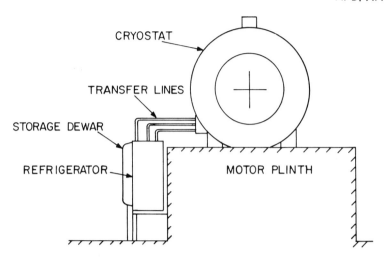

FIG. 37.   Arrangement of Helium Transfer Lines Between Fawley Motor
and Refrigerator.

without having to provide an excessive degree of over-capacity in liquefaction.

The procedure adopted was first to cool-down the refrigerator and
dewar and then to fill the dewar to about 98% full. A level sensor on the
dewar would then switch the refrigerator diverter valve to the high pressure
cryostat position and the cryostat would then proceed to cool down after
initially causing a slight temperature rise in the refrigerator which would
then cool-down again with the cryostat. On reaching a coil temperature
of 5 K, the diverter would again switch to the tank filling position and the
liquefaction rate would be automatically reduced. During this period, the
valve controlling the supply of liquid to the cryostat from the tank has been
manually over-ridden and is now released, allowing liquid to fill the cryostat.

The flow is set up by a small pressure differential maintained between
dewar and cryostat by a pressure bias valve on the return lines in the refrigerator. The pressure differential can be reversed by a manually operated switch should it be required to transfer the liquid from the cryostat
back into the dewar.

D.   Tests at IRD

The machine was completed and ready for test in October, 1969,
and the layout of the test installation is shown in Fig. 36.

The cool-down proceeded without difficulty and it was found convenient to measure the dc resistance of the winding and relate this to its average temperature. Since each of the 18 double pancakes had a voltage tapping
it was possible to study the axial variation of temperature and this was found
to be negligible. Carbon resistors were arranged throughout the winding
and it was found that the maximum temperature differential anywhere in
the winding did not exceed a few degrees. No difficulty was experienced
with the helium liquid level control system. The coil was charged at a
slow rate of a few amperes per minute and a number of observations were
made:

1. The movement of the coil vessel containing the superconducting
winding was measured very carefully and it was necessary to raise the
winding vertically to compensate for the strain in the support cables due
to the iron in the foundations. Movement of the winding in other directions
also had to be compensated as the full rated current was approached because it was found that the mass of the 0.5 MW conventional dc motor used
for the commissioning tests was exerting a pull of a few tons; this was
counteracted by placing iron on the other side of the motor. Further compensation was obtained by placing pieces of iron at the appropriate places
on the motor cryostat.

2. During the charging it was found that a very definite training
effect was present; the instrumentation provided for measurement of the
rate of change of helium pressure was found to be a very sensitive indication of flux jumping; it was found that by discharging the winding after a
rather "bumpy ride" to say, 500 A, the coil could be charged rapidly to
the previous current with no movement on the dP/dt instrumentation. The
full rated current of 725 A was achieved without difficulty after modifications to prevent coil movement had been carried out.

3. With full field, the motor was driven at 200 rev/min by the
0.5 MW motor and an open circuit voltage 2.7% greater than the design
value was obtained. This was due to the presence of iron which had been
ignored in the calculations but which was increasing the total flux of the
machine. At constant speed the voltage versus excitation curve was not
quite a straight line because at the lower flux levels the magnetic iron
shaft robbed some of the useful flux.

4. Using the 0.5 MW machine as a load, the Fawley motor was
operated as a motor up to a maximum load of 0.5 MW. However, by
operating at reduced excitation it was possible to run the motor with full
armature current at the full rated speed. This was the first time that the
segmented slipring concept had been demonstrated at the designed switching
frequency with the full armature current; there was no evidence of sparking
at the brushes. This result confirmed calculations which had indicated that
the reactance of the armature conductors (being in a stainless steel disc)
was about an order of magnitude lower than the level at which sparking
might be expected.

The testing of the motor at IRD was completely successful and sufficient data had been acquired for the tests to be terminated before Christmas, 1969.

E. Tests at Fawley Power Station

The Fawley motor in service duty in the cooling water pump pit at Fawley power station is shown in Fig. 38.

Access to site at Fawley was obtained in April, 1970 and installation began immediately. In spite of being delayed many weeks by an industrial dispute at the power station, installation was completed in September, 1970. The next phase -- cool-down -- began immediately, but difficulties with the helium compressor prolonged the process and it was not until January, 1971 that the machine could be run (on 25% load). Although the helium compressor was still subject to malfunction, 75% full power was achieved in February and 100% in March. The cooling water pump which the motor was driving is connected (hydraulically) in parallel with three other pumps and load on any one pump is taken up by gradually increasing speed. The full load of 3250 hp at 200 rev/min, which was achieved with less than 90% of the rotor field current, was well within the capabilities of the motor. The test program was continued until the formal opening of

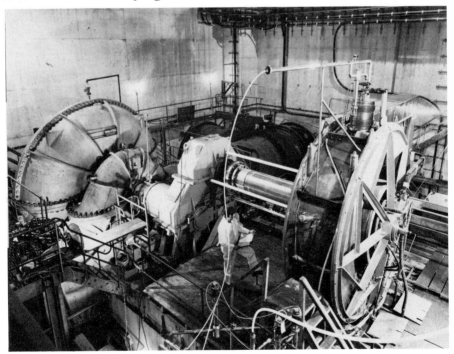

FIG. 38. The Fawley Motor Installed in the Fawley Power Station.

Fawley power station on May 6, 1971, but was subject to some curtailment due to unreliability of the helium compressor. Apart from a small number of defects, none of which was irreparable, the behavior of the motor under all load conditions was excellent; there was no vibration or sparking at the segmented slipring and, notwithstanding a mishap imposing a shock load during transportation, the performance of the superconducting winding was impressive.

The difficulties with the compressor (a reciprocating machine) denied handing over the motor to the CEGB for service. While this was never intended as part of the program, it would have been a gratifying conclusion to the project and a remarkable event for the first prototype of a new class of machine to go into service without modification. Since the power station had now become fully operational, it was not possible to obtain, install and commission a new type of helium compressor in time to avoid putting the output of one of the most modern power stations in Europe at risk.

The successful operation of the Fawley motor has provided a number of benefits for the advancement of superconducting machines. For instance, it has proved that the principles demonstrated with the 50 hp model motor at IRD are valid when the scale of the parameters is increased. It has also provided an immense fund of experience and design data which is available for future machines; these data cover all aspects of cryogenics and helium refrigerators, segmented slipring systems, control and instrumentation. In addition, it has provided the evidence necessary to allow the Ministry of Defence (Navy) to move forward with superconducting dc marine propulsion systems.

Although many of the features of the machines now under construction at IRD are improvements on those provided for the Fawley motor, the basic design approach is the same. The present excellent state of the art with superconducting homopolar machines would not exist but for the Fawley motor funded by NRDC and C. A. Parsons & Co. Ltd.

V.   APPLICATIONS AND FURTHER DEVELOPMENTS

The applications for superconducting dc machines are:
Motors
     Marine propulsion
     Drives to steel rolling mills
     Drives to large pumps and blowers in power stations
     Drives to numerous other industrial process plants including the
         manufacture of cement, paper mills, colliery winders, etc.
     Special purpose drives such as for wind tunnels and possibly airships
Generators
     Power for marine propulsion
     Chlorine production

Aluminum smelting
Copper refining.

As the introduction of these new machines into industry increases, designs will become standardized, costs will fall and the range of applications will increase.

A. Marine Propulsion

Electrical transmission for ships is attractive for a number of reasons but, until the availability of superconducting machines, it has seldom been possible to overcome the problems of weight, space and cost. The power requirement of ships varies enormously but shaft powers have been creeping up beyond 40,000 hp (at about 100 rev/min) and this power level is quite beyond the design limits of conventional dc motors; this is clearly illustrated in Fig. 10. The use of ac motors, whether super-conducting or not, would require the use of a gearbox, and this increases the weight, cost and iron requirements and also reduces the efficiency. The use of an ac system is also restrictive on speed control unless very expensive frequency changing equipment is installed. However, with the development of superconducting dc machines, the necessary power ratings with direct drive to the propeller become available and a range of marine applications may be considered.

The following statistics are of interest.

Approximate Tonnages of Principal Seaborne Cargoes in
World Trade in 1970

Cargo	Million Tons
Crude oil	995
Oil products	245
Iron ore	250
Coal	100
Grain	75
Forest products	55
Iron and steel products	45
Phosphates	35
Bauxite and aluminum	35
Minor bulk cargoes, deep sea	140
General cargoes, deep sea	150
Short sea cargoes	220
	2,345

Oil is clearly the major cargo in world trade and it is interesting to note that legislature may soon make superconducting propulsion systems attractive to tankers; the reason for this is that the practice of filling empty oil tanks with sea water for ballast may become illegal, and this will mean that special clean water ballast tanks must be provided. The effect of this will be that all space will be at a premium and the saving that may be accomplished by employing superconducting machines will be of great significance.

We have already seen that the terminal voltages of the machines may be a few kV, so that the power transmission is not a problem, i.e. it is not necessary to transmit currents of the order of hundreds of thousands of amperes which is sometimes quoted as a disadvantage of homopolar electrical propulsion systems. In strictly volume limited ships it may even be possible to locate the power plant on the deck. The benefits of the electrical propulsion system are:

(1) improved control and manuverability,
(2) improved ship layout with greater cargo space,
(3) easier maintenance of the power unit since it is now much more accessible,
(4) greater efficiency.

The full cost effectiveness of the superconducting propulsion system for tankers and container ships has not yet been fully assessed, but the indications are very encouraging. It is possible to increase the cargo space by between 10% and 20% and the overall efficiency of the transmission system may be as high as 96%. Other benefits that accrue from the electrical system are that a fully integrated ship's electrical system may be designed to give greater simplicity and lower cost, and also that the main power source may be employed for discharge pumps and other loads, thus reducing the installed electrical capacity of the ship. The very high power to weight ratio and maneuverability of superconducting propulsion systems are clearly beneficial to ships such as ice-breakers, tugs and ferry boats.

At the present time there is mounting world activity in superconducting propulsion systems, and at IRD a system is nearing completion with land trials scheduled for 1973.

There are numerous other advantages for navy vessels which are not particularly relevant to commercial ships, but these cannot be discussed at present.

B. Steel Rolling Mills

Large dc motors are employed in the steel industry because of the necessity for very fine speed control, often as in strip mills with a number of motor stands acting in concert. A hot strip mill may have 6 stands with motor ratings of the order of 8,000 hp to 10,000 hp at speeds as low as 30 rev/min. This is a good example of an industrial application for superconducting dc motors and one that is being pursued; a superconducting

motor of 10,000 hp at 30 rev/min will weigh about 40 tons and be capable
of a very rapid acceleration.

A modern steel works complex can cost many millions of pounds
and the penalty of failure of any component which can result in a loss of
production is so high that its probability must be virtually excluded. The
order of priorities becomes:

(1) reliability,
(2) ease of control,
(3) ease of maintenance,
(4) low operating cost,
(5) low capital cost.

The many years of development of superconducting dc machines at
IRD have led to many significant design improvements (some of which are
discussed later) and one of the principal aims in this work has been to increase reliability, reduce the costs, and increase efficiency.

Some detailed studies on an 8,000 hp motor show that efficiencies
of 96% and 97.3% are quite feasible at 40 rev/min, and 90 rev/min respectively; these values compare with about 92% for a conventional dc motor.
The capitalized value of this improvement in efficiency is quite significant.
The power to weight ratio of the superconducting motor is an order of
magnitude higher than the conventional motors and the capital cost saving
is substantial; precise figures cannot be given for commercial reasons.

Other aspects of this particular application which have been studied
in detail, are speed control, power supplies, economy of refrigeration
plant, and maintenance arrangements.

The conclusion reached is that the possible applications of superconducting dc machines in steel mills are:

(a) hot strip mills,
(b) reversing mills,
(c) roughing mills,
(d) finishing mills,
(e) sinter fan drives,
(f) blast furnace blowers,
(g) cold strip mills.

C. Other Industrial Drives

There are a number of applications in which constant speed (or
possibly two speed) ac motors are employed but where the continuously
variable speed of a dc motor would be preferred. The reason for this is
either that a suitable dc motor cannot be manufactured or because it is

too expensive and possibly too large. The availability of the superconducting dc motor changes this position.

In power stations there would be considerable benefit if the boiler feed pumps and large air blowers could have variable speed drives; this particular application becomes more and more relevant to our new machines as the capacity of power stations increases with corresponding increases in the size of the auxiliary drives. An attractive feature, as with a large strip mill, is that a number of machines are in fairly close proximity and it becomes possible to effect economies with the services required by the superconducting machines.

The manufacture of cement requires a massive structure to be rotated at a very low speed, and a common method of achieving this drive is with an ac motor and a rather large gearbox. A modicum of speed control is highly desirable, but more important perhaps is the high capital cost and maintenance cost of the gearboxes. A superconducting dc motor is ideal for this application, although the very low speed does present some problems.

D. Generators

The demand for dc power is growing at about 10% per annum, mainly for aluminum smelting and the production of chlorine; there are other processes, such as copper refining, requiring dc power, but these are not the major users. It is conceivable that as electric cars become more widely used the need for very large battery chargers will create a significant load. All of these applications may be met with the superconducting dc generator. Furthermore this machine will provide the power more cheaply than any other method.

E. Aluminum Smelting

An aluminum smelter requires dc power for the electrolytic reduction of alumina (aluminum oxide), a fine white powder which is extracted from bauxite. To give an indication of the power requirements, at the new Alcan site at Lynemouth in Northumberland, England, a power station with an output of 390 MW produces 120,000 tons of aluminum per annum. The world (excluding the U.S.S.R. and Eastern Europe) output of aluminum in 1972 is estimated at 9×10^6 tons which requires approximately 30,000 MW of installed capacity of dc power. The present method of producing the dc power is by means of transformer rectifier plant with the power being taken either from the grid or generated locally as at Lynemouth, Fig. 39. If a prime mover is employed to drive a superconducting dc generator, the latter replaces the ac generator, transformer and rectifier, and the capital cost saving is substantial; in addition there is an improvement in efficiency so that the operating costs are reduced. It is conservatively estimated that where power is generated on site the

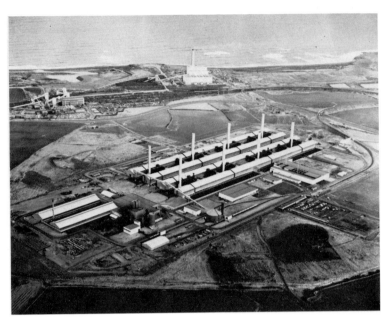

FIG. 39. The Alcan Aluminum Plant at Lynemouth, England.

capital cost of electrical plant is halved with further substantial savings because smaller buildings are required to house the plant. The improvement in efficiency will not be less than 2% which for a 390 MW plant represents a capitalized saving of over £1M (based upon a conservative figure of £150 per kW); the saving on the world scale is clearly very substantial.

F. Chlorine

DC power is required for the production of chlorine and, as with aluminum, there is a growth rate of about 10% per annum. The approximate production data are as follows:

	Million tons
Western Europe	3.5
U.K.	0.9
U.S.A.	7.7

The installed electrical capacity required for this output is approximately 4,500 MW. Comments made with regard to the power supply for aluminum smelters also apply to chlorine.

It may be noted that the voltage requirements for aluminum and

FIG. 40. Resin Encapsulated Superconducting Winding for Generator of a
Marine Propulsion System.

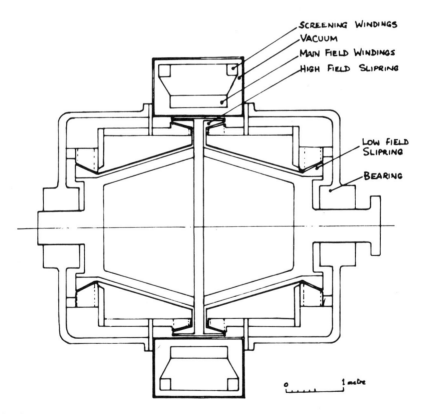

FIG. 41. Possible Arrangement of a Large Motor for Marine Propulsion.

chlorine production may readily be achieved with superconducting dc generators; in both cases a large number of low voltage cells are placed in series and currents in the hundreds of thousands of amperes are required.

G.   Further Developments

In concluding this article on superconducting dc machines, it is appropriate to comment on some recent developments.   Figure 40 shows the superconducting windings for the generator of the marine propulsion system now under construction.   The ability to employ resin encapsulated windings greatly simplifies the design and reduces the cost.   Figure 41 shows the geometry of the motor that may be employed for large commercial ships;   it is seen that the drum type configuration is preferred.

The process of improving the designs of the machines is continuing with considerable success.   At a future date it will be shown how changes to the machine geometry, rearrangement of the refrigeration plant, improved current collection, better access and improved cryogenic systems are all contributing to simpler, cheaper, more robust and more reliable machines.

It is only to be expected that a completely new type of machine such as the superconducting dc motor will take some time to become firmly established in industry.   The benefits to be derived are greater for some applications than others, but there can be no doubt that the heavy industrial power industry is about to enter a new era to the benefit of a large part of the industrial scene.

REFERENCES

1. Appleton, A.D. and MacNab, R.B., Commission 1, London, Annex, 1969, 1 Bull. I.I.R. 261.
2. Appleton, A.D. and Ross, J.S.H., Commission 1, London, Annex, 1969, 1 Bull. I.I.R. 269.
3. Tinlin, F. and Ross, J.S.H., Commission 1. London, Annex, 1969, 1 Bull. I.I.R. 277.
4. Appleton, A.D., Commission 1, London, Annex 1969, 1 Bull.I.I.R. 207.
5. Yamamoto, M., Commission 1, London, Annex 1969, 1 Bull. I.I.R. 285.
6. Clayton, A.E., The Performance and Design of Direct Current Machines, Pitman (1959).
7. Greenwood, L., Design of Direct Current Machines, MacDonald (1949).
8. Appleton, A.D., Shipbuilding and Shipping Record, 122, August 17 (1973).
9. Appleton, A, D, and MacNab, R.B., Proc. Third International Cryogenic Engineering Conference, Berlin, May 1970.
10. Appleton, A.D., Science Journal, 5, No. 4, 41 (1969).
11. MacNab, I.R. and Wilkin, G.A., IEE Journal, Electronics and Power, 18, 8 (1972).
12. Shobert, G.I., Carbon Brushes, Chemical Publishing Co., New York (1965).
13. Baker, R.M. and Hewitt, C.W., Electrical Engineering, (1937).
14. Pattison, R.I.D., Clark, E. and Berger, B., Electrical Times (1965).
15. Appleton, A.D., in Proc. of 1972 App. Supercond. Conf., Annapolis, Md., IEEE Pub. 72CH0682-5-TABSC (1972).

Chapter 5

APPLICATIONS OF SUPERCONDUCTIVITY TO AC ROTATING MACHINES

Joseph L. Smith, Jr., Department of Mechanical Engineering and

Thomas A. Keim, Department of Electrical Engineering, MIT

Cambridge, Massachusetts 02139

I.   INTRODUCTION

A.   Comparison with Conventional Machines

The development of high-field superconducting wire, suitable for fabrication into windings for electric machines, presents the opportunity for significant improvements in rotating electric machines. Previously copper and iron have been the obvious materials to use. In the evolution of the machines since the turn of the century, the most significant changes have been in the insulations and the coolants. The thermosetting resins have completely replaced the older asphalt impregnated insulations which had in turn replaced varnished insulations. Mica has also been used with thermosetting resins for insulations which must withstand high electric stress. Metallurgical improvements have also improved the performance of the magnetic circuits of the machines.

In contrast to the past evolution of machines, the advent of super-conductors should motivate radical changes in the basic construction, especially in the largest machines. The potential of superconductors is best illustrated by examining the limitations of conventional machines and then evaluating how superconductors can advance these limits.

In all electric machines the electromechanical energy conversion results from an average magnetic shear stress between two electromagnetic components with an average relative mechanical velocity. Thus, the machines are limited in power density by mechanical limits on the relative velocity and by electrical limits on the field strengths. In conventional machines the field strengths have been limited by the saturation of the iron which forms the magnetic circuit of the machine. High-field

superconductors promise to extend this field limit by a factor of three
and perhaps a factor of five. In addition, the high current density available in the superconductor allows a large magnetomotive force to be
packed into a small volume without a singificant energy cost.

In a conventional machine the current density in the conductor is
limited by thermal degradation of the electrical insulation (temperature
rise) or by the electrical losses (energy cost). The process of evolving
conventional machines with higher magnetic shear in the air gap has required longer air gaps and correspondingly higher current densities in
the windings. The higher current density has been provided by, first,
hydrogen cooling (rather than air cooling) and second, by direct water
cooling of hollow copper conductors. Most recently direct water
cooling of the conductors has been extended to the field windings on the
rotor of large machines. As a result of the effective cooling, conventional machines may be designed up to an economic optimization of
power loss versus power density (capital cost). Preliminary design calculations indicate that the current density available with high field superconductors may well provide an increase in the economic power density
by a factor of five to ten.

The major limitation on the application of high-field superconductors
in electric machines is the ac loss characteristics. Specifically, the
significant ac loss at power frequencies limits the practical application to
the windings which have essentially a constant flux linkage, i.e., the field
windings. The configuration of a superconducting machine must be a
superconducting field winding (rotating or stationary) and a normally
conducting armature (ac) winding, which also may be stationary or rotating.
The armature winding must be specifically designed to take full advantage
of the high fields which are produced by the superconductors in the field
winding. Ferromagnetic material, if present at all, serves only as an
external flux shield.

If the armature is connected to the power terminals, the machine
is a synchronous ac machine (Fig. 1). If the armature is connected
through a commutator (mechanical rectifier/inverter), the machine is
electrically a conventional dc machine (Fig. 2). A third alternate is the
acyclic machine (homopolar or Faraday disk machine) Fig. 3. In this
machine a field winding produces the field and the electromechanical
interaction occurs between an armature disk (half turn and infinite number
of phases) and a half turn dc reactor winding (current leads to brushes).

A conceptual representation of the electromechanical interaction in
a rotating machine is shown in Fig. 4. The field winding is represented
by a single current loop which produces a dipole magnetic field. The field
is represented by a single vector $\vec{M_F}$. The current in the armature
winding is also represented by a single current loop producing a dipole

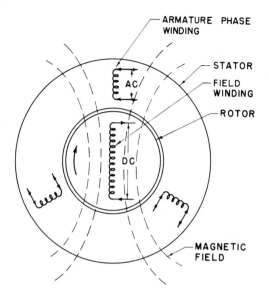

FIG. 1. Schematic for an ac Synchronous Machine.

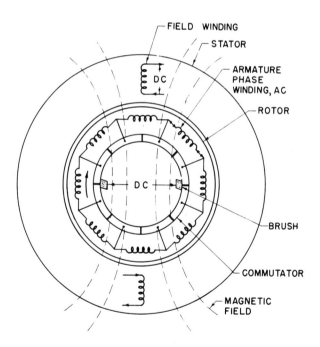

FIG. 2. Schematic for a dc Commutated Machine.

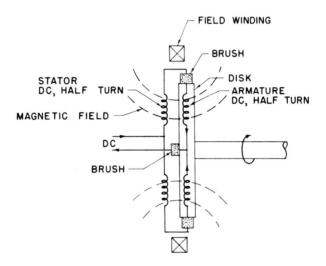

FIG. 3. Schematic for an Acyclic dc Machine.

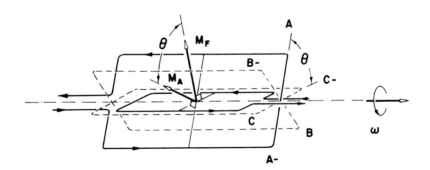

FIG. 4. Conceptual Representation of the Electromechanical
Interaction of a Rotating Electric Machine.

magnetic field represented by $\vec{M}_A$. The current loops are displaced by a torque angle θ about the axis of rotation of the machine so the magnetic dipole vectors are also displaced by angle θ. The magnetic interaction between the current loops is in the direction to force angle θ toward zero. The torque about the axis of rotation of the machine is proportional to the vector product of the two field vectors.

In a rotating field machine, normally an ac synchronous machine, both current loops rotate at synchronous speed. The field current is rotated by mechanical rotation of the conductors and the armature current is rotated electrically by the increase in current in the phase loop ahead, C -, and a decrease in current in the phase loop behind, B -, and in similar sequence around the armature winding. The mechanical rotation of the field-current loop produces mechanical power flow along the machine shaft. The electrical rotation of the armature-current loop produces electrical power flow in the armature terminals.

In a stationary field machine, normally a commutated dc machine, the field is held fixed. The armature conductor loops are rotated forward mechanically and the current loop is stepped around electrically (backward relative to the conductors) by the action of the commutator. The mechanical power results from the rotation of the current loop relative to the armature conductors. The torque angle is held constant by the mechanical action of the commutator.

In each class of machine the power can be increased by increasing the current in the loops or by increasing the intensity of the magnetic field produced by the current loops. However, the strength of the magnetic field produced by armature currents must be limited relative to the excitation field. In synchronous machines too high an armature reaction field produces machine reactances that are so high that the machine will not stay synchronized stably during system transients. In dc machines an armature field which is too high produces commutation problems. In both cases the basic problem is that too significant a fraction of the internal voltage is required to reverse the armature currents.

In machines of conventional copper and iron construction, this limit is complicated by the saturation of the iron magnetic circuit. In large synchronous machines, the limit on the currents has been carried beyond the limits for a short-air-gap magnetic circuit by placing air gaps of up to 5 inches in the machines. This requires high current densities in the field winding and significant exciter power.

The use of superconductors in the field winding of these large synchronous machines provides a significant increase in the field produced by the field winding with negligible exciter power. The armature currents can be increased in proportion to the excitation field without reaching the reactance limits for adequate stability. This would indicate a machine power proportional to B^2; however, it may not be possible to

to increase the armature current to so high a level because of cooling
limitations or power loss (efficiency) limitations. It may well be that the
upper limit on the power density of superconducting machines will be the
strength of the structure, especially under short circuit conditions.

On the other hand, dc machines have limitations not related to the
field strength so the application of superconductors does not increase the
power so drastically. In all dc machines the full power must be transmitted across sliding contacts. In slipring machines the relative velocity
at the brushes is limited by friction and/or wear to perhaps 200 ft/sec.
In the cummutated machine the voltage is limited to several hundred volts
by arcing between commutator segments. In the homopolar machine the
voltage is limited by the half turn windings, and the tip brushes must run
at full air-gap relative velocity. However, multiple disk and segmented
disk configurations and liquid metal brushes have advanced these limits.
The application of superconductors to the field winding of homopolar
machines has been demonstrated and is discussed in detail by A.D.
Appleton in Chapter 4.

The application of superconductors to the field winding of a commutated
dc machine has been studied at the Leningrad Polytechnical Institute and
an experimental machine has been constructed [1]. Unfortunately we have
very little information on this research effort.

In contrast to the dc machines, the ac synchronous machine has the
armature directly connected to the terminals and therefore can utilize high
voltages and modest currents and achieve ratings as large as 1200 MW
with conventional iron and copper construction. These large machines
are mechanically limited by rotor tip velocity (centrifugal stress) and by
rotor lateral flexibility (critical speeds). The use of superconductors in
the field winding offer the possibility of significant increases in machine
rating within the mechanical limitations, since the superconductors extend
the electrical limits. The applications to large machines is most promising
because the cryogenic refrigeration equipment becomes a small part of
the superconducting machine cost [2] as the size of the machine is increased.

B.   Early Superconducting Machines

Work on superconducting alternators was started by Woodson, Stekly,
et al., [3, 4] with the construction of a demonstration alternator employing a stationary superconducting field and a rotating room-temperature
armature, connected through sliprings. This work showed that superconductors performed quite satisfactorily in the field winding of a
synchronous machine.

Early experimental work by Oberhauser and Kinner [5] showed that
superconducting armatures and air-gap shear in a fluid at or near 4.2K
are not desirable features of a superconducting machine. The dissipation

associated with shearing an air-gap fluid (even a gas at low pressure) involves serious problems in maintaining superconducting temperatures and requires excessive power for refrigeration.

The work on alternators at MIT had as its initial objective the demonstration of the practibility of a rotating field winding with no air-gap shear at helium temperature. The combined efforts of the Cryogenic Engineering Laboratory and the Electric Power Systems Laboratory at MIT were brought to bear on the problem. The rotor for the first MIT experimental machine successfully produced a dipole field with a superconducting winding rotating at 3600 RPM. The superconductors, the liquid helium coolant, and the complete dewar vessel all rotate as a unit. The dewar is basically a conventional wide mouth dewar with special modifications to allow high speed rotation.

Experiments with the rotor showed no basic obstacles to the realization of a rotating superconducting field winding. Specifically the spin-up of the liquid helium and mechanical vibrations during rotation do not cause excessive evaporation of liquid helium. As expected, rotation does not adversely affect the characteristics of the superconductor. The experiment did, however, point out the extraordinary effectiveness of natural convection heat transfer at cryogenic temperatures in a 1000 g centrifugal force field.

The armature for the first MIT experimental machine was designed to take advantage of the high field produced by the rotor. The experiments with this machine have demonstrated that an adequately stranded and transposed armature winding can be constructed with the magnet iron serving only as an external flux shield. There was some concern that normal copper conductors exposed to the high ac field might have excessive eddy current and circulating current losses.

The machine was constructed with a vertical axis so that it could be filled with liquid helium and then run until the liquid boiled away. After a series of successful batch-fill runs, a vacuum insulated transfer tube was constructed with a rotatable joint [6]. With this apparatus the machine has been run for several hours and could be run continuously if coupled directly to a helium liquefier.

The steady-state electrical characteristics [7] of the machine were measured in open-circuit and short-circuit tests. These results confirmed the high power density possible in superconducting machines, in spite of the nonoptimum design of the experiment. The field winding achieved 3.2 Tesla both rotating and stationary. This is about 75 percent of the short sample capability of the superconducting wire. More recently the machine has been run as a synchronous condenser connected to the laboratory power supply. The machine has achieved 45 kVA overexcited.

C.   More Recent Machines

After success with the initial experiments with the first machine in 1969, the MIT group began the analysis of transient and fault behavior of superconducting turbine generators connected in an electric power system. The electrical characteristics of full-scale machines of preliminary design were developed [2]. Einstein [8] then used these characteristics to investigate the interactions between the superconducting machine and the power system during a fault on a transmission line and the subsequent action of the circuit breaker to clear the fault. He concluded that the superconducting machine can operate in a stable manner while connected to a power system which experiences normal disturbances.

Luck [9] studied the internal behavior of superconducting alternators during severe fault conditions. He analyzed in detail the currents, magnetic fields, and mechanical forces which result from faults. He concluded that the preliminary designs did not have sufficient mechanical strength to withstand the fault torques which are as high as ten times the full load torque. He devised a rotating, room temperature, cylindrical shield for the rotor which overcomes this problem without introducing an unacceptable heat leak into the cryogenic parts.

The group at MIT is now concentrating on the construction of a second experimental machine, Fig. 5. The objectives of this experiment are: to demonstrate a design and methods of construction which take maximum advantage the high fields produced by superconductors, to verify the analysis of the machine characteristics, to verify practical design equations, and to provide experience and background for a full scale prototype machine.

The size of the machine was selected as the largest experiment within the capabilities of the Laboratory facilities. The preliminary estimate of the machine rating is 2 MVA. The superconducting rotor is eight inches in diameter with a 24-inch active length. With the thermal isolation tubes, seals and bearings the length is about 6 feet, and the outer case is 3 feet in diameter.

The superconductors are bound on the tubular rotor with pretensioned fiber glass in an epoxy matrix. The rotor turns in a high vacuum. The armature is composed of finely stranded conductors held together by epoxy reinforced with continuous-filament fiber glass roving. The winding has a continuous electric-field gradient in the straight length and is forced-cooled with insulating oil. The armature winding fits into the bore of the ferromagnetic flux shield which in turn fits inside the aluminum outer case for the machine. The case can be used as an image-current flux shield when the machine is assembled without the ferromagnetic shield. Data from tests on the machine will provide the basis for selection of the method of shielding; however, preliminary calculations indicate an economic advantage for the image shield in large machines.

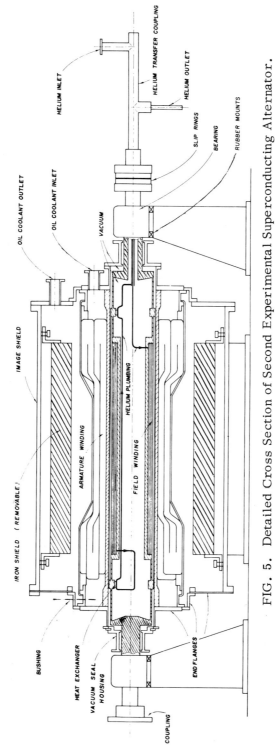

FIG. 5. Detailed Cross Section of Second Experimental Superconducting Alternator.

Since the results of the experiments with the first MIT machine were reported, the activities on synchronous machines with superconducting field windings has increased significantly. Several years ago Westinghouse initiated a program at the Westinghouse Research Laboratory to design and construct a 5 MVA machine [10]. Westinghouse is also designing and building a 5 MVA, 400 Hz airborne superconducting generator for the U.S. Air Force [11]. In England the International Research and Development Co. [12] and GEC power Engineering Ltd. [13] have been studying the potential of large superconducting synchronous machines for utility applications. Their studies present a good case for further development. Work on a small (21 kVA) rotating-armature, stationary-field superconducting machine has also been reported from East Germany [14]. J.T. Hayden at the Cranfield Institute of Technology, Bradford, England, is also constructing a similar 400 Hz machine but has not yet reported on this work. In the Soviet Union a small (100 kVA) rotating-armature superconducting machine has been constructed and tested by the Atomic Energy Agency in Moscow. We also understand that a 1 MVA rotating field superconducting machine is nearing completion at the Leningrad Polytechnical Institute. Only a schematic cross section of this machine is available to us at this time.

In conclusion, the studies by the MIT superconducting machine group have shown that the greatest potential for the application of super-conductors is in the field windings of large synchronous generators. Machines which are optimized to take full advantage of this potential promise the following advantages:

1.   Higher power density will provide smaller, lower cost machines and make possible machines of larger capacity.
2.   Lower machine reactances will improve machine stability in the power system.
3.   Heavy and expensive laminated steel is eliminated from the magnetic circuit.
4.   Higher economic terminal voltages for the machines are possible.
5.   The superconductor reduces exciter power requirements and increases the efficiency.
6.   Reduced temperature rise at the rotor surface during unbalanced fault conditions, increased $(I_2)^2 t$.
7.   Higher rotor critical speeds.

II.   BASIC SYNCHRONOUS MACHINE THEORY

This section presents the basic synchronous machine theory, with special attention to its application to air-core machines. Balanced steady state operation will be described in some detail, and the nature and analysis of transient behavior will be discussed briefly. The computation of machine parameters from a simple electromagnetic model will be described.

A.   Voltage Equation

Figure 6 shows schematically the basic electrical elements of a synchronous machine. It consists of a superconducting field winding f, surrounded by an armature comprised of three phases (a, b, c) with magnetic axes 120 electrical degrees apart. The coil symbols in Fig. 6 represent the actual spatial orientations of the respective winding magnetic axes for a two-pole machine. The analysis and conclusions of this section are applicable to any number of poles.

The system represented by Fig. 6 may be described by a set of algebraic equations relating the flux linkages and currents of the various windings. Because there is no non-uniformly distributed magnetic material in a superconducting machine, the self inductances of all coils are independent of rotor position. In other words, the machine exhibits no saliency. Mutual inductances between the field winding and the armature windings do vary with rotor position. We assume here that they vary sinusoidally with rotor angle φ. With these conditions, the flux-current relationship is simply

$$
\begin{bmatrix} \lambda_a \\ \lambda_b \\ \lambda_c \\ \lambda_f \end{bmatrix} = \begin{bmatrix} L_a & M_{ab} & M_{ab} & M\cos\varphi \\ M_{ab} & L_a & M_{ab} & M\cos(\varphi - \frac{2\pi}{3}) \\ M_{ab} & M_{ab} & L_a & M\cos(\varphi + \frac{2\pi}{3}) \\ M\cos\varphi & M\cos(\varphi - \frac{2\pi}{3}) & M\cos(\varphi + \frac{2\pi}{3}) & L_f \end{bmatrix} \begin{bmatrix} i_a \\ i_b \\ i_c \\ i_f \end{bmatrix} \tag{1}
$$

In the balanced steady state, armature currents are given by

$$
\begin{aligned}
i_a &= I\cos(\omega t) \\
i_b &= I\cos\left(\omega t - \frac{2\pi}{3}\right) \\
i_c &= I\cos\left(\omega t + \frac{2\pi}{3}\right) .
\end{aligned} \tag{2}
$$

In the absence of saliency, such a set of armature currents produces a synchronously rotating wave of magnetic field in the bore of the machine. That is, the field distribution within the machine due to these currents has no time variation when viewed from a reference frame rotating at synchronous speed ω.

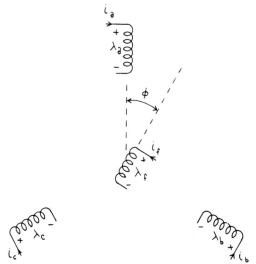

FIG. 6. Schematic Arrangement of Windings in a Synchronous Machine.

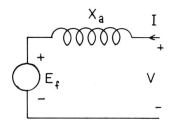

Fig. 7. Steady State Equivalent Circuit.

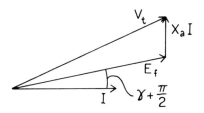

FIG. 8. Vector Voltage Diagram (Motor Reference).

A constant current I_f in a synchronously rotating winding likewise creates a synchronously rotating field distribution, the rotor field being separated from the armature field by some angle γ. The effect of either the armature or the field winding is, then, quite similar to that of a permanent magnet rotated about the axis of the machine. The mechanism of torque production may be simply understood to be the same as the tendency for two such magnets to align with one another.

Neglecting resistance, the phase a output voltage for balanced steady state operation can easily be shown to be

$$v_a = \frac{d\lambda_a}{dt} = -X_a I \sin(\omega t) - E_f \sin(\omega t + \gamma)$$

$$= \text{Re}\left[\left(jX_a I + E_f e^{j(\gamma + \frac{\pi}{2})}\right) e^{j\omega t}\right] \tag{3}$$

where $X_a = (L_a - M_{ab})$

and $E_f = \omega M I_f$.

Phase b and c are identical to phase a except for a time delay.

The second term in Eq. (3) represents the internal generated voltage; the first term represents a reactive voltage drop. This suggests the simply steady-state equivalent circuit of Fig. 7.

Equation (3) also suggests an extremely useful graphical representation for ac circuits. Each of the quantities may be represented by a vector in the complex plane, subject to the laws of vector addition. The resulting picture may be imagined to rotate with time at frequency ω, the projection at any instant of each vector onto the real axis representing the instantaneous value of the corresponding quantity. Figure 8 illustrates Eq. (3) at time $t = 0$.

B.   Power Equation

The instantaneous power input to phase a is given by

$$P_a = v_a i_a = V_a I \cos(\omega t + \psi) \cos(\omega t) = \frac{V_a I}{2}\left(\cos(2\omega t + \psi) + \cos(\psi)\right) . \tag{4}$$

The instantaneous power input to all three phases is

$$P = \frac{3}{2} V_a I \cos(\psi) \tag{5a}$$

or, **if** V_a and I are root-mean-square rather than peak quantities,

$P = 3V_a I \cos (\psi).$

The term $\cos (\psi)$ is called the power factor.

C. Conventions and Terminology

Power engineers accustomed to working with the complex quantities of Eq. (2) often refer to a useful fiction called "reactive power" Q defined by

$$P + jQ = 3V_a I^* \tag{6}$$

and $Q = 3V_a I \sin (\psi).$

The basic unit of power is, of course, the watt, although it is often more convenient to work with kilowatts (kW) or even megawatts (MW). Machine ratings are often expressed in terms of the product $3V_a I = |P+jQ|$ at a given power factor. This rating is always expressed in units of volt-amperes (VA) or kVA or MVA. Reactive power Q is referred to in units of VAR (for volt-amperes reactive) or kVAR or MVAR.

The sign convention illustrated in Fig. 6 is the motor convention. Power Eqs. (4) and (5) give electrical power input. Equation (6) defines reactive power absorbed to be positive if terminal current lags behind terminal voltage. If the positive direction for current in Fig. 6 is reversed (generator sign convention) Eqs. (4) and (5) give the electrical power generated. Equation (6) also holds, but now defines reactive power generated to be positive if current lags voltage. A typical vector diagram for generator operation with generator sign convention is shown in Fig. 9.

D. Another Power Equation

From Eq. (6) and Fig. 9.

$$P = \text{Re}(3V_a I^*) = \text{Re}\left[3V_a\left(\frac{V_a^* - E_f^*}{jX_a}\right)\right] = \text{Re}\left(-\frac{3V_a E_f^*}{jX_a}\right)$$

$$= \frac{3V_a E_f}{X_a} \sin \delta = \omega\left[3MI_f\left(\frac{V_a}{X_a}\right)\sin \delta\right]. \tag{7}$$

The phase angle δ between E_f and V_a is called the power angle. Equation (7) illustrates in very simple terms the permanent magnet analogy described earlier. The power output is the product of the angular velocity and the torque. The torque is proportional to the field strength of the field winding (measured by I_f), to the strength of the field due to armature voltage (measured by V_a/X_a) and to the sine of the angle separating the two field components.

E.   Transient Analysis

In many applications, synchronous machines are operated as components of large interconnected systems of ac apparatus. The transient behavior of a machine connected to such a system is a much more difficult problem to analyze than is steady state performance. A real system may have so many major components that even a simple model of each component leads to a cumbersome system model. Recourse is often made to reduced equivalent circuits to represent the system. The simplest system model is called an infinite bus. This term is used to refer to a source of balanced, constant amplitude, constant frequency voltages.

Even with the system represented as an infinite bus, analysis of transients is much more difficult than the steady-state analysis. The field winding and its associated magnetic field do not in general rotate exactly at synchronous speed. In addition there are magnetic fluxes in the machine not accounted for by the basic flux-current relationship Eq. (1), due to induced currents in the metallic rotor parts. Analysis of transients is facilitated by transforming the machine equations into a coordinate system rotating with the field winding.

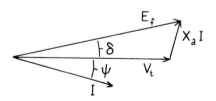

FIG. 9.   Vector Voltage Diagram (Generator Reference).

For transients occurring over a period of time short compared to the field winding time constant, the field winding flux λ_f may be assumed to be constant. If fluxes not included in Eq.(1) are not significant, Eq. (1) may be solved to yield a model much like the one represented by Fig. 9 with a different induced voltage E_f', and a more complicated representation of the reactive voltage drop. The armature reactance of such a model is called the transient reactance and E_f' is called the voltage behind transient reactance.

For transients occurring on a still shorter time scale (during which fluxes due to induced currents in the structure are significant) another model like Fig. 9 may be used, with E_f replaced by the voltage behind subtransient reactance E_f'', and with X_a replaced by a subtransient reactance X_a''. For air core machines, and to a good approximation for cylindrical   rotor iron machines the reactive voltage drop $jX_a''I$ is not more complicated than indicated by Fig. 9.

F.   The Classical Transient Problem

Let us describe a simple solution to a classical synchronous machine transient problem. A generator operating in steady state is connected to an infinite bus by a circuit breaker. At some time $t = 0$, a fault occurs on the line connecting the machine and the bus. The breaker opens, the fault is cleared, and the breaker recloses at $t = t_1$.

Before the fault, Eq. (7) holds, and the steady state $\delta = \delta_0$ is determined by the condition of equilibrium between the mechanical input power P_0 and the output power $\dfrac{V_a E_f}{X_a} \sin \delta_0$ (Fig. 10). When the fault occurs, output voltage and hence output power falls to zero. The input power P_0 is assumed to remain constant. Hence the rotor accelerates and δ increases. When reclosure occurs at t_1, the machine is not in equilibrium. δ_1 is greater than the equilibrium value δ_0 and the rotor has accelerated to a speed above the synchronous speed, in the process acquiring an excess of rotational energy $P_0(\delta_1 - \delta_0)/\omega$, represented by a shaded area in Fig. 10. In this nonequilibrium situation the power output is given by $P = \dfrac{E_f^t V_a}{X_a^t} \sin \delta$. E_f^t and X_a^t are almost universally taken to be transient quantities E_f' and X_{ad}' for iron core machines, but for superconducting air core machines, the subtransient qualities may be appropriate.

In either case, when the breaker recloses, the output power exceeds the input power. The excess is furnished by the rotor's stored kinetic energy. The rotor then decelerates. When the power angle arrives at δ_2

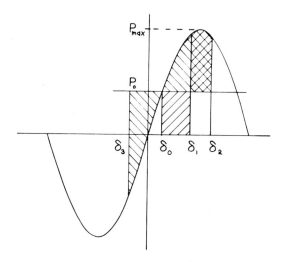

FIG. 10. Simplified Illustration of Transient Problem.

all the rotor's excess energy has been transmitted to the bus, and the
rotor speed is once again at synchronous speed. However, the output
power still exceeds input power, so the rotor continues to decelerate.
When the rotor arrives at the equilibrium value δ_0, the speed is below
synchronous speed, and δ must decrease all the way to δ_3 before the
rotor returns to synchronous speed. The power angle does not oscillate
forever between δ_2 and δ_3 because loss mechanisms, not included here,
eventually consume the store of energy flowing back and forth between
the rotor and the bus, permitting a return to steady state operation at
δ_0.

A more detailed discussion of synchronous machine theory, especially
as it applies to iron machines, may be found in a wide variety of textbooks.
The material presented here is covered in a more thorough but still
elementary way in "Electric Machinery," second edition, by Fitzgerald
and Kingsley.

G.   Calculation of Air Core Reactances

One of the principal differences between air core machinery and
iron core machinery is the distribution of the magnetic field within the
machine. In an iron machine, fields are to a large extent concentrated in
the air gap between the rotor and the armature. In an air core machine,
fields are distributed throughout the active volume. In an iron core
machine, magnetic circuit theory provides a good basic vehicle for

computation of self and mutual inductances. A purely radial magnetic field with no variation of intensity with radius may be assumed. Such computations provide reasonable estimates which need be modified only slightly by computed or empirical corrections to account for deviations of the model from the actual hardware.

A useful inductance model for air core machinery is easily derived from a simple model with some algebra and calculus. Figure 11 shows the basic physical configuration of an air core alternator. The inner-most winding is the field winding, the outer winding is the armature. The entire machine is inside a hollow cylindrical flux shield. This shield may be made of a material with high magnetic permeability or of a material with high electrical conductivity.

Two reasonable assumptions make the model of Fig. 11 amenable to a magnetic field analysis. These assumptions are: 1) that the axial current density J is uniform over the cross section of a winding and 2) that the machine is very long compared to its diameter, so that field variation in the axial direction and the effects of end connections may be neglected. With end effects neglected, the field winding is easy to treat. The only parameters needed to define any winding of this type are three radii, R_{fi}, R_{fo}, and R_s; one angle θ_{wf}; and one number of pole pairs, $\underline{p}$. Each of the three armature phases may be defined by a similar set of parameters. Because the differential equations for magnetic fields are linear, super-position may be employed, and a complete description of the magnetic fields in the bore of the machine may be obtained by addition with appropriate angular displacement of four solutions to the magnetic field problem of Fig. 12.

Even this problem may be broken down into simpler problems. The radial thickness of the winding may be broken down into a series of infinitesimally thick current sheets, each with surface current density $K = JdR$. The angular dependence of current density is periodic and pulse shaped (see Fig. 13) and thus is well represented by a Fourier series

$$K(\theta) = \Sigma K_n \cos(np\theta)$$

Thus, the basic magnetic field problem is illustrated in Fig. 14. A sheet current with constant radius and sinusoidal angular distribution is concentric with a uniform boundary (with $\mu \to \infty$ or $\sigma \to \infty$). The magnetic fields due to such a current distribution are

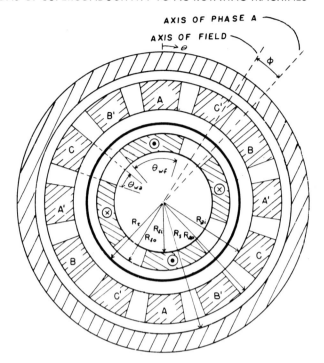

FIG. 11.   Cross Section of a Superconducting Alternator.

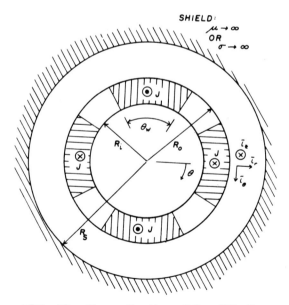

FIG. 12.   Cross Section of One Winding.

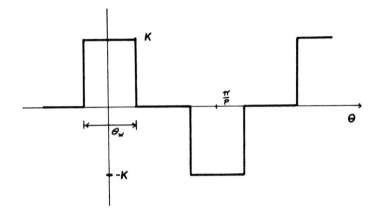

FIG. 13. Angular Dependence of Current Density.

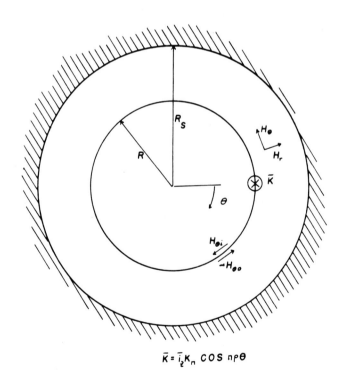

$$\bar{K} = \bar{i}_z K_n \cos np\theta$$

FIG. 14. Basic Magnetic Field Problem.

for $r < R$.

$$H_{ri} = -\sum_n \frac{K_n}{2} \left(\frac{r}{R}\right)^{np-1} \left[1 \pm \left(\frac{R}{R_s}\right)^{2np}\right] \sin(np\theta)$$

$$H_i = -\sum_n \frac{K_n}{2} \left(\frac{r}{R}\right)^{np-1} \left[1 \pm \left(\frac{R}{R_s}\right)^{2np}\right] \cos(np\theta) \; ,$$

(8)

and for $r > R$:

$$H_{ro} = -\sum_n \frac{K_n}{2} \left(\frac{R}{r}\right)^{np+1} \left[1 \mp \left(\frac{r}{R_s}\right)^{2np}\right] \sin(np\theta)$$

$$H_{\theta o} = \sum_n \frac{K_n}{2} \left(\frac{R}{r}\right)^{np+1} \left[1 \mp \left(\frac{r}{R_s}\right)^{2np}\right] \cos(np\theta) \; .$$

Wherever a double sign ($\pm$ or $\mp$) appears, the upper sign applies to the ferromagnetically shielded case ($\mu \to \infty$); the lower sign applies to the conductively shielded case ($\sigma \to \infty$).

After largely disassembling the machine to arrive at the simple model of Fig. 14, it is conceptually straightforward to reconstruct the geometry of Fig. 11. Eqs. (8) are really increments of field due to an entire winding. The field due to a winding can be determined by integrating the appropriate incremental contributions over an interval of radius from R_i to R_o.

The inductances needed to use Eq. (1) can readily be obtained from the field equation by further integration. Each coil may be considered to be composed of a distribution of coil elements, each characterized by a radius r and an angular position ψ. Each coil element, with the corresponding element in an adjacent coil half, links flux which is the integral of $\mu_o H_r$ over the area of the concentric cylindrical surface included between elements, times the numbers of turns in the element,

$$d^2\lambda = \mu_o d^2Nl \int_{\psi - \pi/\gamma}^{\psi} H_r(r, \theta) r d\theta \qquad . \tag{9}$$

The total number of turns per differential element of winding area is

$$d^2N = \frac{\text{Number of turns}}{\text{Area}} \, d\text{Area} = \frac{2N}{p\theta_w(R_o^2 - R_i^2)} \, r dr d\psi \tag{10}$$

Total flux linked by a winding is then the integral of Eq. (9) over the area of the area of the winding:

$$= \frac{2\mu_o 1N}{\theta_w (R_o^2 - R_i^2)} \int_{-\frac{\theta_w}{2}}^{\frac{\theta_w}{2}} \int_{R_i}^{R_o} \int_{\psi - \pi/p}^{\psi} H_r(r,\theta) r d\theta \; dr d\psi \quad (11)$$

If the field expression in the integrand of Eq. (11) is for the winding over which the integration is performed, the result is the self flux linkages. If the field expression is for a different winding, the result is the mutual flux linkages for the two windings. In either case the appropriate inductance is given by the flux linkages divided by the current in the winding generated the flux. This entire magnetic field analysis has been performed in painstaking generality by Professor Kirtley of MIT [15] and in a thesis [16].

III.   EXPERIMENTAL MACHINES AT MIT

Two experimental machines with superconducting rotors have been built and tested at MIT. The first machine, 45 kVA, was built to demonstrate that a machine with a rotating superconducting field winding could be operated at high speed while superconducting. The second machine, 2 MVA, was built to test features that are applicable to very large machines.

A.   First Experimental Machine [7]

The first experimental machine, Fig. 15, consists of a rotating dewar vessel, Fig. 16, which contains the superconducting field winding. The room-temperature stator, Fig. 17, is wound with normal copper conductors and has no magnetic iron teeth. The stator is surrounded by a laiminated iron magnetic shield. Liquid helium is supplied to the rotor through a rotating coupling.

The rotating dewar, Fig. 16, is essentially a wide mouth stainless steel dewar which has been braced with radial support wires so that it will not vibrate when rotating at high speed. The superconducting field is wound on a support structure which is attached to an evacuated neck plug which fits into the dewar. A copper radiation shield in the dewar is cooled by liquid nitrogen stored in the neck plug.

The dipole field coils are wound from a single filament of Nb-Ti alloy wire 0.254mm in diameter. The filament is coated with copper stabilizing conductor to an outside diameter of 0.508mm. The winding has 5500 turns and is scramble wound. The coils are immersed in liquid helium. The maximum current is 54 amp corresponding to a maximum flux density of 3.2 Tesla. The gas cooled current leads to the field coils are composed of two concentric tubes braided from small copper wire. The leads are insulated with woven glass cloth.

FIG. 15.   First Experimental MIT Machine.

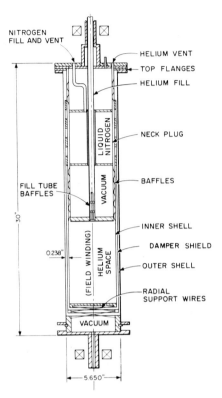

FIG. 16. Cross Section Through Rotor of First Experimental
Alternator.

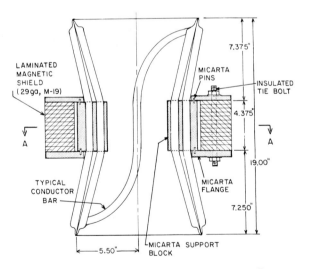

FIG. 17. Axial Cross Section Through Stator.

The air cooled armature coils are wound with litz wire consisting of 133 strands of No. 18 copper wire. The coils are supported on a structure made of phenolic plastic which is attached to the laminated iron magnetic shield which surrounds the armature. Small, well transposed strands are required in the armature conductor because the conductors are fully exposed to the high fields rather than being shielded by iron teeth as in conventional machines.

Liquid helium for cooling the rotor follows the path: storage dewar, stationary transfer rube, rotatable coupling, rotating transfer tube, field winding. The centrifugal acceleration keeps the field coils immersed in the liquid. The cold helium vapor flows out to cool the current leads and the dewar neck tubes. The warm exhaust helium is not collected.

The 45 kVA machine has been subjected to cryogenic, mechanical and no load electrical tests. These tests have demonstrated that:

1. Liquid helium can be contained within a cryogenic vessel while the vessel rotates at 3600 RPM.
2. The superconducting winding is not adversely affected by the rotation.
3. The rotating superconducting winding can operate as the field winding of a synchronous machine.
4. An adequately stranded and transposed armature winding can be constructed.
5. Liquid helium can be supplied continuously to the rotor from a stationary source.
6. The machine can operate as a synchronous condenser to 45 kVA overexcited.
7. The characteristics of the machine can be adequately calculated.

B.   Second Experimental Machine [17, 18]

The goals of the second experimental machine differ considerably from those of the first. Whereas the first machine was built to show feasibility of the concept, the second was constructed to test features that are applicable to very large machines. It is designed to take advantage of the special nature of the superconducting field winding, as described above.

In order to show the most potential, it was decided that the second machine should be as large as possible, limited by the machine tools in our laboratory. This resulted in a machine with a nominal rating of about 2 MVA, Fig. 18.

In order to provide adequate thermal isolation for the cold parts of the machine, the rotor runs in a continuously pumped high-vacuum environment. A stationary epoxy-coated phenolic tube forms both the outer wall of this chamber and the inner support for the armature winding.

FIG. 18. 2 MVA Alternator Ready for Test.

Oil-buffered carbon face seals are used to prevent air leakage along the shaft. The armature, operating at ambient temperature, is oil cooled.

Figure 5 is a cross section of the machine, showing improved design elements.   The rotor mandrel is a 6-in. OD tube of austenitic stainless steel. The superconducting field winding occupies a 1-in. thick annulus around the mandrel. A vacuum tight stainless steel tube covers the field winding and confines the liquid-helium coolant, which enters and leaves the rotor through a rotatable union, Fig. 19, at one end of the machine. Internal helium plumbing and heat exchangers are located within the vacuum space, Fig. 20.

All cold parts of the machine are supported by thermal distance pieces consisting of thin walled tubes of stainless steel. In order to reduce heat leak into the 4.2K region, part of the helium outflow stream is circulated through heat exchangers midway along the thermal distance pieces. In this way, most of the support heat leak is intercepted at an intermediate temperature of about 20K.

Around the outside of the rotor is a thin copper shell which is thermally connected to the intermediate temperature heat exchangers. This is a shield for both thermal radiation and alternating magnetic fields.

Field excitation will be supplied by a controllable solid-state rectifier/ inverter through carbon brushes on copper sliprings. Current leads to the field winding will be cooled by a controllable stream of cold helium gas, part of the exit stream from the field winding space.

The stator consists of an armature winding attached to the phenolic vacuum tube, a laminated iron magnetic shield, and an aluminum outer casing. The armature winding mounts to the aluminum outer case, and is independent of the iron shield. With this design, the machine may be assembled without the iron shield so that the case can serve as an eddy current flux shield. Dimensions of the experimental machine are given in Table 1 and expected characteristics are given in Table 2.

The field winding is constructed from a single length of Niobium-48 percent Titanium superconducting wire with a high purity copper matrix for stabilization. The rectangular wire is 0.125 by 0.050 in and has 24 transposed superconducting strands, each 0.010 inch in diameter. It has a copper to superconductor ratio of 2.6:1. The winding is made up of six layers, with each layer composed of two saddle coils. The winding has a total of 668 1/2 turns.

Fig. 21 is a detailed section drawing of the field winding, illustrating the layered fiberglass support structure and helium coolant passages. Each layer of the winding is held down by a series of bands consisting of continuous filament glass roving and epoxy. These bands were wound in place with the glass roving pretensioned to about 50,000 psi. They serve

TABLE 1

DIMENSIONS
SECOND EXPERIMENTAL ALTERNATOR
(in inches)

Stator	
inner radius of conductors	5
outer radius of conductor	7.675
inner radius of iron shield	12
outer radius of iron shield	16
straight section length	24
iron shield length	40.5
end turn length (total two ends)	16
aluminum shield inner radius	18
Rotor	
inner radius of conductor	3
outer radius of conductor	4
overall length of conductor	38
length of straight section	25.5
radius of electrothermal shield	4.47
thickness of electrothermal shield	0.085

TABLE 2

TENTATIVE PROPERTIES
SECOND EXPERIMENTAL ALTERNATOR

Terminal characteristics	
output (iron shield)	3070 kVA
output (aluminum shield)	2100 kVA
phase currents	740 A rms
Speed	3600 r/min
Number of poles	2
Number of phases	3 (delta)
Synchronous reactance	
iron shield	30%
aluminum shield	29%
Transient reactance	
iron shield	21%
aluminum shield	22%
Subtransient reactance	
iron shield	11%
aluminum shield	13%
Armature self inductance, per phase	
iron shield	1.04 mH
aluminum shield	0.68 mH
Phase-to-phase mutual inductance	
iron shield	-0.47 mH
aluminum shield	-0.3 mH
Number of turns per phase	38
Armature resistance, per phase	22 mΩ
Number of field winding turns	668.5
Field current density	1.25×10^8 A/m^2 rms
Armature current density	2.5×10^6 A/m^2 rms
Losses	
armature resistance	36 kW
iron shield	6 kW
aluminum shield	20 kW
eddy currents, armature	2 kW
Total losses, percent of output	
with iron shield	1.4
with aluminum shield	2.8

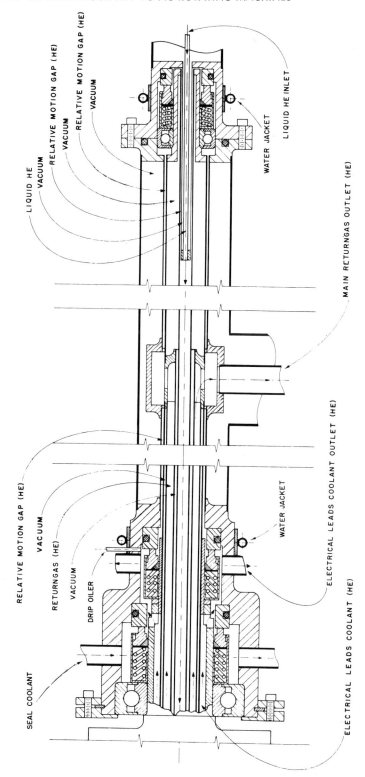

FIG. 19.   Cross Section View of Transfer Coupling.

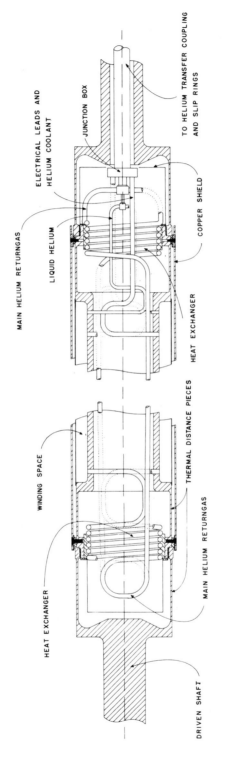

FIG. 20. Cross Section View of Rotor.

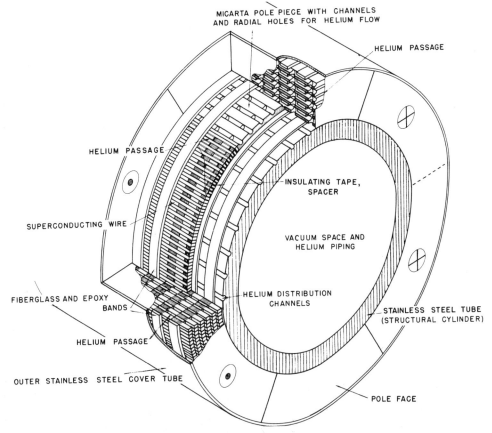

Fig. 21. Details of Field Winding Construction.

Fig. 22. Field Winding During Construction.

to support the field winding against centrifugal and magnetic forces. Spaces in the pole centers and end turns are filled with layers of phenolic plastic the same thickness as the winding. The winding has numerous cooling passages. Radial passages are made by weaving insulating tapes that serve as spacers between the superconducting wires, and by holes drilled in the pole center pieces. Axial passages are cut in the pole centers and stainless steel mandrel. Circumferential passages are formed from the spaces between fiberglass bands.

With the interlayering and epoxy bonding of the prestressed fiberglass structure and the layers of superconductors, it is expected that the field winding will have adequate cooling while achieving the structural advantages of a potted coil. It will produce a field of approximately 2.5T inside the winding for a field current of 800 A, at about 50 percent of the short sample capability of the superconductor.

Fig. 22 shows details of the rotor winding in the end turn region, with temporary supports in the foreground. In this picture the fiberglass banding is white, the superconductor is black. The b-stage epoxy glass tape is woven between turns to space and insulate the turns (b-stage epoxy is essentially "half-cured." Upon heating it first softens and then cures.) When cured it provides turn-to-turn and layer-to-layer bonding.

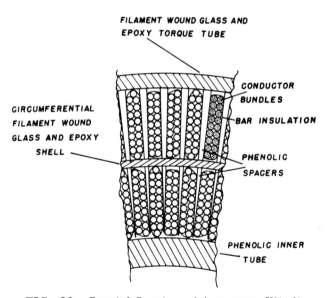

FIG. 23. Partial Section of Armature Winding.

The armature for the machine has been constructed to demonstrate a number of new features which are required to take full advantage of the characteristics of the field winding. About 1/2 of the potential gain from using a superconducting field winding will originate from the increase in the cross section of active conductor in the armature. This results from replacing magnetic iron and ground insulation with a high-strength nonconducting structure. In this design the mechanical structure and insulation are the same.

The armature winding is made up of 228 individual conductor bars arranged around the central phenolic tube as shown in Fig. 23. To keep eddy-current losses low, the individual conductors had to be kept small and well transposed. The bars are made of 18 groups, each of seven strands of film insulated No. 19 (0.036-in.) round wire. Each seven-strand group is twisted once every 0.875 in. The eighteen groups in the bar are transposed one turn in 36 in, or about one and a half times in each bar. To decrease circulating losses, each group is connected separately in the end turns, so that the effective transposed length is the total length of a phase winding.

The seven-strand groups in each bar are bonded together by a b-stage epoxy/glass tape which was served on each group prior to assembly. Similarly, the bars in each layer of the winding are bonded into a mono-lithic structure by b-stage tape, wrapped around each bar prior to assembly into the winding. A wrapping of fiberglass roving and epoxy over each layer completes the monolithic structure. The outer layer of fiberglass roving serves as the main mechanical support for the armature winding. This helically wound tube provides torsional strength to with-stand reaction torque, hoop strength to withstand radial forces, and flexural strength to support the weight of the winding between end flanges.

The winding is cooled by transformer oil, flowing axially, through the passages formed by round conductors in rectangular bars. Since the winding is totally immersed, the oil serves as additional electrical insulation. The outer layer of fiberglass forms a liquid tight covering to confine the coolant.

Since the armature is not mounted in slots in grounded iron, the wind-ing may be constructed so that only turn-to-turn insulation is required, with proper ordering of the conductors, except in the end turn region, where phases must cross over. Here, a simple cylindrical layer of in-sulation serves as phase-to-phase insulation. The end turns of this machine are supported in the same fashion as the straight section. Figure 24 shows the armature winding before the outer layer of fiberglass was applied. Figure 25 shows the furnished armature, with the outer layer and support flanges.

Additional details of the design and construction of the machine have been given in reports to the EEI [19, 20, 21].

FIG. 24. Completed Armature Winding Before Final Structural Wrap
was Applied.

FIG. 25. Completed Armature.

The 2 MVA machine has been subjected to a series of mechanical, cryogenic and electrical tests. Although the test program has been temporarily interrupted by difficulties with the leads to the field winding, the basic characteristics of the machine have been established.

The major mechanical tests of the machine have been to determine the dynamic characteristics of the high speed rotor, both warm and when filled with liquid helium. The long slender rotor of the machine was originally designed to run as a rigid rotor mounted on very stiff spherical-race ball bearings. However, the rotor proved to be more flexible than expected so that the system had a first critical speed at 3000 RPM. Since this resonance was essentially undamped it was impossible to run the rotor over 3000 RPM.

The bearing mounts were tuned by placing them on pads of rubber with significant internal damping characteristics. The pads were sized so that the natural frequencies of the rotor on the tuned mounts were in the range of 1100 to 2000 RPM with the rotor moving as an essentially rigid body. With the flexible mounting, the natural frequency associated with flexure of the rotor is higher since the rotor now vibrates more as a free beam rather than as a beam fixed at the bearings. Data taken with a shaker indicate that this natural frequency is about 5000 RPM.

With the tuned mounts the rotor could be run through the mount resonances with small amplitude vibrations because of the adequate damping of the motion of the bearing mounts. Since no critical speeds are near the operating speed of 3600 RPM, the rotor runs with low vibration levels at design speed.

The rotor was dynamically balanced while at room temperature. The balance did not change as the rotor was cooled down and filled with liquid helium. In fact the vibration levels were somewhat smaller with the rotor at operating temperature. This experience shows that a liquid-helium-filled rotor can be successfully operated above the first critical speed for the system without any vibration problems associated with motion of the liquid.

The field winding has been cooled to operating temperature five times, and the cryogenic performance [22] has been excellent each time. Under steady state conditions, 15 to 20 liters per hour of liquid helium are required to cool the rotor, essentially independent of the rotor speed.

The relatively small mass and efficient thermal design of the rotor provided for rapid cool down with a minimum amount of cryogenic liquids. The rotor may be cooled to about 90K in 3 hours with about 60 liters of liquid nitrogen. The rotor can then be cooled and filled with liquid helium in about one hour with about 25 liters of liquid helium.

The liquid-helium transfer coupling, Fig. 19, for the machine has had good cryogenic performance with no indication of excessive heat leak to the incoming liquid helium. It now is well established that a rotatable bayonet coupling can be made with a small enough clearance so that no rubbing cold seal is required. Clearances on the order of 0.010 appear to be small enough for transfer tubes smaller than about 1/2 inch in diameter. The transfer coupling for the 2 MVA machine did have rubbing problems in the return gas bayonet when the rotor was undergoing excessive vibrations. The face seals in the transfer tube have required lubrication and external cooling to prevent excessive wear and temperature rise.

The high vacuum system for thermal insulation for the rotor has performed well during every cryogenic test. Originally there was considerable concern about the high vacuum shaft seals used on the main shafts of the machines; however, the seals have performed without problems or leakage during the tests. Back streaming from the diffusion pumps and oil leakage from the shaft seals has coated the vacuum spaces external to the rotor with oil. However, the oil has not been detrimental to the performance.

The machine has been subjected to the initial part of the planned test program. Open circuit tests have been performed at low speed (900 RPM) to maximum field current 800 amp. Short circuit tests have been carried out at low currents (500 A). The oil cooling system for the armature has been tested by applying an external current to the armature winding. The armature winding has also been given a surge voltage test (2400 V peak). Since the winding is not in slots in a grounded iron core, the standard high voltage test for an electric machine is not suitable. Figure 26 shows typical test data for the machine together with the expected performance.

The tests have shown that the basic electrical characteristics of the machine are essentially as predicted by the design equations. The armature stranding and transposition has been shown to be effective, keeping the eddy currents and circulating current losses at a low value. In addition the circulating currents in the delta connected armature winding have been shown to be small since the field winding was designed to produce a minimum third harmonic. The armature cooling tests showed the cooling of the winding to be quite adequate; however, the external oil-to-water heat exchanger was found to be marginal. The surface area of this heat exchanger has been doubled. The thyristor exciter which was designed for the machine has supplied maximum field current (800 A) to the field winding while operating as expected from design calculations. Current control has been satisfactory and the voltage ripple does not appear to have had adverse effects on the field winding.

The superconducting field winding has operated to maximum current without quenching. This indicates that the mechanical support structure and the liquid helium cooling system for the winding are satisfactory. During the open circuit run of May 25, 1973, the electrical leads to the

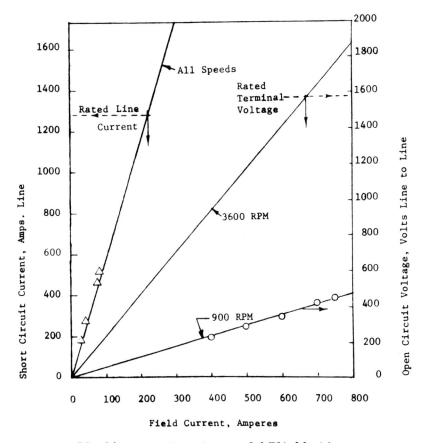

FIG. 26.   Test Data for the 2 MVA Machine.

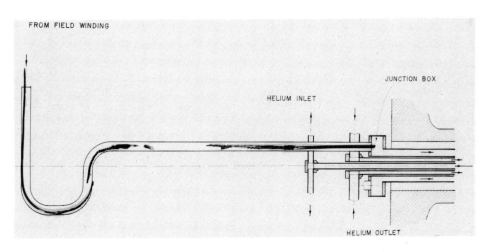

FIG. 27.   Detail of Lead Failure.

field winding failed between the normal to superconducting junction and
the field coils. Figs. 20 and 27. Test conditions were 900 RPM, 800 amp
field current. The failure occurred about 30 seconds after coming to
800 amp. No advance warning of the failure was indicated by the monitoring instruments. Failure was caused by overheating and burnout of a
two inch length of superconductor. Signs of overheating were also
evident in the second, unfailed lead.

Portions of the field winding and the outer stainless steel cover
were also damaged during this failure. This was a result of the discharge of the coil. The inductive field winding maintained an arc between
the ends of the failed lead. When voltage across the arc grew too high,
new arcs were formed to complete the circuit between the ends of the
winding through the stainless steel rotor structure, principally through
the thin (0.030 inch) cover tube outside the winding (Fig. 5, Fig. 20).
The arcs burned holes through the cover tube opposite the lead tubes.
Components of the arc and helium exiting through these holes entered
the vacuum space and resulted in an overpressure. There is no evidence
that loss of superconductivity occurred in the winding proper prior to
the lead failure. Helium flow at the time of failure was 16.6 liters/hour
through the main return and 2 liters/hour along the leads.

The exact location of the failure is shown in Fig. 27 which is a
drawing of the stainless steel lead tube. The failed lead has been placed
on the drawing to illustrate the location of the failure. Undamaged superconductor is evident on both sides of the failure.

Following redesign and repair of the current leads, the test program
will be resumed. The projected tests include: operation at 3600 RPM
to full field current (armature open circuited), operation at 3600 RPM
to full armature current (armature short circuited), operation as a
synchronous condenser on a 13.8 kilovolt bus of the Cambridge Electric
Company. Completion of the tests is anticipated by mid 1974.

IV.   THERMAL ISOLATION AND CRYOGENIC REFRIGERATOR

The superconducting field winding for a synchronous machine must
be thermally isolated from the room temperature surroundings so that
the temperature may be maintained below about 10K. Since refrigeration
below 10K requires a complex refrigerator which has a very low coefficient
of performance, it is very important to have an effective system for
thermal isolation. The problem is complex since the electrical performance
of all practical superconductors increases with reduced temperature.
A well-developed superconducting maching must be a careful compromise
between the conflicting electrical, mechanical, and thermal requirements.
The basic design is a superconducting winding, capable of withstanding
its own self magnetic forces, suspended in a high vacuum enclosure
with carefully designed mechanical and electrical connections to the
room temperature environment.

A.   Gas Conduction and Radiation Heat Leak

The surfaces of the cold parts of the rotor must be insulated by a high vacuum. The vacuum problems are considerably simplified by the cryopumping capability of liquid helium. A normally clean vacuum space which has been pumped to 10^{-3} to 10^{-2} Torr with a good mechanical vacuum pump will perform satisfactorily when sealed and cooled to liquid helium temperature, provided no helium is leaking into the vacuum space.

Even with high vacuum insulation, the heat leak by thermal radiation from a surface at room temperature to a surface at liquid helium temperature is too high to tolerate. The heat transfer by radiation is reduced by the introduction of a refrigerated radiation shield between the room temperature and the helium temperature surfaces. With a single cooled radiation shield the radiation heat transfer is negligible compared to the heat conduction along the mechanical connections to the cryogenic parts of the rotor.

B.   Torque Tube and Mechanical Supports

The unique problem in thermal isolation is to support the conductors against the reaction forces and to provide sufficient lateral stiffness to prevent rotor vibrations without introducing an unacceptably high heat leak. The armature and field windings in rotating machines are placed symmetrically about the axis of rotation of the machine; thus the mechanical connections to the rotor from room temperature need to carry only reaction torque in addition to the weight of the cold parts of the rotor.

Since the thermal conductance per unit length of mechanical support is directly proportional to the cross section, the support should have minimum cross section. For a given torque the cross section will be minimum if the support is a tube of maximum permissible radius made from a material with the highest working stress. In a high speed machine the structural material of the torque tube must withstand the combined centrifugal stress and torsional stress. As the diameter is increased, the strength remaining to carry the torsional moment decreases so that a radius for minimum tube cross section will be reached. For a yield strength of 60,000 lbs/in^2 and a speed of 3600 RPM the radius for minimum cross section is about 18 in.

With the cross section of the torque tube defined, the overall thermal resistance will be directly proportional to the length of the support tube. The torque tube may be lengthened until the flexibility becomes unacceptable.   Thus, the refrigerator power required to keep the superconducting winding at liquid helium temperature is related to the dynamic characteristics of the rotor.

C.   Gas Cooled Supports

However, the power required can be significantly reduced by cooling the torque tube at various points along its length. The effect of the cooling is to increase the temperature gradient near the warm end and to reduce the temperature gradient near the cold end. The refrigerator power is reduced because the work required per unit of refrigeration effect (called the coefficient of performance) is smaller for the intermediate refrigerator than for the helium temperature refrigerator. The absolute minimum refrigerator power for a support of a given area and length will be obtained if the intermediate refrigeration is continuously distributed along the length of the support, for example, cooling by a series of ideal refrigerators each cooling a short length of the support as is shown in Fig. 28.

The optimum distribution of ideal refrigeration effect can be determined by calculus of variations [23]. The variational problem is considerably simplified if the minimization is done in terms of entropy generation and temperature rather than in terms of refrigerator power and geometric variables. The problem reduces to search for the heat flux q in the support as a function of the temperature T which minimizes the entropy generation integral

$$\dot{S}_g = \int_{T_L}^{T_H} \frac{q(T)}{T^2} \, dT,$$

subject to the geometric constraint

$$L/A = \text{const.} = \int_{T_L}^{T_H} \frac{k(T)}{q(T)} \, dT,$$

where k(T) is the thermal conductivity of the material. The resulting heat flux function for absolute minimum refrigerator power is

$$q(T) = [\lambda k(T)]^{1/2} \, T$$

where λ is the Lagrange multiplier.

If the thermal conductivity of the support is independent of temperature, the optimum cooling is for a heat flux which is linear in temperature, and an exponential temperature distribution. However, low-thermal-conductivity construction materials tend to have conductivities which are strongly dependent on temperature. Figure 29 shows the optimum heat flux distribution for a stainless steel support.

The practical way to provide refrigeration distributed along the length of a support is to pass a stream of cold gas up the support from the cold end to the warm end. The gas should flow through a passage with

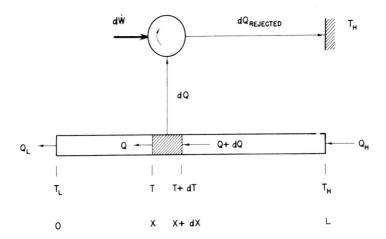

FIG. 28. Mechanical Support with Continuously Distributed Refrigeration.

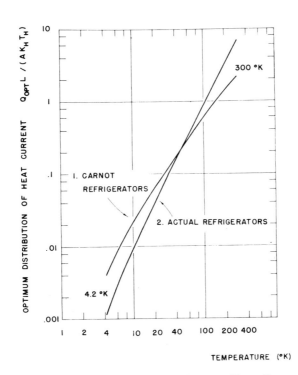

FIG. 29. Optimum Distribution of Conduction Heat Current (for type 304 stainless steel support).

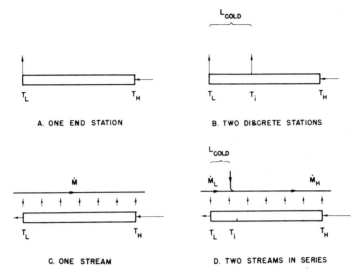

FIG. 30. Practical Methods of Cooling a Mechanical Support.

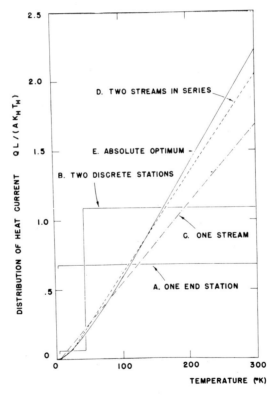

FIG. 31. Heat Current Distributions for Practical Cooling Methods
          (with Carnot Refrigerators and Type 304 stainless steel support).

TABLE 3

Comparison of Various Cooling Schemes Based on the Carnot Refrigerator Power Requirement
(for Type 304 Stainless Steel Support)

Scheme	$\dot{S}\dfrac{L}{AK_H}$	$\dfrac{\dot{W}}{\dot{W}_{opT}}$	T_i (K)	$\dfrac{L_{cold}}{L}$	$\dot{m}_L\dfrac{Lc_p}{AK_H}$	$\dot{m}_H\dfrac{Lc_p}{AK_H}$
a. one end station	47.87	9.015				
b. two discrete stations	10.167	1.915	43	.38		
c. one gas stream	5.84	1.1			1.691	
d. two gas streams (in series)	5.328	1.0034	23.2	.176	.968	2.127
e. absolute optimum	5.31	1.0				

TABLE 4

Comparison of Various Cooling Schemes Based on the Actual Refrigerator Power Requirement
(for Type 304 Stainless Steel Support)

Scheme	$\dot{S}\dfrac{L}{AK_H}$	$\dfrac{\dot{W}}{\dot{W}_{opT}}$	T_i (K)	$\dfrac{L_{cold}}{L}$	$\dot{m}_L\dfrac{Lc_p}{AK_H}$	$\dot{m}_H\dfrac{Lc_p}{AK_H}$
a. one end station	376.00	19.930				
b. two discrete stations	51.87	2.749	37	.43		
c. one gas stream	25.07	1.329			1.709	
d. two gas streams (in series)	19.47	1.032	39	.325	.726	2.853
e. absolute optimum	18.87	1.000				

sufficient surface area so that a small temperature difference will be
required to transfer heat from the support to the gas. In the terms of the
previous analysis, a stream of ideal gas with a constant specific heat
gives a refrigeration effect uniformly distributed in temperature. Figure
30 shows four practical schemes for cooling a support, including two which
employ gas cooling streams. Each of the schemes has been individually
optimized for a stainless steel support and the resulting heat flux distributions are shown in Fig. 31 for comparison with optimum distribution for
distributed ideal refrigerators. Table 3 shows the performance of each
of these schemes compared with the absolute optimum for distributed
refrigerators.

The support cooling problem may also be solved for cooling with
refrigerators with Carnot efficiencies which are less than ideal. Reasonable
values for the efficiency (Work actual/Work ideal) is given by

$$\eta = 0.7 \left(\frac{T}{300K}\right)^{0.427}.$$

The optimum heat flux distribution for distributed refrigerators is shown
in Fig. 29. Table 4 shows the performance of the various cooling schemes
of Fig. 30 when nonideal refrigerators are employed.

The conclusion to be drawn from Tables 3 and 4 is that a
mechanical support cooled with a single stream of gas flowing at the
optimum rate requires only about 30% more refrigerator work than for

distributed refrigerators. Cooling with two gas streams reduces the power
to within 3% of the minimum. More complex cooling arrangements are
obviously not required.

D.   Designing the Refrigerator for the Superconducting Machine

The design of a superconducting machine must be coordinated with
the design of the cryogenic refrigeration system which will provide the
cooling. As we have seen in the previous discussion, it will probably
be desirable to design a refrigeration system which will provide a cooling
effect to the machine at several different temperatures or distributed
over a range of temperature. The design of the refrigerator is significantly influenced by the relative magnitude of these requirements. The
design will be most effective if the refrigerator and the superconducting
machine are designed and optimized together as a unit.

The heat which must be removed from the cold parts of the machine
can originate as electrical losses in the cold areas, as heat conducted
along the mechanical and electrical connections, and as heat transferred
across the vacuum insulation by thermal radiation. The heat which must
be removed from the superconducting winding fixes the required flow of
liquid helium to the winding area, assuming complete vaporization of the
liquid. The cold vapor may then be returned to the refrigerator or it
may be utilized to cool the supports, leads, and radiation shields and
returned to the refrigerator at room temperature. Since the specific
heat of helium vapor (gas) is relatively constant, the refrigeration
available from the vapor is uniformly distributed over the temperature
range from 4.2K to 300K.

If the electrical losses in the superconductor and the parts of the
machine at the temperature of the superconductor are high, then more
helium vapor may be available than can be effectively utilized for cooling
the supports, leads and schields. In this case the design, Fig. 32,
should include a stream of helium vapor at 4.2K flowing back to the
refrigerator. On the other hand, if the electrical losses at 4.2K are
small, then not enough helium vapor will be available and additional flow
from the refrigerator will be required for distributed cooling, Fig. 33.
As we have seen, it will probably be desirable to supply this additional
flow at one or more temperatures above 4.2K. The complexity of employing several cooling streams from the refrigerator to the machine must
be balanced against the reduction in refrigerator power. Since we expect
the superconductors in the machines to be shielded by a room temperature
conducting shell, the electrical losses will be small and a system similar
to Fig. 33 will probably be used.

The next design compromise which must be made between the refrigerator and the machine involves the selection of the operating temperature
for the superconductor. As we have discussed previously, the performance

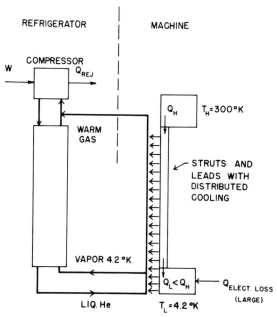

FIG. 32. Refrigeration System for Large Electrical Losses at 4.2K.

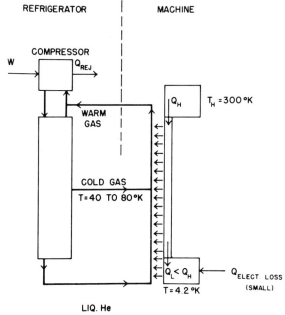

FIG. 33. Refrigeration System for Small Electrical Losses at 4.2K.

of superconducting materials improves significantly with the reduction in the operating temperature. However, a reduction in the temperature increases the refrigerator complexity and size and increases the power requirement.

Figs. 32 and 33 show the superconducting winding cooled by vaporization of liquid helium. The two phase, liquid-vapor, flow which is required may lead to flow instabilities in the cooling system. If this presents a serious problem it may be necessary to utilize forced convection of high pressure liquid helium to cool the superconductor. In that case the cold stream returning to the refrigerator, Fig. 32, will be liquid at a slightly higher temperature than the stream entering the winding. One problem is that about eight times as much flow (1K temperature rise in the liquid) will be required for the same cooling effect as the vaporization of liquid. The c_p of the liquid is about 0.6 cal/gmK and h_{fg} is 4.9 cal/gm. Supercritical state helium (T in the range 5 to 10K, p in the range 3 to 10 atm) is often proposed for cooling superconductors.

In this region the specific heat of helium is high and high heat transfer coefficients are expected. However, the rapid changes in density with temperature at constant pressure can cause flow and heat transfer instabilities in much the same manner as the density difference between phases causes instabilities in the two phase situation. In addition the temperature in this region may well be above the optimum for superconductor operation in a high field.

E.   Basic Refrigerator Design

A brief review of the design of cryogenic refrigerators will show how the operating temperature of the superconductor influences the power requirements and the design of the refrigerator. In addition we will be able to show how distributed cooling requirements influence the design of the refrigerator. More extensive reviews of the design of refrigeration and liquification systems are given in references [24, 25 and 26].

The basic refrigeration unit for a single-fluid system to reach 4.2K is shown in Fig. 34. The unit consists of a compression section, and an expansion section joined by a regenerative heat exchange section. The compressor takes in gas at a low pressure and room temperature and delivers the gas at high pressure and room temperature. If the compressor is to be thermodynamically reversible, the gas must remain at room temperature throughout the compression. In general this is not economically practical because the gas would have to be cooled while in the actual compressor. This would require very slow motion of the compressor and a very large machine. In a practical compressor, the gas is compressed adiabatically and then cooled in a separate heat exchanger by cooling water or air. Even if the adiabatic compression is reversible, the heat exchange in the cooling heat exchanger is still irreversible because of the heat transfer across the large temperature difference (entropy generation).

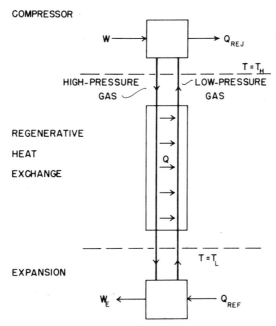

FIG. 34. Basic Refrigeration Unit.

The adiabatic compressor then requires more power than an ideal
reversible isothermal compressor section. The efficiency, w isothermal/w
adiabatic, can range from 50 to 75% depending upon the pressure ratio.
As the pressure ratio approaches one, this loss becomes small; however,
the mass of gas which must be circulated for a given refrigeration effect
becomes very large. Usually the pressure ratio is high enough so that
it is worthwhile to have several compression stages and intercooling
between stages. Three to four stages are commonly employed for pressure
ratios of 15 to 20.

For the same reasons as in the compression section, the expansion
section normally consists of an abiabatic expansion machine followed by
a heat exchanger which absorbs heat from the area being refrigerated and
returns the gas to the heat exchange section at a temperature somewhat
below the temperature of the gas entering the expander. If the simple
single unit of Fig. 34 is cooling a region at constant temperature, a similar
loss (entropy generation) occurs as the cold gas from the expander cools
the fixed temperature load. This loss is also reduced by operating at a
low pressure ratio. When the simple unit of Fig. 34 is used to pre-cool
units at lower temperature, the refrigeration produced in the expansion
machine is used effectively over the temperature range from expander
inlet to outlet temperature.

The heat exchange component of the unit of Fig. 34 is a counter
current heat exchanger, exchanging heat between the two constant pressure

streams of gas. In the simple case of constant and equal specific heats
for the two streams, the first law of thermodynamics requires the temperature difference between the streams to be constant along the length of
the exchanger.

For a fixed pressure ratio, the heat exchanger determines the lower
limit temperature for a single unit, Fig. 34, and the temperature ratio
for the expansion machine is also fixed. Thus, the temperature difference
across the expander is directly proportional to the expander inlet temperature T_L. As T_L is decreased the temperature difference required for
heat transfer in the heat exchanger increases with $T_H - T_L$. The lower
limit T_L is reached when the heat exchange ΔT becomes equal to the
expander ΔT so that the entire refrigeration output of the expander is required to make up the loss due to imperfect heat exchange in the regenerative heat exchange section. The lower limit temperature can be decreased
by use of a larger heat exchanger and by the use of a higher pressure ratio.
The use of a single expansion has not proved practical for refrigeration at
4.2K.

For very low temperatures, the simple system must be modified to
allow better use of the heat exchange surface in the regenerative heat
exchanger. This is done by cooling the heat exchanger along its length
with auxiliary refrigeration so as to decrease the heat exchange ΔT in the
cold region of the heat exchanger.

It is generally recognized that the heat exchanger surface will be
used most effecitvely if the ratio $\Delta T/T$ is uniform along the length of
the heat exchanger. For working gas with a constant specific heat this
can be achieved by an unbalance of the flow in the heat exchanger so that
the flow of cold low-pressure gas exceeds the flow of high-pressure gas.
Fig. 35 shows how a series of expansion machines can be used to unbalance
the flow. Actually five expansion machines are required to span the temperature ratio from 300 to 4.2K since the practical temperature ratio for
one expansion is about 2.

It is interesting to note that the requirement for constant $\Delta T/T$
for minimum heat exchanger loss for a fixed heat exchanger surface
area can be simply demonstrated by the method of minimization of the
entropy generation. The variational technique is applied to ΔT as a
function of T subject to the constraint of constant heater exchanger surface
area.

Fig. 35 shows how a stream of low pressure gas can be taken from
the refrigeration system and used to remove heat from the generator supports and leads over a range of temperature. The extraction of the low
pressure gas from the refrigerator must be compensated for by increasing
the flow through the precooling stages so as to maintain $\Delta T/T$ constant in
the heat exchanger. Fig. 35 also shows how a stream of gas can be
taken from the refrigerator at an immediate temperature and used to increase the cooling available along the warm portion of the generator supports.

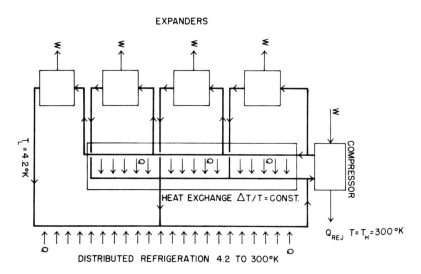

FIG. 35.   Refrigeration System with Refrigeration Effect Distributed
Over the Range of Temperature.

In a practical refrigeration system it may be simpler and less
expensive, but less efficient to use fewer than the ideal number of precooling expansions. In this case the heat exchange $\Delta T/T$ would have to
be greater than the optimum at some points and smaller than optimum
at other points in the heat exchanger. When the specific heat of the gas
is varying rapidly with temperature the heat exchange $\Delta T/T$ cannot be
maintained constant so the expansion machine flows must be adjusted to
keep the average $\Delta T/T$ near the optimum value.

F.   Current State of the Art in Cryogenic Refrigeration

The current state of the art in refrigerators is available in several
review papers [27, 28 and 29]. Strobridge [29] indicates that the costs
of cryogenic refrigeration plants are approximately correlated by the
input power. He suggests that the cost in dollars is given by 6000 times
the input power in kilowatts to the 0.7 power. He also reports that the
reversible (Carnot) refrigerator work is 6 to 20% of the actual work with
the average 10% for plants with a 100 watt refrigeration capacity up to
about 18% for 1000 watt capacity. He found that the refrigerator efficiency
(expressed as a ratio to the reversible work) was more a function of
refrigerator capacity than of the temperature level refrigeration.

A significant factor in the state of the art of 4.2K refrigerators
and liquefiers is the relatively small number of medium and large size
plants which have been constructed. For this reason the compressors

and heat exchangers are taken directly from designs which have been
developed for other applications. Only the expansion machines have been
developed specifically for this refrigeration service. Probably less than
5 well developed expansion machines are available with perhaps one or
two sizes of each basic design. The refrigeration plant designs for single
plants must of necessity be a compromise between the available equipment
and the specific requirements for the plant. If superconducting generators
become a practical reality and a significant number of units are constructed, it should be economically feasible to develop refrigeration
equipment specifically for the generator.

G.   Refrigerator Development

At the present time all practical 4.2K refrigerators and helium
liquefiers employ reciprocating compressors. Since reciprocating machinery does not have a reputation for as high a reliability as turbomachinery,
it is natural to consider turbocompressors for the refrigerator for
superconducting machines. With the current state of the art in turbocompressors this is practical only for very large refrigeration plants [27].
For compression from 1 to 15 atm, a machine would need 32 stages, 7
intercoolers and an input power of 12 MW. Work has been underway for
some time on turbocompressors for small helium refrigeration plants,
but the results have not been too encouraging. The efficiencies are just
too low. The small systems which have been designed [30, 31] have
tended to employ low pressure ratios which reduce the compressor requirements but require a larger and more effective heat exchanger.

Progress in turboexpanders has been more significant. The rotors
of turbomachines are tip speed limited by centrifugal stress. The performance is related to the tip Mach number of the rotor, that is the ratio
of blade tip speed to the local velocity of sound. In the helium compressor
the high velocity of sound in helium (low molecular weight) keeps the
performance low. In the low temperature turboexpanders the velocity of
sound is reduced (proportional $T^{1/2}$) so the expanders have very reasonable performance. The larger turboexpanders have room temperature
oil lubricated bearings and are very reliable. The medium and small
size turboexpanders have room temperature gas bearings which run very
reliably, except they are sensitive to abnormal operating conditions and
thus tend to have accidents. Cold self-actuating gas-bearing expanders
are now operating only in experimental systems.

There will be continued work on turbocompressors, but progress
will be difficult. The axial screw compressor may well provide some of
the advantages of rotating machinery. These compressors are now
being adapted to cryogenic refrigerators by British Oxygen. Gas bearing
or oil bearing turboexpanders will probably be used in the refrigerators
for superconducting machines.

Large capacity cryogenic refrigerators employ standard plate-fin brazed-aluminum heat exchangers. In plants with input powers less than 100 hp, helically wound fin-tube heat exchangers are commonly used. More recently perforated plate and wire screen heat exchangers have been developed for smaller refrigeration systems. The plates or screens are placed transverse to the flow and the flow passages are formed by sealing between the plates or screens with epoxy resin. The development of better heat exchangers is basically a problem of reduction of manufacturing cost so that a larger surface heat exchanger will be economic. The cost of interconnecting and headering the heat exchange is a significant part of the cost, which could probably be subject to a considerable cost reduction if a sizable number of identical units are constructed.

V. PRESENT PROBLEMS, CURRENT DEVELOPMENTS AND FUTURE PROSPECTS

A. Introduction

Significant progress has been made on the realization of practical superconducting ac machines; however, much work remains to be done before these machines will be standardly accepted by the electric utility industry. Small machines have been constructed and operated, to show that it can be done. Machines have not yet been tested under utility system conditions. Most important superconducting machines must eventually be economically competitive with conventional generators. Any estimates of the eventual costs of superconducting machines must be very tentative at this time since many of the basic features of the machines are not yet firmly established. It is, however, generally agreed that the potential of superconducting machines is greater the higher the rating of the machine.

B. Current Problems

The most important problem area in superconducting machines which needs more effort at this time is the behavior of the machines under fault conditions. A superconducting machine which is generating power for a utility system must be capable of operating through the transients which result from short circuits on the transmission system (for example, from the flashover of an insulator). The machine must withstand the large fault currents until the circuit breaker opens. It then must withstand the over voltage, acceleration and the increase in field current until the circuit breaker automatically recloses the circuit. The machine must then decelerate and remain stably synchronized and continue to deliver power to the system. This sequence of events associated with a normal system fault will occur many times during the life of a machine and must be considered an operating requirement for the machine.

An even more severe requirement for a large turbo generator is that it survive an abnormal fault without major structural or electrical damage. These conditions are not expected, but they do occur. The most severe situation is when the short circuit is in the machine, on the bus to the unit transformer or in the low voltage winding of the transformer. This gives the maximum fault currents since the impedance is a minimum. These conditions are so severe that conventional machines are never tested for a full load, full voltage terminal short; however, they are usually guaranteed to come through without major damage.

The construction of a large superconducting generator to withstand a full terminal short is a difficult task. The problem is especially challenging since the measures taken to mechanically strengthen the machine tend to reduce the power density and tend to reduce the transient performance of the machine (the ability of the machine to remain synchronized after a normal circuit breaker operation).

This design problem is complex with many interacting and competing aspects. The following discussion should give some appreciation of the situation. The superconducting field winding must be protected from the ac components of the magnetic field produced by the armature current. (The largest steady fields are produced by unbalanced phase currents in the armature.) In the present machine concepts this is done by placing a conducting shell around the field winding and rotating the shell with the rotor. Eddy currents in this rotor-shield stop the ac fields before they reach the superconductor. When the machine experiences a terminal short circuit very large ac fields are produced by the short circuit currents in the armature. Eddy currents flow in the conducting shell around the rotor as a result of these fields. These eddy currents interact with the magnetic field to produce forces which tend to crush the rotor shield. The forces are equivalent to a pressure of several thousand psi. In addition a large alternating torque is placed on the rotor shield. This torque can be 4 to 10 times the steady full load torque of the machine.

A simple way to think of this situation is to remember that at the time of a fault, the magnetic field is locked into the armature as well as the rotor. As rotation continues, short circuit currents flow to maintain the field locked in the armature and eddy currents flow in the shield to maintain the rotor field in spite of the demagnetizing effect of the armature currents. After 1/2 revolution the two fields are in direct opposition producing maximum currents in the armature and in the shield. This results in severe concentration of the flux between the shield and armature which gives rise to the high magnetic pressure tending to ovalize the shield and the armature.

Thus, the electrically conducting shell which is to shield the superconductor from ac fields will have to take all the transient electromagnetic forces on the rotor and must, therefore, be considerably stronger than the field winding support.

In the experimental machines which have been built the rotor shield has operated at low temperature in order to have high conductivity and effectively shield the superconductor with a thin shell and with a very small dissipation. But the thin cryogenic shield will not withstand the fault loads. If the cold shield is made strong enough to stand fault loads, it will have a rather high heat leak down its torque tube. As a result of these considerations it now seems best to design the rotor shield to operate at room temperature as is shown in Figs. 36 and 37.

As is shown in Fig. 37, the mechanical stress in the shield can be decreased by lengthening the air gap between the shield and the armature, as well as by increasing the thickness of the shield support. Both of these changes reduce the magnetic coupling between the field and the armature. This reduces the reactances of the machine resulting in lower power density and reduced transient performance. In effect the machine will not pull back into synchronism as quickly after a circuit breaker operation to clear a transmission line fault. Clearly then the shield must be carefully designed to the mechanical limit if the superconducting machine is to realize its full electrical potential.

Another factor which complicates the design of the rotor shield is the tendency of synchronous machines to develop self-excited hunting oscillations during steady state operation, especially when run over-excited at low load. During these oscillations the speed of the machine oscillates at a few cycles per second about the steady equilibrium value. In a rotating coordinate system the torsional inertia of the rotor is vibrating against the spring in the magnetic field. In conventional synchronous machines, short circuited damper windings are placed on the rotor to absorb the energy of these oscillations and prevent their being self-excited. In the superconducting machine the shield must serve as the damper winding. If the shield were perfectly conducting, it would be a perfect shield, but it would provide no damping effect to the hunting oscillations. Therefore, the shield must have enough resistance to provide adequate damping yet provide the necessary shielding for the superconductor. These conflicting requirements can be met by using a thick shield with an appropriate resistivity.

An additional complication in the design of the shield for the super-conducting field winding is the requirement for reasonable rapid changes in field current. When the generator is connected to a power system the voltage must be regulated and controlled in response to changes in load and other changes in the power system. Conventional generators have the capability of changing field current and thus terminal voltage at a rate of about 10 percent per second, and an acceptable superconducting machine must at least meet this performance.

When the superconducting field winding is surrounded by a conducting shell, any change in field current will induce currents in the shell which tend to keep the field constant. The voltage of the generator will change only as the eddy currents decay because of the resistance of the

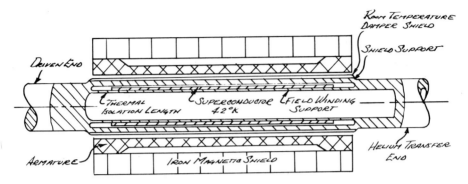

FIG. 36. Superconducting Synchronous Generator with a Room-
Temperature Damper Shield.

1000 MVA Design Rotor Cross Section

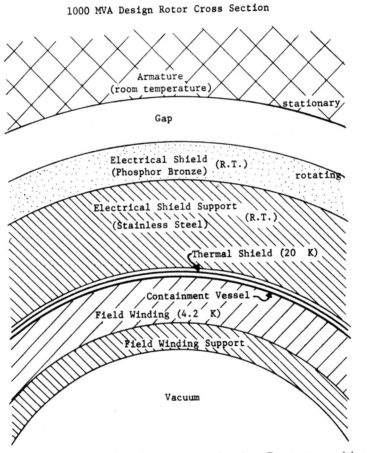

FIG. 37. Partial Section of a Superconducting Generator with a Room-
Temperature Damper Shield.

shield. If the shield has a low resistance for maximum shielding the voltage regulation will be slow. Thus, a trade off must be made between ac shielding and speed of response for voltage regulation. At first it might appear that the controls for the machine could be designed to over drive the field current to achieve good response in spite of the shield. This can produce considerable flux concentration between the field winding and the shield and thus a problem with flux quenching the superconductor.

When a turbine generator that is connected to a power system experiences a fault on the terminals, the only way to remove the fault is to remove the excitation current from the field winding. Since the field winding has an upper limit on voltage, a finite time is required. During this time large eddy currents are flowing in the shield causing a significant heating and temperature rise. The problem in the shield design is to provide enough heat capacity to absorb this Joule heating without melting the surface of the shield or producing other serious thermal damage. In conventional machines the rotor is designed to withstand one per unit negative sequence currents for 5 to 10 seconds without significantly damaging the rotor. Superconducting machines will have to meet this requirement unless they can be designed to be able to reliably reduce the field current in shorter times than is possible with conventional machines.

In summary the rotor shield for a superconducting field winding must provide adequate shielding to prevent quenching the superconductor without unacceptable degradation of the mechinical and electrical performance of the machine.

C.   Superconducting Field Winding

The design of the field winding for a rotating machine involves all of the problems of constructing large superconducting coils, with the additional problems associated with high speed rotation. Any additional weight or size in the coils or their support structure adds very significantly to the centrifugal forces. Thus, the design of the rotor, as in the design of an aircraft, must utilize all space and weight with high effectiveness. The system cannot be made rigid enough to have negligible flexibility, therefore, the designer must carefully avoid resonant vibrations and critical speeds.

One problem in the rotating field coils which is not present in stationary systems is associated with the pressure gradient in the liquid helium as a result of centrifugal force. As a particle of liquid helium moves adiabatically along a radial line through the pressure field caused by centrifugal force it will change pressure at constant entropy. Since liquid helium is compressible it will also experience a change in temperature with pressure at constant entropy. If liquid enters the center of the rotor and flows to the outer radius, it will increase in pressure and temperature as shown in Fig. 38.

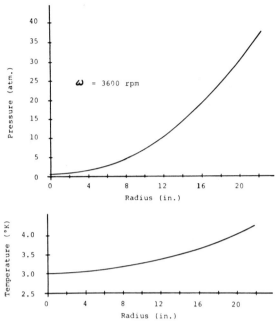

FIG. 38. Pressure and Temperature Distributions for Abiabatic Radial
          Flow of Liquid Helium in a 3600 RPM Rotor.

        A number of methods come to mind to overcome this problem of
temperature rise and the resulting degradation of the performance of the
superconductor. The first, as shown in Fig. 38, is to simply subcool
the incoming liquid sufficiently. This puts severe requirements on the
transfer heat leaks or requires very low pressure (and density) in the
vapor return lines. If a two phase mixture is fed to the coil the temperature rise will be significantly increased. This solution significantly
increases the problems in the cryogenic refrigerator.

        Another method is to use the forced circulation of pressurized
liquid helium, that is at a sufficiently low temperature at the center line
of the coil. The difficulty is that large quantities of helium must be circulated and the refrigeration of the exit stream must be carefully utilized.

        Another method is to utilize conduction through a solid to conduct
heat leaking to the winding at large radius tc the liquid helium which is
confined to areas near the center of rotation. This seems to be practical
for low heat leak designs during steady state operation. However, the
winding will probably have significant heating during swings in the magnetic
field during transients. It appears to be difficult to provide sufficient
thermal conductance to limit the temperature rise in the superconductor
to an acceptable level.

A.O. Lorch [32] has proposed a cooling system for the coils which utilizes a Joule Thomson valve at the outer radius of the rotor. This overcomes the thermodynamic problem but leaves the problem of controlling the valve in the rotor.

Considerable work needs to be done on this problem of cooling with liquid helium in a high speed rotor. In particular the various solutions mentioned or other better solutions need to be experimentally tested.

Some of the work at MIT has focused on the problem of transient heating of the superconducting winding and its support structure during operation of the machine [33]. The results indicate that the instantaneous rate of heating is high enough that it will not be practical to provide the reserve refrigeration requirement with a steady flow of helium. However, the total energy input is a reasonable quantity (because of the short heating time) and the energy is uniformly distributed. The practical solution appears to be to supply a reservoir of liquid helium which can be evaporated (or increased in temperature) to supply the transient refrigeration requirement. We are now working in this area.

The centrifugal force on the helium coolant may also have a significant effect on the thermal stabilization of the superconductor and the propagation of a normalizing zone along the conductor and through the coil. This is the result of the influence of the centrifugal field on natural convection and boiling heat transfer from the superconductor to the helium. The recent experimental difficulty (the lead failure discussed in III) with the MIT 2 MVA machine was apparently the result of an incomplete understanding of these effects.

The mechanical structure for supporting the field coils presents a number of new problems. In most of the superconducting coils which have been constructed, the active field region is inside the coil and no restrictions are placed on the volume available for external support structure. In the superconducting machine the active field is outside the coil and any external support structure limits the utilization of the flux produced by the coil. The most nearly conventional support for the coils would consist of externally blocked and clamped coils within a strong stainless steel tube. The MIT experiment has utilized fiber-glass-reinforced epoxy resin for a structure which is made integral with the coil windings. The basic problem is the new fabrication techniques required and the significant differences in elasticity and expansion coefficient between the fiber-glass-epoxy and stainless steel. In addition the long term strength of this material is not well known when exposed to cylic forces at cryogenic temperature. The high strength and electrical insulating properties of the fiber-glass-epxoy does, however, indicate the potential for a very effective structure in terms of space and weight.

The changes in dimensions which occur as the cold parts of the rotor are cooled down cause serious design problems which are not usually discussed in detail. Since the rotor and its support structure must be rigid as well as strong, many of the techniques usually employed for cryogenic apparatus are not acceptable. A fixed support system must be rigid enough to avoid resonant vibrations and yet flexible enough not to yield on thermal cycling. A contact support must be able to slide during cooldown but must not continually vibrate and wear during normal rotation. The two support systems which have actually been demonstrated have been described earlier; both were fixed support systems. The first was a large diameter tube at one end and radial spokes at the other end. Axial deflection of the spokes allowed for axial contraction of the rotor. Radial contraction of the spokes was allowed by deflection of the attachment rings to an out-of-round shape. The other system was built with tubes at each end of the rotor. Axial changes in the rotor were compensated by motion of the shafts through the shaft seals. The major region of thermal stress in this design occurs at regions of sudden change of the temperature gradient in the tube. (A linear gradient in a tube of constant expansion coefficient does not produce thermal stresses.) Other more effective systems for compensating for thermal changes in dimensions may well provide significant improvement of superconducting machines.

D.   Excitation for the Superconducting Field Winding

The problem of providing the field current for the superconducting machine is quite different from the problem for conventional machines. First there is the possibility of running the superconducting field in a persistent current mode. At this time there has been no practical suggestion for a way to do this and still have control of the voltage of the machine. Even with an external source of field current the superconducting field coils require significant power only when rapid changes in field current are required. For a full scale turbine generator the exciter may well provide only a few kilowatts during steady operation, but may have to supply perhaps 5 to 10 MW during fast changes in field current. Such an exciter will not be of conventional design, and it is not clear how it should be constructed.

The problem of the excitation of the superconducting field has been studied in some detail at MIT [34]. This work concentrated on the problem of the control of the current and voltage applied to the field winding, and concluded that a rectifier-inverter utilizing silicon controlled rectifiers would best meet the special excitation requirements of the superconducting machine. It was further concluded that the voltage ripple from the rectifier-inverter would not adversely influence the field winding. Control circuits were designed to provide a field current proportional to an input signal, with a bipolar maximum voltage limit to prevent excessive rates of change of field current. Work remains to be done to design a machine controller to set the field current (machine voltage) in response to changes which occur in the power system to which the generator is connected.

Work also remains in the selection of a source of ac power to supply the SCR excitation system. The problem is to guarantee exciter power even when the machine is disconnected or when the power system is experiencing abnormal circumstances. In conventional machines these considerations have usually dictated a shaft driven exciter generator.

E.   Armature Winding and Structure

A significant part of the potential gain from the application of super-conductors in the field winding of synchronous machines is realized only when the armature is optimized to take full advantage of the high field produced by the field winding. The armature for the 2 MVA MIT machine has been constructed to demonstrate a number of the unique features necessary in such an armature. As has been described earlier, the experiment has shown that a finely stranded armature winding can be constructed. It remains to be proved that methods can be devised to manufacture full scale armature windings of this general design economically.

There is also a significant problem in scaling the armature design to a full size machine. It remains to be shown that a full scale armature winding can be designed and built with the electrical insulation serving as the major structural elements of the stator. This structure must withstand the normal 60 Hz ovalizing forces for a long time and it must also resist the severe forces during a short circuit. If such a structure can be built it will have a much better resistance to thermal cycling than a conventional winding. The armature will be a monolithic structure with conductors, insulation and coolant all in intimate contact. The conductors will remain at nearly the same temperature as the structure so that there will be no problems of allowing the conductors to expand relative to the support structure (core stack) as in conventional machines.

Although the MIT 2 MVA machine is wound [35] and cooled [36] to illustrate a method of construction which is suitable for high terminal voltage, the modest voltage of the machine does not actually demonstrate the high voltage capability. A machine of reasonable size (perhaps 50 MVA) will be required to demonstrate a significant increase in terminal voltage (perhaps to 69 kV).

F.   Plan of Action

Rapid progress in the realization of practical ac superconducting machines will require major concurrent efforts in four general program areas.

1.   Programs to prove that progress which has been demonstrated in the laboratory is applicable and satisfactory under actual power system conditions.

2.   Programs to establish understanding of the complex compromises between conflicting requirements which are required in the design of superconducting machines. Care must be exercised in avoiding judgments based on previous experience which is not applicable to superconducting machines.

3.   Programs to improve the performance and understanding of individual machine components. Progress here will require iteration with the previous area.

4.   Programs to develop new concepts, configurations and construction methods for superconducting machines. Progress here will lead back to area 1.

G.   New Concepts and Applications

A number of interesting concepts for superconducting ac machines have been devised by considering a rotating superconducting field winding as a source of intense ac magnetic field. This field is then incorporated into various machines in place of the ac field normally provided by an iron magnetic core. In effect such machines are composed of a field winding and two (or more) normal ac windings; that is dual armature superconducting machines [37].

The first such machine (Fig. 39) consists of a freely rotating field winding and two fixed armature windings. Electrically this machine has characteristics similar to a transformer combined with a synchronous condenser. If the two armatures can be changed in relative angular position, external control can be provided for the flow of real and reactive power at the terminal of the machine [38]. Practical realization of this machine will require progress in the construction of high voltage armature windings as will be discussed later.

The second dual armature machine, Fig. 40, consists of a free rotating field winding, a rotating armature winding connected to the machine shaft and a stationary armature. When the rotating armature of this machine has a fixed resistance and is connected to a mechanical load, the machine has characteristics similar to an induction motor, except that it has the high power density associated with superconducting machines. In addition the power factor of the machine may be controlled by means of the excitation current. If the rotating armature winding is connected to an external resistor, the machine operates as a wound rotor induction motor. When the shaft of the machine is connected to a prime mover, the machine is an induction generator. The separately excited field winding overcomes the excitation problems of conventional induction machines operating as generators. In this machine the field winding carries no torque since the two armatures provide the action and reaction with the field providing only excitation. This is similar to the action/reactions in the homopolar machine. This feature may well be especially important

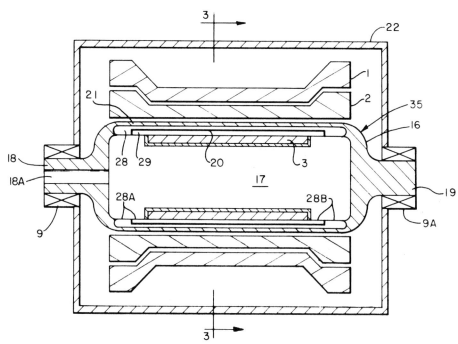

FIG. 39. Superconducting Transformer. See Reference [37] for Key
Numbers.

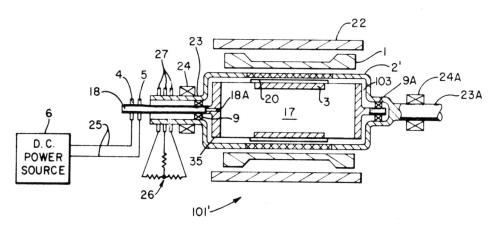

FIG. 40. Superconducting Induction Machine. See Reference [37]
for Key Numbers.

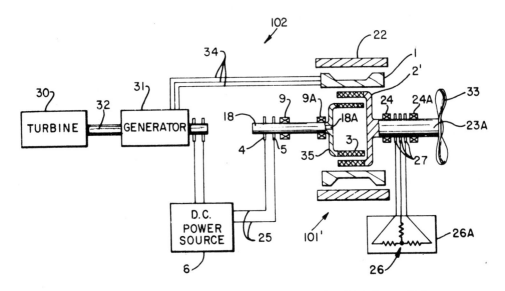

FIG. 41. Superconducting Induction Motor for Driving a Ship Propeller. See Reference [37] for Key Numbers.

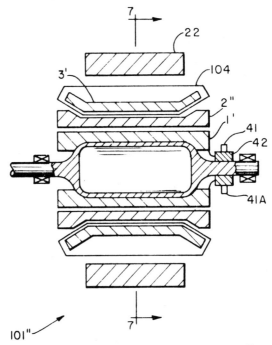

FIG. 42. Commutated Superconducting dc Machine. See Reference [37] for Key Numbers.

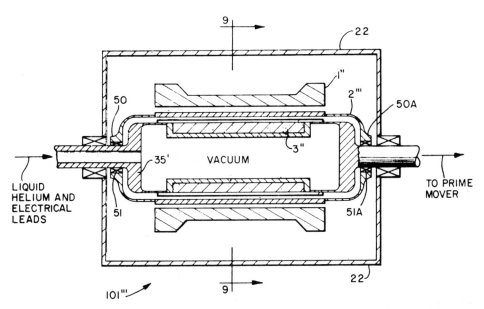

FIG. 43. Superconducting Synchronous Machine with an Inertial Shell for Fault Protection. See Reference [37] for Key Numbers.

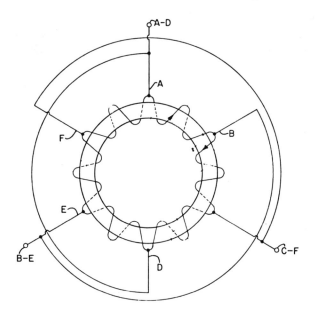

Six Phase Halves are wound in alternating sense
(A-B, C-D, E-F are left hand, F-A, B-C, D-E are right hand)

FIG. 44. Ring-Wound High-Voltage Armature for a Superconducting Machine

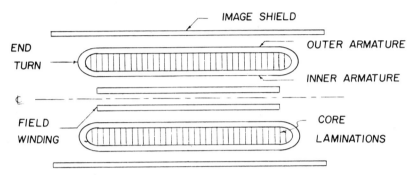

FIG. 45. Cross Section of a Ring-Wound High-Voltage Superconducting
          Machine.

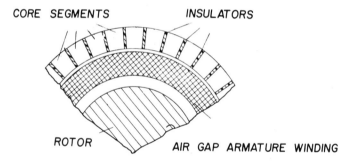

FIG. 46. High-Voltage Armature with a Segmented and Insulated
          Iron Core.

for ship drive motors requiring high torque at low speed and a minimum cryogenic refrigeration. Figure 41 shows a schematic representation of this application. The characteristics of this system are being studied at MIT.

Figure 42 shows how these two winding principles could be utilized to construct a commutated dc machine which required no reaction torque on the field winding. A moving winding and a stationary winding provide action and reaction, with the field winding providing only excitation.

Figure 43 shows a machine (generator) with the second armature turning freely between the field winding and the stationary armature. As long as this machine is operating at steady state the free armature has no function. If the generator experiences a terminal short circuit then the free armature is accelerated by the transient torques. Preliminary calculations indicate that sufficient inertia can be placed in the free armature to absorb the energy of the transient torque and thus shield the rotor and the machine shafts from the high transient torques. This armature can be in the form of a conducting shell, in effect a rotor shield which is free to rotate relative to the rotor. Under steady conditions the main field will magnetically lock the shell to the rotation of the field so that the shell will shield the rotor as previously discussed. The detailed transient behavior of this system is now under investigation at MIT.

Recently several new concepts have been invented to further extend the practical terminal voltage of superconducting machines. Figures 44 and 45 show a ring winding system for a superconducting machine. The winding consists of six sections helically wound, three right hand and three left hand. The sections are arranged in alternating sense to form a toroid. The diametrically opposite reactions are connected in parallel and the three pairs are then delta connected. This symmetric winding provides an absolute minimum gradient in the electric potential with individual turns adjacent, both electrically and mechanically. This system may be capable of potentials up to several hundred kV.

A second idea for a high voltage winding employs a segmented and insulated iron core as is shown in Fig. 46. The iron core is divided into electrically insulated segments which operate at the potential of the adjacent conductors. When this idea is combined with continuous gradient winding [35] it may be possible to construct iron core machines with considerably higher terminal voltages than is now possible; in effect, to construct a generator in the same way that some very high voltage transformers have been built. The segmented core can also be combined with the ring winding to provide more effective flux linkage in the machine.

REFERENCES

1. Kazovskiy, Kartsev and Shaktarin, Superconducting Magnet Systems, Science Publishing Company, Leningrad, 1967, pp. 186-190 (in Russian).

2. H.H. Woodson, J.L. Smith, Jr., P. Thullen, and J.L. Kirtley, IEEE Trans. Power Apparatus and Systems, Vol. PAS-90, No. 2, March/April 1971, p. 620.

3. H.H. Woodson, Z.J.J. Stekly, and E. Halas, IEEE Transactions on Power Apparatus and Systems, Vol. PAS-85, No. 3, March 1966, p. 264.

4. Z.J.J. Stekly, H.H. Woodson, A.M. Hatch, L.O. Hoppie, and E. Halas, IEEE Transactions on Power Apparatus and Systems, Vol. PAS-85, No. 3, March 1966, p. 274.

5. C.J. Oberhauser and H.R. Kinner, Advances in Cryogenic Engineering, Vol. 13, K.D. Timmerhaus, Editor, 1968.

6. P. Thullen, J.L. Smith, Jr., and W.D. Lee, Progress in Refrigeration Science and Technology, Proceedings of the XIII Congress of Refrigeration, Vol. 1, AVI Publishing, Westport, Conn., 1973.

7. P. Thullen, J.C. Dudley, D.L. Greene, J.L. Smith, Jr., and H.H. H.H. Woodson, IEEE Trans. Power Apparatus and Systems, Vol. PAS-90, No. 2 March/April 1971, p. 611.

8. Thomas H. Einstein, Electric Utility Power Generation, Sc.D. Thesis, Department of Mechanical Engineering, MIT, Cambridge, Massachusetts, December 1970.

9. David Lee Luck, "Electromechanical and Thermal Effects of Faults Upon Superconducting Generators," Ph.D. Thesis, Department of Electrical Engineering, MIT, Cambridge, Massachusetts, June 1971.

10. C.J. Mole, H.E. Haller and D.C. Litz, Proceedings of 1972 Applied Superconductivity Conference, IEEE Pub. No. 72CH0682-5-TABSC, p. 151.

11. L.R. Lowry, Proceedings of the 1972 Applied Superconductivity Conference, IEEE Pub. No. 72CH0682-5-TABSC, p. 41.

12. A.D. Appleton and A.F. Anderson, Proceedings of the 1972 Applied Superconductivity Conference, IEEE Pub. No. 72CH0682-5-TABSC, p. 136.

13. H.O. Lorch, Proc. IEEE, 120, p. 221 (1973).

14. D. Eckert, F. Lang, M. Endig, G. Muller, and W. Seidel, Proceedings of the 1972 Applied Superconductivity Conference, IEEE Pub. No. 72CH0682-5-TABSC, p. 128.

15. J.R. Kirtley, Jr., IEEE Winter Power Meeting (1971), paper No. 71-CP-155-PWR.

16. J.L. Kirtley, Jr., "Design and Construction of an Alternator with a Superconducting Field Winding, Ph.D. Thesis, Department of Electrical Engineering, MIT, Cambridge, Massachusetts, August 1971.

17. P. Thullen, A. Bejan, B. Gamble, J.L. Kirtley, Jr., and
     J.L. Smith, Jr., Advances in Cryogenic Engineering, Vol.
     18, p. 372.
18. J.L. Kirtley, Jr., J.L. Smith, Jr., and P. Thullen, IEEE
     Transactions Paper T 73 137-7. Presented at the IEEE PES
     Winter Meeting, New York, Jan. 28 - Feb. 2, 1973.
19. P. Thullen, J.L. Smith, Jr., and H.H. Woodson, Edison
     Elec. Inst., Res. Progress Rep. RP-92, Aug. 9, 1971.
20. P. Thullen, J.L. Smith, Jr., and J.L. Kirtley, Edison Elec.
     Inst., Res. Progress Rep. RP-92, Feb. 15, 1972.
21. P. Thullen, J.L. Smith, Jr., and J.L. Kirtley, Jr., Edison
     Elec. Inst., Res. Progress Rep. RP-92, Sept. 15, 1972.
22. A. Bejan et al., Paper M-3, Cryogenic Engineering Conference,
     Georgia Inst. of Technology, Aug. 1973, (to be published in
     Advances in Cryogenic Engineering, Vol. 19).
23. A. Bejan and J.L. Smith, Jr., (to be published in Cryogenics).
24. R. Barron, Cryogenic Systems, McGraw-Hill, New York,
     1966, Chap. 3 and 5.
25. R.H. Kropschot, B.W. Birmingham, and D.B. Mann, Eds.,
     Technology of Liquid Helium, NBS Monograph 111, 1968,
     Chap. 3 and 4.
26. C.A. Bailey, Ed., Advanced Cryogenics, Plenum Press, London,
     1971, Chap. 7.
27. C. Trepp, Low-Temperature and Electric Power, Proceedings
     of Commission I meeting of IIR, London 1969, Annex 1969-1,
     p. 31, Bulletin of International Institute of Refrigeration, Paris.
28. T.R. Strobridge, IEEE Trans. on Nuclear Science, Vol. NS-16,
     No. 3 June 1969, p. 1104.
29. T.R. Strobridge and D.B. Chelton, Advances in Cryogenic
     Engineering, Vol. 12, 1967, p. 576.
30. F.E. Maddocks, Advances in Cryogenic Engineering, Vol. 13,
     1968, p. 463.
31. R.L. Gessner and D.B. Colyer, Advances in Cryogenic Engineer-
     ing, Vol. 13, 1968, p. 474.
32. H.O. Lorch, Proceedings IEEE, 120, 221, (1973).
33. T.A. Keim, submitted to IEEE.
34. T.A. Keim, "Design and Construction of an Excitation System
     for a Superconducting Alternator," Sc.D. Thesis, MIT,
     January 1973.
35. J.L. Smith, Jr. and J.L. Kirtley, Jr., "Polyphase Synchronous
     Alternators Having a Controlled Voltage Gradient Armature
     Winding," U.S. Patents 3,743,875 July 3, 1973.
36. J.L. Smith, Jr., "High Voltage Oil Insulation and Cooled
     Armature Windings," U.S. Patent 3,743,867, July 3, 1973.
37. J.L. Smith, Jr., "Superconducting Apparatus with Double
     Armature Structure," U.S. Patent 3,742,265, June 26, 1973.
38. J.L. Kirtley, Jr. and J.L. Smith, Jr., 1973 IEEE Winter Power
     Meeting, Paper no. C73-138-5.

Chapter 6

HIGH SPEED MAGNETICALLY LEVITATED AND PROPELLED MASS GROUND TRANSPORTATION

Y. Iwasa

Francis Bitter National Magnet Laboratory, MIT

Cambridge, Massachusetts 02139

I. INTRODUCTION

Most of us who have come to the NATO Advanced Study Institute could come because reliable air flights were available, and because the fare was reasonable. Speed, reliability, capacity, and cost of modern transportation have all made it possible to have international Institutes such as this.

Of course, air transportation is not fast and reliable all the time, as most of us know — by experience. Similarly on highways, traffic congestion is commonplace. The demand for transportation is ever increasing, particularly in industrial nations. Population growth and distribution, new towns and urban patterns, income levels have all contributed to make efficient mass transportation systems essential. In fact we are entering a new era when traditional approaches to solving traffic demands must be modified drastically. Building 5th and 6th lanes when a 4-lane highway gets saturated or building another airport when the 2nd one becomes overcrowded cannot go on forever.

Whatever mode of transportation is proposed to meet this growing demand, it must have a large volume handling capability and satisfy a number of constraints concerning resources, the environment, safety, and land requirements. That is, modern transportation systems can no longer be based purely on technology and economics alone. A consideration of these constraints, some of which are elaborated below, all indicate the desirability of mass ground transportation.

Resources: The efficient use of resources is always desirable and may even become mandatory in the near future. The performance of transportation systems currently in use is shown in Table 1.

TABLE I PERFORMANCES OF VARIOUS TRANSPORTATION MODES

MODE	EMPTY VEHICLE MASS (kg)	NUMBER OF SEATS	VEHICLE MASS/PASSENGER (kg)	FUEL REQUIREMENT AT CRUISING SPEED	ENERGY/PASSENGER-km (10^6 JOULE)
VW (1973)	825	4	200	30 MILES/GALLON (80 km/hr)	0.6
CADILLAC (1973)	2 000	6	330	15 MILES/GALLON (80 km/hr)	1.3
MBTA[a] BUS	975 0	53 (100)[b]	185 (100)[b]	5 MILES/GALLON (40 km/hr)	0.3 (0.15)[b]
MBTA[a] TRAIN	29,000	60 (228)[b]	490 (130)[b]	240 kw (70 km/hr)	0.2 (0.05)[b]
16-CAR NTL[c]	———	1407 (OVER 2000)[b]	———	———————	0.6 (0.45)[b]
BOEING 707	67,000	189	355	10,000 lb/hr (960 km/hr)	1.2
BOEING 747	162,000	490	340	20,000 lb/hr (960 km/hr)	0.9

[a] MASSACHUSETTS BAY TRANSIT AUTHORITY
[b] WITH STANDEES
[c] NEW TOKAIDO LINE

TABLE 2 SPACE PER SEAT OCCUPIED BY VEHICLES

MODE	NUMBER OF SEATS	SPACE OCCUPIED BY VEHICLE (m^2)	SPACE/SEAT (m^2)
		LENGTH(m) x WIDTH(m) = AREA (m^2)	
VW (1973)	4	4.54 x 1.76 = 7.8	1.95
CADILLAC (1973)	6	5.78 x 2.05 = 11.9	1.98
MBTA[a] BUS	53 (100)[b]	12.2 x 2.62 = 32	0.6 (0.3)[b]
MBTA[a] TRAIN	60 (228)[b]	20.4 x 3.05 = 62.3	1.03 (0.27)[b]
16-CAR NTL[c]	1407 (OVER 2000)[b]	400 x 3.38 = 1350	0.96 (< 0.7)[b]
		WINGSPAN (m)	
BOEING 707	189	46.5 x 44.5 = 2070	10.9
BOEING 747	490	70.5 x 59.7 = 4200	8.6

[a] MASSACHUSETTS BAY TRANSIT AUTHORITY
[b] WITH STANDEES
[c] NEW TOKAIDO

From Table 1 it is apparent that the two transportation systems currently used most, the automobile, and the airplane, are the least desirable from the resource viewpoint. They use more energy to transport people than the two other modes of transportation listed, bus and railroad. Equally important, their greater material requirements per passenger mean not only more materials, but more energy as well for processing and manufacturing the equipment. From the resource viewpoint alone, mass ground transportation is clearly the choice for the future.

Safety: Safety is a prerequisite for any system, especially when a great number of people are involved, as in transportation. In this regard it is singularly encouraging to note that there has been no fatal accident in the New Tokaido Line, which has carried nearly 700 million passengers to date since the beginning of its operation in 1964. The annual fatality rate of more than 50,000 persons on the U.S. highways or not so infrequent air accidents of course cannot match such a remarkable record.

Land Requirements: The space requirements of transportation systems are crucial in densely-populated metropolitan areas. One way we can evaluate various modes of transportation in this respect is by examining the space occupied per passenger by the vehicle of a particular transportation mode. Table 2 lists such vehicle space per passenger data for several modes of transportation. Again, private automobiles and commercial airplanes do not fare well against buses and railroads.

These figures are translated roughly into actual land consumption of highways and airports. The newly widened section of the Jersey Turnpike, for example, consumes land at a rate of $0.09km^2/km$ (36 acres per mile) [1]. Present design standards for a 4-land interstate highway in a rural area calls for a land consumption rate of $0.1km^2/km$ (40 acres per mile), excluding interchanges [1] . Similarly, airports consume large areas of land. For example, the Kennedy International Airport in New York occupies $18km^2$ (7.5 square miles, 4800 acres) and the new Dallas-Fort Worth International Airport, $64km^2$ (25 square miles, 16,000 acres). In Japan where land is valued at least one to two orders of magnitude higher than in the U.S., land is utilized more carefully. Nevertheless, the land requirement for the new Narita International Airport, outside of Tokyo, is $10.7km^2$ (4.2 square miles, 2680 acres), which exceeds the $10km^2$ of land consumed by the New Tokaido Line in its entire length of 515.4km [2].

Mass ground transportation should be a primary mode of transportation in the coming decades. The failure of rail service in the U.S., where service is unreliable and train speed is low, and the success of the New Tokaido Line in Japan,where the reverse is true suggest that for such a mass ground transportation system to be successful, it must provide better service at higher speeds.

II. STATE OF THE ART

The ultimate in high speed mass ground transportation based on conventional rail technology was launched in 1964 with the operation of the New Tokaido Line between Tokyo and Osaka. It has been a success story ever since, but has also demonstrated the fundamental limitations of conventional rail technology. Before discussing more advanced systems, let us review the socio-economic impacts of the New Tokaido Line and technical limitations of such a conventional system.

A.   New Tokaido Line

The New Tokaido Line is a high speed ground transportation system which runs through the Tokaido megalopolis, which includes not only Tokyo, Yokohama, Nagoya, Kyoto, and Osaka, but also many other cities with populations in excess of 100,000 located, on an average, every 30km between Tokyo and Osaka. The Tokaido megalopolis occupies 19.2% of the total area of Japan and contains 51.4% of its population. It takes 3 hours 10 minutes to travel between Tokyo and Osaka by the New Tokaido Line, as compared to 7 hours 25 minutes previously. The line was constructed between April 1959 and September 1964 at a total cost of $1.40 billion [Throughout this paper, conversion from the Japanese yen to the U.S. dollar is computed at a current rate of $1 ≈ ¥ 270. Until 1971, it was $1 ≈ ¥ 360.]

Though it is now recognized nationally and internationally as one of the most successful mass ground transportation systems in the world, when it was first proposed in 1957, proponents of the project were severely critized on the grounds of technological and economic feasibility. The critics felt it essential to have low-cost mass transportation rather than high-speed mass transportation. The increase in the standard of living and the "time production effect" which results from shorter travel time were not taken into their calculations.

During the first seven year period, October 1, 1964 to September, 1971, the Tokaido Line carried over 400 million passengers. The beneficial effects of shorter travel time were quickly recognized as important contributions of the New Tokaido Line. It has been computed that the "time production effect," a monetary equivalent to the increase in production because of the saving in travel time, was $195 million in 1967; $450 million in 1970, and is estimated to be $900 million in 1975 [3]. In all, between October 1, 1964 and March 31, 1971, a total of 363 million passengers saved 835 million hours in travel time, an equivalent of $2 billion.

Of the total trips in this period, 70% were for business. The reduced travel time in effect created a 350,000-man skilled labor force. A two-day business trip became a one-day business trip, resulting in an increase in the mobility and productivity of the labor force.

In its first year of operation, the line carried only 60,000 passengers per day (PPD). This figure increased four-fold to 230,000 PPD in 1971. The line has an ultimate annual traffic capacity of 46 billion passenger-km. It carried 26.7 billion passenger-km in 1971 and is expected to reach the saturation figure in 1980. In comparison, the annual travel figure by private automobiles in the U.S. was 384 billion passenger-km in 1967. Considering the 515-km length of the New Tokaido Line and the total projected length of over 65,000 km for interstate highways in the U.S., the New Tokaido Line is much more efficient.

No dramatically new railroad technology was invented in designing the New Tokaido Line. The designers relied entirely on conventional railroad technology, stretching it to its limit. They did introduce many innovations and ideas in fusing these conventional technologies to produce a remarkable end product. In this regard, the New Tokaido Line is very similar to the American Apollo space project.

Often, one need not discover new natural phenomena to produce something daring; this seems particularly so in the field of transportation. All one really needs to do is to innovate and develop existing technology; there exists an abundance of knowledge. Magnetically levitated and propelled transportation is no exception in this regard. Maxwell's equations, superconductivity, linear motors, all have been around for a long time.

The New Tokaido Line may not have revolutionized railroad technology, but it has contributed enormously in achieving a wider public acceptance of and a world-wide interest in the concept of high speed ground transportation (HSGT).

B.   Fundamental Barriers to Conventional High Speed Ground
     Transportation

Speed is an important factor for attracting users to public transportation. Thus among the three basic modes of public transportation, namely air, ground and water, all of which lack the basic conveniences of the private automobile, air transportation is or is becoming by far the most popular mode of public transportation among industrial nations despite its higher cost of travel compared with the other two. It is equally significant to observe that water transportation as a means of intercity transportation has virtually disappeared in the industrial nations. The greater the travel distance and the less significant the cost of travel, the more popular is air transportation. Thus, it is most popular in the U.S., where the distances between major cities are great and the standard of living is higher than any other nation. For example, in New York-Boston travel in 1968, the ratio of the traffic volumes of air to public ground (bus and rails, combined) transportation was about 4, with 74% of the volume of air transportation being for business trips.

A conventional train,   suspended by steel wheels, which in turn are supported by steel rails has fundamental speed limits. At lower speeds friction between the wheels and the rails prevents the wheels from slipping. The friction force is large enough for locomotives to haul a great many cars. Tractive friction decreases with speed and traction through wheels becomes impossible above the speed at which the aerodynamic drag force becomes larger than the tractive force. The worst condition occurs with wet rails in a tunnel and this "traction barrier" limits the wheeled train speed to about 300km/hr.

Wheel vibration is caused by track and wheel irregularities and the peak acceleration associated with vibrations increases with train speed. A severe vibration in the horizontal direction (yaw), for example, can easily result in a derailment. To achieve speeds above 350km/hr in the New Tokaido Line, it has been calculated that the rail width would have to be increased 2 to 3 meters for horizontal stability, the stiffness of the primary suspension spring of the vehicle would have to be reduced by one half, and the amplitude of long-wavelength guideway undulations would have to be limited to 2mm per 10m from the present 3.5mm per 10m. The "suspension barrier" limits the upper speed to about 350km/hr.

At speeds above 350km/hr., a "power collection barrier" appears, preventing a normal catenary overhead electric power-collection system, because vibrations and wave propagation along the line become excessive at these high speeds.

These three barriers clearly indicate the fundamental importance of both suspension and propulsion mechanisms in HSGT.

III. PRINCIPLES OF MAGNETIC LEVITATION

The idea to use a magnetic force to suspend vehicles has been around for over half a century, [4]. The possible application of superconductivity has renewed an interest in the idea [5] . The idea seems new simply because it has never been put to practical use; wheels have always proved unbeatable. Wings have proved to be superior to wheels, but only at speeds above 200km/hr or so. Even the airplane relies on wheels below its take-off speed. These facts suggest rather strongly that the best chance of success and eventual public acceptance of a magnetic levitation system lies in a speed range above 250/km/hr, a range that is currently not possible by conventional wheel suspension systems.

We shall, therefore, examine magnetic levitation systems in this content. Magnetic levitation, or for that matter any other new system of levitation, which can operate in the same speed range as that of conventional wheel systems is not considered in this review paper.

A.   Magnetic Force

A magnetic force results from the Lorentz interaction of current and magnetic intensity, $F = IxB$ (Newton/m). Current can be provided in a variety of ways: 1) the Amperian current in a permanent magnet, (2) an induced current in a ferromagnetic material due to a static magnetic field or in a conductor due to a time-varying magnetic field, and (3) a transport current in a conductor. The magnetic force can also be thought of as a magnetic pressure, similar to that of pressurized gas in a confined volume. At a field strength of 1 tesla (10,000 gauss), the magnetic pressure is about $4 \times 10^5 N/m^2$ (57.5 psi), which is about the tire pressure of a Greyhound bus.

When a permanent magnet is used to provide the current, the force can be either attractive or repulsive. That is, the permanent magnet is either attracted to or repelled by the rest of the system. The force between a ferromagnetic material and a source of magnetic field is always attractive while the force between a normal metal and a source of time-varying magnetic field is always repulsive. For the case of two pieces of metal carrying transport current, the force can be either attractive or repulsive.

B.   Attractive System

As mentioned above, the magnetic force between a ferromagnetic material in a magnetic field and the source of the field is attractive. In the attractive Maglev systems, the weight of a vehicle is supported by this attractive force. A simplified schematic of such a system is shown in Fig. 1 [6] . Basic features of this system, currently investigated for use in mass ground transportation systems by Messerschmitt-Bolkow-Bolhm (MBB) and Krauss-Maffei of Germany and Rohr Industries of the United States, are:

1.   Because of the presence of ferromagnetic material, the necessary magnetic field can be generated by normal-metal electromagnets.

2.   The use of normal-metal electromagnets necessitates a small gap ($\sim$ 1cm) between the ferromagnetic material and the magnets. Even with ferromagnetic material used to reduce the magnetic reluctance, magnetic flux is more expensive than with superconducting magnets.

3.   The magnetic force increases as the gap gets smaller, and decreases as the gap gets greater. That is, the attractive system is inherently unstable — the system has a "negative" spring constant — and is stabilized by a feedback mechanism which controls the magnet current.

Although no hasty negative conclusion can be made on the attractive system, there are at least two disadvantages to this system, if it is to operate above 250km/hr. The first is the small gap with which it must

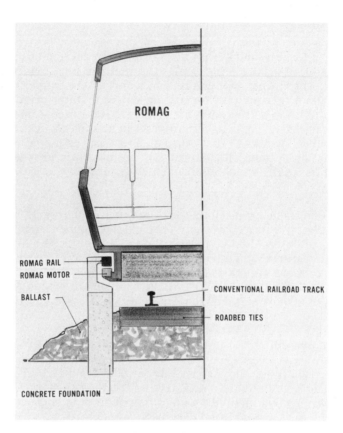

FIG. 1.   Attractive Maglev System being Developed by Rohr Industries, California [6].

operate. A fundamental reason that the conventional wheel-rail suspended train can not travel beyond 300km/hr is because its vertical position has to be maintained to within ± 0.2cm over a 10m distance. The second reason is that the system is inherently unstable with respect to vertical motion. These two disadvantages, if they do not make high-speed operation impossible, result in a great power requirement for maintaining a correct gap distance at speeds beyond 250km/hr. It has been suggested that the normal-metal electromagnets could be replaced by superconducting magnets, thereby making it possible to increase the gap distance. This advantage is balanced by the difficulty of controlling their currents, which is necessary for stabilizing vertical position.

C.   Repulsive System – General Theory

By the so-called Lenz law, a current is induced in a conductor in such a way as to maintain constant flux. Thus by applying a time-varying magnetic field, one can induce a current in a conductor. The direction

of the induced current is such that a magnetic field of the opposite polarity
to that of the externally applied magnetic field is created, hence resulting
in a repulsive force between the magnet and the conductor.

Another way of looking at the same phenomenon is as follows. Time-
varying magnetic flux lines generated by a current loop in the absence of
a conductor, at one instant of time, are shown in Fig. 2a. When the conductor is brought near the loop, the flux lines get compressed as shown in
Fib. 2b, and "magnetic pressure" is developed between the loop and the
conductor. Note that the system is inherently stable in the vertical direction because the repulsive force increases as the distance between the loop
and the conductor gets smaller.

For a given excitation field, the induced current increases with the
excitation frequency but levels off asymptotically at higher frequencies.
The asymptote is reached when no flux penetrates the conductor. This
means that the repulsive force also increases with the excitation frequency
initially but reaches an asymptotic value at high frequencies.

One other important aspect of the repulsive system is the power dissipation which arises in the conductor. It is a simple resistive loss
due to the finite conductivity of the metal. As in induction heating, this
power dissipation depends on the frequency of excitation and has a
maximum at a single frequency, decreasing to zero at higher frequencies.

The most important feature of the repulsive system as applied to mass
ground transportation is the use of superconducting magnets to provide
the necessary magnetic field. Superconducting magnets make it feasible to
generate a large magnetic field over a large volume and the use of superconducting magnets has profound effects on the overall design of the
system. The most important features are:

1. The gaps between the magnets and the conductor can be at least
   one order of magnitude larger than those possible in the attractive
   system. This is of fundamental importance to the design and
   operation of high speed vehicles.

2. A large magnetic field generated over a large volume by the magnets
   can be incorporated easily into a propulsion mechanism, making the
   suspension and propulsion mechanisms compatible.

Unless more detailed investigation reveals otherwise, it appears, at
least to this author, that there are no fundamental technical problems
for the repulsive system. There are, however, many technical innovations needed before the system can be put into practical use.

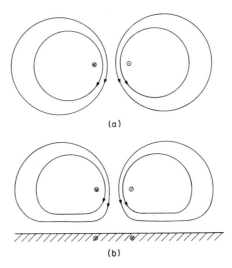

FIG. 2. Time-Varying Magnetic Flux Lines Generated by a Current
Loop: (a) in the Absence and (b) Pressure of a Conductor.

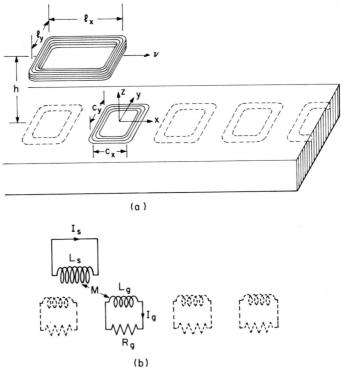

FIG. 3. Single Superconducting Magnet Moving above a Guideway
Composed of a Series of Normal-Metal Coils.

D.   Theories of Lift and Drag Forces

The repulsive force and power dissipation discussed above are examined for the case of a vehicle moving over a metallic guideway. The vehicle is provided with a number of superconducting magnets, located at its bottom and energized with a constant current. The magnetic field generated by these magnets is essentially dc in the moving coordinate system of the vehicle; to a stationary point in the guideway, however, it appears time-varying. The repulsive force thus created by the motion of the vehicle levitates the vehicle and is usually called the lift   force. The lift force is zero when the vehicle is stationary. Just as with the airplane, therefore, the repulsive maglev vehicle must also be suspended by wheels at low speeds. Power dissipated in the guideway appears as a drag force against the motion of the vehicle, necessitating an external propulsive force to maintain the vehicle motion.

We wish to derive approximate expressions for the lift and drag forces for a repulsive system utilizing superconducting magnets. The purpose is not to obtain precise expressions of these forces applicable to a particular vehicle-guideway configuration, but rather, by using simple approaches, to point out the essential parameters which affect the lift and drag forces. There are a number of methods one can use to derive these expressions [7, 8, 9, 10]. We present here two approaches: one an energy point of view and the other a field point of view.

1.   Energy approach

Let us consider a simple vehicle consisting of a single superconducting magnet energized with a current and operated in a persistent mode. The vehicle moves with a speed v(m/sec) at a distance h(m) above a guideway into which a series of normal-metal coils are placed in the direction of the vehicle motion. Let us examine an interaction between the superconducting magnet and one of the guideway coil (Fig. 3a). We define:

L_s = self inductance of the superconducting magnet,
L_s = current in the superconducting magnet,
Φ_0 = total flux through the superconducting magnet at rest,
M = mutual inductance between the superconducting magnet and the guideway coil,
R_g = resistance of the guideway coil,
L_g = self inductance of the guideway coil,
I_g = current in the guideway coil,

A lumped circuit representation of the system is shown in Fig. 3b.

Initially when the superconducting magnet and the guideway coil are far apart, we have:

$$L_s I_s = \Phi_o \, ,$$
$$M = 0,$$
$$L_g I_g = 0.$$

For the sake of simplicity, let us first assume $R_g = 0$. We can treat the case $R_g \neq 0$ as a perturbation to the $R_g = 0$ case. In general when $M \neq 0$, we have:

$$L_s I_s + M I_g = \Phi_o \, , \tag{1}$$

$$M I_s + L_g I_g = 0 . \tag{2}$$

Solving for I_s and I_g, we get:

$$I_s = \frac{\Phi_o}{L_s} \left(\frac{1}{1 - M^2/L_s L_g} \right) = I_o \left(\frac{1}{1 - M^2/L_s L_g} \right) \tag{3}$$

$$I_g = - \frac{M}{L_g} I_s = - \frac{M}{L_g} I_o \left(\frac{1}{1 - M^2/L_s L_g} \right) \tag{4}$$

where $I_o = \Phi_o/L_s$, the initial current in the superconducting magnet when the vehicle was at rest or far away from the guideway. In practice the current in the superconductor is not I_o but $I_o/(1 - M^2/L_s L_g)$ which should not exceed the critical current of the superconductor. It should also be noted that the current in the superconducting magnet oscillates between these two values as the vehicle travels.

The total instantaneous energy of the system consisting of the super-conducting magnet and the guideway coil is given by:

$$E = \frac{1}{2} L_s I_s^2 + M I_s I_g + \frac{1}{2} L_g I_g^2 \tag{5}$$

Inserting Eqs. (3) and (4) into Eq. (5), we obtain:

$$E = \frac{L_s I_o^2}{2} \left(\frac{1}{1 - M^2/L_s L_g} \right) . \tag{6}$$

The mutual inductance M between the superconducting magnet and the guideway coil, of course, depends on the distance which separates them. More specifically, if we define a Cartesian coordinate system located at the center of the guideway coil as shown in Fig. 3a, the x-axis in the direction of the vehicle motion, the z-axis in the vertical direction of the vehicle, and the y-axis in the horizontal direction of the vehicle, the mutual inductance $M(x, y, z)$ is a function of $x, y,$ and z. The quantitative dependence of $M(x, y, z)$ is shown in Fig. 4.

The force in any direction n is given by:

$$\vec{F}_n = - \nabla_n \vec{E} \, .$$

(7)

We can thus obtain expressions of force in the three directions:

$$F_z = - \frac{I_o^2}{L_g} \left(\frac{1}{\left[1 - M^2/M_s L_g \right]^2} \right) M\left(\frac{\partial M}{\partial z} \right)$$

$$F_x = - \frac{I_o^2}{L_g} \left(\frac{1}{\left[1 - M^2/L_s L_g \right]^2} \right) M\left(\frac{\partial M}{\partial x} \right)$$

$$F_y = - \frac{I_o^2}{L_g} \left(\frac{1}{\left[1 - M^2/L_s L_g \right]^2} \right) M\left(\frac{\partial M}{\partial y} \right)$$

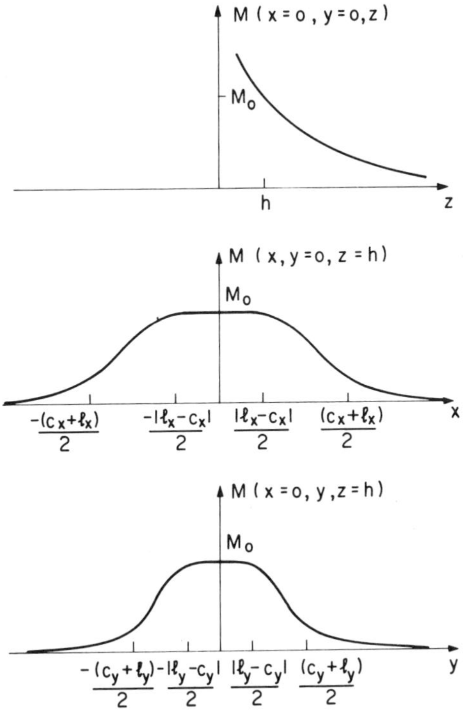

FIG. 4. Qualitative Dependence of the Mutual Inductance M between the Superconducting Magnet and the Guideway Coil.

Based on the dependence of M on x, y, z shown in Fig. 4, the qualitative dependence of F on x, y, z can be plotted.

In Figs. 5a and 5b are shown, respectively, the x and y dependences of the lift force. When contributions from all the guideway coils are summed, the lift force becomes nearly independent of x. The lift force ripple can be minimized by adjusting ℓ_x, c_x, and the distance between adjacent guideway coils [11]. In the case of a continuous metallic sheet guideway, the lift force is uniform. We can also see, from Fig. 5b, that the vertical displacement of the vehicle is stable.

In Fig 5c is shown the force which acts on the vehicle in the direc-of motion. For x < 0, the force is negative. That is, an external force must be applied in the direction of motion to overcome this repulsive force and keep the vehicle in motion. The mechanical energy supplied by the external source during this process is converted into a magnetic energy and the total magnetic energy of the system increases as the vehicle approaches x = 0. (Note that the quantity $1/(1-M^2/L_s L_g)$ in Eq. (6) gets greater as M increases.) For x > 0, the force is positive, meaning the vehicle is pushed forward in the direction of motion. This force has an average value of zero for $R_g = 0$, and corresponds to energy converting back and forth losslessly between mag-netic and mechanical energy. This conclusion can also be drawn easily by observing the fact that the initial energy (at x = $-\infty$) and the final energy (at x = $+\infty$) of the system are identical.

The lateral force profile is shown in Fig. 5d. It shows that when the vehicle is centered with respect to the y- axis the system is unstable. That is, a slight horizontal push in either direction pushes the vehicle out of the guideway. This unstable vehavior of the vehicle is obvious if one considers how the energy of the system varies with y, being maximum at y = 0. If not constrained by an external force(s), a system always seeks the minimum energy state, and for this particular case the minimum energy state occurs at y = $+\infty$. This problem did not arise in the z-direction because of the weight of the vehicle. That is, in the case of the displace-ment in the z-direction, not only the magnetic energy, as given in Eq. (6), but also the gravitational potential energy of the system, mgz, enters into the total energy of the system.

This horizontal instability of the system must be corrected, by providing restoring, or constraint forces in the horizontal direction. It might be interesting to point out that a simple wheel-rail mechanism would also be unstable in the horizontal direction, were it not for the flange attached to the rim of the wheel.

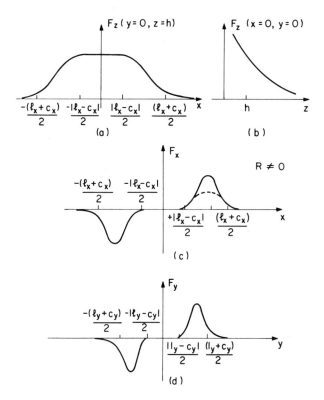

FIG. 5.   Qualitative Dependence of Force on x, y, and z.

We now discuss what happens when R_g is not zero. The lift force is no longer independent of the vehicle speed. We must deal with two characteristic frequencies: one associated with the characteristic frequency of a current induced in the guideway coil due to the vehicle motion and the other associated with the guideway coil circuit. The former is proportional to v, the latter to R_g/L_g. It can be shown that as long as the ratio of these two chracteristic frequencies, $\sim v/(R_g/L_g)$, is very large, the $R_g = 0$ approximation presented above is valid. When this ratio is not large, the lift force is the value computed for zero resistance, weighted by a factor which is purely a function of this ratio. More specifically, the lift is zero at v = 0 and gradually increases and levels off towards an asymptote at high speed, as illustrated in Fig. 6a. A physical description of this behavior has been given previously.

Another important change when $R_g \neq 0$ is the appearance of a drag force in the direction of the vehicle motion. In describing F_x in Fig. 5c above, we have noted that $F_x = 0$ for $R_g = 0$. This was energitically possible because of the absence of power dissipation within the system consisting of the vehicle magnet and the guideway.

When the guideway coil is resistive, energy is dissipated during the time the vehicle magnet enters and leaves the vicinity of the guideway coil. The net result is the necessity of power flow, equal to that dissipated, < P >, into the system if the vehicle is to travel at a constant speed, and this appears as a drag force from the relationship:

$$\vec{F}_D \cdot \vec{v} = <P>. \tag{8}$$

A qualitative modification of the force profile due to the non-zero R_g is indicated by the dotted curve in Fig. 5c.

We describe qualitatively the dependence of the drag force on the vehicle speed. As noted above, energy is dissipated during the time the vehicle is in the vicinity of the guideway coil. The rate of dissipation depends on the circuit decay time constant, L_g/R_g. Therefore, for a given value of L_g/R_g, the faster the vehicle moves, the less energy is dissipated. This means that above a certain speed, at which point the current induced in the guideway has become sufficient to cancel the excitation field, the total energy dissipated in the guideway, and consequently the drag force, should decrease with the vehicle speed. More specifically, the right-hand side of Eq. (8) approaches a constant value at high speeds and $F_D \sim 1/v$. At still higher speeds, as flux is excluded from the bulk of the guideway conductor and the induced current flows only at the surface of the conductor ("skin depth"), the resistance of the coil increases, and < P > varies as $(v)^{1/2}$ and $F_D \sim 1/(v)^{1/2}$.

At low speeds when flux is still distributed uniformly within the guideway, the amount of current induced depends directly on v, and < P >

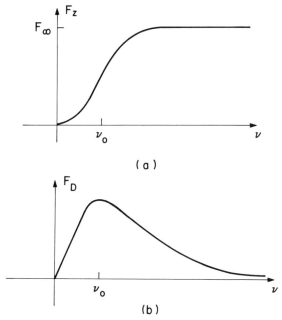

FIG. 6.   Velocity Dependence of (a) Lift and (b) Drag Forces.

varies as v^2, and $F_D \sim v$. For a given combination of guideway size and material, there is a certain vehicle speed at which the process of power input becomes "in phase" with that of dissipation and a maximum amount of power is dissipated. At this speed, the drag force is maximum.

From these observations of the drag force described above, we can draw a general velocity dependence of F_D, and it is shown in Fig. 6b. If one ignores the "skin depth" effect at very high speeds, it can be shown that in general,

$$\frac{F_L}{F_D} = \frac{v}{v_o} \tag{9}$$

for the entire velocity spectrum. Here v_o is a characteristic velocity, corresponding to the characteristic frequency mentioned above, associated with the guideway size, material parameters, and the configuration. Typically, v_o has a value of $5 \sim 20 km/hr$ for a full-scale system.

Since the lift force must be equal to the weight of the vehicle, we can express, from Eq. (9), the magnetic drag force as:

$$F_D = \frac{v_o}{v} \, W , \tag{10}$$

where W is the vehicle weight. If a vehicle travels at a constant speed $\vec{v}$ for a distance ℓ, the total energy expended to overcome the magnetic drag force will be simply:

$$E_m = \vec{F}_D \cdot \vec{\ell} = \frac{v_o}{v} W\ell .\tag{11}$$

That is, the energy expended to overcome the magnetic drag force decreases with operating speed.

However at high speeds, energy expended to overcome an aero-dynamic drag force accounts for the majority of the total energy expenditure, and it increases as the square of velocity.

2.   Field approach

The energy approach presented above, because it is based on a lumped circuit description, offers little insight into how the geometry and configuration of a magnetic system affects the lift and drag forces, although their importance can be anticipated by noting the dependence of the lift force on $M \dfrac{\partial M}{\partial z}$ and the drag force on M^2 . For this discussion, let us use a simple magnetic system shown in Fig. 2.

From Newton's law, the lift force acting on the primary current ring is equal and opposite to the force acting on the induced current ring in the conductor. (The current induced in the conductor is really distributed over the entire conductor, but most of it is concentrated in the form of a ring right beneath the primary current ring, as shown in the figure.) The lift force can, therefore, be given approximately by:

$$F_L \approx B_t I_i (2\pi a),\tag{12}$$

where B_t is the tangential component of magnetic field, generated by its primary current ring, acting on the induced current ring; I_i the induced current, and "a" the radius of the induced ring. The induced current I_i is proportional to the total flux linked by the induced ring, which in turn is proportional to the normal component of the applied magnetic field. By the definition of self inductance, we have:

$$LI_i \approx \pi a^2 B_N .\tag{13}$$

Eq. (13) is valid only for the case $R = 0$ or $v = \infty$. Inserting Eq. (13) into Eq. (12,) we obtain an expression for the lift force:

$$F_L \approx \left(\frac{4\pi\mu_0 a}{L}\right) (\pi a^2) \left(\frac{B_N B_t}{2\mu_0}\right).\tag{14}$$

The quantity in the first parentheses in the right-hand side of Eq. (14) is dimensionless and of the order of unity; the second quantity is the area enclosed by the ring; and the last quantity is a magnetic pressure. Note that the pressure is a product of the normal and tangential components of the magnetic field due to the primary currents.

From Eq. (8) the drag force can be computed readily once the average power dissipation of a guideway is computed. The power dissipation is given as

$$<P> \approx RI_i^2 . \tag{15}$$

From Eqs. (13) and (15), we obtain:

$$<P> \approx R\left(\frac{\pi a^2}{L}\right)^2 B_N^2 , \tag{16}$$

and the drag force becomes:

$$F_D \approx \frac{R}{v}\left(\frac{\pi a^2}{L}\right)^2 B_N^2 . \tag{17}$$

By combining Eqs. (14) and (17), we obtain an expression for the ratio of lift to drag:

$$\frac{F_L}{F_D} \approx \left(\frac{2LB_t}{RaB_N}\right) v \tag{18}$$

Equating Eq. (18) with Eq. (9), we can identify v_o:

$$v_o \approx (2a)\left(\frac{R}{L}\right)\left(\frac{B_N}{B_t}\right) , \tag{19}$$

or more generally:

$$v_o \approx \frac{1}{(\text{length})}\left(\frac{\rho}{\mu}\right)\left(\frac{B_N}{B_t}\right) \tag{20}$$

We have here an expression for v_o, the single most important parameter in the repulsive magnetic system, in terms of physically identifiable quantities: guideway size, guideway material parameters, electrical resistivity, ρ; magnetic permeability, μ; and the vehicle-magnet and guideway configurations. Equation (20) shows that, for a given choice of guideway size and material, one can improve the lift-to-drag ratio by designing vehicle magnet configurations such that a small value of B_N/B_t results. Attempts to obtain as small a value of v_o as possible have resulted in a number of ingenious, sometimes impractical, designs for the vehicle magnet and guideway configurations.

In one variation of the "null-flux" system, proposed by Danby and Powell, [12] a guideway is placed in a magnetic "cusp", created between two oppositely energized vehicle magnets, one above and the other below the guideway. In such a configuration, the normal component of the field is zero at the midpoint between the magnets and a very large value of B_t/B_N results. However, due to the vehicle weight, the position of the guideway is not at midpoint. In addition, to support a given vehicle weight in the null system, because B_N is small or from the lumped circuit viewpoint because M is small, a very large current is required in the vehicle magnet. This is not a serious limitation if superconducting magnets are used.

E.   Experimental Techniques

For a magnetic levitation system, as in any other system, two immediate concerns are that the vehicle be both levitated and propelled with a reasonable consumption of material and power. Only if levitation and propulsion seem favorable does one worry about vehicle heave motions, horizontal stability, cryogenics etc. It is understandable, therefore, that most of the theoretical and experimental work on magnetic levitation which began intensively shortly after the proposal of Powell, has dealt with the functional dependence of lift and drag forces on speed for various configurations of vehicle magnet and guideway. This concentration on lift and drag forces has had the unfortunate effect of obscuring the importance of formulating an integrated magnetic levitation-propulsion system.

A typical [13, 10] experimental set-up and results obtained by it are shown in Figs. 7 and 8 respectively. The rotating wheel in the set-up simulates relative linear motion between the vehicle magnets — here represented by a single magnet — and the guideway. Another common configuration is to use a large turn table.

The lift and drag dependence on speed can also be investigated by the impedance method [7, 14]. In this method, the impedance Z of a network (Fig. 9) consisting of a model vehicle magnet and guideway is measured as a function of frequency. By noting that the real part of the impedance is proportional to the average power dissipation $<P>$, one can investigate the speed and spatial dependence of the drag force. The spatial dependance is obtained by varying the position of the magnet with respect to the guideway. Similarly, the imaginary part of the impedance is proportional, in the absence of electric energy, to the average magnetic energy stored in the network. That is, by measuring the imaginary part of the impedence, one can investigate the frequency and spatial dependences of the forces.

One fundamental shortcoming of these experiments using scaled-down models is that a projection of performance at full scale is not always

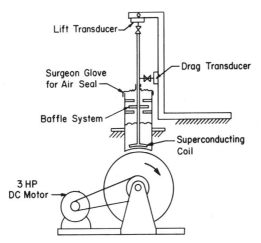

FIG. 7. Typical Experimental System to Investigate Velocity Dependence
of Lift and Drag Forces [8].

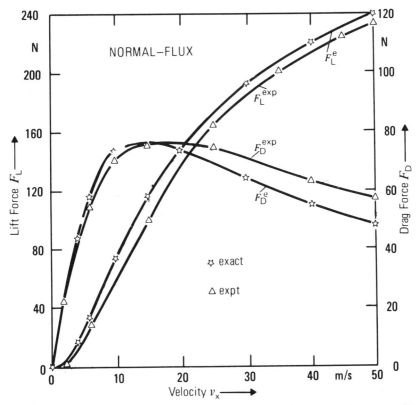

FIG. 8. Comparison Between Theory and Experiment on Velocity Dependence
of Lift and Drag Forces [10].

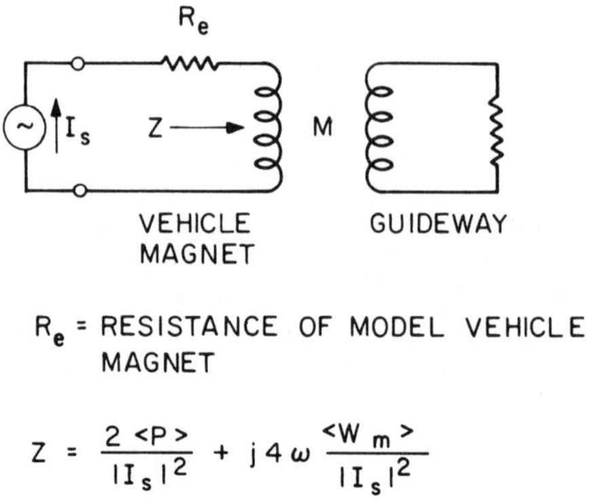

R_e = RESISTANCE OF MODEL VEHICLE MAGNET

$$Z = \frac{2 <P>}{|I_s|^2} + j4\omega \frac{<W_m>}{|I_s|^2}$$

FIG. 9. Circuit Representation of Resistive Vehicle Magnets and Normal-Metal Guideway [14].

straightforward. This is especially true when one investigates the dynamics of the vehicle and considers the effects of magnetic damping. Scaling is not straightforward basically because the characteristic time associated with the vehicle motion and the magnetic time associated with the circuit depend differently on dimensions:

$$\tau_{vehicle} \sim (length)/v.$$

$$\tau_{circuit} \sim (length)^2 \mu/\rho.$$

Thus, if one scales down every dimension by 25, for example, $\tau_{vehicle}$ is reduced by a factor of 25, and $\tau_{circuit}$ by a factor of 625. This point is also clear from the fact that v_o depends on system dimensions (Eq. 20), while v does not.

IV. PRINCIPLES OF LINEAR MOTOR PROPULSION

Power collection is one of the most crucial problems in HSGT. As mensioned previously, the conventional method of power collection fails at a speed of about 350km/hr, the power-collection barrier. One approach, therefore, is to eliminate the need for power collection altogether with a propulsive power source on board the vehicle. Although carrying the power source poses no problem, generation of power on board is usually not desirable ecologically. The author feels that jet engines, though seriously suggested by some, will never be accepted as a source of propulsive power for a HSGT vehicle because of excessive levels of noise and air pollusion. Mechanical systems such as pressure difference, air streams, or gravity, all requiring stationary power source, have also been proposed as alternatives to the conventional method of power collection. However, linear motor propulsion seems best suited to magnetically levitated systems.

In both ordinary induction and synchronous motors, ac power is fed into the stator coils in such a way as to generate a rotating magnetic field. In the induction motor, the rotor is a conductor and eddy currents are induced in the rotor by and interact with the rotating field of the stator, generating a torque to drive the motor. The torque is produced at any speed of rotation, irrespective of the rotation speed of the field, although, good efficiency is achieved only near synchronism. In the synchronous motor, the rotor is provided with dc currents in the form of the Amperian current of a permanent magnet, the transport current of an electromagnet, or the induced current of a ferromagnetic material. Because the direction of the rotor current is independent of the orientation of the stator with respect to the rotor, and because torque is dependent on orientation, only certain orientations create a torque. To produce a torque continuously, the rotor must always maintain a correct orientation, meaning a synchronous rotation.

The linear motor can be thought of as a rearrangement of the conventional rotary motor, cut along a radius, unrolled, and laid out flat. The air gap between the primary and the secondary remains, permitting relative linear motion between the two. One of the members must be lengthened in the direction of travel so that motion can continue.

At first glance, because it can tolerate speed variation and because one of its two members can simply be a conductor, the linear induction motor, (LIM) appears much more attractive than the linear synchronous motor(LSM). In fact, the LIM has received most attention in the past as a possible propulsive device in high-speed trains. When subjected to conditions prevalent in speeds above 250km/hr , however, the LIM appears distinctly inferior to the LSM for the following two reasons:

1.   If the primary winding of the LIM is to be on board the vehicle, then one is faced with the same problem of power collection. By laying the primary winding in the guideway and having a passive reaction plate on board the vehicle, one can eliminate the problem of power collection, but one now is faced with a high cost of energizing the guideway as well as a task of removing heat, generated by the induced currents, from the passive plate.

2.   The gap distance between the reaction plate and the primary windings must necessarily be small, which is difficult at high speeds.

On the other hand, the LSM appears the most suitable means of propulsion for the repulsive magnetic system using superconducting magnets for levitation, for the following principal reasons:

1.   A large magnetic field generated for the purpose of levitation by
      superconducting magnets can also be used as the field of the LSM.
      The suspension and propulsion mechanisms are thus integrated.

2.   The gap distance can be at least one order of magnitude greater than
      that possible in the LIM.

One major aspect of LSM's, considered disadvantageous by critics
of LSM's, is the need for an active guideway to generate traveling magnetic waves. Compared to a conventional electrified rail system, the
guideway of a LSM System, of course, is much more complicated; but
as will be shown later, the cost of such an active guideway is not
prohibitive and the overall guideway cost is quite competitive.

Two basic requirements of a LSM system are that: (1) the vehicle
must be kept synchronized with the traveling magnetic wave, whose speed
is proportional to the frequency of the ac currents supplied to the guideway,
and (2) to keep the power requirement minimum, the entire length of
guideway must be divided into a number of blocks and only the block which
contains a vehicle is to be energized. As the vehicle moves to the next
block, the power must be switched to the next block. To accomplish these
basic functions, a sophisticated control system is required.

A.   Basic Theory of Synchronous Propulsion

As with the lift force considered above, the magnetic thrust can be
understood in terms of the magnetic energy. The guideway windings are
assumed to be composed of three-phase windings, each phase displaced
1/3 of a pole pitch apart from the next (Fig. 10). We also assume that each
phase winding carries sinusoidal current of amplitude I_o and 120° apart in
time. Namely,

$$I_1 = I_o \cos \omega t ,$$
$$I_2 = I_o \cos(\omega t - 120^o),$$ (21)
$$I_3 = I_o \cos(\omega t - 240^o).$$

The total thrust acting on a vehicle is equal to the total thrust acting
on a single superconducting vehicle magnet multiplied by the number of
magnets on board the vehicle. If we assume the current in the superconducting magnets to be constant, we obtain an expression for the thrust acting
on one superconducting vehicle magnet [15]:

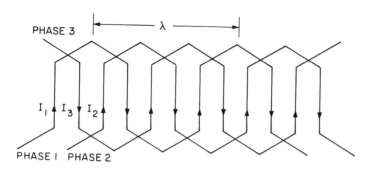

FIG. 10. Guideway Windings Composed of Three-Phase Windings.

$$F_x = I_v I_o \left[\sum_{n=0,\,3,\,6} \frac{\partial\, M_n(x_W - \frac{n\lambda}{3})}{\partial\, x_W} \cos \omega t + \sum_{n=1,\,4,\,7} \frac{\partial\, M_n(x_W - \frac{n\lambda}{3})}{\partial\, x_W} \cos(\omega t - 120^\circ) \right.$$

$$\left. + \sum_{n=2,\,5,\,8,} \frac{\partial M_n(x_W - \frac{n\lambda}{3})}{\partial x_W} \cos(\omega t - 240^\circ) \right] \qquad (22)$$

where I_v is the current through the superconducting magnet, M_n the mutual inductance between the superconducting magnet and nth winding, and λ the wavelength of the winding. The coordinate x_W is measured with respect to a fixed reference point on the winding and is related to the moving coordinate system, x_v, measured with respect to a fixed point on the vehicle by:

$$x_W = x_o + vt + x_v, \qquad (23)$$

where x_0 is the arbitrary separation distance between the two reference points at $t = 0$ and v the velocity of the vehicle.

The mutual inductance is periodic in the x-direction and, therefore, M can be expressed by the Fourier series:

$$\sum_{n=0,3,6} \frac{\partial M_n}{\partial x_W} = \sum_{m=1,3,5\ldots} b_m \cos m \frac{2\pi}{\lambda} x_W \; ,$$

$$\sum_{n=1,4,7} \frac{\partial M_n}{\partial x_W} = \sum_{m=1,3,5\ldots} b_m \cos m \frac{2\pi}{\lambda} (x_W - \frac{\lambda}{3}) \; , \tag{24}$$

$$\sum_{n=2,5,8} \frac{\partial M_m}{\partial x_W} = \sum_{m=1,3,5\ldots} b_m \cos m \frac{2\pi}{\lambda} (x_W - \frac{2}{3}\lambda) \; .$$

Combining Eqs. (22) and (24), we obtain a simple expression for the dominant term (m=1) of the thrust acting on a single superconducting vehicle magnet:

$$F_x(m=1) = \frac{3}{2} I_v I_0 b_1 \cos(\omega t - \frac{2\pi}{\lambda} x_W) \; . \tag{25}$$

That is, the thrust is composed of two traveling waves, one traveling in the positive x-direction with speed $v=\omega\lambda/2\pi$ and the other traveling in the negative x-direction with speed $v=-\omega\lambda/2\pi$. The speed of a vehicle is, therefore, directly related to the frequency of excitation and the pole pitch of the windings. Note the peak thrust is given by:

$$F_x = \frac{3}{2} I_v I_0 N b_1 \cos\left(\frac{2\pi x_0}{\lambda}\right) \tag{26}$$

where N is the total number of magnets carried by the vehicle. The quantity $2\pi x_0/\lambda$ is the analogue of the torque angle defined in ordinary synchronous motors. It can be shown that higher harmonics of thrust are either zero or very small compared to the dominant term when excitation currents are sinusoidal and equal in amplitude in all three phases.

Reaction forces in the other two directions can be derived essentially the same way, except that $\partial M/\partial z$, when expressed by the Fourier series, contain sine rather than cosine terms, resulting in peak reaction forces which vary as $\sin 2\pi x_0/\lambda$. That is, by choosing $x_0=0$ and thereby maximizing thrust, we minimize vertical and transverse forces.

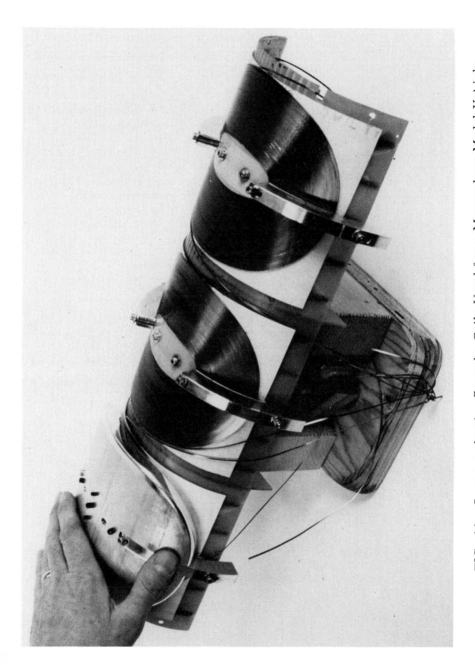

FIG. 11. Superconducting Pancake Coils Used for a Magneplane Model Vehicle.

B.   LSM Control System For a Model "Magneplane" Vehicle

To illustrate the basic control features involved in the LSM, we
describe here briefly the control system designed, constructed, and
tested for a model "Magneplane" vehicle under development at MIT.
The system proposed has features which resemble an airplane rather than
a train and hence is called magneplane [16]. It is guided electromagnetic
flight at low altitudes.

Two 1/25th scale models of a full-scale vehicle along with several
hundred feet of matching guideway have thus far been constructed. The
first vehicle uses sammarium-cobalt permanent magnets as the motor
field. Ten rows of magnets are attached, in alternating polarity, on an
iron plate made from a 120° section of a cylinder. The pole pitch is 3
inches and the total vehicle weight is 35 pounds. In the second vehicle, the
permanent magnets are replaced with three 4-layer pancake superconducting coils, which when energized with the same polarity, generate a field
distribution similar to that generated by the permanent magnets. The pole

FIG. 12.   Model Guideway of Magneplane.

pitch is 3 inches and the total vehicle weight is around 40 pounds. The coil windings are shown in Fig. 11.

The guideway consists of two parts, one for suspension and the other for propulsion, as shown in Fig. 12. Until the vehicle reaches its cruising speed of about 60km/hr, it is suspended by four wheels which rest on curved non-conducting strips, located at both ends of a 120° section. In the cruising section of the guideway, these strips are made of aluminum. The active guideway is located in the middle. It is a 3-phase armature"meander" winding laid in a linear direction, similar to that shown in Fig. 10.

A block diagram of the control system used to accelerate the model vehicle is shown in Fig. 13. Three-phase, 400-Hz power is fed into a cycloconverter which converts to the desired voltage, frequency, and phase across the guideway windings. The vehicle starts off at rest and as it begins to accelerate, the frequency of the current in the guideway increases synchronously with the vehicle. A Hall probe on board the vehicle senses the magnetis field of the guideway. The probe is placed such that it would read zero field when the relative position of the vehicle with respect to the traveling magnetic wave results in the maximum propulsive thrust. For small displacements of the vehicle from the optimal operating position, the probe output is proportional to the displacement. The position information is transmitted and received by a ground control station and processed, using a standard communication technique as indicated in the block diagram of Fig. 13. Frequency control of the guideway current is accomplished by creating a reference signal whose frequency is derived from the fed-back signal and which the cycloconverter attempts to duplicate at a higher power level.

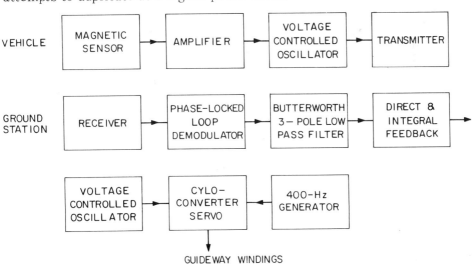

FIG. 13. Block Diagram of the Magneplane Control System [17].

V.   OTHER IMPORTANT CONSIDERATIONS FOR REPULSIVE MAGNETIC
      LEVITATION SYSTEM

A.   Ride Quality and Vehicle Dynamics

If speed attracts users to mass ground transportation, a comfortable ride keeps them. Ride quality is an important parameter in vehicle specifications and is related intimately to vehicle dynamics.

In any system in which energy is stored in more than one form, the system undergoes resonant modes unless dissipative mechanisms of the system quickly act on additional energy inputs. Thus, in the repulsive magnetic levitation system, where energy is stored in two forms, inertia and magnetic, and where dissipation is designed to be minimal, resonant modes are inevitable. The question is how resonant modes affect the ride quality and how the quality may be improved.

Let us consider, as an illustration, a vehicle, levitated at a height h_0 above the guideway, traveling with a constant speed v. Suppose the vehicle is pushed up momentarily from its equilibrium height by a amount δ by an external disturbance. In the absence of dissipation, its subsequent height above the guideway will be given by:

$$h(t) = h_0 + \delta \cos 2\pi \nu t, \tag{27}$$

where ν is the resonant frequency of the system. The resultant vertical acceleration the passengers experience is given simply by the second derivative of Eq. (26):

$$\ddot{h}(t) = -\delta(2\pi\nu)^2 \cos 2\pi \nu t. \tag{28}$$

Let us obtain an order-of-magnitude peak vertical acceleration for a typical vehicle. The resonant frequency ν is given by:

$$\nu = \frac{1}{2\pi} \sqrt{\frac{k}{m}} , \tag{29}$$

where k is the "spring constant" of levitation and m the mass of the vehicle. In order to estimate k, we first relate the weight of the vehicle to the magnetic force,

$$mg = kz_0, \tag{30}$$

where z_0 is a natural deflection of the "magnetic" spring force due to the weight. If the vehicle is to be maintained at a constant height of h_0, it is resonable to assume that z_0 is a fraction of h_0. Let us say $z_0 = 0.1h_0$. Then from Eq. (30), we have:

$$k = \frac{mg}{z_o} \sim \frac{10mg}{h_o} \quad . \tag{31}$$

Combining Eqs. (29) and (31), we obtain

$$\nu \sim \frac{1}{2\pi} \sqrt{\frac{10g}{h_o}} \quad , \tag{32}$$

which for $h_o = 0.25m$ results in a natural frequency of about 3Hz.

An amplitude of oscillation of 2cm, due to an external disturbance, is not unlikely, and, therefore, if we insert $\delta = 2cm$ and $\nu = 3Hz$, we obtain in Eq. (28):

$$\ddot{h}(t) \sim - g \cos 6\pi t \quad . \tag{33}$$

This is comparable to the motion one experiences when one is jumping rope. It is not a comfortable experience if one wants to concentrate on reading at the same time.

There are several criteria for specifying ride quality. One of these is the Janeway criterion, shown in Fig. 14, which specifies, as a function of frequency, the maximum vertical acceleration humans can tolerate without feeling uncomfortable. The above motion clearly violates this criterion.

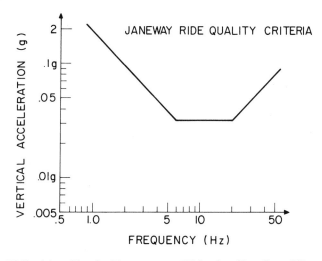

FIG. 14. Single Frequency Ride Quality Specification.

Another vertical acceleration which affects ride quality and is simple to analyze is that caused by a long-wavelength guideway undulation. Suppose that a guideway is a sinusoidal undulatory track, having a vertical undulation amplitude of ϵ with a wavelength λ.   If a vehicle travels with a constant speed of v, the vertical acceleration generated by this guideway undulation is given by:

$$a = \epsilon\left(\frac{2\pi v}{\lambda}\right)^2 \sin\frac{2\pi v}{\lambda} t. \tag{34}$$

Taking the New Tokaido Line, as an example, ϵ is limited to 3.5×10^{-3}m for a λ of 40m. Therefore, if we take these values as the maglev guideway specifications and insert them into Eq.(34) together with the vehicle speed of 500km/hr ($\sim$140 m/sec),we obtain a peak vertical acceleration of about 0.2g at an oscillation frequency of 3.5Hz, which violates the Janeway criterion. (Note that at the maximum speed of 210km/hr of the New Tokaido Line, the corresponding value is 0.03g at 1.5Hz, which satisfies the criterion.)

As the built-in damping mechanism of the repulsive magnetic system is not adequate to meet ride quality specifications, either a secondary suspension or some form of active control mechanism must be introduced. But in order to minimize vehicle weight, we should try to avoid a secondary suspension system. One obvious approach of the active control system would be to introduce an independent set of magnets, separate from superconducting magnets. The current through these auxiliary magnets is controlled to damp oscillatory motion. Preliminary investigations of the Ford group have shown that the additional power required for such a control system to make ride quality comfortable would be on the order of one kilowatt per ton of vehicle weight [18]. Another approach would be to induce vertical damping forces in the vehicle by changing the phase of currents in the stationary propulsion winding. Vertical forces on the order of 0.1g are available free in as much as use of the propulsion windings adds no vehicle weight, and the components are already part of the system.

B.   Cryogenics

Cryogenics could make or break the repulsive magnetic levitation system. Among the many disciplines this new system requires, cryogenics is perhaps the least advanced, yet it is one of the most important. Cryogenics is one area where innovations are clearly needed.

It was only after helium was liquefied that superconductivity could be discovered, and superconductivity remained a curiosity for a few physicists for a long period of time chiefly because a lack of adequate amount of liquid helium prevented others from participating in superconductivity research. The phenomenal growth of applied superconductivity which started in the 1960's is a direct result of the large-capacity liquefiers developed by Prof. S. Collins in the late 1940's and 1950's. We now have means to produce large amounts of liquid helium, but our method of keeping the magnet superconducting, by direct immersion in a bath of liquid helium, is the same technique that Kamerlingh Onnes used in 1914 when he tested the first superconducting magnet. Recently an approach based on forced circulation techniques has drawn considerable attention among magnet designers.

Basically, the cryostat for maglev must be lightweight and capable of transmitting a large lift force across a great temperature gap with minimum heat conduction.

Helium must be recycled in a maglev system as freon is recycled in home refrigerators. The amount of helium required to operate a maglev system is not really important, neither is the power requirement for refrigeration, which is relatively small. What is important is the size and weight of the on-board refrigeration system.

It is estimated that a maglev system, operating in the northeast corridor in 1985, would carry about 4,000 passengers per hour in one direction, or a total of 8,000 passengers per hour, requiring about seventy 120-passenger vehicle every hour [19]. Each vehicle will probably be equipped with at least ten superconducting magnets, each weighing about 500 kg and requiring refrigeration of at least 2 watts, or 20 watts per vehicle. This amounts to a total refrigeration load of about 1400 watts for the system. One watt of refrigeration load at 4.2K is roughly equivalent to a refrigeration power requirement of 1,000 watts at room temperature. Thus, the 1400-watt load means a total system refrigeration power requirement of 1.4 megawatts, a minute amount compared to what is needed for propulsion.

Figure 15 shows the weights and volumes of stationary and projected lightweight 4-K refrigerators as a function of heat load [20]. The curves for the stationary refrigerator are based on the actual weight and volume of existing refrigerator systems, while the curves for the lightweight refrigerators are projected, based on the limited data on lightweight airborne refrigerators currently available. For a 20-watt load, the stationary system requires a volume of $6m^3$, which is about 3% of the volume of a maglev vehicle, while the projected lightweight system requires only 1/10 the volume of the stationary system.

We examine the advantages of a forces circulation technique compared with the direct immersion technique. With the forced circulation technique, it is, of course, best to wind a magnet with a hollow conductor through which helium is circulated. Helium could be operated either in the supercritical state, [21], which exists at pressures above 2.26 atmospheres at 4.2K, or in the conventional two-phase mode. The basic advantages are:

1.   Improved magnet stability, as every section of the magnet comes in direct contact with the fluid.

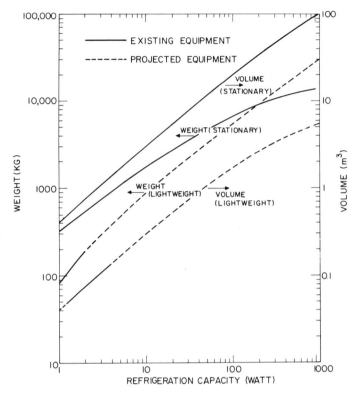

FIG. 15. Refrigeration System Weight and Volume Versus Capacity [20].

2.   Improved strength of the magnet, since no empty space, such as
      that created by spacers, is needed in the magnet for liquid helium to
      flow.

      These advantages are particularly evident for large magnets and the
technique has been used successfully in large-volume magnet systems
[22, 23].

      Another possible area where new ideas could be tried is in the area
of magnet energizing. The conventional technique of energizing a superconducting magnet calls for a pair of current leads. When a magnet gets
to be the size of a vehicle magnet, it is normally charged with a current at
least 1000A. A well-designed pair of 1000-A leads conducts at least 2
watts of heat into the liquid helium. Since the vehicle magnets, once charged,
are put into a persistent mode, it is desirable that the leads be removed
from the liquid helium environment to eliminate heat conduction through
the leads. If each magnet is to be charged independently, there must be
ten pairs of such removable current leads for each vehicle carrying ten
magnets. One possible simplification would be to charge all of the ten

magnets simultaneously by connecting them in series and thereby reducing the number of leads to just one pair. To keep the magnets independent while they are in a persistent mode, each magnet can be equipped with a persistent-mode switch, as illustrated in Fig. 16. In Fig. 16(a) is shown an equivalent circuit while the magnets are charged. Each persistent-mode switch which shunts the magnet is in a resistive mode. Once the magnets are fully charged, the persistent-mode switches are turned off, leaving all the magnets in a persistent mode (Fig. 16b).

The magnets can, in principle, be charged without even a single pair of leads. In such a scheme, a "charging" magnet which produces a sufficient magnetic field is placed near the vehicle magnet while its circuit is held resistive by means of a persistent-mode switch. Once the field has penetrated completely into the vehicle magnet, the vehicle magnet can be switched into a persistent mode. Now the charging field can be turned off. The amount of current which is induced in the vehicle magnet, I_v, is given simply by:

$$I_v = I_o \frac{M}{L_v}$$

where M is the mutual inductance between the vehicle and charging magnets, L_v the self inductance of the vehicle magnet, and I_o the original current in the charging magnet. Because M/L_v is quite small in practice (~ 0.1 at most), I_o must be quite large, making this procedure not as simple as it might sound.

C.   Magnetic Shielding

Magnetic shielding is an important matter in the repulsive magnetic levitation system using superconducting magnets [24]. We must consider two types of shielding, one pertaining to dc shielding to shield passengers in a vehicle from stray dc magnetic fields coming from the superconducting magnets and the other pertaining to ac shielding to shield the superconducting magnets from ac magnetic fields generated by the heave motion of the vehicle and the propulsive windings in the guideway.

1.   DC shielding

There is at present no data which enables one to predict the human tolerance of magnetic field intensity. An excellent review of the biological effects of magnetic field by H. Aceto, Jr., et al. [25] gives a number of examples of magnetic field affecting biological systems. Since oxygen is a paramagnetic molecule, exposure to magnetic fields might alter some bio-chemical interactions. One can conclude from the review by AcetoJr., et al that, in general, pronounced effects may occur in fields above 0.1T. We might, therefore, want to design for peak fields of less than 0.005T in the passenger cabin. By comparison, the earth's magnetic field is on the order of 10^{-4}T.

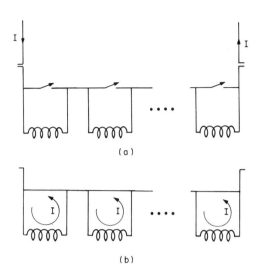

FIG. 16. Current-Lead Arrangements for Energizing Vehicle Magnets:
(a) A Circuit During a Charging Process. (b) Each Magnet is in
a Persistent Mode.

Aside from biological effects which we must at present assume may
be deleterious, there are other reasons to keep the level of magnetic field
in the passenger cabin low. Magnetically sensitive materials, such as
magnetic tapes and electronic devices, would be affected and any ferromagnetic object would be subjected to a magnetic force.

In dc magnetic shielding the following parameters are important:
(1) the level of field to which the space in question is subjected without
shielding, (2) the level to which the field in the space in question must be
brought by shielding, and (3) the volume of the space to be shielded.
Depending on the sizes of these parameters, different approaches are chosen
to provide satisfactory shielding with acceptable consumption of material
and power. It should be noted that there always exists a force between the
source of the original field and whatever provides shielding and that this
force can become quite large.

Basically there are three approaches to dc shielding: (1) passive
shielding with high-mu materials or superconducting sheets, (2) active
shielding with compensation magnets which produce a cancelling field,
and (3) a combination of the above two. Using a very simple geometry, we
present here the basic concept of passive dc shielding with high-mu materials. Based on this example, we can conclude that when the volume of the
space to be shielded is large, the use of high-mu material results in an unacceptably large increase in the weight of the overall system.

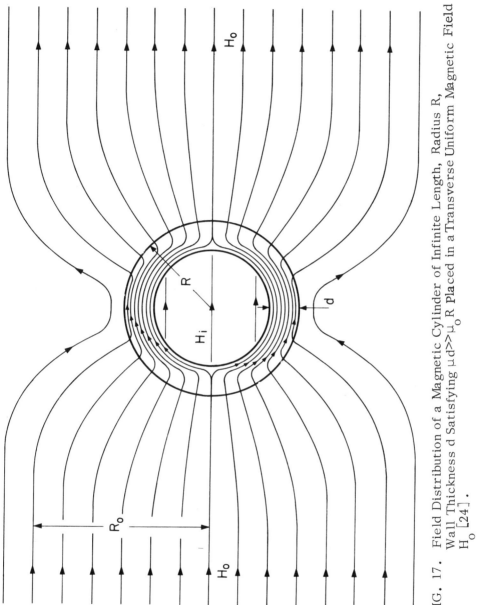

FIG. 17.   Field Distribution of a Magnetic Cylinder of Infinite Length, Radius R, Wall Thickness d Satisfying $\mu d \gg \mu_o R$ Placed in a Transverse Uniform Magnetic Field H_o [24].

A magnetic cylinder of infinite length, radius R, and wall thickness d is placed in a transverse magnetic field H_0 which is uniform at large distances from the cylinder. Assume the cylinder material is homogeneous and isotropic, and its magnetic characteristic is linear for H less than H_s, and saturated above H_s. That is, the magnetic induction B is related to H according to the relationships:

$$B = \mu H, \qquad\qquad\qquad 0 < H \le H_s$$
$$B = B_s \qquad\qquad\qquad H \ge H_s \qquad\qquad (35)$$

If $\mu \gg \mu_0$, where μ_0 is the permeability of free space, we can say that in-coming magnetic flux lines passing within an unknown distance R_0 of the cylinder axis are attracted to and flow through the cylinder wall, as shown in Fig. 17. The conservation of magnetic induction, $\vec{\nabla} \cdot \vec{B} = 0$, results in the following approximate relation:

$$\mu_0 H_0 R_0 = \mu H_w d , \qquad\qquad\qquad (36)$$

where H_w is the peak magnetic field through the wall of the cylinder when the material is not saturated. To keep the material from saturating, the

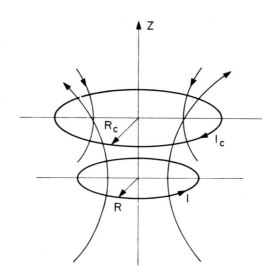

FIG. 18. Active Shielding Compensation Dipole of Radius R_c Carrying a Current I_c [24].

wall thickness of the cylinder must be such that $H_w < H_s$:

$$d = \frac{\mu_o H_o}{\mu H_w} R_o \geq \frac{\mu_o H_o}{\mu H_o} R_o = \frac{B_o}{B_s} R_o \, (m) \tag{37}$$

The minimum thickness is thus given when $H_w = H_s$. Since in the absence of a surface current, the tangential component of a magnetic field must be continuous across the boundary, the inside field is equal to H_w. For iron the inside flux density is less than 0.005T when the shielding is un-saturated. The above example can be solved exactly and its solution agrees with Eq. (37) if $R_o = 2R$ for the limiting case, which is $\mu d \gg \mu_o R$.

Let us generalize Eq. (37) and draw a conclusion on a case existing in a maglev vehicle. Eq. (37) states that by means of a high-mu material one can divert and concentrate flux, spread over a large space, into and through high-mu material, provided it is sufficiently thick. This means one can use Eq. (37) as long as one chooses appropriate values for B_o and R_o. The value of B_o is that flux intensity which exists in the unshielded passenger cabin, is a maximum ($\cong 0.05T$) at the floor. R_o is the size of flux spread and since the vehicle has a width of about 3m, its order of magnitude value would be 1m. Inserting $B_o = 0.05T$, $R_o = 1m$, and $B_s = 2T$ into Eq. (37), we obtain,

$$d \sim \frac{0.05}{2} \, m$$

Thus, the thickness of the plate, which can be used as a part of the floor, would have to be about 2cm. For a 120 passenger vehicle with passenger dimensions of 3m by 30m, a 2-cm thick iron floor weighs roughly 20,000kg, about 30% of the total vehicle weight. The actual weight increase would be somewhat less than 20,000kg, because: (1) the entire floor need not be covered with the full 2-cm thickness, (2) the shielding material, which does not have to be in plate form, can be used for structural purposes, and (3) the low-reluctance return flux path provided by the shield reduces the requirements on the superconducting material. The actual weight increase, therefore, might be closer to 10,000 kg. The 10,000-kg additional weight just for the purpose of shielding is still too much. An active shielding approach described below appears more attractive for magnetic flight.

In active shielding, a compensation magnet is placed above the main magnet. In selecting the compensation magnet, there are three parameters to adjust: the size, the height above the main magnet, and the amplitude of the current. As this compensation magnet will reduce the magnetic field everywhere, including at the guideway, it is desirable to place it as high above the main magnet as possible. Cryostat design restrictions, however, favor a lower location.

A preliminary analysis [24] of a very simple system consisting of one current ring, representing a main magnet, and another current ring, representing a compensation magnet and being comparable in size and placed right above the first ring (Fig. 18), indicates that the magnetic field at the floor of the passenger cabin can be brought below a level of 0.005T. The compensation magnet would also be wound of superconductor. Slightly more superconductor would be required for the main magnet, since the compensation magnet reduces the field at the guideway. The active shielding approach would require about 30% more superconductor. Total vehicle weight would be increased by not more than 2000 kg, suggesting that active shielding is preferable to passive shielding with high-mu material. Because the space to be shielded is large, passive shielding by superconducting sheets appears impractical.

2.   AC Losses in superconducting magnets

High-field, high-current superconductors, such as the alloys of Nb-Ti which are most likely to be used for the vehicle magnets, are not truly superconducting under ac conditions. That is, they exhibit ac losses when subjected to a time-varying magnetic field and/or when they are carrying an alternating current. More detailed accounts of ac losses in superconductors and superconducting magnets are given by Dew-Hughes (Chapter 2) and Brechna (Chapter 3). For our present purpose, we simply point out various sources of ac power dissipation.

One obvious source of dissipation is due to the propulsion windings of a linear synchronous motor propulsion system. Theoretically the magnetic field generated by the propulsion windings appears essentially dc to the vehicle magnets as the vehicle moves synchronously with the traveling field. The phase unbalance and the nonsinusoidal nature of the traveling wave, causes ac magnetic fields to appear at the vehicle magnets and they are sufficient to cause rather serious dissipation. Without proper shielding, the dissipation can reach as high as 25 watts for a typical vehicle magnet. Such a dissipation is devastating not only to the performance of the magnet itself but to the refrigeration requirement.

Dissipation is also caused by the vehicle heave motion and/or by levitation conductor nonuniformity [26]. As was pointed out in connection with Eq. (3) the persistent current is not a constant current. That is, the vehicle magnet current fluctuates as the mutual inductance between the magnet and the guideway varies.

Since the mutual inductance depends strongly on the distance between the magnet and the levitation conductor, even with an ideally smooth conductor, the heave motion of the vehicle can cause the current to oscillate, thereby generating dissipation. If the levitation conductor is periodic-structured, then the mutual inductance undergoes a periodic modulation, causing dissipation.

These dissipations can of course be reduced by shielding against ac magnetic fields. As long as the frequency of oscillation is reasonably high, at or above 100 Hz, a highly conductive plate, such as an aluminum plate of thickness 2mm at 4.2K, can be quite effective. At low frequencies ($\cong$ 1Hz) the metallic plate becomes less effective. Recent studies[27, 28] have shown that ac losses due to the vehicle heave motion, however, is not as great as was previously estimated [26] and may amount to about 1 watt per vehicle.

VI.   COST ESTIMATES FOR A MAGNEPLANE SYSTEM

Ground transportation systems demand a very large fixed investment. Thus, an established system is little disposed to change and the implementation of a new system is very difficult to realize.

As with any commodity which is based on a large sum of investment, the success or failure of maglev depends on traffic volume. In Japan, where the New Tokaido Line is a financial success story thanks entirely to an overflow of users, the decision to build a new system is not as much financial as it is political. In the U.S., where rail transportation has been declining steadily over the years and is suffering its worst financial crisis, economics is clearly a major issue.

Compared to this fixed investment cost, the operating cost is small. An estimate of the fixed investment cost for the proposed "Magneplane" system which is to serve the northeast corridor follows. Some of the estimates are based on those made for a TACV system, which are included here for comparison [29].

A.   The Northeast Corridor Traffic Forecast and Magneplane Characteristics

The total length of a Washington, D.C.-Boston route is about 700km. The corridor's predicted total transportation demands between 1975 and 1995, based on a study by the Department of Transportation [19] are listed in Table 3 together with the traffic share of a TACV or Magneplane system. The total demand is assumed to grow at an annual rate of 4%, while the share of the TACV or Magneplane grows at a 3% rate. The experience of the New Tokaido Line, which showed an average annual passenger growth rate exceeding 20%   in the first 6 years of operation, suggests that the growth rate of 3% for the TACV or Magneplane is clearly pessimistic. A figure of 10% does not seem unreasonable. The 10% annual growth rate will make the traffic share for TACV or Magneplane 104 million in 1995.

TABLE 3   TRANSPORTATION VOLUMES IN THE NORTHEAST CORRIDOR

	1975	1985	1995
BY ALL MODES [a] (MILLION PASSENGERS)	203	300	444
BY TACV OR MAGNEPLANE [b] (MILLION PASSENGERS)	——	40	54
BY TACV OR MAGNEPLANE [c] (MILLION PASSENGERS)	——	40	104

[a]   BASED ON ANNUAL GROWTH RATE OF 4 %
[b]   BASED ON ANNUAL GROWTH RATE OF 3 %
[c]   BASED ON ANNUAL GROWTH RATE OF 10 %

The 40-million passenger figure predicted for the new system in 1985 means a system capacity of about 4,000 passengers per hour in one direction is required if a 15-hour day operating schedule is assumed; it increases to about 5,000 in 1995.

The Magneplane is designed to operate at a maximum speed of about 430km/hr. The maximum headway distance is set for about 7km, or 60 seconds at the maximum operating speed. Unlike conventional trains, a Magneplane will consist of one vehicle. One most obvious advantage of such a one-vehicle system is a tremendous reduction in peak propulsive power demand and point-to-point scheduling flexibility. The vehicle will have a 120-seat capacity. The Magneplane system will, therefore, have a maximum system capacity of about 7,000 seats per hour in one direction, or 77 million passengers annually based on a 15-hour day operation.

B.   Capital Costs

Of several categories which account for capital costs, the most important is guideway construction, followed by land acquisition and roadbed preparations. Costs for such items as communication systems, terminals, yards and ships, and vehicles constitute a small part of the total costs. A large error in the estimate of these small items will probably be insignificant compared with a small error in the estimate of guideway costs.

1.   Guideway

The biggest difference in fixed investment cost between the Magneplane and TACV comes in the cost of the guideway. The Magneplane guideway, whose section is basically trough-shaped, requires a great deal of aluminum for both levitation conductors and propulsion windings [30] . The principal levitation conductors are tubular and each conductor weighs about 20kg per meter. Propulsion windings are made of aluminum bars, 1-cm thick and 12.5 -cm wide. One proposed winding scheme is shown in Fig. 19. Guideway members are made of non-magnetic metal, principally aluminum, and of electrically insulating material where required. Magnetic material is kept at least 60cm away from any conductor. It is estimated that one kilometer of guideway for the Magneplane would cost roughly $3.5 million, while the figure would be roughly $1 million for the TACV: cost breakdowns are listed in Table 4 [31]. Based on these figures, total guideway costs of the Magneplane and TACV for the entire northeast corridor route would, respectively, be $2.3 billion and $700 million.

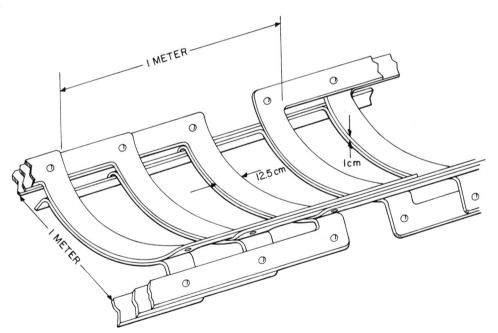

FIG. 19.   General Arrangement for Prototype Magneplane Propulsion
          Conductors [30].

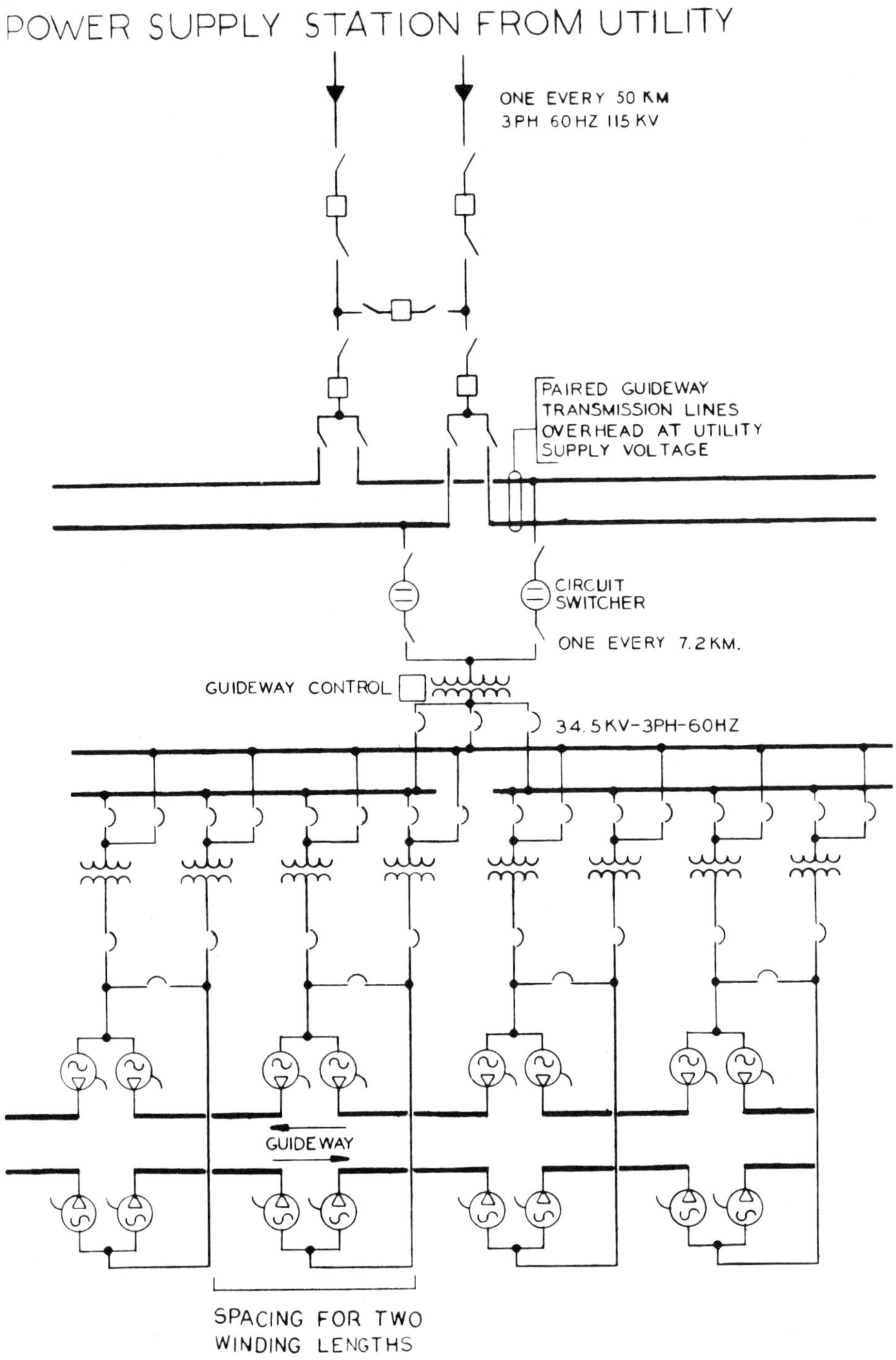

POWER SUPPLY STATION FROM UTILITY

FIG. 20. Magneplane Power Distributor [30].

The cost of land acquisition for the entire route is estimated to be roughly $600 million. Roadbed preparation is expected to reach about $300 million. In addition, it is expected to cost another $500 million for construction of tunnels in the route.

2. Guideway Power Supply System

The system supplying power to the Magneplane guideway LSM windings is illustrated in a circuit diagram shown in Fig. 20. Power is taken every 50km at utility supply voltages which ranges from 69kV to 161kV at 60Hz. Supply power is assumed to be a double-circuit at each point and is transmitted along the guideway over a bi-directional double-circuit[30]. The control of power to the LSM windinss is by back-to-back phase controlled thyristor power conditioning units. Each winding is energized only during the period a vehicle is in the winding.

TABLE 4   TWO-WAY GUIDEWAY COSTS PER 1km
FOR MAGNEPLANE AND TACV

MAGNEPLANE

1. GRADE	$ / 1 km
ALUMINUM	1,000,000
CONCRETE	100,000
ANCHOR BOLTS	300,000
GRAVEL & EXCAVATION	100,000
PROPULSION CONDUCTOR	600,000
SUB TOTAL	2,100,000
OMISSION, CONTINGENCY, CONTRACTOR OVERHEAD, etc.	1,000,000
TOTAL	3,100,000

2. ELEVATED	
ALUMINUM	1,600,000
CONCRETE & REINFORCING STEEL	200,000
FORMWORK, EXCAVATION, etc.	100,000
PROPULSION CONDUCTOR	600,000
	2,500,000
OMISSION, CONTINGENCY, CONTRACTOR OVERHEAD, etc	1,100,000
	3,600,000

TACV

GRADE	
CONCRETE	400,000
TWO LIM RAILS	330,000
ASPHALT SHOULDERS, etc.	30,000
	700,000

To reduce the cost of insulation and surge protection for power conditioning units, the voltage is stepped down to a level of 34.5kV with transformers every 7km. At this spacing the entire system will be loaded continuously when the system is operating a 60-second headway.

One of the most important variables which affects the cost of the power supply system is the length of LSM winding which will be energized by each power conditioning unit. The length of LSM winding section affects the levels of steady state, peak, and rms voltages and currents, which in turn influence the ratings of equipment and conductors. It also determines the number of power conditioning units needed along the guideway.

The ratio of vehicle transit time to headway time, for example, increases with winding length, thereby increasing the ratio of rms current to peak current. The winding dissipation and thus the operating costs will be higher if the winding is longer. The inductance of the winding will also increase, which means a higher rating for power conditioning unit and more vars of series capacitor compensation. The longer winding section, consequently, means more costly but fewer unit of power supply and control equipment. For the proposed Magneplane system, a winding length of 2.5km was chosen. At this length the cost of system electrification for the entire northeast corridor route is estimated to be $260 million. This figure is almost identical to the figure estimated for a TACV system.

C.   Comparison with a TACV System

As has been noted above, the greatest difference in capital costs between the two systems is in the guideway cost. Other costs are either identical, by virtue of the same problems involved, or very close. The differences in any case is insignificant compared to the difference in the guideway cost. The total capital costs for the Magneplane and TACV are estimated to be roughly $5 billion and $3.5 billion, respectively; breakdowns, rounded off to $50 million, are listed in Table 5.

VII.   CURRENT STATUS OF NEW TRANSPORTATION SYSTEM RESEARCH AND DEVELOPMENT IN VARIOUS NATIONS

A number of industrial nations are currently engaged in research, development, and testing of new transportation systems. There are basically three approaches to high speed ground transportation systems: (1) to build entirely new lines based on new technologies such as magnetic or air-cushion levitation; (2) to build new lines based on existing railroad technologies; and (3) to upgrade existing lines. Here we will focus our attention on the first category, in which Japan and West Germany are leading, followed by the U.S., France, and Canada.

TABLE 5 CAPITAL COSTS OF MAGNEPLANE &
TACV SYSTEMS PROPOSED FOR THE
NORTHEAST CORRIDOR ROUTE (700 km)
(MILLIONS OF 1972 DOLLARS)

	MAGNEPLANE	TACV
LAND ACQUISITION	600	600
ROADBED PREPARATIONS; ALLOCATION OF HIGHWAY, etc.	450	450
GUIDEWAY	2300	700
SYSTEM ELECTRIFICATION	250	250
BRIDGES	200	200
TUNNELS	500	500
SWITCHES	100	NEGLIGIBLE
COMMAND, CONTROL, COMMUNICATION	100	100
TERMINALS	200	200
YARDS & SHOPS	50	50
VEHICLES FOR INITIAL YEAR OF OPERATION	100	200
TOTAL	4,850	3,250

A.   Japan

As with other research and development programs related to railroad technologies, the maglev program is under the direction of the Japanese National Railways (JNR). Universities and industries participate in these programs under contract from JNR. Four major industries, Hitachi, Mitsubishi Electric Corporation, Toshiba Electric Corporation, and Fuji Electric Company, have been participating actively in the maglev program.

The maglev program is the latest in a series of research and development programs that JNR has been conducting. Since 1963 JNR has actively been investigating linear motor propulsion schemes. The ever-increasing traffic volume of the New Tokaido Line has been a major factor in making JNR plan a third trunk line in the Tokaido corridor. In 1970 JNR tentatively decided to build this new line using the repulsive magnetic levitation driven by linear induction motors. In 1972, it decided to use linear synchronous motors. The line is scheduled to be operational in 1985.

The cumulative result of maglev research and development programs to date has been the first public demonstration of a model vehicle conducted at the Japanese National Railways Technical Laboratory in Tokyo on September 12, 1972. The vehicle, weighing 3.5 tons and carrying four 250,000 ampere-turn superconducting magnets, was levitated to about 5cm and propelled at a maximum speed of 60km/hr by means of a linear induction motor drive over 200 meters on a 480-meter long test guideway [32].

The next major goal has been set for 1975 to finish building a 7-km test guideway on which to operate a test vehicle to a maximum speed of 500km/hr. Prototype vehicles are expected to be built by 1977.

The Ministry of Transportation recently announced a 5-year program to develop a "pollution-free" mass transportation system beginning in 1974, separate from the current maglev project. Instead of aiming for high speed, its principal aims will be quiteness and vibration-free performance, designed to travel at speeds less than 200km/hr suspended by magnetic fields or by air-cushions. The government and industries are expected to spend up to $15 million for the program.

One distinct and important aspect of railroad-related research programs in Japan is that all programs are under the firm control and direction of one agency, the Japanese National Railway. Each program is directed by an advisory committee composed of representatives from participating groups and chaired by JNR officials, and all jobs are divided and allocated to these groups. Duplication of effort is thus absent.

B.   Germany

Four separate industrial groups working on maglev have joined forces in conjunction with the Federal authorities to form a group called GBI (Gesellschaft fur Bahntechnische Innovationen) to make the maglev effort more unified and centralized. Under this plan, a central test facility will be constructed with Government funds and joint test programs will be carried out.

The four industrial groups working on maglev are: Messerschmitt-Bolkow-Blohm (MBB), Krauss-Maffei (KM), Krupp, and a consortium comprised of the major electrical industries, AEG (Allgemeine Elektrizitats Gesellschaft), Brown, Boveri and Companie, Telefunken, and Siemens. Of these, MBB and KM are developing attractive systems, Krupp a permanent magnet suspension system, and the Siemens group is developing a repulsive system with superconducting magnets.

The first public demonstration of an MBB vehicle was on May 7, 1971 on a 670-meter test guideway. The 5.2-ton vehicle with a rated speed of 85km/hr is the first experimental vehicle to be suspended, guided, and propelled by magnetic fields alone. The first prototypes are scheduled to run on a 70-km test guideway in 1977.

On October, 1970 Krauss-Maffei presented a 1/30 scale functional model of a high-speed transportation system called Transrapid. The vehicle is of a multi-purpose type with the following possible combinations of suspension and propulsion: (1) magnetic suspension and control with LIM propulsion, (2) air-cushion suspension and control with LIM propulsion, and (3) air-cushion suspension and control with jet-engine propulsion. This was followed by the public demonstration on October 11, 1971 of the Transrapid 02, a magnetically suspended and guided, LIM-propelled vehicle roughly 12 meters in length and weighing 1000 kg. A speed of 130km/hr was reached on a 1000-m test guideway in December 1971.

One significant recent development has been the announcement, by the prime minister of Ontario, Canada, of a $16 million contract with Krauss-Maffei for a test demonstration of an attractive maglev system, called Personal Rapid Transit (PRT). The PRTs will travel at 80 km/hr over elevated concrete guideways on an oval-shaped 4-km loop at Toronto's lakefront. The system is expected to be ready for testing in late 1974 and to be in operation in 1975. The vehicles will be 6 meters long, weigh 6 tons, and carry 12 passengers seated and eight standing. They will run, propelled by LIM, at intervals of 15 to 20 seconds, using four stations. The system will ultimately be expanded to 90km long around Toronto. The system is not high speed but its success should greatly enhance the public acceptance of maglev systems.

The Siemens group is the only German group investigating repulsive maglev with superconducting magnets. The group has constructed a 280-meter diameter circular test guideway with a C-shaped concrete guideway banked at 45° in Erlanger. A more detailed review of German Maglev programs is given by Bogner (Chapter 12).

C.   United States

As in France and Great Britan, a major emphasis was placed at first on the concept of air-cushion suspension in the United States. From 1966 until 1973, about $41 million was spent by the U.S. government on research and development of air-cushion suspensions, and another $36 million on related areas such as system engineering, propulsion, guideway evaluation, tunnel technology, and communication. The project is known collectively as the TACV project. Despite these expenditures, TACV has not met with much success.

The maglev program, on the other hand, has been moving slowly in the U.S., about $750,000 having been spent to date. Ford Motor Company and Stanford Research Institute have been funded by the Department of Transportation to evaluate the repulsive maglev system. The next phase of this program, announced in April 1973, calls for full-scale, 500 km/hr rocket sled tests, to be conducted on a 1 km test track, to choose between attractive and repulsive maglev. Incidentally, both MBB and KM have used rocket slet tests to study problems associated with levitation at high speed.

Another group engaged in repulsive maglev in the U.S. is at Francis Bitter National Laboratory at MIT. Called the "Magneplane" project, it is

financed by the National Science Foundation and involves two industrial
concerns as well: Raytheon Company and United Engineers and Construc-
tors, Inc.

Another maglev project currently underway is that being developed
privately by the Rohr Corporation of California. It is a ferromagnetic
attractive systen but differs from the two German systems. In the Rohr
system, levitation is accomplished by time-varying fields and the same
fields produce LIM propulsion. That is, levitation and propulsion are
integrated, which might prove to be superior to the German systems for
a narrow-gap, low-speed system by reducing on-board weight.

Unlike in Japan or West Germany, the U.S. program unfortunately
does not, at least presently, have any long-range planning and is subject
to yearly fluctuations in the budget.

D.   France

Since October 1965 when the French government granted $880,000 to
Companie Aerotrain, a developer of air-cushion suspended vehicles, the
research and development program of the Aerotrain has been sponsored
by the government. Since then several prototype vehicles have been tested
at various test tracks, but the Aerotrain program has not progressed as
rapidly as was first anticipated.

The first large-scale Aerotrain line will be a government-financed
$60 million, 26-km line outside Paris, connecting Cergy-Pontoise and
Defense. One or two-car, linear-motor propelled, 40-passenger Suburban
Aerotrains running at 180 km/hr will be serviced every two minutes during
rush hours. The construction has not begun,however.

E.   Canada

Besides the Toronto plan, in which Krauss-Maffei is involved, there
is at present no national goal to establish high speed ground transportation
systems using maglev or air-cushion suspension. However, the Canadian
Institute of Guided Ground Transportation of Queen's University in Ontario
is engaged in the study of repulsive maglev. It is supported by the
Transportation Development Agency of the Ministry of Transportation.

VIII.   CONCLUSIONS

We have covered this highly complex field of high speed ground
transportation in grossly broad generalization and with many oversimplifi-
cations. The maglev system is fascinating and exciting particularly to

those who are engaged in the application of superconductivity. As has been demonstrated, it requires cooperative efforts involving the resources of many disciplines. Therefore, although one may be involved in research in only one technical aspect of a solution, it is desirable that one appreciate the vast complexity of the problem. Only with a sound understanding of the overall picture, can one responsibility advocate the large scale application of superconductivity to transportation systems.

There are still many technical questions unresolved in the repulsive maglev system comprising of superconducting magnets. Some of these questions summarized are:

A.   Levitation System
    1. Guideway: discrete (coils) versus continuous (sheets).
    2. Dynamic behavior and stability.
    3. Interactions of mechanical and magnetic systems.

B.   Superconducting Magnets
    1. Reliability.
    2. ac losses and magnetic shielding.
    3. Weight reduction.
    4. Biomagnetic effects and dc shielding.

C.   Cryogenics
    1. Refrigeration system: refrigeration cycle, lightweight
       refrigeration system.
    2. Lightweight cryostat.

D.   Propulsion
    1. LIM versus LSM
    2. Control and instrumentation
    3. Switching.

E.   Guideway Economics

In spite of these unresolved technical questions, if transportation demand continues to grow for the next few decades, the only logical and sensible solution is to develop high speed mass ground transportation.

ACKNOWLEDGEMENTS

The author wishes to thank H. Kolm, S. Foner, B. Schwartz, S. Brown, and J. Schultz for critical reading of the manuscript and many suggestions for improvement. He also wishes to express his thanks to Prof. M. Tinkham of Harvard University for bringing his attention to T.K. Hunt's latest measurement of low-frequency ac losses in superconductors.

REFERENCES

1. M. Miller, M. Cheslow, N.T Ebersole, J. Gerba, and D. J. Igo, "Recommendations for Northeast Corridor Transportation. Main Report. Volume 2," Department of Transportation, Washington, D.C. Office of Systems Analysis and Information.(September, 1971).

2. Japanese National Railways, Facts and Figures, 1970 Edition (in Japanese), Japanese National Railways, Tokyo, Japan.

3. T. Oku, Y. Kyotani, and T. Sanuki, Super High Speed Line, (in Japanese), Chuokoron-sha, Tokyo (1971). [An English edition, translated, edited, and supplemented by Y. Iwasa, will be published by the M.I.T. press in 1974.]

4. G.R. Polgreen, New Applications of Modern Magnets, Boston Technical Publishers, Inc., Newton, Massachusetts (1970).

5. J.R. Powell, American Society of Mechanical Engineers, Paper 63-RR-4 (1963).

6. J.A. Ross, Proc. of the IEEE 61, 617 (1973).

7. C.A. Guderjahn, S.L. Wipf, H.J. Fink, R.W. Boom, K.E. MacKenzie, D. Williams, and T. Downey, J. Appl. Phys. 40, 2133 (1969).

8. J.R. Reitz, J. Appl. Phys. 41, 2067 (1970).

9. P.L. Richards and M. Tinkham, J. Appl. Phys. 43, 2680 (1972).

10. L. Urankar and J. Miericke, to appear in Applied Physics (October, 1973).

11. E. Ohno, M. Iwamoto, and T. Yamada, Proc. of the IEEE 61, 579 (1973).

12. J.R. Powell and G.T. Danby, Cryogenics 11, 192 (1971).

13. R.H. Borcherts and L.C. Davis, J. Appl. Phys. 43, 2418 (1972).

14. Y. Iwasa, J. Appl. Phys. 44, 858 (1973).

15. R.S. Kasevich, "Linear Synchronous Motor Theory," in an unpublished report Magneplane Linear Synchronous Motor Study by Raytheon Company, Wayland, Mass. (1973).

16. H.H. Kolm and R.D. Thornton, IEEE Conf. Record, IEEE Cat. No. 72 CHO 682-5 TABSC, (1972).

17. J.H. Schultz, "Electric Control of Linear Synchronous Motor," S.M. Thesis, MIT, unpublished, (August, 1973).

18. J.R. Reitz, R.H. Borcherts, L.C. Davis, T.K. Hunt, D.F. Wilkie, "Preliminary Design Studies of Magnetic Suspensions for High Speed Ground Transportation," a report by Ford Motor Company prepared for The Office of Research, Development and Demonstrations, Department of Transportation, Washington, D.C. (March, 1973).

19. "High Speed Ground Transportation Alternatives Study," U.S. Department of Transportation, Washington, D.C. (January, 1973).

20. Z.J.J. Stekly, T.A. de Winter, J.A. Vitkevich, J.M. Tarrh, and A.E. Emanuel, "Design Study of Superconducting Magnetic Levitation Pads," A report prepared by Magnetic Corporation of America for Ford Motor Company (April, 1973).

21. H.H. Kolm, M.J. Leupold, and R.D. Hay, Advances in Cryogenic Engineering, Vol. 11, Ed. K.D. Timmerhaus, Plenum Press (New York, 1966).

22.   Mr. Morpurgo, in Proc. of the 1968 Summer Study on Superconducting
       Devices and Accelerations, Ed. A.G. Prodell (Brookhaven National
       Laboratory, Upton, New York 1969), p. 953.
23.   R.L. Bailey, B. Colyer, and G.J. Homer, Rutherford Laboratory
       Report RHEL/R 258 (December, 1972).
24.   Y. Iwasa, Proc. of the IEEE 61, 598 (1973).
25.   H. Aceto, J., C.A. Tobias, and I.L. Silver, IEEE Trans, Magn.
       Vol. MAG-6, 368 (1970).
26.   M. Tinkham, J. Appl. Phys. 44, 2385 (1973).
27.   T.K. Hunt, "ac Losses in Superconducting Magnets at Low Excitation
       Levels," submitted to J. Appl. Phys.
28.   T. Satow, M. Tanaka, and T. Ogama, "ac Losses in
       Multifilamentary Superconducting Composites for Levitated Trains
       under ac and dc Magnetic Fields," presented at The 1973 Cryogenic
       Engineering Conference, Atlanta, Georgia (August, 1973).
29.   P.F. Dienemann and J.P. Large, "Cost Analyses for Northeast
       Corridor Transport Project, Volume 1. High Speed Ground Modes,"
       Department of Transportation, Washington, D.C., NECTP-22 (December
       1969).
30.   D.W. Jackson, G.S. Chawla, and P.A. Wheeler, "Report on a Study
       of Magneplane Power System and Guideway," An unpublished report
       prepared for MIT Francis Bitter National Magnet Laboratory by
       United Engineers & Constructors Inc. Boston, Massachusetts (June,
       1973).
31.   M. Yano, "Magneplane Cost Analysis," S.M. Thesis, Massachusetts
       Institute of Technology, unpublished, (August, 1973).
32.   K. Oshima and Y, Kyotani, "High Speed Transportation Levitated by
       by Superconducting Magnet," presented at the 1973 Cryogenic
       Engineering Conference, Atlanta, Georgia (August, 1973).

Chapter 7

TRANSMISSION OF ELECTRICAL ENERGY BY SUPERCONDUCTING CABLES

G. Bogner

Research Laboratories of Siemens

AG Erlangen, Germany

I. INTRODUCTION

The growth of the world's population, the general rise in the standard of living and the gap to be closed in this respect by the developing countries lead to a constant increase in the demand for power. Figure 1 shows the expected trend of the world's power demand up to the year 2000 [1]. The present consumption of about 7.5 billion tons of hard-coal units (HCU) [1 t HCU = 8,120 kWh] will then have risen three-fold, i.e. to 22 billion tons HCU. While nuclear energy today represents a minute fraction compared with the other primary sources of energy such as coal, petroleum and natural gas, Fig. 1 shows an expected increase to about 35% at the turn of the century.

The demand for electrical power is rising even more sharply than the general demand for energy. At present its share is about 27% of the total demand and it is expected to increase to about 50% [2] by the year 2000. The annual growth rate for electrical power consumption in the major industrialized countries of the world now stands at about 7%, meaning consumption will double over a period of 10 years. Figure 2 shows the expected demand for electrical power up to the year 2000, both for the whole world and a number of industrialized countries. Figure 2 indicates an estimated demand of about 40,000 billion kWh for electrical power throughout the world in the year 2000.

No definite forecast is possible as to how long energy consumption will continue to grow at such rates and whether there will be any letup before the year 2000. Ecologists have already voiced their concern about limited growth. They see dangers lying in the continued exploitation of our limited energy sources and an increasing effect on the environment

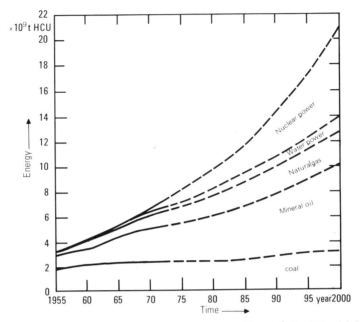

FIG. 1. Prospective Energy Consumption of the World [1].

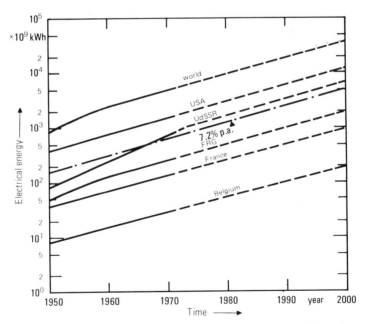

FIG. 2. Consumption of Electrical Energy in the World and in Selected
Countries [2].

from mounting energy conversion. As a clean form of energy, electricity, especially when produced on an increasing scale in nuclear power stations, allows the necessary sociologic developments which result in an increase in power consumption, and also takes sufficient account of ecological considerations. Nevertheless, the main object of the electricity supply industry will always be to meet power requirements fully and economically at all times.

Economical power generation necessitates power station units of ever increasing size, regardless of whether conventional, thermal or nuclear power stations are involved. The latter, because of their capital-intensive nature, are especially under the dictate to utilize the heavy depreciation allowed on investment cost for stations of large unit ratings. Power stations with ratings of 1000 - 2000 MW are under construction or have already been completed. Schemes have been conceived for power station pools with ratings up to 10,000 MW.

Figure 3 shows the increase in turbo-generator unit ratings associated with the increase in station unit size [3]. Present unit ratings for water-cooled generators are already as high as 1500 MW for 4-pole machines, and 1300 MW for 2-pole machines. The output limits of water-cooled generators are expected to be just below 2000 MW for 2-pole machines and around 2500 MW for 4-pole machines, with such unit ratings being required around 1985 [4]. It is hoped to realize still higher unit ratings with the aid of superconductors.

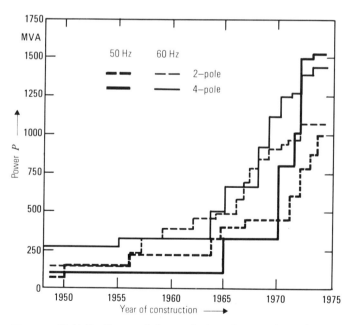

FIG. 3. Power Unit Ratings of 2- and 4-Pole Turbo-Alternators [3].

Only about one-third of the initial cost of power, referred to the location of the load, is accounted for by power generation, while the major portion is made up of power transmission and distribution. In order to provide an economical and uninterrupted supply of electrical power, the capacity and efficiency of future transmission systems will have to be increased at the same rate as the power generation equipment. This can be accomplished by using higher transmission voltages or artificial cooling to achieve a marked increase in the unit ratings of conventional means of transmission, or by using entirely new high-capacity transmission media of the kind now under development. The latter primarily include superconducting cables, in addition to gas-spacer and cryogenic cables. From the technical point of view, superconducting cables certainly represent the most complex system, but in the long run they will probably be the choice for underground transmission at very high power levels in the areas with high load densities.

The realization of superconducting cables still requires much research and development work. Companies and governments throughout the world have put numerous laboratories to work on solving the problems involved, and work is still in the component development stage. Before we discuss the state of development in detail and draw conclusions from the results obtained so far, the capability and the limits for conventional and other new kinds of transmission systems will be considered.

Finally, it should also be noted that, in addition to the familiar "electrical solution", another form of power supply is receiving serious attention for the future, namely that using hydrogen gas ("hydrogen economy"). The power system based on hydrogen gas involves high temperature nuclear power stations located a few kilometers offshore, which split sea water thermally or electrically into hydrogen and oxygen, pumping the former in the gaseous state through pipelines to load centers. Part of the gas is burnt for direct conversion to heat, another part is used for chemical processes and the rest is fuelled in gas turbines for the generation of electrical power. The cost of transporting hydrogen gas through pipelines would be many times less than that for transmitting the equivalent of electric power through overhead lines. The power transmission capability of a hydrogen-gas pipeline of 1 m dia. is about 10,000 MW. One problem which still has to be solved with this concept is that of economical hydrogen production [5].

II. POWER TRANSMISSION SYSTEMS IN EXISTENCE AND IN PROCESS OF DEVELOPMENT

A. Overhead Lines

About 95% of all electrical energy has been transmitted in the past and up to the present by means of overhead lines. This is due to the great advantages which overhead lines provide in various ways. These advantages are:

     1. Low-cost construction. The cost of an overhead line is 1/9th to 1/26th of the cost of underground cables of equivalent capacity [6].

     2. The ease of access allows rapid repairs to be made when faults occur.

     3. Low dielectric losses, since air is used as the insulation.

     4. By adopting very high voltage ranges, very high transmission capacities per system are possible. Several circuits are possible for each tower.

In the West-European network the highest voltage level at present is 420 kV. In the rest of the world, particularly in the USSR, the USA, Canada, Australia and South America, 500 kV systems of considerable extent — in the USSR alone there are 9000 system kilometers — have been in operation for 15 years [2]. In Canada, as far back as 1965, a 735 kV transmission system was erected by Hydro-Quebec. Between the power stations on the St. Lawrence River and Montreal a power of more than 5000 MW is transmitted over a distance of 600 km by three 735 kV lines. At the present time a further project of this kind is being carried out in Canada. With the aid of three 735 kV lines a further 5225 MW are to be transmitted over a distance of about 1200 km. In an initial stage a partial system with 950 MW was put into operation in 1971. Ultimate extension of the system is planned for 1975 [7]. Since 1966 a 750 kV experimental transmission line, 88 km in length, has been in operation in the USSR (Konakowo - Moscow). American Electric Power (AEP) is at present installing a 765 kV system with a highest system voltage of 800 kV and a single circuit length of 2000 km. Figure 4 shows the development of the transmission voltages of three-phase systems. The mean value curve of the step-function corresponds with a mean growth of 3% per annum.

Energy transmission is carried out almost without exception via three-phase systems. Only in special cases where, for example, large distances are bridged without intermediate stations, is high-voltage direct current transmission (HDCT) employed. An example of this is the HVDC transmission of Cabora-Bassa (Mozambique) to Apollo (near Johannesburg) carried out as a joint venture by German, French and Italian firms. Here 2400 MW are being transmitted at a maximum voltage of ± 533 kV over a distance of 1400 km via two monopolar DC overhead lines run on separate routes [8]. The first section of the transmission line will go into service in 1975.

The transmission capacity of very long overhead lines is determined by the natural load and in the case of short lengths by the thermal rating. The natural load is given by:

$$P_{nat} = \frac{U^2}{Z}$$

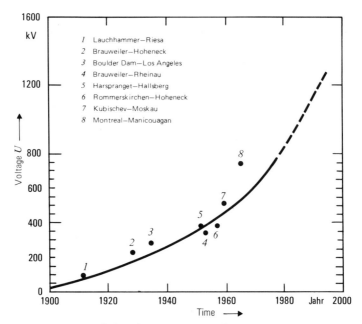

FIG. 4. Development of the AC System Voltage U in Germany, Canada, Sweden, USSR, and USA from 1900 to 1970 [2].

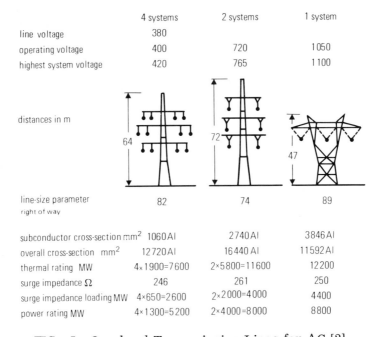

	4 systems	2 systems	1 system
line voltage	380		
operating voltage	400	720	1050
highest system voltage	420	765	1100
distances in m	64	72	47
line-size parameter right of way	82	74	89
subconductor cross-section mm²	1060 Al	2740 Al	3846 Al
overall cross-section mm²	12720 Al	16440 Al	11592 Al
thermal rating MW	4×1900=7600	2×5800=11600	12200
surge impedance Ω	246	261	250
surge impedance loading MW	4×650=2600	2×2000=4000	4400
power rating MW	4×1300=5200	2×4000=8000	8800

FIG. 5. Overhead Transmission Lines for AC [2].

where U is the transmission voltage and Z is the surge impedance of the overhead line. The surge impedance for high-capacity overhead lines is 200 to 400 Ω, depending upon the number of conductors and the transmission voltage.

Consequently, to increase the transmission capability of overhead lines, two courses are being pursued: First, the increase in transmission voltage in the range of 1000 - 1500 kV. There are even investigations in progress (joint research programs of American Electric Power and ASEA Sweden) to clarify the technical and economic limits of ultrahigh voltage transmission up to the range of 2000 kV transmission voltage [9]. It is envisaged that system voltages of 1000 kV and over will be introduced by certain countries even in this decade and by further countries in the eighties. In addition, the thermal rating of the systems is being increased which is greater than the natural load by a factor of about 3 to 7. This increase is achieved by enlarging the conductor sizes and by using a greater number of circuits per tower. In this way considerable increases in the thermal rating of a transmission line are possible without increasing to a marked extent the required width of the route. By employing this solution, problems concerning the erection of new lines, such as lack of space, heavily built-up areas and property rights can more easily be solved by the electric utilities. Half of all the newly constructed or planned 400 kV transmission lines are, for example, fitted with four or six circuits; in the remainder two-circuit systems are the rule [2]. Concentration of this kind naturally restricts freedom of operation, makes assembly and repair work difficult and in many cases calls for the provision of equipment not otherwise required until later.

Figure 5 gives a number of the most important geometrical and electrical data for three-phase high-voltage ac transmission lines [10]. It is seen that when adopting transmission voltages of 1000 kV and above, transmission capacities of about 10,000 MW and more are possible per transmission line. By using several circuits per tower considerable blocks of power can be transmitted, e.g., 5000 - 10,000 MW at voltage levels of 400 kV and 720 kV. Therefore, overhead lines are a method of transmission which, from the technical aspect, will be able for a long time to come to fulfill economically the transmission of the largest loads.

Great problems however are occurring to an increasing extent in conurbation areas. Here the difficulties of insufficient space and excessive costs for procuring land occur in an escalated form. Disturbances in radio and television reception are to be expected with higher operating voltages and also unacceptable noise levels. In addition large local concentrations of overhead lines aggravate the amenity problem.

In congested areas it is therefore becoming increasingly essential to change over to underground high-capacity connections. The same must also apply to areas in which there is considerable impairment to energy transmission through frequent atmospheric disturbances (e.g., in the USA hurricanes, tornados). Moreover the tendency is becoming apparent in the USA of resorting to the laying of cables under public roadways instead of creating new routes for overhead lines [11].

B. Underground High-Power Cables

The proportion of underground high-power cable circuits in relation to the total length of high-power circuits is at present only about 4 to 5%. The reasons for this are the high costs of manufacturing and laying the cables which, as already stated above are about ten to twenty times higher than for corresponding overhead lines. In particular in congested areas, an increasing use of cables is to be expected. However, cables are also gaining in importance in or around large power stations and in the vicinity of substations where overhead lines are not practicable because of the clearances required or because of height restrictions around airports.

1. Conventional ac high-voltage cables

Nowadays the functions of underground power transmission are almost exclusively fulfilled by paper-insulated oil-filled cables found reliable for many years. As the most important types of these cables, the following are available:
— External gas pressure cable in a steel pipe,
— Internal gas pressure cable in a steel pipe,
— Low-pressure oil-filled cable of the single conductor type,
— High-pressure oil-filled cable in a steel pipe.

Internal gas pressure cables can be constructed for voltages of up to about 110 kV and rating of up to 150 MW and gas external pressure cables for voltages of up to about 220 kV and ratings of up to 250 MVA. For higher transmission voltages and powers, only low-pressure and high-pressure oil-filled cables can be considered. Figures 6 and 7 show the stepped samples of such cables [12].

The dielectric of the low-pressure oil-filled cable consists of paper and an oil of low viscosity, which normally is under a pressure of 1.5 to 2 bar. The oil fills up all voids in the space beneath the metal sheath. When the cable is carrying current and consequently its temperature rises, the oil expands and flows through the hollow duct in the conductor into expansion vessels from which it is automatically returned into the cable when the system cools off.

In the case of the high-pressure oil-filled cable (Oilstatic cable) the three shielded cable cores without metal sheaths are in a common steel pipe which is filled with low viscosity oil at a pressure of 15 to

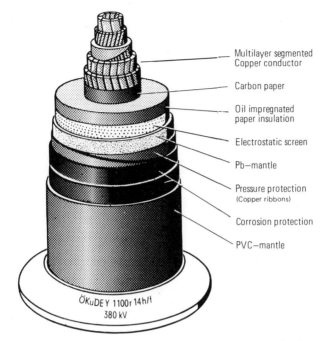

FIG. 6. Low-Pressure Oil-Filled Cable [12].

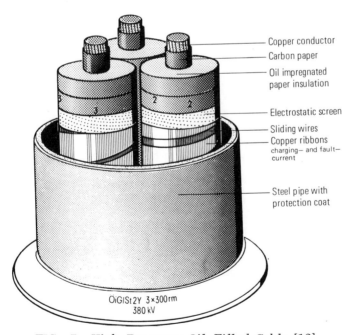

FIG. 7. High-Pressure Oil-Filled Cable [12].

17 bar. The quantity of oil in the pipe is adapted to the thermal loading
of the cable by a safety valve   and an automatically operating pump.

Low-pressure oil-filled cables for 400 kV have been installed in
England to a considerable extent. In the USA the first 525 kV low-pressure oil-filled cable is to be put into operation in 1973 and in Europe the
development of a 750 kV cable of this type is already under way (Pirelli,
Italy).

345 kV high-pressure oil-filled cables in a steel pipe have been in
operation in the USA for many years and have given satisfactory performance. For some time trials have been carried out with 500 kV cables.

In the Federal Republic of Germany 110 kV cable installations
have so far been mostly in use together with a few short 220 kV cable
connections, these being for the greater part single conductor oil-filled
or gas-pressure cables. The first 380 kV installations are scheduled
for operation in 1974.

The maximum possible transmission voltage of impregnated-paper
oil-filled cables is dependent upon the dielectric losses of the insulation.
With the papers which are in vogue today the dielectric loss factor tan δ
is between 1.7×10^{-3} and 2.5×10^{-3}. The dielectric losses increase as
the square of the transmission voltage and at high transmission voltages
form a considerable part of the total losses. As the power losses that
can be dissipated to the surrounding soil are restricted, the transmission
capacity of a cable system decreases as the dielectrical losses increase.
The conditions are shown in Fig. 8 where the loss factor is the parameter. Figure 8 reveals that with the use of normal paper (tan $\delta \approx$
2×10^{-3}) the transmission voltage limit lies at 750 kV.

The maximum power transmission capability of conventional cables
is determined by the restricted dissipation of the heat losses generated
in the cable to the surrounding ground and the permissible continuous
temperature for the dielectric which with an oil paper dielectric is 85°C.
The cable losses are composed of : the I^2R losses which in the case of
three-phase cables are further increased by the skin and proximity effect, losses through induced currents and eddy currents in the cable
sheaths and, possibly, hysteresis losses in the latter, and the dielectric
losses already mentioned. Assuming a maximum conductor temperature of 85°C and a partial drying out of the ground up to an isothermal
of 30°C the following possible maximum continuous loads (load factor 1)
are obtained for 400 kV impregnated-paper oil-filled cable systems with
natural heat dissipation [12]:
    – Low-pressure oil-filled cable of the single conductor type: 450 MW,
    – High-pressure oil-filled cable in a steel pipe: 350 MW.

As Fig. 9 shows these power ratings are obtained with a copper
conductor cross section of about 1000 mm^2. Even if the cross section

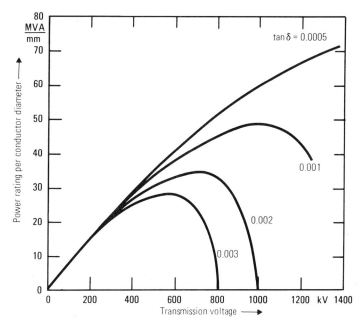

FIG. 8. Maximum Transmission Voltage as a Function of Dissipation Factor [12].

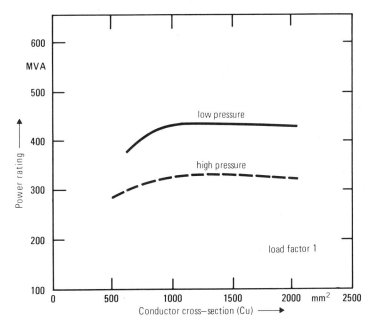

FIG. 9. Power Ratings of Natural Cooled Oil-Filled Low- and High-Pressure 400-kV Cables [12].

of the conductor were increased — for technical manufacturing reasons up to 2500 mm^2 are feasible — no further increase in the maximum power can be achieved as is also shown in Fig. 9. The laying of many parallel cable systems is restricted by the mutual thermal effects and by the amount of ground space required. An increase in the transmission capa- city can be obtained by using special backfill (e.g., lean concrete, sand- gravel mix) in the immediate vicinity of the cable [13]. This method has already frequently been used in England. Drastic increases in the sys- tem transmission capacities may, however, be obtained by artificial cooling of the cables.

2. Artificially-cooled impregnated paper oil-filled cables

For the forced cooling of supertension cables, the following methods can be considered:

     a) Direct or indirect cooling of the cable surface by water and oil,

     b) Direct cooling of the conductor in the cable by the insulating oil.

With the indirect cooling method of the cable surface (lateral cooling) pipes through which water is circulated as the cooling medium are laid in the earth parallel to the cable. The heat from the cable is dissipated through the surrounding soil to the water cooling pipes. The heat absorbed by the water over a certain length is carried away into cooling stations placed along the cable route. This method has so far been used chiefly in England for 275 kV transmissions systems.

When employing the direct cooling method (integral cooling), the cable surface is in direct contact with the cooling medium (water or oil). The cooling principle can be used both with high-pressure oil-filled cables and also — particularly effectively — with low-pressure single- conductor cables. Figure 10 shows for both types of cable the increase possible in the transmission capacities with integral cooling. This capacity increase is the higher, the lower the outlet temperature of the cooling medium. Moreover, the maximum conductor cross-section can be utilized with this type of cable cooling. With a maximum cooling- medium temperature of 40° C and a conductor cross-section of 2000 mm^2, the possible transmission capacities are 1500 MW for a 400 kV single- conductor system and 1200 MW for a 400 kV high-pressure cable.

Examples of the application of direct cooling of the surface of single-conductor oil-filled cables with natural flowing water (cable troughs with gradients and weirs) are the well-known three tunnel in- stallations in England: Southampton Water, Woodhead - Dunford - Bridge and the Thames sub-crossing. In Germany a forced-cooled 400 kV low-pressure single-conductor system has also been proposed (BEWAG - Berlin) with a transmission capacity of 1 GW per system.

Direct cooling of the conductor with insulating oil (internal cooling)

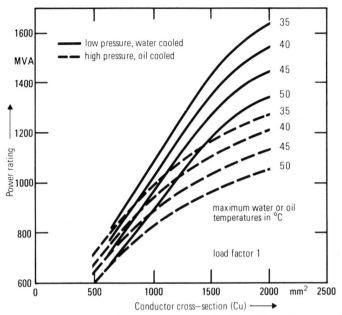

FIG. 10. Power Ratings of Forced Cooled Oil-Filled Low- and High-Pressure 400-kV Cables [12].

is particularly suitable for single-conductor oil-filled cables. For efficient cooling, even over long sections, the resistance to flow should be as low as possible, i.e., hollow conductors with comparatively large diameters must be employed which again lead to greater cable dimensions. With cable installations of this kind transmission capacities of around 3500 MW could be achieved at voltages of 400 kV, and of over 4000 MW at 750 kV. Efforts in this direction have particularly been made in England [14]. An even more efficient internal cooling of the conductor would certainly be achieved with water, the thermal capacity of which is more than twice as great as that of insulating oil. Special difficulties arise here due to the fact that the water to be recirculated is at high-voltage potential.

In conclusion it can be said that forced cooling enables enormous increases in the transmission capacities of oil-filled cables to be made. These increases can be obtained by combining well-known techniques (oil-filled cable and cooling system).

It remains to be seen how this solution will fare economically compared with other competing concepts.

3. High-voltage dc cables

dc cables have the following advantages as compared with ac cables:
- no reactive charging power is needed,
- no dielectric losses occur,
- no skin effect, therefore full utilization of the conductor cross-section,
- no losses in the metal sheaths.

The first advantage means that any reactive power compensation, which imposes restrictions on the power transmission capability, is unnecessary and consequently power transmission over great distances is possible. The second advantage has the consequence that for the same dielectric higher transmission voltages are possible. This, in conjunction with the other two advantages, results, for the same expenditure, in a considerable increase in the transmission capacity as compared with ac cables. The advantages mentioned must, however, be outweighed against the disadvantage of costly converter installations and stations.

Existing HVDC cable installations are almost exclusively long sea cable links. For power station connections and system interties land cables may in future also gain in importance. An example of this kind is an installation with low-pressure oil-filled cables which was put into operation sometime ago in England (Kingsnorth - London).

4. SF_6 - gas - spacer cables

This type of cable has been developed from metal-clad SF_6 insulated high-voltage switchgear. The pipe cables consist, in the single-phase type, of an outer pipe made of aluminum alloy, in which the bare aluminum conductor supported by insulating spacers is arranged. In these cables sulphur hexafluoride gas (SF_6) at a pressure of 2.5 to 4 bar is used as the insulant. Figure 11 shows the construction of an SF_6 pipe cable for 400 kV. The outer sheath is used as the earth wire (return conductor) [15]. SF_6 pipe cables have the following advantages:

a) Because of the comparatively large distance between the conductor and the outer sheath and the low dielectric constant of the insulating gas ($\epsilon_r \approx 1$), the capacitance and consequently also the capacitive charging power of pipe conductors of this kind is small compared with conventional cables. (The capacitance is about five times greater than in the case of overhead lines and about five times smaller than with conventional cables.) Consequently the maximum possible transmission lengths can be considerably increased.

b) Dielectric losses are negligible.

c) The comparatively good heat conductivity of the insulating gas provides satisfactory heat transfer from the conductor to the outer sheath. The maximum possible temperature of the conductor depends upon the temperature stability of the spacers which should be at about 100° C.

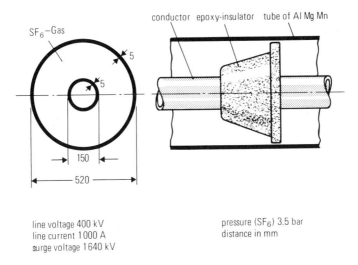

conductor   epoxy-insulator   tube of Al Mg Mn

SF$_6$–Gas

line voltage 400 kV
line current 1 000 A
surge voltage 1 640 kV

pressure (SF$_6$) 3.5 bar
distance in mm

FIG. 11. Construction of a SF$_6$–Gas–Spacer Cable for 400 kV [12].

Mention must, however, also be made of the following disadvantages:

a) The dielectric strength of the insulating gas is smaller than that of an oil paper dielectric. Moreover in the area of the spacers, the dielectric strength of the insulating gas is in general not achieved when the spacer is not cast onto the inner conductor [16]. The latter is however technically difficult to achieve. These circumstances lead to comparatively large pipe diameters which in turn result in wide trenches to accommodate the pipes (for 500 kV systems a diameter of 600 mm is a normal figure for single-phase pipes). The dielectric strength may be considerably reduced through impurities in the gas or contamination of the surface of the insulating spacers.

b) The laying of rigid pipe conductors is possible only in limited lengths (maximum 20 m). This means many joints. Moreover, to compensate for thermal expansion, expansion joints must be incorporated. In addition to the cleanliness required, there is a considerably higher expenditure for assembly and laying compared with conventional cables. For working voltages up to 220 kV corrugated pipes can also be used for the conductor and outer sheaths which enable conductor lengths of about 250 m to be manufactured. These could then be wound on drums.

It is expected that when laying pipe cables underground a transmission capacity of 2500 MW can be achieved at a transmission voltage of 400 kV. In the case of 500 kV systems this capacity can be increased to 3500 MW. By employing forced cooling by gas circulation inside the

outer sheath, or water cooling of the surface of the outer sheath, it is expected to attain transmission capacities of up to about 10,000 MW [17].

The first commercial cable section employing SF_6 gas was installed
in New York in 1969. It is a 345 kV single-conductor system with a transmission capacity of 2000 MW and a length of 180 m. In the USA several
pipe cable systems of a relatively short length (up to 2000 m) have been
ordered for operating voltages of 138 kV - 345 kV. The first installation
in Europe will be carried out in the Federal Republic of Germany in 1974.
It will be a 400 kV cable installation with two parallel systems for a
pumped-storage power station in Southern Germany.

Finally it should be mentioned that SF_6 gas-spacer cable designs
have already reached a notable level of development and in the future could
probably serve as an effective means of transmitting large blocks of
power. They will doubtlessly be in strong competition with forced-cooled
oil-filled cables.

5. Cables with plastic insulation

Cables with extruded polyethylene insulation have the advantages
of a small tan δ of $\leq 5.10^{-4}$, a simple method of production and a simple
construction (elimination of pressure and expansion tanks which are essential with oil-filled cables). As short samples of polyethylene have in
addition a high breakdown strength, cables of this kind could be expected
to be ideal for high transmission voltages. However, it seems to be very
difficult to verify the high dielectric strength with geometrically extended insulations. It is greatly influenced by the purity of the raw materials
and the conditions of the manufacturing process in relation to impurities,
freedom from voids and mechanical stresses. As the thickness and length
of the insulation increase, technological difficulties arise to more than a
proportional extent. Work is still going on to investigate and solve the
problems.

So far 110 kV and 132 kV cables have been laid to a considerable
extent and with them the most varied operating experience has been gained.
In France a short length (5 km) was laid for operation at a voltage of 225
kV [18]. With large transmission capacities artificial cooling would also
have to be employed. Since, however, no data can be provided at the
moment concerning the maximum possible transmission voltages for these
cables forecasts cannot be made either as to the possible transmission
capacities.

The difficulties with cables having extruded polyethylene insulation
should be eliminated to a large extent with a wound plastic film insulation.
Films made from the following materials have been proposed for wound
insulations and have also partly been tested:

- low pressure polyethylene (also irradiated),
- polypropylene biaxially stretched,
- polycarbonate (e.g. Macrolon BASF),
- polyphenylene oxide and polysulphone.

Film insulations, in a similar way to the paper insulations of conventional cables, have to be impregnated in an insulating liquid or an insulating gas. Normal insulating oil, silicone oil and SF_6 have been proposed [11]. Problems exist here in that the plastic films are normally impervious to the impregnating liquid and the latter may react with residues of plasticisers remaining in the film. If use is made of perforated films to improve impregnation, a dielectric strength is obtained for the insulation which corresponds essentially with that of the impregnating liquid. From the above mentioned materials, polycarbonate and polysulphone must probably be excluded because of their poor elastic performance which spoils the bending behavior. Sulphur hexafluoride has also proved to have disadvantages as an impregnating liquid. As compared with oil, decreases in the dielectric strength were clearly noticeable which is to be attributed to the great differences in the dielectric constants. In order to obtain a better impregnation mixed windings of polypropylene and cellulose or synthetic paper and cellulose have also been investigated.

For economic reasons cables of this kind are attractive only for working voltages above 400 kV where the small loss factor becomes an advantage. The development, however, is only in its infancy so that no final conclusions can yet be drawn as to the possible use and transmission capacity.

6. Sodium cables

Since sodium as a conductor material has only 1/3.25 of the weight compared with copper calculated on the same conductor resistance and on the other hand the price of sodium is only about 30% of that of copper, the cost for cable conductors would be reduced to about 1/10 if sodium were used.

Figure 12 shows a cross-section which reveals the simple construction of this type of cable. During manufacture, special measures have to be taken. When extruding the sodium the plastic sheath (polyethylene) must be extruded around this at the same time so as to avoid the entry of air (combustibility). The advantages of the cable are:
- cheaper conductor material,
- greater flexibility of the conductor and thus of the cable,
- low weight,
- a semiconducting layer between the polyethylene and sodium is not necessary because of the good bonding of sodium to polyethylene.

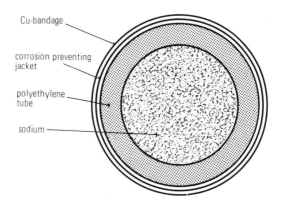

Cu-bandage

corrosion preventing
jacket

polyethylene
tube

sodium

FIG. 12. Cross Section of a Sodium Cable.

    In the USA 50 sodium cables for sub-transmission voltages have
already been laid. A cable for 138 kV has been constructed by the
American Electric Power Corporation which apparently has remarkable
properties in relation to the breakdown strength (high breakdown voltage
with a low variation of the breakdown values). On the other hand, terminations and connectors have still to be developed for this type of cable.

    In the meantine, however, development of this cable has in general
been discontinued in the USA because there is a danger that on heating
the cable above the melting point of the sodium (98°C) overheating (thermal runaway) may occur through the sudden increase in resistance
of the conductor.

    A further possible source of faults revealed in tests carried out in
Germany (Siemens AG) consists of cavities in the conductor material
where the film of sodium adhering to the polyethylene does in fact maintain the electrical contact but the small cross-section of the conductor
under current load results in an excess temperature of the polyethylene
wall. Thus there is the danger of sodium exuding from the cable [19].

    At the present state of the art, considerable doubt must be ex-
pressed whether sodium cables will ever become of importance for high-
power transmission.

7. Microwave power transmission

Waveguides for power transmission are generally known in com-munication engineering where, for example, antenna leads are used with transmission capacities of about 1 MW. It was therefore appropriate to investigate whether waveguides of this kind are also suitable for high-power energy transmission. Microwave transmission is likely to be economically attractive only for covering great distances because the re-quired main-frequency/high-frequency converters in the transmitting and receiving stations, as in the case of HVDC, are very expensive. Con-sequently, this approach has to be compared with high-voltage overhead lines. The power region for which microwave transmission has been con-sidered lies above 3 GW.

Two wave types have been investigated:
– the H_{01}-wave in the circular waveguide, and
– the H_{10}-wave in the rectangular waveguide.
(H denotes the main mode of oscillation of such waves at which axial mag-netic field components occur, i.e., $H_z \neq 0$ and only transverse electrical field components, i.e., $E_z = 0$. In the case of a rectangular waveguide, the indices indicate the number of current modes along the wide and narrow sides of the tube and in the case of the circular waveguide, the number of zero positions of the Bessel functions describing the wave type or their derivatives.)

With microwave energy transmission, the attenuation of rectangular waveguides is too great so that they can only acquire importance as short coupling elements. More attractive is the attentuation of circular wave-guides with the H_{01}-wave and consequently these alone appear suitable for microwave transmission. To enable them to compete with conven-tional transmission lines as regards losses, the attenuation must be $\lesssim 10^{-3}$ dB/km.

The first difficulties begin to appear when selecting the operating frequency. If the meter wave and decameter wave ranges (VHF and SW range) are employed, techniques are available which would enable the production of high frequency energy to be obtained in the Gigawatt range. On the other hand waveguides then assume physical dimensions which are prohibitive. To obtain reasonable dimensions for the components, an operating frequency of 3 to 10 GHz (centimeter waves) was generally selected for microwave transmission [20, 21]. Microwave oscillators must, however, still be developed for these frequencies and outputs in the Gigawatt range.

Figure 13 shows for circular waveguides the maximum transmission capacity and attenuation as a function of the diameter of the hollow con-ductor: the parameter is the frequency. Accordingly, if it is desired to transmit power with losses similar to those of extrahigh voltage lines (i.e., $\alpha \lesssim 10^{-3}$ dB/km), a tube diameter must be selected of 2.3 m at 3 GHz or 1.3 m at 10 GHz. The capacities which can then be transmitted

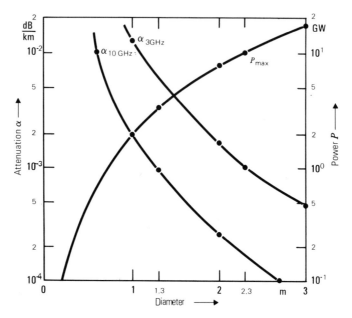

FIG. 13. Attenuation α and Power-Handling Capability P of the Circular
Waveguide [21].

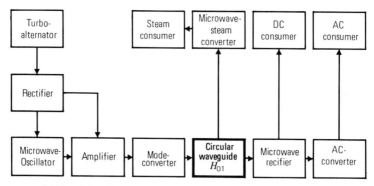

FIG. 14. Power Transmission by Microwaves.

are 10 GW or 3.4 GW respectively.

An attentuation of 10^{-3} dB/km corresponds to a specific power loss of about 200 kW/km GW, i.e., if 5 GW are to be transmitted, losses of 1000 kW/km are to be expected. In order to keep losses at this level, exacting demands must be made on the accuracy in the manufacture and installation of the hollow tubes. Thus changes in diameter and displacements of the axial centre must be $\leq 10^{-3}$ and the same also applies to the ellipticities of the tubes. At the flanges where the individual tube sections 5 to 20 m in length are joined up, care must be taken to see that the bend angles are kept at a minimum ($\varphi < 10^{-3}$ rad). The radius of tube bends must not be less than 1000 m. If the tolerances are not maintained there is a danger that other wave types will be excited which are attenuated more strongly than the H_{01}-wave and the specific losses will increase considerably.

In addition to the attentuation it is mainly the excitation of H_{01}-waves at high powers and also their coupling out at the end of the conductor that cause great problems. At the heavy local power densities to be expected ordinary excitation devices and mode converters will burn even when they are water cooled [22]. For the production of microwaves and also their rectification or conversion into 50 Hz no insuperable difficulties are to be expected.

Figure 14 shows the block diagram of a complete installation for microwave power transmission. A conventional turbogenerator for 50 or 60 Hz supplies a microwave oscillator and microwave amplifier. The wave guide is coupled to the amplifier via a mode converter. In order to improve the efficiency of the installation the whole of the incoming high frequency energy is not converted into power-frequency energy at the load side but a large part is passed on direct steam production and a further part in the form of dc power to dc loads.

Even taking a long view, the efficiencies of the transmitting and receiving stations combined with that of the hollow wave-guide will most likely not be high enough to come even near the efficiency of ac or dc transmission systems. The use of microwave transmission is not considered a practical possibility in the foreseeable future because of the state of development and the disadvantages described.

8. Cryogenic normal-conductor cables

Proceeding from the forced-cooled conventional cables dealt with above, this type of cable represents the next logical step forward. In place of the conventional cooling medium such as water or oil, low-boiling cryogenic liquids are to be used with this cable. Here not only the heat formed in the cable is conducted away but by using ultrapure conductor materials which increase their conductivity at low temperatures, a high power density is obtained.

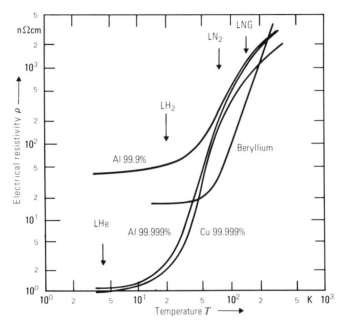

FIG. 15. Temperature Dependence of the Electrical Resistivity of Al, Be, and Cu of Different Purity [113].

Figure 15 shows the variation of the resistivity of the three conductor materials in question, Cu, Al and Be in relation to the temperature. The following have been considered as coolants: liquid natural gas (boiling point ≈ 110 K), liquid nitrogen (80 K), liquid hydrogen (20 K) and liquid helium (4.2 K).

Direct conductor cooling with natural gas can hardly be considered because the increases in conductivity at temperatures around 110 K are still not very pronounced. Nevertheless, it has been proposed [23] to transport liquid natural gas in the radiation shield of a cryogenic cable parallel to the electrical energy. Nitrogen and beryllium would represent a very good combination because the latter at 80 K exhibits a decrease in resistance of almost two orders of magnitude. Unfortunately, the technology and processing of Be is extremely difficult and costly so that at present only Cu and Al can be seriously considered for use as conductors. At 80 K both materials exhibit a decrease in resistance of 10. A marked increase in the conductivity of Cu or Al can be obtained by cooling with liquid hydrogen. With pure materials (99.999%) the conductivity at 20 K increases by a factor of 1000: aluminum is preferred because it is easier to obtain and cheaper. The use of liquid hydrogen, however, conflicts in particular with the existing safety regulations. Liquid helium must be excluded as a coolant because the increase in the conductivity of Cu and Al at 4.2 K, as compared with 20 K, is small and on the other hand the cost of cooling at 4.2 K is considerably higher than it is at 20 K. Intensive comparisons of the various concepts have shown an aluminum conductor with liquid N_2 cooling [24] to be technically the simplest and also an economical combination. And it is adequate if the degree of purity of the aluminum is 99.9 - 99.99%.

Development work on aluminum-conductor cables cooled with liquid nitrogen is at present being carried out by the General Electric Co., USA, by Hitachi and Furukawa, Japan and previously by the Simplex Wire and Cable Co., USA. While the first three firms mentioned are working on a cable which contains flexible Al conductors comprised of subconductors, the last mentioned firm has developed a cable with rigid aluminum tubular conductors (see Figs. 16a and b) [25-28]. In the case of the flexible phase conductors, several single conductors are arranged on a round flexible hollow element. By selecting a suitable diameter for the single conductors and by transposition of the insulated conductors, the skin and proximity effects are coped with, i.e. the conductor losses are minimized. (The penetration depth in the case of pure Al at 80 K is 3.4 mm.) Because of the skin effect, the diameter of the rigid tubular conductor must in general be made greater for the same working current than that of flexible conductors. If excessively large pipe dimensions are to be avoided, larger losses must be accepted than in the case of flexible conductors.

The electrical insulation of the two types of cables also differs. In the case of rigid tubular conductors a vacuum serves as the high-voltage insulation, the conductors being held by suitable supports in the vacuum pipe, which at the same time serves as an electromagnetic shield. For the flexible conductors an insulation wound from synthetic paper (Tyvek) and impregnated with liquid N_2 has been proposed [27]. In Fig. 16a the insulating fluid is at the same time the cooling fluid. For this type of insulation a dielectric strength of 16 kV/mm is assumed [24].

Since aluminum conductors still produce losses despite cryogenic cooling, the highest possible transmission voltages, e.g., 500 kV, must also be employed with these kinds of cables to obtain transmission capacities in the Gigawatt range. To enable dielectric losses to be restricted to a reasonable value, the loss factor of the insulation must be $\leq 10^{-4}$. It has been shown in experiments [29] that the tan δ values of a Tyvek film insulation impregnated with nitrogen may be in the range of 10^{-5}.

In both of the types of cables, the three phase conductors are placed in an electromagnetic shield, i.e., aluminum pipe, which prevents eddy currents, and consequently losses, from being generated in the outer cable pipes. In the case of the flexible cable this pipe in addition to the flexible hollow conductors has nitrogen flowing through it. In the rigid type of cable the cooling liquid flows only in the conductor tubes. Single phase nitrogen at an increased pressure (sub-cooled liquid) is particularly suitable. According to investigations made so far, the cooling stations for recooling the warmed nitrogen are spaced at invervals of about 10 km, the assumption being made that the warmed nitrogen is

Liquid nitrogen-cooled flexible
cable system

Liquid nitrogen-cooled rigid
tube system

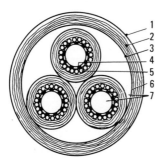 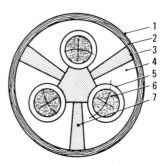

1 outer casing
2 super insulation
3 inner casing and E.M.shield
4 perforated tubular form (nonmetallic)
5 stranded conductors
6 fluid saturated insulation
7 Cryogenic fluid

1 outer casing
2 super insulation
3 inner casing and E.M.shield
4 vacuum insulation
5 rigid tube conductor
6 cryogenic fluid
7 spacer

FIG. 16. Cryoresistive Cables [24, 27].

TABLE 1. Characteristic Data of Liquid Nitrogen Cooled Al-Cryo-resistive Cables 500 kV, 3500 MW [24]

	Flexible type	Rigid type
Conductor area cm^2	26.45	13.03
Strand diameter mm	1.3	119
Resistance ratio conductor/shields	1/10	1/10
Tan δ	3×10^{-4}	0
Cable outer Diameter cm	17.3	20.6
Pipe/Shield inner diameter cm	43.5	64.5
Costs per cent		
Conductor metall	4 ⎫	2 ⎫
Rest of cable	8 ⎬ 16	14 ⎬ 19
Shield	4 ⎭	3 ⎭
Stainless steel pressure pipe	6 ⎫ 28	10 ⎫ 46
Cryogenic pipe	22 ⎭	36 ⎭
Refrigerator	9 ⎫ 25	10 ⎫ 34
Power charge	16 ⎭	24 ⎭
Terminals	8 ⎫ 31	6 ⎫ 24
Installation	23 ⎭	18 ⎭
Total	100	123
Refrigerator input kW/km		
Conductor and eddy losses	653	1019
Dielectric	75	---
Shield	367	640
Pipe	59	83
Pump	21	52
Return pipe		108
Total	1214	1902

returned in a separate pipe to the refrigerator. In both
types of cable the thermal insulation consists of a vacuum space between
the electromagnetic shield and the outer cable pipe. The vacuum space
contains a reflecting multilayer superinsulation, an insulating powder
filling or even foam plastic, such as, for example, polyurethene foam
(see also Section VII).

Table 1 shows by way of comparison a number of important data
for both types of cable [24]. The comparison is based on a 500 kV three-
phase system with a transmission capacity of 3500 MW. It is quite clear
that the cable with the rigid pipe conductor has a greater outer diameter
and also considerably higher losses than that with the flexible conductor.
It will be seen later on that even the losses of the flexible type which are
about 360 kW/km GW are relatively great as compared with the other
underground transmission systems. The outer diameters of the two types
of cables of about 45 cm (flexible) and 65 cm (rigid) respectively are
remarkably small. The manufacturing costs, including laying in the case
of the rigid cable are about 25% higher than for the flexible type so that
the latter is to be preferred in all respects (losses, dimensions, costs).
It will be shown below that the specific capital costs are to be set at
rather a higher than a lower figure as compared with other types of
cables.

The state of the art is perhaps best illustrated by a 12 m long single
phase test circuit of the General Electric Company, USA [28]. It was
fitted with a high-voltage termination and was principally used for testing
the dielectric strength of a wound Tyvek film insulation impregnated with
liquid nitrogen. With an insulation thickness of 21.6 mm a breakdown
occurred at 435 kV corresponding to a mean breakdown field strength of
20 kV/mm. The tests are being continued in particular to improve
electrical insulation.

From the facts described, it must be concluded that cryogenic
normal-conductor cables are comparable in their transmission capacity
with forced-cooled oil-filled cables and with SF_6 pipe-type cables. Since
it must be assumed that, with the present-day state of the art the capital
costs and transmission losses will be at least as high as they are with
the latter, they represent no progress compared with these. A higher
power density alone, i.e. smaller cable dimensions hardly justify the
considerable outlay for development.

III. A GENERAL COMPARISON BETWEEN SUPERCONDUCTING DC
     AND AC CABLES

The employment of cryogenic fluids for cooling cables results in
definite advantages only if the conductor losses are reduced to zero or
practically to zero by making use of the phenomenon of superconductivity.
This property of superconductivity makes it possible to increase the

efficiency of power transmission, which today stands at around 95%, to about 99.5%.

Shortly after the discovery of the so-called high-field superconductors with their high critical current and flux densities at the end of the fifties and beginning of the sixties, an exceedingly optimistic forecast was given in many publications [30-32] on the possible applications of superconductivity in electrical engineering. Initial assessments showed that enormous increases in power, improvements in efficiency and savings in weight and volume are obtained for almost all of the important components in power generation and transmission such as generators, transformers and cables. These assessments were based on the false premises that high-field superconductors are able to carry both direct current and alternating currents without losses. These views had to be corrected by 1963 after it had clearly been proved [33, 34] that high-field superconductors in particular, but also niobium and lead, exhibit hysteresis losses in magnetic alternating fields. With high-field superconductors hysteresis losses are so great that their use at main frequencies (50 or 60 Hz) and large alternating field amplitudes ($\mu_0 H > 1$ Tesla) must be ruled out for economic reasons. On the other hand it was also shown then that with small flux density amplitudes (0 to 0.2 T) the losses, in particular with lead and niobium but also with hard superconductors, could be so low that use of these superconductors in three-phase cables is possible. Consequently superconducting ac power cables were being proposed relatively soon [35, 36] in addition to superconducting dc cables. As an introduction to the descriptions which follow, the essential differences between superconducting dc and ac cables will first be discussed in brief, an outline given of their advantages and disadvantages and their potential applications will be shown.

Superconducting dc cables have a number of advantages as compared with ac cables. The construction of the conductor system is simpler because, to begin with, the number of conductors is smaller in the ratio 2:3. As in addition, under normal operation, the conductor carrying direct current does not produce any magnetic alternating fields which might produce eddy currents in normal-conducting cable components, superconducting shields are unnecessary, such as are required with ac cables (see Section IV). In the case of dc cables, the choice of conductor materials is much wider, practically all superconductors being suitable which have high current densities in the flux density range of 0.1 T - 0.5 T and thus, in particular, the high-field superconductors. With three-phase cables, on the other hand, it is only practical to use superconductors which have low hysteresis losses at 50 or 60 Hz. Consequently, as we shall see, the maximum flux density amplitude and the maximum linear current density (r.m.s.) of an ac conductor is confined to 0.1 T and 560 A/cm respectively, and the diameter of ac conductors for the same working current is in general greater than that of dc superconductors. In addition dc superconductors produce no losses under normal working conditions. Similar conditions apply as far as the electrical insulation is concerned when the most promising way, that is to say a wound, multi-layer foil insulation is considered. With dc cables a selection can be freely made according to the criteria: dielectric strength, workability and costs since dielectric losses do not occur. For ac cables only materials with a small loss factor can be employed. However, the fact that the dielectric strength of wound insulation systems is much higher at room temperature with a dc voltage load than with alternating field voltages [37, 38] appears to apply only to a limited extent at low temperatures. Final proof of this assertion must still be established.

The advantages described so far of the dc cable result in smaller cable dimensions and thus in lower thermal losses when equivalent loads are to be carried. Taking into account the absence of electrical losses, considerably smaller total losses are thus obtained. A quantitative comparison shows for example that with a dc cable, practically twice the power can be transmitted [39, 40], i.e., 5 GW instead of 2.5 GW, at the same transmission voltage of 120 kV and with the same overall cable diameter of 0.5 m. The total losses with the dc cable are about one half less than with the ac cable. The specific capital costs (DM/km. MW) for the dc cable with the same transmission capacity are less by a factor of 2 to 3.

With regard to possible transmission lengths and charging power, similar differences exist in principle as in the case of conventional dc and ac cables but, strictly regarded, conditions are somewhat different with superconducting cables. With superconducting dc cables there are practically no limits to transmission lengths from the point of view of line and leakage losses. The capacitive charging power of superconducting ac cables is smaller than it is with conventional cables of the same power rating in particular because of the comparatively low transmission voltages (due to the high transmission currents). The low transmission voltage has also the result that superconducting ac cables, in contrast to conventional cables, must in general be operated above their natural load (U^2/Z_W). The transmission lengths are then restricted not through reactive power losses in the conductor but through the inductive voltage drop along the cable. For greater distances, several phase conductors with loads close to the natural one must be connected in parallel in order to keep the voltage drop within limits.

From what has been said so far, it follows that the superconducting dc cable is superior to the ac cable in practically all respects. Nevertheless its possible application is restricted due to the fact that at its ends space-consuming and costly static converter installations are required. The proportional costs of the static converter in the transmission system are so high (e.g. for 500 kV: 180 DM/kW), that dc cables are economically attractive only for power transmission over great distances; for short lengths ac systems are preferable. Figures given on the break-even distances, above which dc transmission would be more economic than ac transmission, vary widely and range from 50 to

800 km [41-44]. They depend upon the power being transmitted and the transmission voltages and naturally also upon what costs are assumed for the two types of cables. The small distances refer to low transmission voltages (100 kV) because for the same capacities, the static converter costs for these are smaller than for high voltages. In comparisons of this kind reference is frequently made also to the fault-current limiting effect of the static converter installations. By using dc transmission systems for power-station infeeds or system interties, fault currents in the power systems can be limited.

Our own estimates have shown that for transmission lengths of <200 km, ac cables represent the more economic solution in every case. They are therefore feasible in particular for the supply of conurbation areas. Over greater distances superconducting dc cables are in competition with the overhead line and will only gain over the latter when boundary conditions make underground transmission necessary.

IV. SUPERCONDUCTING CABLE CONCEPTS

A. Mechanical Construction

In the case of superconducting cables, a distinction can roughly be made between two features, namely thermal insulation and electrical conductor system. As far as the mechanical construction is concerned, there are generally three different cable concepts:

a) The rigid type of cable (single-phase model shown in Fig. 17a): the thermal insulation and conductor are made of rigid pipes. One of the main difficulties encountered with this design is that the maximum transportable manufacturing length is about 20 m which results in a large number of cable joints. A further problem consists in controlling the thermal contraction of the cold components of the cable. With metals this is about 0.2 to 0.4% and with plastics up to a few percent for temperatures less than 80 K. To compensate for longitudinal contraction, use must be made of corrugated components both in the thermal insulation and in the conductor and these are best fitted in the area of the cable joints. When using pipes made of special alloys with low coefficients of expansion (e.g., Invar: $\Delta \ell / \ell \approx 0.05\%$) as mechanical supports for the conductors and as the material for the cold components of the thermal insulation, corrugated components are only required at wide intervals (several hundred metres). When Invar is used corrugated components may even be completely dispensed with if axial forces occurring during cooling of about 6 kg/mm^2 are accepted.

b) The semiflexible type of cable (single-phase model shown in Fig. 17b): the thermal insulation is again made of rigid pipes with corrugated components to compensate for thermal contraction. The conductor, however, is flexible and consists either of a corrugated pipe or of strips or wires which are wound helically on a hollow cylindrical support.

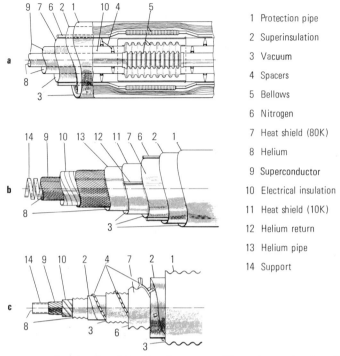

1	Protection pipe
2	Superinsulation
3	Vacuum
4	Spacers
5	Bellows
6	Nitrogen
7	Heat shield (80K)
8	Helium
9	Superconductor
10	Electrical insulation
11	Heat shield (10K)
12	Helium return
13	Helium pipe
14	Support

FIG. 17. Mechanical Cable Designs. a, Pipe-type; b, Semiflexible; c, Totally flexible.

These conductors can be made in lengths of 200 to 500 m and transported on drums. In these lengths they are also drawn into the previously laid rigid thermal insulation. With this cable the spacing between conductor joints is very great while the thermal insulation joints, as in solution a, are spaced at about 20 m.

c) The completely flexible type of cable (single-phase model shown in Fig. 17c): here the thermal insulation is flexible as well. It is constructed of corrugated pipes [45] so that there are no problems regarding transport and thermal contraction. The conductor may again consist of a corrugated pipe or of helically wound strips. The whole cable can be made and transported in relatively large lengths (≈ 200 m) so that the cable joints are also spaced relatively widely apart.

In the rigid and semiflexible types of cable, all of the conductors can be accommodated in a common rigid thermal envelope which has a favorable effect as far as the thermal losses and the required width of the cable route are concerned. In the case of the fully flexible cable, on the other hand, the outer diameter is restricted to about 250 mm for manufacturing and transport reasons (transportable length about 200 m), so that the same can only be used as a single-conductor cable for larger transmission capacities (i.e., each conductor lies in its own thermal insulation). This results in higher thermal losses and in the need for a

greater amount of ground space. On the other hand, the problems with
transport and laying are considerably simplified. Apart from a few
exceptions solutions b and c are today preferred.

B. Conductor Configurations

1. Three-phase conductor systems

The conductor configurations proposed so far for three-phase ac
cables are based on the fulfillment of two requirements:
— Restriction of the electromagnetic fields to the areas between
the superconductors to prevent the occurrence of undesired eddy-current
losses in the normally conducting components of the cable.
— Low magnetic flux densities on the superconductor surfaces to
minimize losses in the superconductor.

Round hollow conductors are mainly used as conductors which
provide both good flow and cooling conditions for the helium. Figs. 18
to 22 show the principal conductor arrangements.

With the concentric conductor arrangement [46] shown in Fig. 18
complete field compensation appears at first sight to be achieved by a
symmetrical loading of the phases. To this end, one phase must be
divided into two auxiliary phases (Fig. 18), so that in the two succeeding
coaxial pairs of conductors $(R, S_2$ and $S_1, T)$ the currents are in each case
displaced in phase by 180°. The distribution of the field over the surface
of the superconductor is uniform and with a symmetrical arrangeement
no forces occur between the conductors. In reality, complete field compensation is not obtained because of the different impedances of the
individual phase conductors. This negative effect is further intensified
by unbalanced loading between the individual phases. In addition, the
provision of the auxiliary phases necessitates additional phase-shift
arrangements.

Figure 19 shows the pairs of conductors in the coaxial arrangement
[47]. The three-phase conductors or a multiple of the same have the
same impedance so that in addition to the advantages of the concentric
arrangement, namely regular distribution of the field on the surface of
the conductor and freedom from forces with symmetrical position of the
conductors, complete field compensation is also provided. A disadvantage of this arrangement is the greater cost for superconductor
material and the higher ac losses caused through the larger surface. A
possible electrical connection for the phase conductors is shown in
Fig. 19. The star point is brought out from the transformer and via the
return conductors moved forward to the cable input (generator side).
The star point may be earthed additionally at the cable input end. By
means of this arrangement complete field compensation is obtained even
with unbalanced loads.

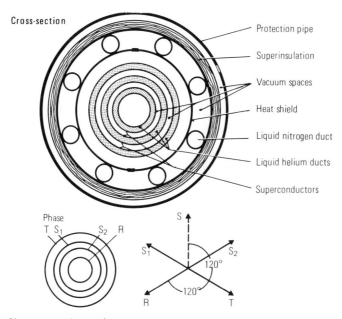

FIG. 18. Superconducting AC Cable Design with Concentric Conductors [46].

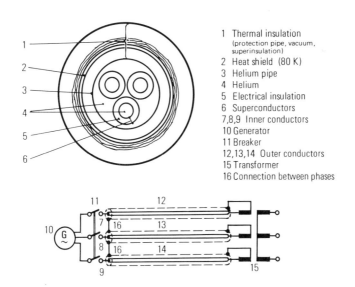

FIG. 19. Superconducting AC Cable Design with Coaxial Conductor
          Pairs [46].

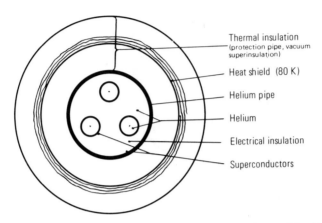

FIG. 20. Cross Section of a Superconducting AC Cable Design with
Common Superconducting Screen [48].

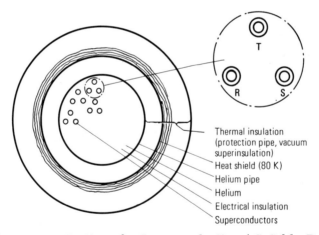

FIG. 21. Cross Section of a Superconducting AC Cable Design with
Conductors in Multiphase Arrangement [49].

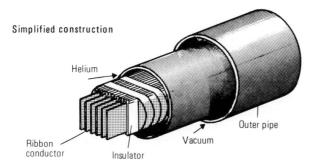

Simplified construction

Helium

Ribbon conductor

Insulator

Vacuum

Outer pipe

Electrical field control and cooling configurations

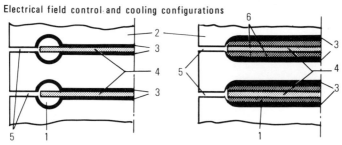

1 Helium 2 Solid dielectric 3 Conducting layers 4 Superconducting ribbons
5 Helium coolant channels 6 Spacers

FIG. 22. Interleaved Ribbon Cable Design [43].

Fig. 20 shows a design in which the three phase conductors [48] or a multiple of the same are contained within a common superconducting shielding. Inhomogeneous magnetic and electrical field distributions occur on the surface of the conductors. The conductors must be insulated to withstand the full transmission voltage and cannot be arranged so as to be free of forces. Compared with a coaxial arrangement, a larger conductor diameter is necessary for a given transmission current in order not to exceed a certain critical field value, e.g., H_{C1} due to the increase in the magnetic surface flux density. Apart from this an increase in ac losses must also be expected.

In the multiphase system [49] of Fig. 21 three phase conductors always are arranged as closely as possible so that an almost complete field compensation is obtained without any additional shielding. A disadvantage of this arrangement is that each one of the many individual conductors must be insulated for the full transmission voltage.

Consequently, only low transmission voltages are possible, and to obtain high transmission capacities, heavy currents are necessary. These however, involve considerable losses at the cable termination and a greater expenditure of superconducting material.

Similar problems occur with the interleaved ribbon cable [43] shown in Fig. 22. The outgoing and the return conductors of a phase lie parallel so that a practically complete field compensation is obtained. Here edge effects occur for the magnetic and electrical field which, particularly in the case of the electrical field, are not easy to control and which result in a limitation of the transmission voltage. A reduction of the increase in the electrical field is to be obtained by specially formed conductive layers embedded in the electrical insulation (Fig. 22). A further disadvantage of this design is the increased capacitance which results in higher charging currents and dielectric losses. Owing to the large surface of the superconductor, low magnetic flux densities can be obtained on the surface of the ribbon so that this conductor arrangement is particularly suitable for hard type II superconductors such as Nb_3Sn.

2. DC conductor systems

Most of the conventional and also some of the proposed superconducting dc transmission systems consist of two separately arranged single conductor cables the potential of which is symmetrical to zero. Thus with a high transmission voltage the electrical insulation has only to be designed for half of the voltage and in case of emergency (damage to a conductor) half of the power can still be transmitted.

For cooling reasons hollow conductors are generally recommended also for a dc cable. They are contained either coaxially [40] or in parallel [38] in a common thermal insulation (Fig. 23) or as mentioned before, are single-conductor cables with their own thermal insulation. For conductors arranged in parallel in a common thermal insulation the transmission voltage is symmetrical to zero as in the case of the single-conductor cable, while with the coaxial conductor arrangement the electrical insulation must be designed for the full transmission voltage. The helium pipes for the parallel conductor arrangement shown in the figure are made of Invar. In this way the thermal contraction is reduced and practically no forces are present between the two conductors because the highly permeable Invar pipes in each case shield the magnetic field of the other conductor.

With a dc cable it is, however, also possible to use a solid conductor of small dimensions. This enables extremely high powers to be transmitted. The cable in Fig. 24 is designed for a load carrying capacity of 100,000 MVA at ± 100 kV and 500 kA [50]. The current could be transmitted in a single solid wire with a diameter of about 2.5 cm. The high losses to be expected during transient phenomena, in particular switching operations and load fluctuations, make it however essential to divide up the solid conductor into transposed single conductors. The disadvantage of this arrangement of solid conductors is the high expenditure of material required for the superconductor. The reason for this is the relatively low attainable current density of about 10^5 A/cm^2 as a consequence of the high flux densities which occur.

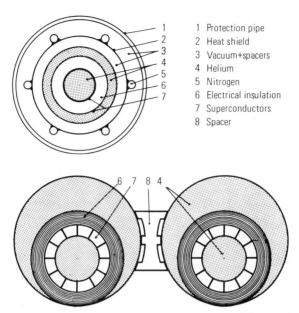

1 Protection pipe
2 Heat shield
3 Vacuum+spacers
4 Helium
5 Nitrogen
6 Electrical insulation
7 Superconductors
8 Spacer

FIG. 23. Cross Sections of Superconducting DC Cable Designs [38, 40].

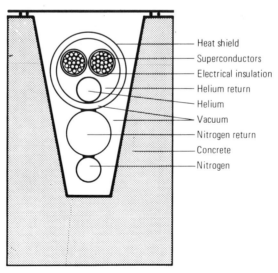

Heat shield
Superconductors
Electrical insulation
Helium return
Helium
Vacuum
Nitrogen return
Concrete
Nitrogen

FIG. 24. Cross Section of Superconducting DC Cable Design Proposed
by R. H. Garwin and J. Matisoo [50].

It can be shown quite generally in theory [51], that in a coaxial pipe conductor arranged as in Fig. 23, the superconductor material can be utilized more economically than in the noncoaxial arrangements, due to the low and evenly distributed magnetic flux density. Minimizing of the capital costs for a dc cable with a coaxial conductor arrangement in relation to the radius of the cable gives optimum values for the magnetic field. These values vary only slightly with the transmission capacity and are about 1. to 2. x 10^5 A/m [40]. In addition, the superconductor is arranged as in the case of the coaxial ac conductor (the inner conductor is backed with the superconductor on the outside and the outer conductor on the inside) so that transient phenomena and in particular the residual ripple of the current do not produce any eddy-current losses in the normal conductor metal. Electromagnetic forces between the coaxial conductors are only present in the case of an asymmetrical arrangement and they attempt to reduce this asymmetry.

The different conductor arrangements discussed have widely differing inductances. The inductance is considerably smaller for the coaxial conductor arrangement than for the two parallel conductors. On the one hand, a high conductor inductance has a further damping effect on the rise of the fault current and causes additional smoothing of the dc current. On the other hand, however, the large amount of magnetic energy stored up during the short-circuit in the dc cable poses certain problems. Without additional measures (such as the dissipation of the energy through shunted resistances), the whole of the stored-up energy would discharge via the fault point and result in considerable damage to the cable.

The conductor systems constructed from coaxial pairs of tubular conductors feature considerable advantages both for ac and dc transmission. Consequently, most of the schemes now being developed are based upon this conductor configuration, particularly for ac cables.

3. Other cable concepts

To obtain higher transmission voltages it is possible, especially with single-conductor dc cables, to arrange the electrical insulation, which is normally at the temperature of the helium, also for operation at nitrogen or room temperature. In one concept shown in Fig. 25, the higher dielectric strength of liquid nitrogen as compared with helium is utilized [52]. Liquid nitrogen is also used for the thermal insulation of the cable. With the other arrangement shown in Fig. 25 the same high dielectric strength as in the case of conventional cables is obtained at room temperature by using paper insulation impregnated with oil [53]. With the first arrangement it is also possible, in the event of a failure of one conductor, to transmit half of the power for a short time through the earthed outer pipes. The ohmic losses occurring in them are to some extent carried away by the liquid nitrogen. The evacuation of the vacuum spaces which are at a high-voltage potential presents however, a

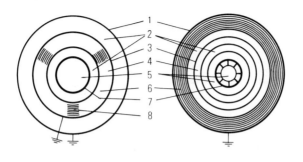

1 Protection pipe   5 Helium
2 Vacuum   6 Electrical insulation
3 Heat shield   7 Superconductor
4 Nitrogen   8 Spacers

FIG. 25. Cross Sections of Superconducting DC Cable Designs [52, 53].

formidable problem in these schemes, unless the vacuum produced before putting the cable into service can be maintained without pumps during operation. This however, requires a high degree of sealing, in particular of the pipes for the cryogenic liquids.

V. CABLE CONDUCTORS

A. Superconductors

1. DC superconductors

Up to the present, dc superconductors have been used primarily for the construction of high field magnets. In connection with the latter, interest has centered on the performance of the superconductor in high magnetic fields at comparatively low current densities. If, however, superconductors are employed in cables for dc transmission, their properties at low flux densities and high current densities are of particular importance.

When selecting the superconductor for a dc cable, two aspects are of importance. It should have high critical current densities at low magnetic

flux densities (0.2 to 0.5 T) so that the superconductor material required and the dimensions of the conductor can be kept low. In addition
a high transition temperature is desirable so that the cable can be
operated at the highest possible temperature where the efficiency of
helium refrigerating plants is better. For this reason high-field superconductors are primarily suitable for use as dc cable superconductors,
namely the alloys NbTi and NbZr and also the compounds Nb_3Sn and
V_3Ga. For economic reasons pure niobium can also be considered [54].
It is less expensive, easier to process and attains the same high critical
current densities at flux densities of 0.2 T as the aforementioned materials. PbBi also has been proposed for use as a superconductor mainly
because of its low price [55]. Attention has even been turned to lead
[56] but, in our opinion, it should be excluded from consideration because of its low critical flux density (0.05 T).

At the present state of the art, Nb_3Sn and NbTi have the best prospects for employment in dc cables. The advantage of the former lies in
particular in its high transition temperature (18.4 K). NbTi on the other
hand is easy to handle, very stable in the form of multi-strand conductors and has been tested and found satisfactory for many applications.

In the case of high-field superconductors the lower critical flux
density $\mu_0 H_{C1}$ is relatively small (0.01 to 0.02 T) so that under the
normal operating conditions of the cable the current flows in the volume
of the superconductor. With high-field superconductors the following
relationship applies approximately concerning the dependence of the
critical current density j_c on the magnetic flux density B [57]:

$$j_c = \frac{\alpha(T)}{B + B_0} \qquad\qquad (1)$$

where $\alpha(T)$ is a proportionality factor dependent upon the temperature
and B_0 is the constant which determines the critical current in the self-
field of the conductor. Fig. 26 shows the curve of the critical current
density as a function of the magnetic flux density B determined by experiment in the case of Nb44 wt % Ti round wires. Here B is an external
field acting on the conductor.

Despite Eq. 1 and a large number of experimental j_c vs. B curves,
it is difficult to give exactly the current densities attainable from superconductors when used in cables. This is due on the one hand to the fact
that practically all known j_c vs B curves have only been measured for
flux densities ≥ 1 T (the range of interest for high-field magnets). On
the other hand, in the low-field range the critical values are greatly
dependent upon the respective conductor configuration owing to the great
influence of the self-field. An additional factor is that at high critical
current densities, the experimentally measured values may also be
dependent upon the degree of stabilization and the cooling conditions. The
maximum obtainable superconductor current densities should therefore

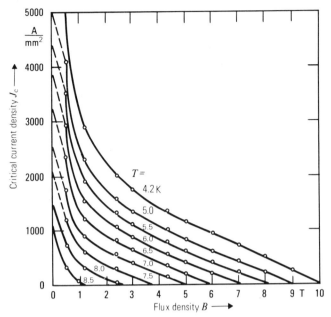

FIG. 26. Field Dependence of Critical Current Density for Nb−44 wt% Ti [58].

be determined experimentally for the respective conductor configuration under the conditions envisaged. If it is assumed that the flux density present at the cable conductor shall not exceed 1 T, current densities of about 5×10^5 A/cm^2 can be expected for NbTi and about 10^6 A/cm^2 for Nb$_3$Sn.

For the design of the cables it is also essential to know the variation of the critical current densities with temperature. Here a linear trend can be assumed in good agreement with experimental findings, (Figs. 27 and 28 [58, 59]), viz:

$$j_c(T) = j_c (4.2 \text{ K}) \frac{1 - T/T_o}{1 - 4.2/T_o} . \tag{2}$$

In the case of small flux densities, T_c can be directly substituted for T_o; with greater flux densities the decrease of T_o with an increasing magnetic field must be taken into account.

Superconductors in dc cables are also exposed to alternating fields. These are caused continuously by a certain residual ripple in the direct current and from time to time by transient phenomena (load changes, switching operations). It is known [60] that in high-field superconductors

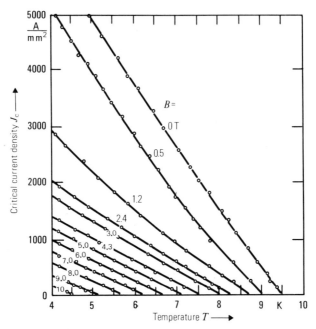

FIG. 27. Temperature Dependence of Critical Current Density Nb−44 wt%
Ti [58].

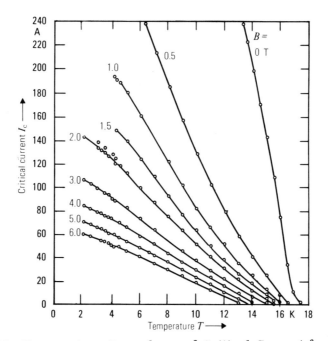

FIG. 28. Temperature Dependence of Critical Current for Nb₃Sn [59].

even with surface fields $> H_{C1}$, screening currents may flow at the surface which correspond to field changes of $\pm \Delta H$ ($\mu_0 \Delta H \approx 0.010$ T). When ripples occur with field changes less than $\pm \Delta H$ the superconductor volume is screened from flux changes and losses become nil [61]. During transient phenomena, with large field changes, appreciable losses occur momentarily which on the average, however, are of no consequence. It must be ensured through suitable stabilization measures that the cable conductor does not permanently become normally conductive because of transient phenomena.

2. AC superconductors

Pure niobium has chiefly been investigated as superconductor material for three-phase cables and more recently also Nb_3Sn. The advantages of niobium are clearly revealed in the low ac losses already obtained so far, the relatively low cost and the simplicity in the manufacture of the conductor. Nb_3Sn on the contrary offers the possibility of higher working temperatures and perhaps a better short-circuit current performance.

The high lower critical flux density $\mu_0 H_{C1}$ of niobium enables the working current to be chosen in such a manner that the self-field of the conductor does not exceed this value. The superconductor is in the Meissner state the current flows only in a thin surface layer and the losses are purely surface losses. If a maximum surface flux density B_{max} of 0.1 T is selected for the conductor, the maximum possible working temperature of niobium is fixed at ≤ 6 K (Fig. 29) due to the temperature restrictions of H_{C1}.

Since H_{C1} is much lower with Nb_3Sn than with niobium, the superconductor at a flux density of 0.1 T is in the mixed state. The current flows in the volume and causes greater volume losses. Higher working temperatures, however, are possible, e.g., of 8 K and more (Fig. 29).

Pb has also been proposed occasionally as ac conductor. It is cheap and easy to process but more susceptible to corrosion than Nb. Moreover the low values for T_c and $\mu_0 H_c$ (7.16 K, 0.05 T) discourage the use of Pb. NbTi as a multistrand conductor must also be ruled out because of the high additional eddy current losses in the matrix material.

At power frequencies the losses both in the Meissner and in the mixed state are of the hysteresis type. An important characteristic of the losses is their linear increase with the frequency. The surface losses are mainly brought about by defects such as cavities, roughness and impurities in the region of the surface [62]. Owing to the field increase at these defects, local flux penetration takes place resulting in losses. With higher fields the magnetic flux then completely penetrates the superconductor volume and the volume losses which then occur are

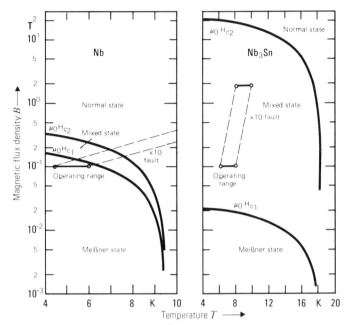

FIG. 29. Temperature Dependence of Critical Fields.

influenced by the following factors:

a) Meissner currents which prevent magnetic flux penetration for external fields having field strengths less than H_{C1};

b) Surface currents which shield the superconductor volume in the case of changes in the externally applied field with a field strength of less than ΔH [60];

c) The critical current density which determines the flux distribution in the superconductor volume [63, 64].

The following components may contribute to the surface currents mentioned under (b):

– The surface barrier [65] due to the interaction of flux lines with the superconductor surface (Bean-Livingston barrier);

– The surface layer which is responsible for superconductivity between the upper critical field strengths H_{C2} and H_{C3} [66] and which also exists for smaller field strengths down to H_{C1}[67];

– Flux pinning in heavily deformed areas near the surface [68].

In conductors fabricated on a large scale, it is to be expected that the effects associated with the surface barrier and the surface layer, as a consequence of the relatively rough surface, become small or completely vanish and that eventually it is only the pinning of the flux lines in the surface which make any contribution to the surface currents [69].

The following assumption was made by Melville [70] concerning the calculation of <u>surface losses</u>: the flux penetration (and consequently the losses) at points of surface roughness take place from zero field strengths onwards, while the volume is shielded by Meissner currents until the external field exceeds the value of H_{C1}. In the area of surface roughness peaks, because of the considerable increase in the field strength, no shielding effect is provided by currents in the superconductor, and the magnetic flux density in this area has a homogeneous distribution. Consequently we get the following expression for the losses per cycle and surface:

$$L = (\mu_0 \, sf^{-1} \, Q \, H_{C1}^{-w} \, H_{C2}^{-\epsilon}) \, H_m^{w+2+\epsilon} \qquad (3)$$

where μ_0 is the vacuum permeability, w a form factor for the surface profile, H_m the amplitude of the surface field strength and s denotes approximately the depth of the surface roughness. Small values for w denote sharp-edged, and large values rounded projections in the surface; e.g., w = 1 for a triangular profile and w = 2 for a parabolic or sinusoidal profile. The constants Q and ϵ depend in a complicated way on the surface profile and on the pinning force in the surface layer. The exponent ϵ assumes values between -0.2 and +1.6 and Q values between 10^{-2} and 3.10^{-1}. The higher values of Q correspond with smaller values of w. The constant f depends upon the form factor w and where there is a symmetrical surface profile it has the value $f = 2^{1-w} \approx w^{-1}$. Surface losses (Eq. 3) are thus largely influenced by the nature of the surface while the pinning forces have a less pronounced effect. They also vary to a greater extent with H_{C1} than with H_{C2}. A decrease in the purity of the material generally gives rise to an increase in H_{C2} and to a decrease in H_{C1} so that this contamination of the surface may result in a considerable increase in the losses.

The temperature dependence of the losses is mainly determined by the temperature dependence of the critical field strengths H_{C1} and H_{C2}. For H_{C1} the following relation is valid:

$$H_{C1}(T) = H_{C1}(0) \left\{ 1 - \left(\frac{T}{T_c}\right)^2 \right\}. \qquad (4)$$

In the case of niobium in particular, measurements on samples of pure material [71] show that the otherwise parabolic temperature relationship of H_{C2} for $T \leq 6.5 \, K$ must be somewhat modified [70] so that

$$H_{C2}(T) \approx H_{C2}(0) \left\{ 1 - \left(\frac{T}{T_c}\right)^2 \right\}^2. \qquad (5)$$

If the expressions for H_{C1} and H_{C2} are inserted in Eq. 3, we see that the surface losses are related to the temperature as

$$L(T) = L(0)H_m^{w+2+\epsilon} \left\{ 1 - \left(\frac{T}{T_c}\right)^2 \right\}^{-w-2\epsilon} \tag{6}$$

where $L(0)H_m^{w+2+\epsilon}$ are the losses for $T = 0 \, K$.

The volume losses have been calculated by Dunn and Hlawiczka [61] based on a modified Bean-London model [63, 64]. By extending the "critical state model", complete exclusion of the field is assumed for superconductor volumes at field strengths $H_m \leq H_{C1} + \Delta H$.

For the particular case where the current density is independent of of the field (j_c = const), and where ΔH is constant, we get for the volume losses per unit of surface

$$P = \frac{2}{3}/\mu_o \frac{\nu}{j_c} (H'_m)^2 (\frac{3}{2} H_{C1} + 3\Delta H + H'_m),$$

and $H'_m = H_m - (\Delta H + H_{C1}); \quad H_{C1} + \Delta H \leq H_m \leq H^*.$ \hfill (7)

Here ν is the frequency and H^* the field strength at which the magnetic flux has just penetrated into the whole superconductor volume. For $H_m \gg H_{C1} + \Delta H$, Eq. 7 becomes converted into the simple Bean-London relationship [63, 64] and results in an increase in losses to the third power of H_m.

Volume losses begin at $H_{C1} + \Delta H$ and rise steeply as the field increases. This starting point of volume losses is shifted towards smaller field strengths, in particular, through the temperature dependence of H_{C1} (Eq. 4) as the temperature increases. Measurements on niobium samples stabilized with normal conductor metal show a parabolic temperature dependence also [72] for the surface current,

$$\Delta H \approx \Delta H(0) \left\{ 1 - \left(\frac{T}{T_c}\right)^2 \right\}, \tag{8}$$

so that with Eqs. 4, 7, and 8 the temperature dependence of the volume losses is established. If in Eq. 7 the field strengths H_{C1} and ΔH are disregarded in relation to H_m, such as is possible in particular for the hard type II superconductors, we obtain for the increase in loss with rising temperature

$$P(T) \approx P_o (1 - T_o/T_c)/(1 - T/T_c), \tag{9}$$

where P_o are the losses at temperature T_o.

3. Experimentally determined ac losses

Based on the thermal losses to be expected, maximum ac losses of $10\,\mu\,W/cm^2$ at 4.2 K are tolerable for niobium. In a large number of measurements the required low values for losses were obtained [73-76, 38]. In our laboratory, extensive measurements were carried out on Nb samples which were specially developed for use in cable conductors [69, 72]. The stabilized conductor samples were prepared by engineering methods based on the latest knowledge concerning surface finishes and to some extent already in larger lengths. Consequently as far as niobium is concerned, this paper will deal exclusively with our own experimental results.

Measurements on commercial Nb_3Sn strips give considerably higher losses [43]. These also confirm measurements on Nb3Sn vapor-deposited strips from our own production. Since it is apparently possible to work at somewhat higher operating temperatures with Nb3Sn, higher losses are permissible within certain limits.

Figure 30 shows the curve determined experimentally at 4.2 K for the 50 Hz losses of two niobium samples and a Nb3Sn sample as a function of the amplitude of the surface flux density $\mu_0 H_m$. For smaller flux densities, i.e., within the range of surface losses, the losses, in the

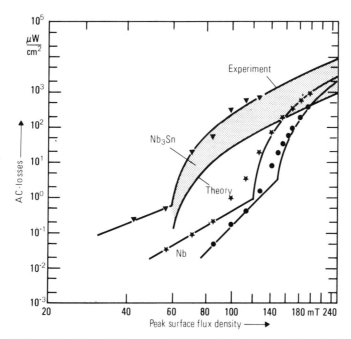

FIG. 30. 50 c/s Losses of Nb and Nb_3Sn at 4.2 K [69].

logarithmically divided diagram, increase linearly with the field with the slope $w+2+\varepsilon \approx w+2$ (Eq. 3). In the present case the values for the slope are 4.4 or 7.5 for niobium and 3 for Nb_3Sn. Generally, values of 3 to 7.5 were determined with niobium for $w+2+\varepsilon$ depending upon the quality of the surface. A corresponding spectrum of values for Nb_3Sn is not yet available.

In the range of volume losses, the measured values rise steeply and are well given by Eq. 7. The extrapolated transition between the two branches of the loss curve lies approximately at the field strength $H_{C1} + \Delta H$. The deviations in the measured points from the straight line begin, however, at a somewhat smaller field strength H_p. That is to say from H_p onwards the magnetic flux penetrates locally into the super-conductor volume. The theoretically predicted tendencies, viz., the increase in the losses for decreasing values of w and H_{C1}, are confirmed in the range of the surface losses of Nb. Adaptation of Eq. 7 to the mea-sured points provides the value for the critical current density j_c. For Nb_3Sn the value thus found lies below the measured short sample value (2.5×10^6 A/cm^2) by a factor of 10. For this reason, the loss curve calculated with the short sample value of the critical current density in accordance with Eq. 7 has also been plotted. It is assumed that the high experimental values were caused by magnetic instabilities (flux jumps) and that these could be made to approach the theoretical curve by stabili-zation with normal metal of good conductivity.

As shown in Fig. 30, losses of $0.2\,\mu$W/cm^2 are to be expected with niobium samples at surface flux-density amplitudes of 100 mT and at 4.2 K. The losses of Nb_3Sn are higher by more than two orders of magnitude. The losses in the niobium samples are not influenced to a great extent when they contain correspondingly prepared welded seams in the longitudinal or cross direction.

An important aspect when fixing the maximum possible working temperature for a three-phase cable is the variation of the superconduc-tor losses with temperature. Fig. 31 shows the relationship for a large scale fabricated aluminum wire with an Nb layer. The field at which the transition takes place from surface losses to volume losses ($\approx H_{C1}$) decreases as the temperature rises according to a parabolic curve. Nevertheless at a working temperature of 6 K and a surface flux density of 0.1 T it is possible to remain in the range of the low surface losses, in agreement with Fig. 29. A prerequisite for this is that the niobium layer is as smooth as possible but heavily worked (H_p then lies closer to $H_{C1} + \Delta H$ and ΔH is 10% of H_{C1}). A quantitative comparison of the losses at 4.2 K and 6 K and at a surface flux density of 0.1 T shows that the losses at the higher temperature are greater by approximately one order of magnitude but still remain just under 10 μW/cm^2. The ex-perimentally determined parameters w and ε lie within the range of theoretical predictions ($w+2+\varepsilon \approx 3.6$; $w \approx 1.6$ and $\varepsilon \approx 0$).

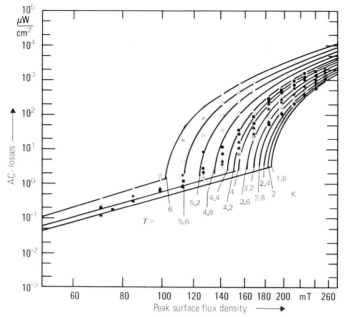

FIG. 31. 50 c/s Losses of Niobium at Various Temperatures [72].

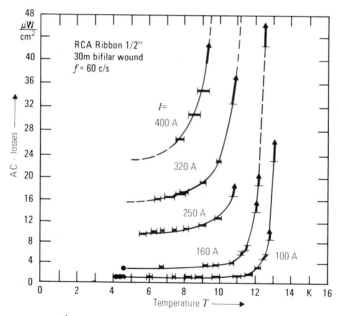

FIG. 32. 60 c/s Losses of Nb_3Sn at Various Temperatures [77].

Results on similarly accurate measurements on Nb_3Sn have not been published so far. The results shown in Fig. 32 obtained on the Nb_3Sn vapor-deposited strip of the RCA must be regarded as provisional ones because they were determined with an improvised apparatus [77]. They however confirm qualitatively the assertion made in Eq. 9 that with Nb_3Sn in the temperature range of 4.2 to 10 K the losses do not vary much with the temperature. Moreover the results appear to support the view that with Nb_3Sn maximum working temperatures of 7 to 8 K are possible.

4. Measuring methods

For measuring the ac losses, use was made both of calorimetric and electrical measuring methods. With the latter, the alternating current is coupled directly to the sample being measured or inductively through a magnetic alternating field. The easiest way is to measure the losses by means of the amount of helium being vaporized. The power dissipation capable of detection in practice is, however, restricted to about 3×10^{-4} W [78]. A considerably greater measuring sensitivity up to 10^{-7} W can be obtained by measuring the increase in temperature of a thermally insulated sample suspended in the calorimeter [79]. Electrical measuring methods have the advantage of shorter measuring times. Here the losses are in principle determined through a current-voltage measurement (the considerably greater inductive component of the voltage signal must be suppressed by a compensation technique [33, 62]) or by direct determination of the effective resistance of the sample being measured by means of a resistance bridge [56].

The losses of the Nb_3Sn strip were measured calorimetrically (Fig. 30) or electrically (Fig. 32). The measured results shown above on niobium samples (Figs. 30 and 31) were determined electrically by a method proposed by Buchhold [62]. This method, further developed in our own laboratory, is described in greater detail below.

The cylindrical samples (tubes or wires) are subjected to an alternating magnetic field parallel to their axes, $H = H_m \sin \omega t$. A pickup coil wound directly on the sample measures the time variation of the magnetic flux penetrating into the surface. The purely inductive component of the voltage signal due to the finite wire thickness of the pickup coil and the depth of penetration of the superconductor is suppressed by a compensation technique. The losses per unit surface are obtained from the mean time value of the Pointing vector as

$$ P = \frac{1}{SN} H_{eff} U_{eff} \cos \varphi, \tag{10} $$

where S is the circumference of the sample, N the number of turns of the pickup coil, H_{eff} the value of the effective field strength, U_{eff} the effective value of the potential of the pickup coil: $Nd\Phi/dt$ and φ the phase angle between H and U.

Figure 33 shows the principles and diagrammatic construction of the measuring arrangement [69]. The field coil is part of a resonance-tuned oscillatory circuit so that the magnetic field is free of harmonics. The tube samples are fitted at the ends with niobium caps. In this way edge effects and premature flux penetrations are prevented. The pick-up and compensation coils are connected together in series. The amplitude of the compensation coil can be adjusted with a potentiometer. Complete balance is obtained with the potential of the small balancing coil which can be altered in magnitude and phase. The lock-in amplifier reference input is a potential proportional to the magnetic field strength. The amplifier output is the value $U_{eff} \cdot \cos\varphi$ and the losses are obtained directly with the factors from Eq. 10. The measuring sensitivity is restricted by the background noise of the input amplifier and is between $10^{-8}\,W/cm^2$ and $10^{-9}\,W/cm^2$ depending upon the circumference of the sample and the number of turns of the pickup coil.

To measure the variation of the niobium losses [72] with temperature, the samples are thermally insulated and suspended in a calorimetric vessel (Fig. 34). The vessel is evacuated and is placed in a helium bath. The test temperature is varied either by heating the sample or by pumping on the outer helium bath and it is measured with two Ge resistance thermometers. These are arranged separately in the field-free space of the sample and in good thermal contact with it. The losses are also determined by the electrical method described above.

5. Conductor losses of ac cables

If it is assumed that the results obtained with laboratory samples of niobium are in general made worse on going over to large production lengths and additional processing methods, mean conductor losses of $\leq 5\,\mu W/cm^2$ may be expected with flux density amplitudes at the surface of the conductor of up to 100 mT (or surface current densities $j_{eff} \approx 560$ A/cm) and working temperatures of 4.4 K at the start of the cable and of 6 K at the end of the cable. Based on these values, the resulting conductor losses are 20 Watts per kilometre of cable and per Gigawatt transmitted at 4 to 6 K, for 120 kV ac cables with pairs of co-axial niobium conductors (see Fig. 39).

If the losses of Nb_3Sn, in accordance with Fig. 30, are taken as $50\,\mu W/cm^2$ at a working temperature of 6 to 8 K and a surface flux density of 100 mT, the electrical losses for a corresponding cable with Nb_3Sn conductors are 200 Watts per kilometre of cable and Gigawatt transmitted.

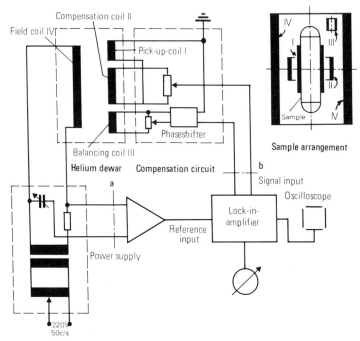

FIG. 33. Sample Arrangement and Block Diagram of the Circuit for Determining the AC Losses [69].

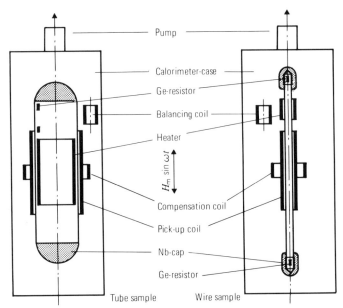

FIG. 34. Sample Arrangement for Determining the Temperature Dependence of the AC Losses [72].

Even taking into consideration a higher efficiency for the refrigerator, with which the total losses (electrical and thermal) of an Nb_3Sn
cable are removed by cooling, the losses of Nb_3Sn are still too high. They
must be reduced by a factor of at least 2 to 3 in order that the power input
at ambient temperature (refrigerator input) is not greater for a cable with
Nb_3Sn conductors than it is with a corresponding niobium cable. The way
to achieve this is a further improvement of j_c, which may possibly be
brought about by other manufacturing methods (vapor deposition in vacuo [80]).

B. Transient Behavior of Cable-Conductors

1. Stabilization of cable-conductors

An important point is the stability of the conductor to transient increases in temperature, which lead to a local loss of superconductivity.
Transient phenomena of this kind may be caused by flux jumps (electrical
instabilities in the superconductor) or through conductor movements
under the influence of electromagnetic or thermal forces (frozen in stresses
through contraction). The aim of stabilization is to prevent the occurrence
or spread of these areas of elevated temperature.

The flux jumps in the superconductor release magnetic energy
which is high in the low field range, and may be suppressed or considerably damped in their effect by the following stabilization methods:
a) Combining the superconductor with a stabilizing material   of a
high specific heat so as to prevent a critical rise in temperature (en-
thalpic stabilization).
b) Combining the superconductor with a stabilizing material of
good electrical and thermal conductivity in order to retard, through
eddy-current damping, the process of energy release and to increase
heat removal (dynamic stabilization).
c) Reducing the dimensions of the superconductor so that the
energy of a flux jump remains sufficiently small (intrinsic stabilization).
For this purpose thin superconductor strands are generally used in a
normal conductor matrix. (The matrix also fulfills a stabilizing function
in accordance with **b.**)

To prevent intolerable heating up of conductors through movements,
rigid positioning is necessary, particularly in the case of flexible types.
Nevertheless motion of the conductor cannot be completely excluded.
This is because the thermal contractions of the components of a flexible
cable conductor vary too much, and in the case of a short-circuit,
particularly with three-phase cables, the magnetic forces are very large.
In order to render them harmless it is consequently advisable to stabilize
the cable conductors fully or cryostatically. (In the case of cryostatic
stabilization the superconductor is combined with sufficient normal metal
of good conductivity so that the stabilized current is able to flow in the

normal metal and the Joule heat generated is transferred to the helium
coolant without any appreciable temperature increase.) This means that
if the superconductor becomes normal as the result of motion of a conductor, it is able in a short time to revert back to the superconducting
condition. Unfortunately, as we shall finally see, it is most unlikely
for economic and engineering reasons that in the case of three-phase
ac conductors complete stabilization can be obtained for fault currents
so that, in general, it can only be applied to working currents and certain
smaller overcurrents. The normal metal, however, protects the superconductor from destruction under faulty conditions. A fully stabilized
conductor has the additional advantage that, despite minor, locally
restricted faults in the superconductor or at the cable (thermal leak), it
can remain completely functional.

Preferred stabilizing materials are high-purity copper and aluminum.
Of these two, aluminum recommends itself, in particular, due to its
easily attainable high resistivity ratios, its low price and its low weight.
To a satisfactory performance of the stabilization, an intimate metallic
bonding between the superconductor and the normal metal is absolutely
necessary. In the case of niobium conductors stabilized with copper or
aluminum and NbTi multistrand conductors with a copper matrix, this is
achieved during the manufacturing process (coextrusion). Fig. 35 shows
three samples of coextruded copper-niobium or aluminum-niobium

FIG. 35. Copper and Aluminum Stabilized Nb Samples. 4 cm φ Copper
        Tubes with 50 μm Thick Nb Layer Outside; 2 mm φ Aluminum
        Wire with 50 μm Thick Nb Layer Outside [69].

conductors. Nb_3Sn which at present is only available in strip form is stabilized by subsequently soldering on copper or aluminum strips. Currently, Nb_3Sn and V_3Ga multistrand conductors with a normal metal matrix are being developed which will very likely be suitable for use in dc cables. The stabilizing metal in the conductors with thin Nb or Nb_3Sn layers (thickness: 10 - 100 μm) serves at the same time as a mechanical support while in the case of multistrand conductors the normal metal matrix is required also for manufacturing reasons. With ac super-conductors the normal metal must be arranged in such a manner in the cable conductor that it is not penetrated by the magnetic field in order to prevent eddy-current losses or the metal layers must be so thin that the losses are negligible.

2. Behavior of cable conductors under fault conditions

In ac transmission systems, fault currents of about 100 kA and clearing times of about four cycles must be expected in the future be-cause of the close interconnection of the systems through transformers. These currents must be carried by the stabilized conductor until dis-connection without the cable being destroyed or its conductivity adversely affected due to electrical overheating of the conductors or because an excessive rise in pressure of the coolant. In addition, the increase in temperature of the conductor or the coolant must be kept as low as poss-ible so that a short while after disconnection the cable can again carry its rated current. Fig. 36 shows the time curve of a 10 times fault current with dc offset for the relatively improbable case of a failure of the first circuit-breaker (four cycles) so that the fault has to be cleared by another backup circuit-breaker only after 0.5 s. While the maximum initial short-circuit currents (t = 0) in 400 kV systems are at present around 50 kA, maximum fault currents of about 100 kA will have to be dealt with in the future.

The fault-current behavior of superconducting ac cables has not yet been finally clarified. This is due, on the one hand, to the fact that the corresponding experiments are complicated and have not yet been carried out and, on the other hand, quantitative theoretical data are not yet available due to lack of knowledge of important parameters (e.g., transient heat transfer figures around the critical point of helium). The following statements concerning the fault-current behavior of three-phase cables must consequently be considered as provisional ones. A number of material techniques (Nb, Nb_3Sn) have been proposed which will serve to provide superconductor ac cables with an acceptable fault-current behavior.

In the case of a niobium cable the surface flux density at the Nb conductor under normal operation is 0.1 T. Under fault conditions, this flux density rises to about 1 T, i.e., the superconductor becomes (as shown in Fig. 29) normally conducting and the whole current must be

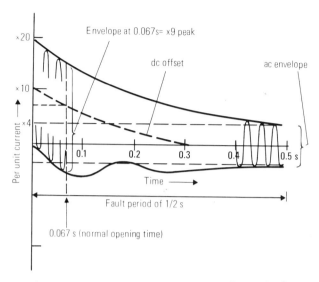

FIG. 36. Fault-Current Waveform [43].

carried by the stabilizing material. The useful current cross-section is restricted by the skin effect and in the case of a round, hollow conductor the resistance per unit length (Ω/m) is given by:

$$R' = \frac{2}{D} \sqrt{10^{-9}\, \rho\, \nu} \qquad\qquad (11)$$

where ρ is the resistivity (Ω cm), ν is the frequency (Hz) and D is the dia-meter (cm). In order to keep the ohmic losses small, a stabilizing ma-terial with the largest possible resistivity ratio must be used. The values of ordinary commercial copper are hardly adequate for this and conse-quently high-purity aluminum which can easily be obtained in large quantities is preferred.

If use is made of aluminum with a resistivity ratio of 2500, a sur-face power density of about $1\,\mathrm{W/cm^2}$ on the cable conductor must be reckoned with in the case of a fault current of 100 kA at the beginning of the fault. As Fig. 37 reveals, this power flow could, in the steady condition, just be removed by circulating supercritical helium without the increase in temperature of the conductor being greater than 1 K. In this case the superconductor would be cryostatically stabilized with respect to the fault current and, after interruption of the current (after four cycles $\simeq$ 80 ms), would be able to carry the full load current again immediately such as in the case with conventional cables.

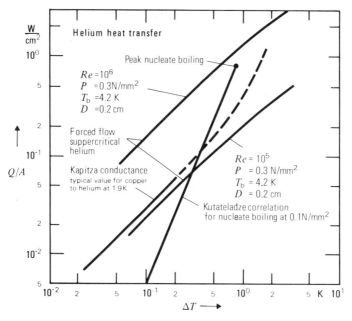

FIG. 37. Comparison of Various Modes of Helium Heat Transfer [118].

The heat-flow values given in Fig. 37 are steady-state values which cannot be applied to the transient events in question. Even with turbulent flow a certain time is required for the heat pulse from the wall of the pipe to become evenly distributed over the helium, the so-called helium mixing time, which with cable pipe arrangements will probably be of the order of magnitude of 10 s. This presents a completely new aspect: the conductor and the adjacent helium layer heat up to about 25 K during the 80 ms which elapse until the interruption of the fault current. Only after a few helium mixing periods does the temperature of the conductor become lower to such an extent that it will revert to the superconducting state. In the equilibrium condition a temperature or pressure rise of, say, 0.3 K and 0.7 bar is obtained at the end of the cable. To enable the original condition to be restored again in a shorter time, the cable must be cooled momentarily with additional cooling capacity. At present it has not been determined to what extent the full load current may again in the usual way be applied to the conductor while the latter is still in the normal conducting condition. If this is not permissible, a longer recovery time must be allowed for the conductor ($\approx$ 2 to 3 helium-mixing time constants). In the unlikely case of the current being interrupted not after 80 ms but only after 0.5 s the maximum conductor temperature does not become very much higher, i.e. about 35 K. The increase in pressure and in particular in temperature at equilibrium then however assumes values which after the full load current has again been restored

results in considerably greater electrical losses in the superconductor.
It was reported by R.W. Meyerhoff at the CEC 1973 in Atlanta, that a
niobium-copper composite tube conductor recovered from a 10 times
fault-current within 0.1 s.

Because a Nb_3Sn conductor has a high upper critical field H_{C2} (see
Fig. 29), it could carry the fault current in the superconducting state.
The only drawback is the increased losses in the superconductor which
result in a power density per unit surface of about $0.05 \, W/cm^2$. Relatively thick layers are however required ($\approx 100 \, \mu m$) which, if they can be
manufactured at all, are very unstable magnetically and mechanically.
It is therefore planned to construct conductors of this kind in the form of
multilayer sandwiches in which Nb_3Sn layers about 10 μm in thickness
alternate with thin Cu or Al strips. Under normal operation the current
is carried only in one Nb_3Sn layer, while under fault conditions it distributes itself over all of the layers of the sandwich. For such conductors it has been stated [43] that even with a 10 times fault current
with dc offset, (see Fig. 36) and a fault clearing time of 0.5 s the rise
in temperature is not more than 2 K, e.g., from 7 to 9 K. In the equilibrium condition the rise in pressure and temperature of the coolant
is, for example, 0.2 bar and 0.15 K. The Nb_3Sn conductor should therefore be able to carry the full load current again immediately after this
unlikely fault condition.

Finally, a composite superconductor stabilized with normal metal
has been proposed by Taylor [81]. On a copper substrate there is a
thicker layer of high-field superconductor such as NbZr or NbTi and
over this a thin Nb layer. Under normal operation the current flows in
the niobium layer and causes slight losses, but under fault conditions
the current transfers into the high-field superconductor layer without its
going normal.

The problem with the two last-named types of conductors (sandwich
and composite conductors) lies in the fact that high-field superconductors,
particularly Nb_3Sn, are prone to become normally conducting when there
are rapid changes in flux density such as occur in the case of a fault
(dB/dt $\approx$ 100 T/s). A further danger of a transition to the normal state
is, as mentioned above, the possibility of conductor movements under
the impact of a fault current. If the possibility that a high-field superconductor goes normal cannot be ruled out, the same situation arises
as in the case of the aluminum stabilized Nb conductor, that is to say
the fault current must be carried by the stabilizing material. Then the
cheaper aluminum stabilized Nb conductor is to be preferred to the other
two types which, and this applies in particular to the Nb_3Sn sandwich
conductor, are far more costly to fabricate and more difficult to handle.

What has just been said has clearly demonstrated that a number of
unsolved problems still exist with regard to the fault-current behavior of
superconducting ac cables. It is therefore absolutely necessary, and this

has already been planned, that the unsolved questions concerning heat transfer be resolved with the aid of simulation tests (pulse heated pipe walls cooled by single phase helium). In addition tests made in the near future will without doubt provide information on whether it is possible to carry the fault current in hard superconductors.

In connection with the problems discussed here, it is of interest to make reference to the general tendency to reduce the fault currents in power systems in the future by current-limiting devices which are arranged at the system infeed points in series with the transformers. Series-resonance circuits which would be able to perform this task are too expensive. In connection with superconducting cables, i.e., with the refrigerating plant which is to be provided anyway, a superconducting current limiter would be of particular interest. Current-limiting installations have already been proposed and it has been shown that they should in principle, be able to limit the fault current to twice the value of the normal current [82-84]. For this application use is made of the transition of a superconductor from the superconducting into the resistive normal conducting state. However, some problems must still be solved before superconducting current-limiting switches of this kind can be realized in practice.

With dc cables the fault current is considerably less of a problem than it is with ac cables. Through the current-limiting effect of the static converter installation, the fault current is limited to about twice the nominal current. There are two possibilities for conductor construction by means of which the fault current can be controlled [40]:
  – The cross-section of the superconductor and of the stabilizing metal is selected to suit the working current. The fault current can then flow in the stabilizing metal until the instant of current interruption whilst heating and an increase in pressure of the helium take place.
  – The cross section of the superconductor is selected to suit the fault current. The cross-section of the normal metal is sufficient for the cryostatic stabilization of the same. If the conductors are not moved, no losses occur in the superconductor during an increase in current apart from short period eddy current losses in the normal metal and hysteresis in the superconductor.
It is clear that the first solution leads to a much smaller expenditure for the superconductor and the normal metal conductor. The potential application is however restricted by the increase in pressure and temperature of the helium when faults occur in the succession.

C.   Conductor Configurations

So far, arguments concerning the configuration of the cable conductors were mostly carried out under the aspect of cooling conditions, their manufacturing costs, cable assembly and installation and compensation for their longitudinal contraction at low temperatures. As already

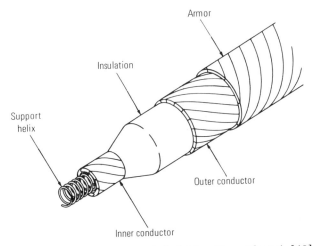

FIG. 38. Flexible Coaxial Cable (One Phase) [43].

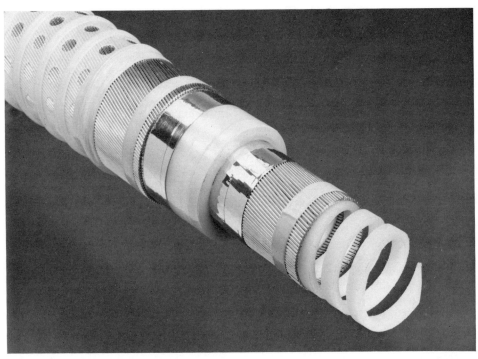

FIG. 39. Flexible Coaxial Cable Conductor (Model Fabricated by Siemens A.G.).

stated in Section IV, a coaxial arrangement of the tubular cable conductors is in many cases preferred for economic, refrigeration engineering and electrical reasons. Both rigid pipes and flexible hollow conductors made of wires, strips and multistrand cables appear suitable for use. Rigid pipes have the advantage, for example in the case of Nb conductors, of considerably lower manufacturing costs and smaller ac losses. On the other hand, additional compensating features (concertina contraction joints) are required to accommodate thermal longitudinal contractions, or expensive measures are required — the use of noncontracting support pipes (e.g. made of Invar), a helical arrangement of the pipes [38] or bending the pipe at discrete places into Ω-bends [85]. The conductor lengths which can be transported are restricted to a maximum of 20 m which leads to a great number of joints with all of their difficulties.

Flexible hollow conductors, as shown diagrammatically in Fig. 38, are certainly somewhat costlier to make and have among other things higher ac losses (deviation from an ideal cylindrical surface in the case of the wire conductor or edge effects on strip conductors backed on one side with a superconductor [38]). However, they have the advantage that they can be manufactured in longer lengths (several hundred metres) and transported on reels, which results in a considerable reduction in the number of joints. Moreover they require no additional compensating components. During thermal contraction compensation for the length of the conductor takes place automatically when the lay angle β of the conductors wound spirally on a core is arranged in such a manner that its contraction coefficient α_L and that of its core α_T are related by

$$\beta = \text{arc sin } (\alpha_L/\alpha_T)^{1/2}. \tag{12}$$

Fig. 39 shows the model of a flexible one-phase conductor constructed of wires. Conductors of this kind, as in the case of conventional water-cooled high-capacity cables, may be drawn into a rigid thermal insulation previously installed in the field or laid with a flexible thermal insulation as a complete single-conductor cable.

Corrugated tubular conductors made of copper or aluminum with Nb or Nb$_3$Sn layers have advantages similar to flexible hollow conductors-manufacture and transport in large lengths, automatic length compensation. When using high-purity aluminum as the stabilizing material, the low mechanical stresses permissible for avoiding drastic increases in resistance, result in comparatively large corrugations [85]. This means higher costs for superconductor material, a greater conductor diameter and consequently greater cable dimensions and larger thermal losses. In addition, the electrical insulation will probably present great difficulties.

VI. ELECTRICAL INSULATION

The dielectric strength of the electrical insulation of a superconducting cable is of crucial importance for its transmitting capacity and thus
for its economy. Moreover it naturally plays an important role in relation to the operational reliability of such cables. For the reasons mentioned above, the current densities with which superconductors, in particular those used for ac may be loaded, cannot be selected as high as
one would desire. If the dimensions of the conductors and thus diameters
of the cables are to remain within certain limits for the power loads to
be transmitted, the transmission voltages must certainly lie within the
range $\geq$ 100 kV. What maximum transmission voltages it is possible to
use cannot yet be stated because comparatively little is known about the
properties of the various electrical insulation systems at low temperatures. In addition, it is not to be expected that, like in conventional
cables, the insulation of dc cables may be stressed considerably higher
(factor 3 to 4) than that of ac cables because thermal effects probably
do not play such an important role with superconductor cables. DC
cables do, however, offer advantages due to the fact that because of the
absence of alternating electrical loads during steady operation a much
wider choice of solid dielectrics is available than with ac cables. The
insulation of dc cables can therefore be chosen entirely under the aspect
of the dielectric strength.

The electrical insulation of a superconducting ac cable must have,
in addition to a high dielectric strength, a low dissipation factor $\tan \delta$
because   the dielectric losses with all of the cable concepts considered
have to be carried away at low temperatures and the helium refrigeration plants used for this have a comparatively low efficiency. For the
dielectric losses in the three phases of an ac system the following is
valid:

$$P_d = \omega\, C\, U^2 \tan \delta \tag{13}$$

where ω is the angular frequency, C is the capacitance of one phase,
U is the transmission voltage (phase-to-phase voltage), and $\tan \delta$ is the
dissipation (or loss) factor.

As Eq. 13 shows, the dielectric losses are proportional to
$(\omega\, C\, U^2)$ the capacitive wattless power of the cable.

Tan δ is generally measured with the Schering bridge as the ratio
of the dielectric losses to the charging capacity. With this instrument
sensitivities up to 10^{-6} have been obtained [86]. For the $\tan \delta$ determination in the temperature range of liquid helium, calorimetric measurement methods may however also be used [87].

Basically the following dielectrics can be considered for use in
superconducting cables:

— Vacuum,

— Helium, liquid, subcooled liquid or supercritical, and

— Wound foil insulation impregnated with helium or foil insulation in vacuum.

A. Vacuum Insulation

As an advantage of electrical insulation with a vacuum, the possibility is recognized of employing it simultaneously as thermal insulation. The vacuum is, in principle, the insulation with the highest dielectric strength. In theory, breakdown in a vacuum gap should only occur at field strengths in the region of 10^7 V/cm when field emission begins at the cathode. In practice, the breakdown occurs already at field strengths which are smaller by two orders of magnitude, that is to say which lie in the range of 100 kV/cm. In addition, under voltage loads a current flows from the cathode to the anode long before reaching the breakdown voltage, the so-called prebreakdown current. This prebreakdown current is due to field emission of electrons from the cathode at points of highest gradient caused by projections on the cathode surface.

The prebreakdown current of a vacuum insulation represents both for ac and dc cables a source of loss which cannot be disregarded. With a transmission voltage of 120 kV and a current density for the prebreakdown current of 10^{-6} A/m^2 losses of 7 μW per cm^2 surface of the phase conductor occur in a three-phase system. They are of about the same magnitude as the hysteresis losses of a niobium conductor system. Care must therefore be taken that the prebreakdown current is kept as low as possible. Improvements in this direction are possible if the electrodes are given a protective dielectric coating. L. Jedynak [88] was able, for example, to reduce the prebreakdown current by two to four orders of magnitude by applying SiO or epoxy coatings. At room temperatures he attained current densities of 3×10^{-8} A/m^2 (field strength ≈ 60 kV/mm; 5 mm gap). Anodic oxidation of the electrodes also results in an improvement [89]. A lowering of the temperature of the electrodes has also a decidedly beneficial effect on the prebreakdown currents as has been shown by Looms et al. [90] on pure niobium electrodes. On a decrease of the temperature from room temperature to 4.2 K the prebreakdown current is reduced to 3×10^{-8} A/m^2 at a field strength of 15 kV/mm (electrode gap 1 mm) Fig. 40. It is therefore to be expected that the prebreakdown currents can be reduced to a tolerable degree by a suitable surface treatment of the electrodes, i.e., conductor surfaces and by the low temperature effect.

The dielectric strength of vacuum gaps is a complex function of many parameters. It depends to a large extent on the surface condition of the electrodes. Coatings of insulating layers generally improve the breakdown strength by up to a factor of 2 [88]. Cooling the electrodes to 4.2 K also results in a similarly great improvement as shown by [90].

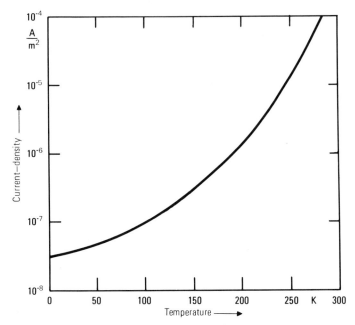

FIG. 40. Variation of Pre-breakdown Current with Temperature at
15 kV/mm [90].

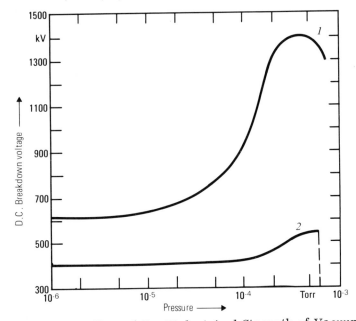

FIG. 41. Pressure Effect of the Dielectrical Strength of Vacuum,
Measured with Steel Electrodes and Gaps of 20 cm (1) and
5 cm (2).

In order to obtain the maximum possible breakdown voltage for a system, conditioning is required in which several breakdowns at a reduced energy are triggered at time intervals through the electrode arrangement. Generally, breakdown voltages of a few to several 100 kV may be expected with a gap width of 1 cm and it has been shown that the breakdown strength increases less than linearly with the gap width. The following empirical relation has been derived for the breakdown voltage [91];

$$V = kd^n \quad \text{where } n = 0.5 \text{ to } 0.7. \tag{14}$$

The breakdown voltage is practically independent of the electrode material and also of the frequency in the frequency range of 0 to 60 Hz that is of interest for power cables so that no essential differences are to be expected in relation to ac and dc cables. Generally speaking, the vacuum should have a pressure of $\leq 10^{-4}$ mbar, but it has been established, in particular through dc voltage measurements [92], that the maximum of the breakdown voltage in vacuum is at pressures of a few 10^{-4} mbar to a few 10^{-3} mbar (Fig. 41). At higher pressures a steep decline in the dielectric strength occurs. This dc voltage-pressure effect was not observed by Graneau [93] in tests at 60 Hz.

In arrangements as they are used for superconducting cables — long, cylindrical (coaxial) electrodes — spacers are placed between the electrodes to support the conductors. It is known [94] that the dielectric strength of vacuum gaps with spacers is primarily determined by the surface dielectric strength of the latter and that it is in general considerably smaller than the dielectric strength of a blank vacuum gap. The flashover strength of spacers in vacuum is in general smaller by the factor of ten than the breakdown strength of the spacer material. In the following the known facts will be presented which lead to the optimization of the flashover strength of spacers in vacuum [94, 95].

The breakdown strength of the spacer material should be as high as possible. Surface coatings with a high surface resistivity generally improve the flashover strength and the service life. Rough surfaces are usually better than smooth ones. The angle which the surface of the spacer forms with the surface of the electrodes is of great importance. Surfaces resting vertically on each other (cylindrical spacers) result in very low flashover values; they can be increased up to a factor of five if angles between 30° and 50° are chosen.

A factor of decisive importance to the dielectric strength is the configuration of the cathode-spacer junction because it is primarily in this that the charge carriers are produced which initiate the flashover process. The insulator-anode junction is of minor importance. With simple bearing contacts between the spacers and the cathode there is the danger of voids at the cathode - insulator - vacuum triple interface which, because of the differing dielectrical constants of the vacuum and the spacer material and the gap shape, result in field increases.

For this reason it is in any case advisable to use for the spacers a
material with a small ε_r. Proposals to sink the spacers into the cathode
so as to reduce the electric field at the junction [94] cannot be realized
in practice with cable arrangements. Solutions, which are successful
with normal vacuum installations, such as bonding spacers to the cathode
by means of thermoplastic or thermosetting dielectrics, will also meet
with difficulties (thermal contraction) particularly in the case of three-
phase cables (alternating polarity). For dc cables, however, a firm
cathode-spacer bonding should be possible. Good experience has also
been made with spacers the surfaces of which were formed in such a way
that they contain barriers for surface discharges and also reduce the field
strength at the cathode-insulator junction. Fig. 42 shows examples of
such spacers [96]. In all other respects a vacuum fitted with spacers
behaves similarly to a blank vacuum. The breakdown voltage can be
increased by as much as 100% through conditioning. For dc voltages
the maximum of the flashover voltage was found at pressures within the
range of a few 10^{-3} mbar. The flashover voltage increases less than
linearly with the spacer length.

On the basis of the results discussed so far, it can be expected
that the breakdown strength of insulating vacuum with correctly selected
spacers and with electrode gaps such as are usual in cables and at 4 K is

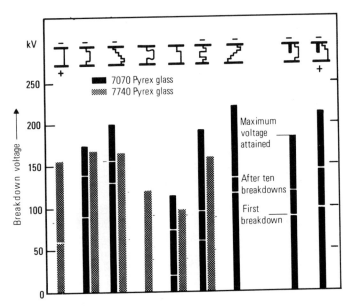

FIG. 42. Effect of Insulator Geometry on Breakdown Voltage [96].

about 200 kV/cm. This assumes that by cooling to a low temperature an improvement by a factor of 2 is obtained, which is probably, but has not been proved. Good spacer materials appear to be epoxy resins, nylon and polyethylene. In this connection consideration should be given to the fact that the conditioning of longer lengths of cable will not be possible because of their large charging capacities. Perhaps a heating process may be used in its place.

In addition to the more general results given above on voltage breakdown in vacuum, results of test carried out specially in rela- tion to vacuum-insulated, refrigerated cables were published and pro- vided very interesting information. Tests by Graneau [93] on a nitrogen- cooled single-phase coaxial arrangement three metres in length indicated that with a correct design of the spacers, namely, the use of embedded metal screens in the spacers (so-called ion screens) the breakdown in vacuum should take place without damage to spacers and to the conductor electrodes. The ion screens in the spacer are to prevent exchange of the charge carriers between the negative and positive electrodes and thus lead to rapid self-extinction of an electric arc. In this way fault currents are prevented from flowing via the insulation which would result in its destruction. Unfortunately these favorable properties of a vacuum dielectric have not been confirmed by other experiments.

Swift [97], on the other hand, found that in electrode arrangements similar to those in cables, electromagnetic forces may act on a struck arc which drive it away from the point of origin and cause it to run along the cable at a great velocity (e.g., 10 km/s). During a power follow- through current, large cable sections may consequently be destroyed. Swift's results also show, however, that under certain circumstances it may be possible to construct the spacers of the vacuum insulation as plasma traps which restrict the electric arc to the cable sections be- tween the spacers.

The above details show that vacuum insulation for superconducting cables will lead to very costly spacer designs. The junction of the cathode-spacer insulator poses a special problem because the generally employed bonding techniques cannot be directly applied to cables be- cause of the differing coefficients of thermal expansion of the spacers and the metal electrodes. This results in difficulties particularly in the case of ac cables with alternating polarity. Another problem lies in maintaining the vacuum over long sections in relatively narrow gaps through the spacers even when extremely good vacuum is not required. The problem is further complicated by the fact that the medium adjoining the insulation vacuum is helium which cannot be frozen. An extremely tight sealing of the tubular conductors is a prerequisite for the satisfac- tory performance of the vacuum insulation. It will be readily understood that vacuum insulation can only be used for rigid pipe conductor cables or corrugated pipe conductor cables.

B. Helium Insulation

An idea which is obvious and also desirable for design reasons is to use the cooling medium helium simultaneously as the electrical insulating medium for superconducting cables. The operating temperatures of superconducting cables lie in the range of 4 to a maximum of 10 K. As an insulating medium for the cable itself only liquid, subcooled or supercritical helium are therefore suitable for use. If use is made of He gas cooled current supply leads at the cable terminations, the dielectric properties of helium gas are however also of interest at temperatures up to 300 K.

Helium gas at room temperature is a very poor insulator, its breakdown strength with large electrode gaps lies at around 0.1 kV/mm, Fig. 43 [98]. At first sight this is somewhat surprising, since helium as an inert gas has a comparatively high ionization energy. It is understandable, however, when it is considered that helium is a monoatomic gas, in which only ionization producing inelastic collisions are possible, in contrast to diatomic gases where inelastic collisions can result in molecular vibrations. Therefore even at relatively low electrical field strengths the electrons can reach such high energies that ionization occurs. As Fig. 43 further reveals, the breakdown strength of helium gas increases considerably as the temperature is reduced. At 4.4 K and 1 bar it is at least 60 times greater than at room temperature [99]. As the reciprocal temperature is proportional to the density of the gas in the first approximation, Fig. 43 confirms that for helium gas Paschen's law applies, which states that the breakdown voltage is proportional to the product of the electrode gap length and the gas density. According to Gerhold, Paschen's law holds good up to gas densities of about 20 kg/m^3 [100].

As can be further seen the breakdown strength in the immediate vicinity of the boiling point increases more than proportionally. In liquid helium at 4.2 K and 1 bar at gaps of 1 mm, breakdown gradients of 20-40 kV/mm can be expected which are better than the values for transformer oil. In the case of cables the breakdown field strengths at greater electrode gaps are of interest and it must be expected that, as in the case of a vacuum as dielectric, the breakdown field strength decreases as the electrode gap is increased. Fig. 44 shows the corresponding curve of the breakdown field strength of liquid helium at 4.2 K and 1 bar for electrode gaps up to 10 mm and an ac voltage of 50 Hz. The decrease in the field gradient from 1 mm to 10 mm is here about 30 % [101]. This decrease in the dielectric strength of liquid helium with increasing gap spacing was also reported by a number of other authors [102, 38]. However, no gap relationship was found in other measurements with smaller electrode gaps up to a maximum of 3 mm and direct current [102, 103]. The measured values in liquid helium are subject to a relatively high degree of scatter which is about 15 to 30 % for a 5-10 mm electrode gap.

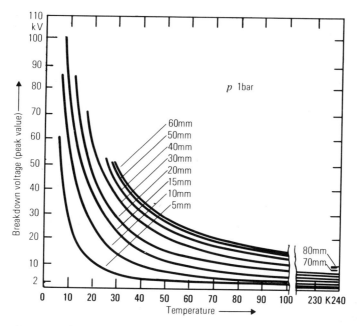

FIG. 43. Breakdown Voltage of Gaseous Helium as a Function of
Temperature and Gap Width [98].

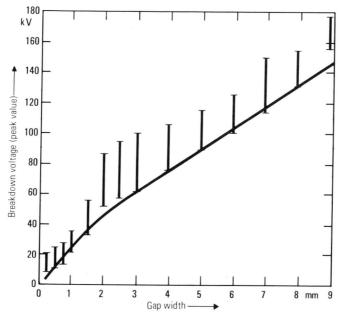

FIG. 44. Dielectric Strength of Liquid Helium at 4.2 K as a Function of
Electrode Distance [101].

Increases in the pressure and temperature have practically no influence on the breakdown values of liquid helium if the liquid is kept in the saturated condition, i.e., in the range of 4.2 K and 1 bar up to the critical point ($\approx$ 5.1 K and 2.2 bar). The scatter in the measured values, however, decreases sharply as the critical point is approached [100]. However, in the case of subcooled liquid helium (i.e., T < 5.1 K and a pressure greater than the respective saturation pressure) the dielectric strength increases with rising pressure without the deviation range of the measurement values altering [100]. At 4.2 K and 4 bar the breakdown strength is, for example, 15 to 20 % higher than with normal liquid helium. The same applies to supercritical helium with which an improvement in the dielectric strength of 65% compared with liquid helium can be obtained by an increase in pressure to 10 bar (T = 5.2 K) as Fig. 45 shows [102]. The range of deviation in the measured values in the region of the supercritical helium is negligibly small. On cooling the helium to the λ point an increase in the breakdown strength of about 50% was established [104].

The results of the investigations reported so far on the dielectric strength of liquid, subcooled and supercritical helium can be explained very well with the aid of a model hypothesis by J. Gerhold [100] concerning electrical breakdown in liquid helium. This hypothesis is based on the fact that with the aid of impurities which are polarized and accelerated in the electrical field, gas bubbles occur in the helium which initiate the disruptive discharge through corona formation. Based on existing knowledge, a zone in the pT-diagram of helium can be given in which reasonable values for the dielectric strength of the insulation of superconducting cables are to be expected. This zone has been drawn in Fig. 46. However, it has been assumed that the temperature of the helium is not higher than 10 K (because of the critical temperatures of the superconductors) and that for mechanical reasons the pressure is not higher than 20 bar. Absolute values cannot be given yet in this zone because it has not yet been clarified whether a "gap" effect occurs with supercritical helium, in which bubble formation cannot occur. It is also not yet clear whether any differences occur in the dielectric strength for ac and dc voltages. While theoretical arguments and certain experimental results [103] point to the fact that there is no difference, other results [38, 104] lead to the conclusion that the dc voltage values are higher than those at 50 Hz. It is safe to assume that in the case of helium it will be possible to obtain dielectric strengths of 10 kV_{eff}/mm at the electrode spacings which will be used for cables (d $\geq$ 10 mm).

The electrode material has little influence on the dielectric strength of helium. Thus the same results were obtained for example with Nb electrodes as with steel electrodes [100]. Impurities on the surface of the electrode, such as thin oxide films, also do not appear to matter to any great extent, whereas mechanical irregularities of the surface, such as scratches, drastically reduce the dielectrical strength. The electrode surfaces must be as smooth as possible. Impurities in the helium as

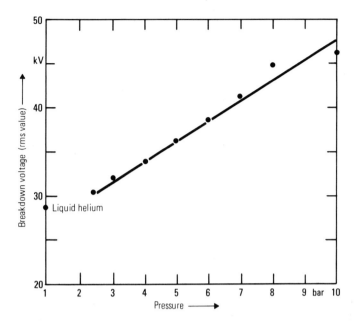

FIG. 45. Dielectric Strength of Supercritical Helium as a Function of the Pressure (Gap Width 1.7 mm) [102].

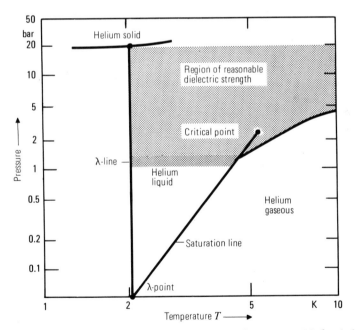

FIG. 46. Temperature and Pressure Region of Proper Dielectric Strength of Helium (For Cable Application) [100].

far as frozen air particles are concerned, have practically no influence
on the dielectric strength whereas slight traces of oil reduce the dielectric strength considerably [100]. The production of additional charge
carriers as a result of increased radiation also reduces the dielectric
strength as was established by experiments carried out with iridium 192
[102].

Measurements of the dielectric constants and of the dielectric loss
angle tan δ were, for example, carried out by Mathes [104]. His results
are shown in Fig. 47 for temperatures ≤ 4.2 K. It is found that ϵ_r lies
at around 1.05 and tan δ at 10^{-6}. Consequently helium, as in the case of
vacuum, has no dielectric losses. The latter occur only in the spacers
but they can be kept small by selecting the right material for the
spacers.

As in the case of a vacuum, great difficulties must, however, be
reckoned with in the region of the spacers as regards the dielectric
strength. Although no measurements in this respect are as yet available,
it is expected that especially the cathode-spacer junction will raise problems. The same applies here as in the case of a vacuum namely, that
owing to the different dielectric constants of helium and the spacers,
gradient increases may occur in the gaps at the junction resulting in a

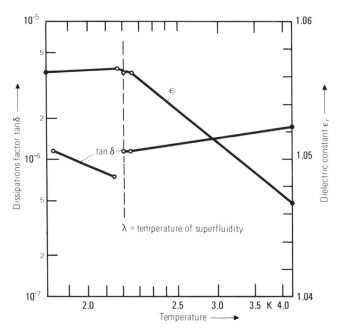

FIG. 47. Dissipation Factor and Dielectric Constant of Liquid Helium
versus Temperature [104].

premature disruptive discharge. To eliminate this danger as far as possible, similarly expensive junction techniques and shaping of the spacers are required as in the case of vacuum. However, as in the case of vacuum insulation little is known as to whether thermal contraction will not present an insoluble problem to the spacer designs.

Helium, like a vacuum, will be mainly used as insulation in rigid pipe conductors. It remains to be considered whether the flowing helium coolant should be used simultaneously as the dielectric or whether separate nonflowing helium should be employed for the latter. For electrical reasons separate nonflowing helium is to be preferred be- cause, on the one hand, the danger of impurities is removed, and on the other hand, at least with helium gas at room temperature, it has been established that flowing gas has a smaller dielectric strength than non- flowing gas [105], not to mention the flow effects at the spacers. Main- taining the condition of separate insulating helium results, however, in a more expensive arrangement.

Regarded in its entirety, only subdued optimism exists with regard to the use of vacuum and helium as dielectric for rigid pipe conductor systems. The problems associated with the spacers appear to outweigh the initially mentioned advantages of a common vacuum for electrical and thermal insulation or of the use of helium for both cooling and in- sulation.

C. Wound Foil Insulation in Helium or in Vacuum

Because of the difficulties to be expected with vacuum or helium insulation, attention was turned at an early stage to wound foil insulation made of paper or synthetic material. The materials selected were those which are also used in other fields of electrical engineering. The pos- sibility is provided of impregnating these insulation systems with liquid or supercritical helium or of using them in a vacuum environment. Foil insulation is not intended for use with rigid coaxial pipe arrangements, they are particularly suitable with flexible cable systems.

In addition to the dielectric strength, the dissipation factor of the insulating tape used is of particular interest in ac cable insulation sys- tems, because the losses which occur must be dissipated at a low tem- perature, i.e., at a poor efficiency. The dielectric constant is also of great interest because, apart from the losses, it determines such im- portant cable data as the surge impedance and the capacitive wattless power. Fig. 48 shows the temperature curve of the dissipation factor and the dielectric constant of nonpolar plastic materials − (polyethylene (PE), polypropylene (propylex) and Teflon PTFE [86, 106]. Nonpolar plastics are characterized by the fact that they have no permanent dipole moments. Fig. 49 gives the corresponding graph for slightly polar plastics and Figs. 50 and 51 for polar plastics and dry or wet normal

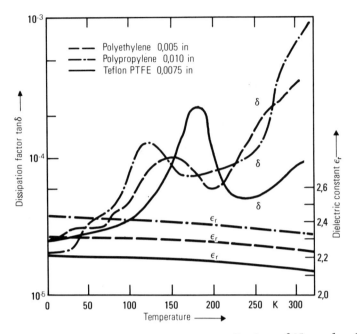

FIG. 48. Dielectric Constant and Dissipation Factor of Nonpolar Materials versus Temperature at 75 c/s [86].

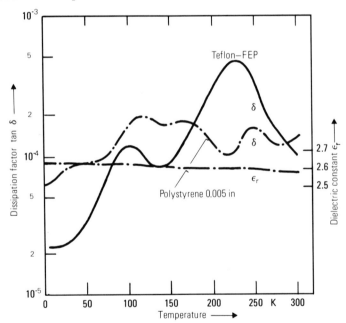

FIG. 49. Dielectric Constant and Dissipation Factor of Slightly Polar Materials versus Temperature at 75 c/s [86].

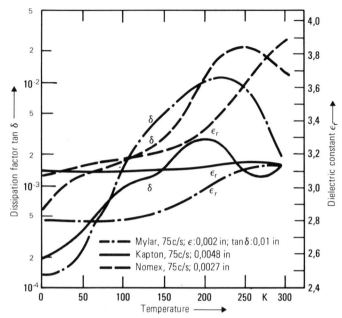

FIG. 50. Dielectric Constant and Dissipation Factor of Polar Materials versus Temperature [86].

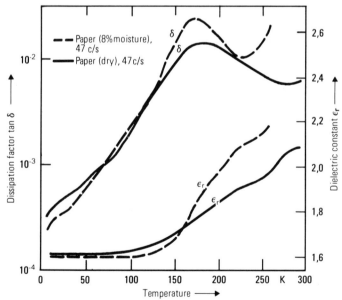

FIG. 51. Dielectric Constant and Dissipation Factor of Polar Materials versus Temperature [106].

insulating paper. The values were determined by electrical bridge measuring methods on thin foils.

Figure 48 indicates that the loss angle of nonpolar plastics decreases by about one to two orders of magnitude between room temperature and 10 K and below.

The values determined at 4.2 K for power frequencies lie in the
region of 10^{-5}. Values have been given by P.S. Vincett which lie even
considerably below 10^{-5}, namely at 10^{-6} [87]. The loss maxima occurring at intermediate temperatures are to be attributed to resonance phenomena; they are not of interest in connection with superconducting cables.
The dielectric constants remain practically constant in the whole temperature range. As Figs. 49 and 51 show the loss angles with slightly
polar and polar plastics and also with insulating paper also decreases considerably at low temperatures. For the former, values of almost 10^{-5} have
been obtained and around 10^{-4} for the latter. While the dielectric constants
for the slightly polar plastics also remain constant, those of the polar
plastics, except those of Kapton, decrease by a maximum of 20 % as the
temperature is reduced. The decrease with dry and moist paper is
about 30 %.

Figures 48 to 51 show that for ac cables only nonpolar, and to a
restricted extent slightly polar plastics are suitable for use if the dielectric losses are to remain tolerable compared with the other losses
(thermal losses and conductor losses). Consequently, the choice of
insulating materials is considerably narrowed down and cannot be based
indiscriminately on the material with the highest dielectric strength.
In contrast to this (without taking into consideration mechanical aspects),
a choice can be made in the case of dc cables based on the maximum dielectric strength.

Figure 52 shows comparative measuring results of the 50 Hz dielectric strength of 0.1 mm thick plastic foils in liquid helium or in a
vacuum. This shows that the dielectric strength of foils in vacuum is
considerably greater than in He. This fact is explained by the greater
dielectric strength which a vacuum has compared with helium. Partial
discharges in the helium reduce the dielectric strength of the foils.

For the engineering design of cables, measurements on individual
foils are not so much of interest as they reflect only relationships and
tendencies, but measurements on wound multilayer foil sections from which
extrapolation to the actual cable can more safely be made. Fig. 53
shows the respective preliminary results of the breakdown field strengths
of wound multilayer insulations made of paper or plastic foils placed in
helium of different states or in vacuum [107, 38, 108]. To begin with, as
in the case of the pure impregnating materials helium and vacuum, a
marked decrease in dielectric strength can be observed for an increase
in thickness of the wound foil sections. Moreover, the dc voltage values

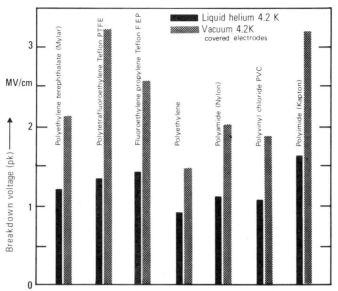

FIG. 52. Comparison of 50-Hz Dielectric Strength of Film Polymers as a Function of the Immersion Medium (Film Thickness 100 μm) [J. C. Bobo].

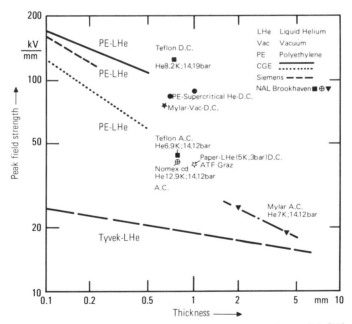

FIG. 53. Compilation of Breakdown Measurements on Multilayered Plastics in Helium or Vacuum [107, 38, 108].

are considerably higher than the ac voltage values. It is also confirmed that the foil sections in vacuum as a surrounding medium have a higher dielectric strength than in helium. Wound foil sections in vacuum presuppose the use of helium-tight pipe conductors (rigid or corrugated). They are used predominantly in French cable designs [38] which also assume that the vacuum must have only values between 10^{-3} and 10^{-4} mbar.

As can be clearly seen in the case of polyethylene and Tyvek, the dielectric strength values of compact foils are higher than for paper-like foils of the same material. Measurements, which are not shown in Fig. 53, reveal that the dielectric strength of paper-like plastic foils of this kind can be improved by calendering. In a similar way the dielectric strength values increase with helium as the surrounding medium as the pressure increases and the temperature falls similarly to the behavior of pure helium. The as yet very incomplete measuring results indicate that with the materials investigated so far, and with the insulation thicknesses usual for cables, effective breakdown field strengths of about 10 kV/mm for ac voltage and of about 20 kV/mm for dc voltage may be expected. Consequently, transmission voltages in the 100-200 kV level are quite possible with ac cables and in the 200-400 kV level with dc cables.

In the ac voltage measurements made so far on wound foil sections, intensive corona phenomena (= partial discharges) could be observed in the foil sections. Depending upon the type of foil and with He as the impregnating agent, these begin partly at voltages which are only a relatively small fraction of the breakdown voltage. These partial discharges occur in the triangular or rectangular gaps in the wound foil section which are formed at the overlap or jointing positions of the foils and which are helium or vacuum filled. Because of the smaller dielectric constant of helium and vacuum as compared with the foil material and the respective geometrical form of the gaps, a considerably higher electric field strength occurs in these (e.g., four times greater) than in the rest of the wound foil unit. Particularly, if helium is used as the filler material, predischarges occur in the gaps because of its smaller dielectric strength even at moderate voltage gradients. Since these predischarges result in losses and to the gradual destruction of the foil materials in the vicinity of the gaps and thus to a reduction in the dielectric strength of such foil sections, they must be kept low or be eliminated as far as possible. Further intensive investigations are necessary in this respect.

In addition to the dielectric properties, mechanical properties in particular also play a decisive role in wound foil insulation. In order to assure the required reversible flexibility with flexible cables, the surfaces of the foils must exhibit good sliding properties. The heavy thermal contraction ($\approx 1\%$) and embrittlement of plastic foils at low temperatures present a special problem. Experience shows that with an incorrect choice of foil material the wound foil insulation can be mechanically destroyed when it is cooled to 80 K or even merely electrically loaded at 4 K. In general, paper-like foils exhibit better mechanical properties but smaller dielectric strengths so that the choice of the correct foil insulation will represent a compromise between me-chanical and electrical properties.

D. General Aspects of Electrical Insulation and Test Methods

The results of measurements of the dielectric strength of the vacuum, helium and wound foil insulation were obtained on so-called laboratory samples. It is known from conventional cables that the di-electric strength however is subject to a geometrical and size effect whereby the shape of the electrodes and the size of the electrode surface exerts a decisive influence on the dielectric strength. Consequently breakdown measurements should as far as possible be carried out with the actual shapes of the electrodes. Since the dielectric strength in general decreases with an increase in electrode surface, test pieces as long as possible should be measured. Finally, because of corona phenomena long-term effects must also be taken into consideration.

These general aspects find their expression in corresponding test instructions for conventional cables. If, for example, VDE Specifica-tions relating in the GFR to conventional three-phase cables with voltages up to 275 kV are applied to superconducting cables, the following test procedures should be used or the following voltages be applied:

a) Short lengths of cable ($l \leq 16$ m); for 24 hours 2.5 times the line to neutral voltage U_0, i.e., for a cable with a line-to-line voltage of 120 kV, the test voltage is 175 kV. The cable fittings, joints and cable terminations must also be subjected to this high-voltage test.

b) Longer lengths of cable ($l \geq 16$ m): Twice the line-to-neutral U_0 for 15 minutes.

c) With a complete cable: The line-to-line voltage for 15 minutes.

AC cables are frequently connected to overhead lines from which lightning overvoltages travel into the cable and may damage the insula-tion. This stressing must be simulated by an impulse test by which a voltage is produced which rises in 1-$2\,\mu$s to the peak value (front time τ_s) and which reaches half of the maximum value after 40-$60\,\mu$s (tail to half value time τ_r). The peak value is about five times the value of the line-to-line voltage.

At the present time a number of experiments are in progress in different laboratories in which the high-voltage strength of larger test lengths (20 to 30 m) of superconducting cables and fittings is being tested. The tests are both ac and dc voltage tests. In this connection it should however be mentioned that at present test regulations for dc cables similar to those for ac cables do not yet exist.

In summing up it can be said that, particularly with regard to the preferred flexible coaxial types of conductor, a wound insulation made of paper or plastic foil is to be preferred. To enable final judgement to be made concerning the performance of insulations of this kind, the required measurements must be extended to include larger cable lengths. The results of tests so far suggest that with the insulation thicknesses realizable mechanically, transmission voltages of up to about 200 kV should be possible for ac and of about 400 kV in the case of dc. An advance into still higher voltage levels would call for the use of new types of material.

VII. THERMAL INSULATION

A. Heat Sources

While in the case of conventional cables, the problem arises of dissipating the heat produced in the cable to the surrounding ground as efficiently as possible, the contrary is the case with the superconducting cable. Here measures must be taken to ensure that the minimum amount of heat penetrates from the outer (steel) pipe which is at ambient temperature to the helium-cooled conductor system because helium has a very small cooling capacity and the efficiency of helium refrigerating plants is very low. The heat is conveyed through conduction (convective gas currents, the solid-state heat conduction of the spacers) and through radiation. The thermal insulation of a superconductor cable extends from the outer pipe, which is at ambient temperature, to the helium tube which contains the superconductors. Instead of a single helium tube, several such tubes may be used.

B. Methods for Reducing the Heat Inflow

1. Vacuum

Gas convection is prevented by evacuating the space between the outer pipe and the helium tube to at least 10^{-4} mbar. Evacuation of cables can be achieved in various ways. In every case each method should meet with the expressed wishes of the user as to the shortest possible pumping times at the beginning of operation or during repair work. On the other hand, the cost of the vacuum plant from the point of view of its installation and maintenance must remain within economic limits. The procedures for producing and maintaining the required vacuum also depend upon the specific construction of the thermal insulation. As will be shown further on, the latter may consist of one vacuum chamber only or of several vacuum chambers separated from each other. Moreover, operation is possible with or without cooled shields at intermediate temperatures.

In the case of a single vacuum space, a very simple method of evacuation would appear at first sight to suggest itself, that is to say the freezing out of the air filling during the cooling of the inner parts of the cables to the temperature of liquid helium with the aid of the cryo-sorption effect of the cold surfaces. This method, however, entails the following difficulties:

— Too great an inflow of heat in the first part of the cooling phase,

— The residual gas pressure of the unfrozen helium is too high,

— The thickness of the air layer adsorbed on the helium tube is so great (~ 200 μm), that the reflectance of the surface of the tube becomes too small for the incident heat radiation.

Two basic methods of evacuation are possible for long lengths of cable:

a) The generation of the required vacuum of about 10^{-5} m bar in the warm condition of the cable with the aid of high-vacuum pumps which must be placed at relatively small distances from each other to secure tolerable pumping times; maintenance of the vacuum is then obtained also with these pumps despite small leakages of helium.

b) With the aid of vacuum backing pumps an initial vacuum of about 10^{-2} mbar is created. At this pressure the cooling of the cable is commenced and the required end-vacuum is obtained through the cryo-sorption effect of the cold surface of the helium tube. The inflow of heat from the outer pipe during the initial cooling process is tolerable at pressures $\lesssim 10^{-2}$ mbar. In this case far greater pump spacings are possible than in case a). Maintenance of the vacuum against leakage of helium is obtained with the aid of high-vacuum pumps, which must have only a relatively small suction capacity. Their spacing corresponds to that of the backing pumps. An essential condition for this mode of opera-tion, however, is that at ambient temperature, the leakage of the helium tube (or tubes) over its entire length should be just above the detection limit of helium leakage test instruments. Inward diffusion of hydrogen through the outer steel pipe is prevented by a suitable protective coating.

In addition to the leakproof properties of the helium cooling system, attention must be given in the case of a) and b) to the leak-free properties of the outer pipe and any pipes conveying nitrogen. Although the gases which penetrate as a consequence of leaks of this kind are frozen through the 4.2 K cold surface, the adsorbed layers must be prevented from be-coming too thick, and hence the reflectance of the helium pipe surface from decreasing to an unacceptable extent.

When the thermal insulation has two separate vacuum chambers separated by a vacuum-tight nitrogen-cooled shield, an initial vacuum of at least 10^{-3} mbar must be created in the outer vacuum space (outer pipe - N_2 shield) before cooling when employing method b). The reason for this is the smaller cryo-pumping effect of the N_2 shield. Moreover in this case the two vacuum chambers must be pumped with different pumps, which may also be differently spaced, both for method a) and for method b).

After these qualitative data concerning vacuum formation a semi-quantitative review will be given of the possible pumping distances and the pumping times to be expected. The pumping times may be calculated approximately by the following equation:

$$\tau = \frac{V}{S_{eff}} \ln \frac{p_o^- \left(p_E + \frac{W_{ein}}{S_{eff}} \right)}{p_{max}^- \left(p_E + \frac{W_{ein}}{S_{eff}} \right)} , \qquad (15)$$

where

V	is the volume of the cable to be evacuated in m^3,
p_o	is the initial pressure at t = 0 in N/m^2,
p_{max}	is the required maximum end pressure of the cable installation (at that point which is farthest from the adjacent pumps),
p_E	is the minimum pressure of the vacuum pumps employed,
W_{ein}	is the rate of the gas inflow, e.g. in N.m/s, and
S_{eff}	is the effective suction capacity of the arrangement (pump and cable, in m^3/s.

For S_{eff} the following applies depending upon the arrangement of the pumps:

$$\frac{1}{S_{eff}} = \frac{1}{S_p} + \frac{1}{n.L_R} \qquad \text{where n = 4 or 8,} \qquad (16)$$

where

S_p	is the suction capacity of the pump, and
L_R	is flow coefficient of the section between two pumps in m^3/s.

For which arrangement of the pumps n = 4 or n = 8 should be applied is shown in Fig. 54. The flow coefficient of air in m^3/s at room temperature is:

$$L_R = 12.1 \times 10^{-5} \frac{d^3}{\ell} \times \alpha . \qquad (17)$$

In this equation d, the real or equivalent diameter of the installation to be pumped, must be expressed in cm and the distance ℓ between two adjacent pumps in metres. α is a correction factor which takes into account the influence of the mean free path of the gas particles. Its dependence upon pressure will be seen from Fig. 55 [109].

The gas inflow W_{ein} may consist of the following components: air flowing in via leaks in the outer pipe; hydrogen which diffuses in through

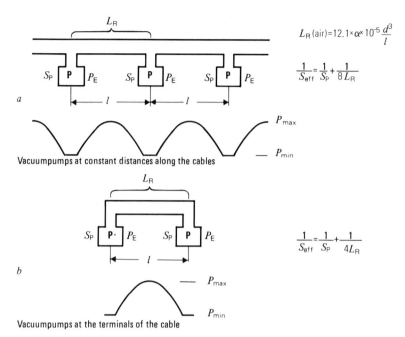

FIG. 54. Evacuation of Superconducting Cables.

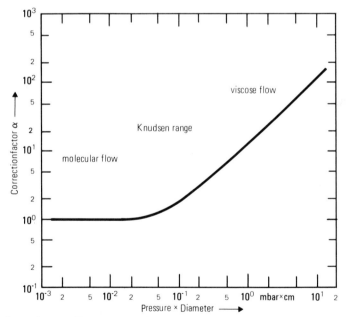

FIG. 55. Correction Factor α for the Calculation of the Conductance of Cylindrical Ducts [109].

the outer pipe; water vapor adsorbed from the walls of the pipe; gas evolved from the thermal insulating material and from the suspension and support materials made of plastic; gas evolved from lubricants and greases of bushings. The inflow of gas is difficult to account for mathe-matically because it is not constant as to location and time.

For the thermal insulation of a 2.5 GW ac cable having an external diameter of 0.5 m and only one vacuum space, the pumping times in ac-cordance with Eq. 15 were calculated approximately for both of the evacuation methods in accordance with a) and b). In the case of method b) (initial vacuum + cryovacuum) it was assumed that vacuum pumps are installed only in the terminal stations of a 10 km long cable section, i.e., the arrangement in accordance with Fig. 54b applies. When there is a maximum pressure p_{max} in the centre of the cable of $2.5. 10^{-2}$ mbar (the minimum pressure p_{min} at the ends of the cable then being about 2.10^{-3} mbar), the pumping time is about 3 days when a suction power S_p of 1000 m^3/h and an end pressure p_E of 1.3×10^{-3} mbar was adopted for both pumps. The additional time required to achieve the end vacuum through cryo-pumps is identical with the cooling time of the cable (see Section VIII). The heat inflow through residual gas convection at the beginning of the cooling process is tolerable and is only about 1 W/m^2 on reaching the boiling temperature of liquid N_2 (80 K).

If the same cable in the warm condition with pumps spaced 1 km apart is pumped to an end vacuum of 10^{-5} mbar (method a), Fig. 54a) about three days are again required with the initial vacuum of about 10^{-3} mbar being reached after several hours. As pumping units, high vacuum pump stations consisting of backing pump and oil diffusion pump (S_p = 1000 m^3/h, p_E = $6.5. 10^{-7}$ mbar) were selected. In both cases the time dependence for the inflow W_{ein} was taken similar to that confirmed experimentally.

The above examples indicate that with evacuation method b) it is completely possible to locate the vacuum pumps at intervals of 10 km and still to obtain satisfactory evacuation times. Attention must then be given of course to the maximum possible vacuum tightness of the cable system. It will, however, be better to install the vacuum pumps at smaller intervals, as for example 1 km, along the cable route and to adopt either evacuation method a) or b). This then provides greater reliability against the occurrence of small leaks.

Suggestions have also been made of locating the vacuum units at the position of the joints which have a maximum spacing of 500 m [110].

A further interesting process for shortening the evacuation times has been proposed by Klaudy [56] and others. In this process the spaces to be evacuated are first scavenged with a scavenging gas which has a solidi-fying temperature which is higher than the boiling temperature of liquid N_2, as for example CO_2. When the air has been completely displaced, the pressure of the scavenging gas is reduced to a few mbar and then the cooling of the cable is started preferably initially with liquid N_2 and then continued with liquid helium. The scavenging gas then freezes and the required high vacuum is obtained.

2. Cooled intermediate shields – superinsulation

A well known method of effecting a considerable reduction in the heat radiated and conducted through supports or spacers to the tube carrying the helium is the use of actively cooled intermediate shields at higher temperature levels at which the heat can be led away with a better efficiency. Many cable designs therefore incorporate an intermediate shield cooled with nitrogen (80 K) between the outer and the helium pipe. If the operation of the superconducting cable is restricted to tempera - tures $\le 6\,K$, as in the case of niobium, it might be advisable to intro- duce a further shield between the N_2-cooled shield and the helium pipe containing the conductor, which is cooled to a temperature of 6 to 8 K by the helium return flow (see also Section VIII).

Various proposals have been made for the design of the N_2 inter- mediate shields. Annular channels completely filled with liquid nitro- gen, which are formed from two coaxial metal pipes; these divide the space between the outer pipe and the helium pipe into two separate vacu- um chambers which could be a disadvantage for the evacuation of the cable (see above). In this respect the use of single pipes, which are cooled by smaller cooling pipes soldered or welded on to the former, is an advantage. These arrangements can be provided with openings which connect the spaces between the outer pipe and the N_2 shield on the one hand and between the N_2 shield and the helium pipe on the other hand without causing any radiation short circuit. To reduce radiation losses it is recommended that highly reflecting materials be used for the inter- mediate shield such as Al or Cu or when using other materials such as stainless steel or Invar to provide these with highly reflecting surface coatings.

The energy radiated from the outer pipe of the cable per unit of surface is given by:

$$Q = \sigma \epsilon T^4 \tag{18}$$

where σ is the Stefan-Boltzmann constant ($5.77 \times 10^{-8}\ Wm^{-2}K^{-4}$), ϵ is the emissivity of the pipe and T its temperature. If we take for the outer pipe made of normal steel with a correspondingly corroded sur- face an emissivity of 0.3 we get for its power radiated at 300 K a value of 140 W/m^2. The maximum of the emitted radiation occurs at a wave length of 9.6 μm. In order to reduce the irradiation on the N_2-cooled shield, the space between it and the outer pipe is filled with a large number of highly reflecting layers – so called superinsulation. Its radiation losses are then represented by

$$\dot{Q} = \frac{\sigma}{\left(\dfrac{1}{A_1} + \dfrac{1}{A_2} - 1\right)} \cdot F_2 \; \frac{T_1^4 - T_2^4}{N+1} \tag{19}$$

where F_2 is the surface of the N_2 shield and T_2 its temperature, T_1 is the temperature of the outer pipe, N is the number of reflecting layers, and A_1 and A_2 are the corresponding absorptivities.

This superinsulation may consist of a multilayer arrangement of Mylar foils with a vapor deposite of aluminum on one or both sides, or aluminum foils which are separated from one another by intermediate layers made of glass-fibre cloth, nylon netting etc. Table 2 gives the thermal conductivities of a number of multilayer insulations of this kind with 300 K and 20 K as the surface temperatures [111]. The different kinds of superinsulation only become fully effective when the residual gas pressure of the space in which they are located is between 10^{-4} and 10^{-5} mbar. Fig. 56 shows the relationship between the thermal conductivity of various superinsulations and the residual gas pressure [112].

When selecting the superinsulation for long cable sections, not only does the thermal conductivity of the superinsulation in question play an important part, but also its suitability for processing and its weight, which make themselves felt in the manufacturing costs. From the point of view of workability and density, superinsulation made of Mylar foils with Al vapor coating on one side is the most favorable material. Consequently, although it has not the best thermal conductivity, it is frequently used for cable insulation. Aluminum foils with glass paper layers are very difficult to use because of the liability of the glass paper to tear.

It is preferable not to fill the space between the helium tube and the N_2-cooled shield with superinsulation since radiation from the latter on the helium tube is small owing to the T^4-radiation law. Consequently the full annular gap is available for evacuating the cable.

Suggestions have also been made that the thermal insulation be made without actively cooled intermediate shields and that thicker layers of superinsulation, i.e., a greater number of N of reflecting layers be used for this. This solution leads automatically to a greater radial extension of the thermal insulation and to a rise in radiation losses and conductivity losses through the spacers. Losses must then be anticipated which are about 5 times greater than in the case of insulations with N_2 cooled intermediate shields. It is also questionable whether the certainly much simpler cable design and the absence of a cooling cycle compensate for the considerably greater losses and the larger cable dimensions. In the comparison, the reliability in operation of the two

TABLE 2. Thermal Conductivity of Multilayer Insulations*

Material	Thickness μm	Density g/cm³	Thermal Conductivity, 7 N/m² compression W/cm K x10⁻⁷	Thermal Conductivity, 0,1MN/m² compression W/cmK x10⁻⁷
1145-H19 Aluminium & Nylon Mesh	50.8 177.8	0.24 0.048	0.62 0.84	230 260
Double-Coated, Aluminized Polyester Film & Nylon Mesch	6.3 6.3 177.8	0.048 0.048	0.84 0.84	260 260
Soft Aluminium & Fiberglass Cloth (3 layers)	127.0 25.4	0.16	1.2	476
Soft Aluminium & Fiberglas Mat	12.7 355.6	0.08	1.4	63
1145-H19 Aluminium & 1/8 x 1/8 Vinyl-Coated Fiberglas Netting	50.8 508.0	0.256	2.0	–
Soft Aluminium & Fiberglass Mat	25.4 76.2	0.048	2.0	110
Crinkled, One-Side Aluminized, Polyester Film	6.4	0.0224	2.7	591

* Values are for boundary temperatures of 300K-20K

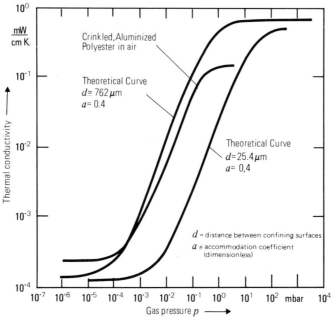

FIG. 56. Effect of Gas Pressure on Thermal Conductivity of Multilayer Superinsulation (T_{cold} = 80 K and 20 K) [112].

variants must also be included as a further parameter and here sufficient experience will only become available after experiments with test lengths of cable have been made. In any case the solution becomes more promising when the possible working temperatures of the cable conductors are increased, e.g., with Nb_3Sn conductors.

In addition to the techniques described above there are also other possible solutions. Thus it has been proposed for example [38] that the space between the outer pipe and the N_2-cooled intermediate shield be filled with an evacuated powder insulation so that supports or spacers between the two pipes can be dispensed with. The thermal conductivity of fillings of this kind in the temperature range in question (300 to 70 K) is greater by a factor of at least 5 than is the case with the above mentioned multilayer superinsulations, which result in losses at the N_2 shield which are higher by a factor of 3 to 4.

3. Pipe suspensions and spacers

The pipe carrying the helium and the actively cooled intermediate shield must be suspended in the outer pipe or be supported against the latter. In order not to introduce too much heat through the suspensions or spacers, materials must be chosen for them which have a low integral heat conductivity at the temperature range in question and high mechanical strength. Table 3 gives the integral heat conductivities and mechanical strengths of a number of materials. In this connection, the quotient $\sigma_{0.2}/\int \lambda dT$ is of interest.

For suspensions, ropes made of fine wires, e.g., stainless steel are suitable. Synthetic ropes, e.g., of nylon have even better values for $\sigma_{0.2}/\int \lambda dT$. They have however the disadvantage of a greater thermal contraction and among other things of evolving gas and becoming embrittled in the high vacuum. Fibre glass reinforced epoxy resins which, in addition to a high $\sigma_{0.2}/\int \lambda dT$, exhibit a relatively low gas discharge rate and great stability in a vacuum, are particularly suitable as spacers.

From the suggestions made so far, suspensions are clearly in the majority. With them the forces occurring during the cooling or warming up processes can be controlled well. In the case of spacers, buckling forces and axial displacements must be taken into account. As already mentioned above, in the case of powder fillings, supports or suspensions can be dispensed with either completely or in part.

C. Mechanical Construction

Apart from small variations, there are essentially two types available, constructed from rigid pipes or from flexible corrugated tubing. The advantage of rigid insulation lies in the fact that any desired diameter

TABLE 3. Integral Heat Conductivity $\int_{T_1}^{T_2} \lambda\, dT$ and Mechanical Yield Strength $\sigma_{0.2}$ of Low-Temperature Structure Materials

	Temperature K	Titan Ti-5,Pt-2.5,Sn	V$_2$A	Nylon	Teflon	Epoxy-Fiberglass*
Integral heatconductivity W/cm	300/4.2	16.5	30.6	0.895	0.702	0.9/1.1
	300/80	14.3	27.1	0.753	0.563	0.75/0.9
	80/4.2	2.2	3.5	0.142	0.139	0.15/0.2
Mech.yield strength $\sigma_{0.2}$ tension and pressure N/mm²		$\sigma_{\text{tension}}0.2\%$		$\sigma_{\text{pressure}}0.2\%$		
	300	580	350	77	7	300
	80	1100	750	245	18	700
	4.2	1300	1000	300	27	800
$\dfrac{\sigma_{0,2}}{\int \lambda d T}$ $\dfrac{N}{Wcm}$	300/80	$4{\times}10^3$	$1.3{\times}10^3$	$3.2{\times}10^4$	$3.2{\times}10^3$	$8{\times}10^4$
	80/4,2	$5.9{\times}10^4$	$2.8{\times}10^4$	$2.1{\times}10^4$	$1.9{\times}10^4$	$4.5{\times}10^5$

* D. KULLMANN

can be constructed and the dc conductor or ac conductor system can always be accommodated in one thermal insulation. Because of the good mechanical stability of rigid pipes a smaller number of spacers for the individual pipes are required which, in conjunction with what has already been asserted, results in the fact that insulation made from rigid pipes is the most favorable as regards thermal losses. What does constitute a disadvantage is the fact that rigid pipes can be transported only in maximum lengths of 20 m, and an N$_2$ intermediate shield and the helium pipe must be provided with additional axial compensation means because of the linear contraction occurring with respect to the outer pipe during cooling, or materials must be used for the cold pipes with a low coefficient of expansion (e.g., Invar $\Delta 1/1$ [300 K $\rightarrow$ 4 K] $\approx 0.04\%$).

It is envisaged that the rigid thermal insulation is prefabricated in transportable lengths and assembled in the field into longer lengths. Then the rigid tubular conductors are inserted from one end or flexible cables are pulled in (as in the case of water-cooled cables).

A flexible thermal insulation made of corrugated piping has the advantage that it can be wound on reels and therefore transported and laid in greater unit lengths (e.g., 200 m). Moreover, no additional measures have to be taken against the variation in thermal contraction of different cable pipes.

The following disadvantages should be mentioned. The outer diameter of an insulation with corrugated piping is restricted for manufacturing and transport reasons so that with the power ratings under consideration only single-conductor cables can be used. Apart from the greater expenditure to provide inner supports for the corrugated pipes (e.g., in the form of plastic spirals) this results in an increase in thermal losses and in a larger amount of ground space.

With corrugated-pipe insulation provision is made for the conductor to be drawn in before transportation and for the cable to be laid complete as a single-conductor cable.

Finally, it is also possible to employ a rigid pipe for the warm outer jacket and corrugated tubes for the inner tubes to be cooled.

Figure 57 shows, in conformity with the above examples, a number of the main possible designs for thermal cable insulation.

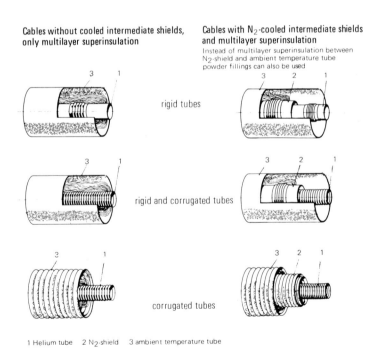

Cables without cooled intermediate shields, only multilayer superinsulation

Cables with N_2-cooled intermediate shields and multilayer superinsulation

Instead of multilayer superinsulation between N_2-shield and ambient temperature tube powder fillings can also be used

rigid tubes

rigid and corrugated tubes

corrugated tubes

1 Helium tube   2 N_2-shield   3 ambient temperature tube

FIG. 57. Construction of Thermal Cable Insulations.

D. Thermal Losses of Different Designs

As mentioned above, in a thermal insulation constructed with rigid pipes, and an N_2-cooled intermediate shield, multilayer superinsulation and rope suspension, the lowest thermal losses are obtained for the smallest radial dimensions. The following losses were measured [113]:

At the temperature level of the helium pipes (4-6 K): $\leq 100 \, mW/m^2$. At the temperature level of the N_2 shield (80 K): $\leq 2 \, W/m^2$.

In a design with flexible corrugated tubes (for all components) the following are obtained approximately [114]:

At the N_2 shield (80 K): $\leq 3.5 \, W/m^2$, at the helium shield (4 - 6 K): $\leq 0.5 \, W/m^2$.

Since, in the power range under consideration, only single conductor cables are feasible for these designs, which, in addition, have a considerably greater surface, the ratio of the losses of this design to the former becomes even less favorable.

If a thermal insulation is employed which does not make use of an N_2 intermediate shield but only of a multilayer superinsulation, losses increase by a factor of approximately 5 as compared with the above mentioned designes with increasing radial dimensions. (Here a comparison was made of the total losses at ambient temperature; with an efficiency of 10 % for the N_2 refrigerating plant and of 0.25 % for the He refrigerating plant).

If a powder insulation is used between the outer pipe and the N_2 pipe instead of superinsulation, the losses at the liquid N_2 level rise to approximately four times the value of the former design, i.e., $8 \, W/m^2$, apparently independently of the fact whether rigid or corrugated pipes are used. The losses on the helium side are not influenced.

VIII. CABLE COOLING

The cooling of long sections of superconducting cables leads to a number of specific cryogenic as well as economic problems which are not relevant in other systems or are of a secondary order. In a cable, the geometric dimensions of the parts to be cooled are disproportionately greater than with other superconducting plants such as magnets or machines. Although relatively simple in their mechanical design, the ducts carrying the coolant contain numerous weld seams, requiring these to be executed with great care for maximum tightness. The object being cooled is practically inaccessible. As we will see in the following, the amount of helium contained in major stretches of superconducting cables is very large and constitutes an appreciable cost element of the cable, meaning that the economic efficiency of the cable is uniquely determined by the price of helium. As will be seen in Section XII, the preservation of the expensive helium filling in the event of a cable defect raises considerable problems. Optimizing cable cooling on the basis of numerous aspects is therefore absolutely necessary, but not yet fully possible, since little theoretical or experimental information is available on such data as the unstable range of helium close to its critical point. For this reason, the following results based on optimum operating conditions of superconducting cables should be regarded as tentative.

A. Magnitude of Losses

While practically only thermal losses occur in dc cables, ac cables also produce conductor losses and dielectric losses. If we assume that the thermal cable insulation includes a nitrogen-cooled intermediate shield at a temperature of about 80 K, the following thermal losses at 80 K and 4 K will result for the various types of cable:

4 K: 0.1 W/m^2 for a rigid thermal insulation,
      0.4 W/m^2 for a flexible thermal insulation.

80 K: 2   W/m^2 with wound superinsulation and rigid thermal insulation,
      4   W/m^2 with wound superinsulation and flexible thermal insulation,
      8   W/m^2 with powder insulation.

At a peak flux density of 0.1 T corresponding to 560 amperes per centimeter (r.m.s.), Section V indicates the electrical losses to be $5\,\mu$W/cm^2 for niobium and $30\,\mu$W/cm^2 for Nb_3Sn at the mean operating temperature involved (5 K for Nb and 7 K for Nb_3Sn). The value for Nb_3Sn has not yet been attained by experimental means, but instead represents a target value. Thus the electric losses per unit length of a 120 kV ac cable with three coaxial conductor pairs and a transmission capacity of 2500 MW, are about 50 mW/m and approximately 300 mW/m for conductors of Nb and Nb_3Sn, respectively. Assuming a dissipation factor of 2×10^{-5} for the electrical insulation and a dielectric constant of 2, the dielectric losses for the same cable will be around 30 mW/m. For the first approximation, the flow losses in the conductor system may be neglected.

B. Stationary Nitrogen Cooling Circuit

The coolants, both helium and nitrogen, are used in single-phase form instead of liquid form, thus avoiding the well known hydrodynamic problems occurring with two-phase flow. Unfortunately in this case the heat of evaporation of the liquid is not available for cooling. This especially is disadvantageous because of the low heat capacity of helium.

Nitrogen is used as a subcooled liquid. With cable types having rigid thermal insulation, the nitrogen shield generally consists of a cylindrical metal jacket. The shield is split down the middle, and both halves of the cylinder are thermally insulated from each other. The nitrogen flows in one direction through ducts which are brazed or welded to one half of the shield, and returns to the refrigerating unit through corresponding ducts on the other half of the cylinder. With flexible thermal insulations of corrugated tubes, where it is not possible to use the two-way flow system just discussed, the nitrogen flows through annular ducts. At higher capacities, several single-conductor flexible cables are laid side-by-side, providing a convenient two-way arrangement.

For a 2500 MW ac cable with wound multilayer superinsulation, the minimum heat inleak at 80 K is about 2 W per meter of cable length. If nitrogen at 77 K and 4 bar is used at the cable entry end, a flow of about 2 l/s will be quite sufficient to cool a cable section 10 km long [115]. The temperature rise of the liquid coolant returning to the refrigerating unit is 84 K. The pressure drop and the resulting flow losses can be kept low (e.g., 1 bar) or within the range of 1 % of thermal losses if the correct tube diameter is chosen.

It should be noted here that intermediate cooling with liquid N_2 has not been specified for all of the cable concepts that have been developed so far. It has been proposed to use a similar system as with exhaust-helium cooled cryostates [116] for cables having rigid thermal insulation. Six copper cylinders cooled with helium are arranged between the helium duct and outer tube of the cable. At the cable joints, the helium is taken from the cooling circuit and passed through a helical duct arrangement which is connected (brazed) to the six copper cylinders kept at a low temperature between the joints (maximum distance: 20 m) by heat conduction. This particular cable concept requires only a single helium circuit. However, a separate pipe has to be provided to collect the helium gas issuing from the sleeves.

C. Stationary Helium Circuit

The operating pressure and temperature of the helium circuit of a superconducting cable depend both on the superconductor material used, the magnitude of the losses and, of course, on the properties of the helium. All cooling concepts devised so far are based on an operating pressure of the helium circuit which is above the critical helium pressure (p = 2.24 bar). The operating temperature is primarily determined by the superconductor material used. If the superconductor material is

niobium, the maximum operating temperature possible is 6 K, as we know
from Section V, since the hysteresis losses rise inadmissibly at higher
temperatures. However, to have a safe margin (temperature rises following overcurrents or short-circuit current), the maximum conductor
temperature chosen for normal operation will only be around 5.6 K. Even
where NbTi is used as a dc cable conductor, the temperature should not
be above 6 K, since the current density of the superconductor will then
drop to about 50 % of 4.2 K. This is why an interval of only 4.4 K to 5 K
has been proposed for the operating temperature of NbTi dc cables [40].
Operating temperatures of 6 - 10 K **are** envisaged for both dc and ac
cables using Nb$_3$Sn for the conductor.

Figure 58 shows a temperature-entropy diagram of helium [117]
in which possible working ranges have been marked for ac and dc cables.
Also entered in Fig. 58 is the curve of maximum specific heat, the so-
called transposed critical line, which intersects the boundary curve of the
two-phase area at the critical point of helium. The shaded boundary
lines indicate areas of similar helium properties (areas 1 to 4). There is
distinct phase separation (liquid meniscus) in the equilibrium state in
area 2, while no phase separation is possible in area 1, the so-called
pseudo-liquid area. In area 4 and in the upper part of area 3, helium is

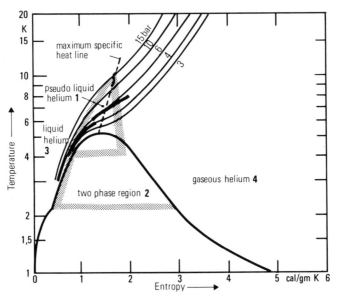

FIG. 58. Temperature – Entropy Diagram for Helium Showing the Regions
of Similar Property and Heat Transfer Behavior.

considered to be gaseous, while it is regarded to be a liquid in the lower part of area 3.

The curve of maximum specific heat is characteristic in that major properties of helium such as heat capacity, density, viscosity and heat conduction change considerably when the curve is intersected or approached on temperature changes. This is a characteristic similar to the transition from the liquid to the gaseous phase or vice versa. As with a two-phase liquid, the result is the danger of pressure and temperature oscillations in area 1 and especially near the transposed critical line. Moving away from the critical point, e.g., in the direction of higher operating pressures, reduces the tendency for instability. Operation in the neighborhood of the 15-isobar should be free of oscillations.

Operation at high pressures is undesirable for two reasons: The helium ducts have to be mechanically stable, which is uneconomical, while at the same time the heat capacity of the helium drops noticeable with increasing pressure, which results in correspondingly higher mass flow rates for carrying away the heat inleak. In order to keep the helium flow rate and the resulting flow losses small, operating pressures (cable entry) of 4 to 6 bar have been suggested. At the dimensions involved, the pressure drops in the helium ducts would be quite insignificant. Fig. 58 assumes the entry pressure of an Nb three-phase and an NbTi dc cable to be 4 bar (temperature 4.4 to 5.6 K), and that of cables with Nb_3Sn conductors to be 6 bar (temperature 6 to 8 K). On the basis of the above, the hypotheses presented make it reasonable to assume that Nb_3Sn cables are more susceptible to oscillations in actual operation than Nb or NbTi cables. It was found in experimental cooling circuits that it should be possible to maintain both operating conditions without disturbing oscillations.

The pressure and temperature operating ranges selected for superconductor cables also have to be considered under another aspect, namely the Joule-Thomson effect. Depending on whether operation involves the area of negative or positive J-T coefficients, an additional temperature rise or drop of the coolant will result from the pressure drop along the helium ducts. Fig. 59 shows the curves of constant J-T coefficients of He as a function of temperature and pressure [118]. On average, the curves indicate a positive J-T effect for both modes of operation, i.e., the expansion is likely to produce cooling, with the effect being greater in the case of Nb_3Sn cables.

For establishing the operating data of the helium circuit, and in addition to the need of avoiding disturbing oscillations and of utilizing a positive Joule-Thomson effect, the dielectric strength of helium may also play a decisive role when used for electric insulation or as an impregnating compound for a wound tape insulation, as in the case with flexible or semiflexible cable types. As we have seen in Section VI, the dielectric strength rises with increasing density, i.e., with increasing

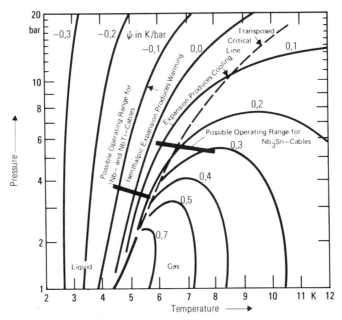

FIG. 59. Contours of Constant Joule–Thomson Coefficient, as a Function of Pressure and Temperature [118].

pressure. The dielectric strength of helium appears to be adequate in the operating range favorable for cooling. However, final statements on the operating range actually suitable can only be made following cryogenic and dielectric experiments on long cable sections.

As with the N_2 circuit, a return line has to be provided for the He circuit. A number of more or less acceptable proposals on the design involved has been made. With single-phase cables laid separately, one cable may be used for helium return, as may be one of the helium ducts where there are several phases and helium ducts in a thermal insulation. However, all systems must ensure that the helium input and return ducts are well insulated from each other thermally, since a thermal short-circuit may produce temperatures in the middle of the helium routes higher than that at the helium outlet [119]. Such temperature rises are inadmissible, since they lead to an increase in conductor losses and to a reduction in the critical current densities of the superconductors, as well as to a less efficient coolant circuit. The corresponding calcula-tions show the mean thermal conductivity between the helium flows have to be $\leq 1.8 \times 10^{-5}$ W/cm K [43].

Accordingly, a number of cable concepts calls for the helium return to be run separately and vacuum-insulated in the cable. The return line may then be used to cool an additional thermal shield which interrupts the heat flow from the thermal insulation to the conductor system. This arrangement is particularly favorable with conductors having a low upper operating temperature (Nb and NbTi), but involves some complexity in the cable design.

D. Transient Cooling

Transient cooling is defined as the processes taking place when a cable cools down or warms up or after the occurrence of an overcurrent, in particular a heavy short-circuit current. The first process is equally important for dc as well as for three-phase cables, while the latter, as we have indicated earlier, is especially important for three-phase cables.

It will be readily appreciated that the pressure and flow conditions for an optimal transient cooling process differ from those for normal operation. While large mass flows at small pressure drops may be achieved in normal cooling operation, considerable pressure drops due to the high flow resistance of the coolants at ambient temperature occur in the cooling process at relatively small mass flows. This causes the cooling periods of major cable lengths to be comparatively long. The cooling periods which operators of superconducting cables would like to see as short as possible, depend on many parameters: type and length of the cable, the flow cross section of the coolant, the magnitude of the masses to be cooled, the additional heat inleak from ambient, the available capacity of the refrigerating system and of the cooling procedure itself. The components of the refrigerating systems should be so designed as to ensure both optimum stationary operation and the shortest cooling periods possible.

Generally speaking, it is possible to cool a section of cable with the cooling system required for stationary operation if normal reserve capacity is available, and the components of the system are suitable for low mass flow and high initial pressures. However, the cooling periods will be relatively long, especially with cables having low cooling requirements in the stationary state. This gave rise to plans to reduce the cooling periods by using additional coolant reservoirs and compressor units.

The method proposed for cooling long stretches of cable is the same as the one used for cooling long transfer lines handling low-boiling liquids [120, 121], where the coolant enters the cable at a temperature close to operating temperature. From what has been found so far, the following process occurs in the cooling of long cable tubes: the tubes cool down in the manner of a clearly defined cold front growing from the inlet end. Ahead of the cold front, the tube walls assume the coolant temperature, with ambient temperature prevailing at the tube walls and in the gas stream behind the cold front. The heat absorption of the coolant from initial temperature to ambient temperature occurs in the cold front. The actual expansion of the cold front is about 10^3 times the diameter of the tube, meaning that it is small compared with the length of the cable (approx. 10 km). The cold front grows at a speed proportional to the flow of coolant fed per unit time.

This speed is limited by the speed at which the warm gas flows out at the end of the cable; this speed cannot be greater than the speed of sound. The maximum speed of propagation theoretically possible for the cold front is therefore only a small fraction (in the order of 1/100) of the speed of sound of the warm gas. The speed of propagation decreases with increasing length-to-diameter ratio of the tube and increases with an increase in the ratio of entry pressure to exit pressure. The entire pressure drop over the length of the cable occurs practically in the warm gas flowing behind the cold front.

In actual practice, the minimum cooling period attainable is primarily determined by the maximum permissible pressure drops over the length of the cable tubes and by possible temperature and pressure oscillations during the cooling process. Pressure drops of more than 10 to 20 bar are not possible for mechanical and economic reasons. Especially the first phase of the cooling process requires careful, that is to say, very slow cooling in order to avoid dangerous oscillations. Specific difficulties in the cooling of long flexible tubes might arise if the oscillation frequency of the coolant resonates with the natural frequency of the flexible tube.

Let us assume a cable having a helium and nitrogen circuit. The nitrogen shield is best cooled down by subcooled nitrogen. The helium ducts and the conductor system are cooled down by cold helium gas at high pressure, the gas in the cooling down phase being precooled in a liquid N_2 bath to 80 K via heat exchangers.

For example, in the case of a 120 kV, 2500 MW ac cable of 10 km length, the amount of heat which the cooling down process requires to be absorbed from the nitrogen shield is about 1.3×10^7 kJ, and that absorbed from the conductor system including helium ducts, is about 7×10^7 kJ; of the latter 95% arise at temperatures greater than 80 K. At a maximum pressure of 20 bar and a maximum possible He mass flow rate of 400 g/s, the time required for cooling this cable to 4.4 K is estimated at 2 to 3 weeks. Similar values have been indicated by other authors [122]. Together with the evacuation period (Section VII), total startup time is about 3 weeks, with about the same period applying for shutdown. It should be noted, however, that these estimates are based on assumptions which still have to be verified experimentally in tests involving fairly long stretches of cable. Such experiments will also have to provide the answer to the question whether the wound tape electrical insulation favored for flexible cable types will come out unaffected by the cooling process.

During the cooling down process, nitrogen and helium have to be continually fed at a rate at which the temperature of the circuits drops. The helium content of the helium circuit of cable sections 10 km long and rated at several 1000 MW is 200 to 300 m^3 of liquid or pseudo-liquid helium. This corresponds to a storage volume of about 150,000 to 220,000 m^3 required at atmospheric pressure; even at a pressure of 200 bar, high pressure storage vessels would still have a volume of 750 to 1,100 m^3, which corresponds to 15,000 to 22,000 standard pressure cylinders. Storage of such a volume of gas at the refrigerating station, or delivery as compressed gas during the cooling process is not feasible. The delivery of liquid helium in large road tankers of the kind already existing is considered a practicable solution. If a cable is put out of service, the above is applied in reverse, i.e., the helium is filled into road tankers and delivered to other load centers. However, this solution assumes that superconducting cables are integrated in a system of large scale helium users.

In addition to technical problems, the large helium content of superconducting cables also present economic questions. The helium filling accounts for an appreciable percentage of the cable cost, which would thus be largely determined by the price of helium. Warnings to the effect that a helium shortage might lead to an increase in the price of helium in the future cannot be completely disregarded. A sensible helium conservation programme as existed in the USA, would ensure a continued supply [123]. There are also considerations of financing developments aimed at an economical production of helium from the air. It has already been pointed out that superconductor cables have a good helium storage capability [43].

Much less of a problem than the cooling down and warming up process are the transient processes occurring at overcurrent or especially short circuit currents where only the equilibrium state resulting after a certain time (several He mixing times) is considered (see also Section V). From the point of view of the coolant circuit, this approach is permissible, since no action can be taken from the refrigerating station anyway to influence the actual transient process.

Assuming a 2500 MW ac cable as described in Section XIII, a short circuit current of 100 kA cleared after 80 ms will produce a brief temperature rise of up to 20 to 30 K in the cable conductor. In the equilibrium state, the mean temperature rise of the conductor or the helium charge is about 0.3 K, corresponding to a pressure increase of 0.7 bar in the helium. In order to reestablish the original operating condition after such a short circuit, the refrigerating system will have to have a temporary capacity margin above rated. The recovery period is equivalent to the time in which a definite volume of helium passes through the cable both ways (transit time). For a stretch of cable 10 km long, the transit time of helium is about one day (see Section VII, E), meaning the capacity margin required would be as small as about 100 W. If the

short circuit is cleared after 0.5 s, there will be no change in recovery
time, only the capacity margin will have to be increased by the factor
2 - 3. In this case, the maximum helium equilibrium temperature will
still not exceed 6 K ensuring 100 % availability. of the cable even after
this rather unlikely case. The general overcurrent capacity of superconducting cables, i.e., sustained overcurrents to cover peak loads, is
considerably limited by the long transit time of helium.

E. Refrigeration Requirements and Refrigerating Systems

The manner in which the outlet of the refrigerating unit is connected
with the cable ends depends primarily on the operating temperature of
the cable. For the magnitude of the power ratings in question, the basic
refrigeration cycle of the refrigerating unit consists of a two-stage
Claude cycle with expansion turbins and liquid nitrogen precooling.

Where operating temperatures of ≥ 6 K are possible, as is the case
with cables having Nb_3Sn conductors, the outlet of the refrigerating unit
may then be directly coupled to the cable, see Fig. 60 a. Circulation of
helium through the cable is effected by the main compressors at ambient temperature. If operating temperatures ≤ 6 K are required, as in
Nb and NbTi conductors, the refrigerating unit may be connected to the
cable, using an additional turbine or a J.T. valve and a heat exchanger
or a liquid tank, see Fig. 60 b. In this case also, the main compressors
circulate the helium through the cable. An indirect connection of the
cable with the refrigerating unit results from the use of a He pump,
See Fig. 60 c, where the helium gas stream of the refrigerating unit and
of the cable are separated from each other and connected only by way of
a heat exchange or a liquid tank. In this particular example, the pressure
of the system is controlled from the outside, the pump having to overcome
only the pressure drop caused by the flow resistance.

The cooling arrangement using a pump as in Fig. 60 c undoubtedly
represents the most flexible solution. On the other hand, the additional
losses of the pump cause a decrease in efficiency, but on the other hand
temporary changes necessary in the operation of the cable have no
effect on the operating conditions of the refrigerating unit, especially if
a liquid tank with buffer action is used as a coupling element. The refrigerator may be set to an optimum operating condition which should
result in a good Carnot efficiency. Since there are also helium pumps
with an efficiency greater than 90 % [124], the Carnot efficiencies of the
refrigerating systems in Figs. 60 a), b) and c) should be roughly the
same. For reasons of greater flexibility, solution c) is preferable.

Figure 61 a is a schematic diagram showing the refrigerating unit
connection to cable stretches of about 10 km or more, where several
refrigerating units are arranged along the cable. For stretches of
cable about equal to or less than 10 km, the coolant is fed in at one end

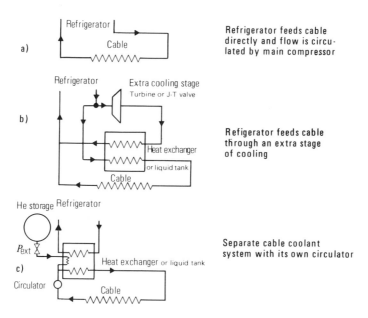

FIG. 60. Three Arrangements of a Cable as the Refrigerator Load.

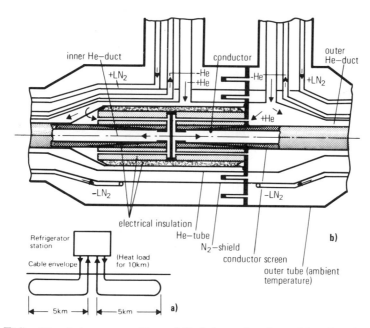

FIG. 61. Interconnection of Refrigerator to Cable [110].

through the terminal   (see Section X, Fig. 68). Fig. 61b shows the practical implementation of such a connection as applied to a flexible phase of a three-phase cable [110]. Part of the helium is admitted into the conductor bore by way of an axial duct through the wound tape electrical insulation. The transition from earth potential to high-voltage potential is accomplished by the use of transverse potential dividers which linearize the potential gradient along the helium duct. That part of the helium which cools the phase conductor from the outside remains at earth potential, and therefore poses no problem. The same method of connection is used if the helium is admitted through the termination or also where the helium passes from the inside of the conductor to the return duct which is also at earth potential. The type of connecting piece (cooling sleeve) as shown in Fig. 61b also permits the vacuum spaces to be sectionalized, in addition to separation of the He coolant circuits.

Before we refer to several examples to demonstrate the refrigerating capacity required for superconductor cables, the efficiency of the machines involved has to be given some further consideration. The ratio of the refrigerating capacity $\dot{Q}(T)$ available at temperature T to the input $W(T_0)$ required by the refrigerating unit at ambient temperature T_0 is expressed by:

$$\frac{\dot{Q}(T)}{W(T_0)} = \eta \cdot \frac{T}{T_0 - T} \ . \tag{20}$$

The second factor of the product is the familiar ideal Carnot cycle efficiency and η the so called Carnot efficiency which indicates the fraction of ideal efficiency actually achieved by the given refrigerating system. The Carnot efficiency practically depends only on the refrigerating capacity, as shown in Fig. 62   and not on the cooling temperature [125]. At the capacity of refrigerating systems needed for cable sections of 10 km in length, η can be expected to be around 0.20. For a helium refrigerating plant whose working temperature is 4.2 K, this would result in an efficiency of about 1/350, and about 1/10 for a nitrogen refrigerating unit operating at 80 K.

According to Section VIII A, total losses of around 180 W/km can be expected for a 120 kV/2500 MW ac cable with coaxial niobium conductors at helium temperatures, and 430 W/km for a cable with coaxial Nb_3Sn conductors. Table 4 shows the major cryogenic data of these cables and input required by the helium refrigerating plants. Two variants have been assumed in each case; they differ in that the one variant provides for the thermal losses to be arrested by a helium shield cooled by return helium, while the other one does not.

Unusual are the low mass flow rates which lead to low pressure and flow losses (<10 % of total losses). They occur mainly in the helium

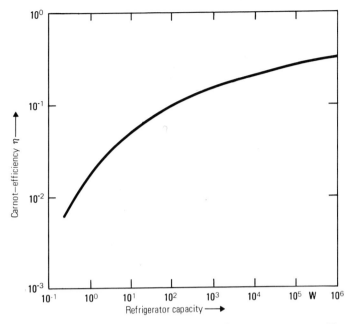

FIG. 62. Efficiency of Low Temperature Refrigerators as a Function of Refrigeration Capacity [125].

TABLE 4. Helium-Refrigerator-Input Powers for 120 kV, 2500 MW Coaxial AC Cables (Cable length, 10 km)

Temperature range of conductor K	Superconducting Material	Total ΔT go and return K	Inlet pressure bar	Total pressure drop bar	Massflow rate g/s	Line heat load W	Helium shield load W	Total heat load W	Hot work at 20% Carnot kW
4.4–5.6	Nb	1.7	4	0.17	130	800	1000	2000*⁾	550
4.4–5.6	Nb	1.1	4	0.58	260	1800	–	2000*⁾	630
6 – 8	Nb₃Sn	2.6	6	0.25	180	3300	1000	4300	855
6 – 8	Nb₃Sn	2.0	6	0.38	225	4300	–	4300	920

*⁾ + 10% for Helium Pump

return duct. The low mass flow rates result in low flow rates as well.
In superconductor cables, they are 0.1 to 0.3 m/s; they are higher in
the helium return duct, depending on its dimensions. As already mentioned, these low average flow rates result in very long helium transit
times on long cable stretches; for a 10 km section, the time is about
one day.

Table 4 also shows that despite the favorable loss figure assumed
for Nb_3Sn (30 $\mu W/cm^2$ at 7 K) and the improved refrigerating plant efficiency at a higher mean operating temperature, the input required
for the Nb_3Sn cable at ambient temperature is about 50 % above that
needed for the Nb cable. Nevertheless, the specific input required for
both cables, i.e., per unit length and power is naturally much less than
with nonsuperconducting cables. It can also be clearly seen that the use
of an additional helium shield which prevents the inleak of thermal losses
from affecting the conductor system, leads to reduced refrigerating plant
input. However, the decrease in input is not so impressive as to warrant the extra mechanical outlay in every case.

Assuming the heat inleak affecting the nitrogen circuits of the above
cables to be 2 W/m at 80 K, the input required by the refrigerating unit
to absorb the heat from a cable section 10 km long will be about 200 kW.
These losses must be taken into account when stating the specific cable
losses.

Far lower are the specific losses of dc cables. A coaxial
120 kV/5000 MW cable with flexible conductor of NbTi copper composite,
flexible helium duct, rigid outer pipe and nitrogen shield, requires a
helium refrigerating plant input of around 30 kW per km of cable length
at an operating temperature of 4.2 to 5 K and a mean working pressure
of 4 bar [40]. Thus the specific losses of a dc cable are only 25 to 35 %
of the figure for ac cables of the same voltage. Assuming that the insulating materials used for a dc voltage can take a higher dielectric
stress than when used for an ac voltage, the ratio in favor of the dc
cable will be even better.

There are already helium refrigerating systems of several kilowatt input at 4.4 K. The low pressure drops occurring in the cable
tubes at normal operation make it appear realistic to have refrigerating
stations spaced at intervals of 10 to 20 km in the case of ac cables, and
at even greater intervals in the case of dc cables. Limits on the intervals might be imposed by the cooling down process, that is if the operators
of such cables specify fixed periods for startup and repairs.

Another interesting aspect is the weight and especially the volume
of refrigerating systems. Fig. 63 shows these factors as a function of
refrigerating capacity with regard to helium refrigerating systems. The
volume of refrigerating systems suitable for cooling a cable section
10 km in length, is several hundred m^3. The weight and volume of such

systems may be reduced at any time, but this might lead to greater costs and, definitely where volume is reduced, to more difficult maintenance and repair conditions. The cost of refrigerating systems for a temperature range of 1.8 to 90 K may be approximated using the relation

$$C = 6000\,P^{0.7},$$ (21)

where C is the cost in dollars and P the installed input in kilowatt.

The operation of superconducting transmission systems requires maximum reliability and availability of refrigerating systems. This may be achieved by installing double the number of refrigerating plants required. It will also be necessary to have machines designed to permit the major components and those most likely affected to be readily and quickly replaced. Liquid reservoirs acting as buffers from which the cable sections may be cooled over short periods of time, also increase reliability. These measures are necessary although the reliability of cryogenic equipment will continue to improve as more experience is gained in this field. For example, there have been reports of small automatic 4 K helium refrigerating plants operated without interruption for more than 20,000 hours [126].

IX. CABLE JOINTS

The basic problem with cable joints is the conductor connection technique. It plays a decisive role particularly in rigid pipe conductors because with these, as already stated, cable joints at maximum intervals of 20 m are present, i.e., a long cable has very many joints. For the electrical connection of rigid pipe conductors, a sliding contact connection with a corrugated fitting has been proposed by Roger, Slaughter and Swift [127] to obtain the required vacuum or helium tightness (Fig. 64). This connection solves primarily the problem of thermal contraction on cooling. The resistance of a contact of this kind between superconducting surfaces is not known but should be of the order of $10^{-6}\,\Omega$. Consequently the losses per contact point for an effective current density of 565 A/cm and a conductor diameter of 7 cm are about 150 W compared with about 0.1 W for the ac losses of a 20 m conductor. With a normally conducting connection of the backing material (Cu or Al) through brazing or welding a resistance of about $10^{-8}\,\Omega$ is to be expected with losses of about 1.5 W. The examples shown reveal that the connection of rigid pipe conductors must be carried out superconductively because of the large number of cable joints.

For stabilized Nb conductors this is possible. During manufacture the pipes are fitted at the end with Nb rings so that on jointing the pipes pure Nb can be welded. These welded seams are superconducting. Because of the small areas of the welded seams a heavy mean increase in the losses of ac cables (referred to the cable as a whole) is not to be

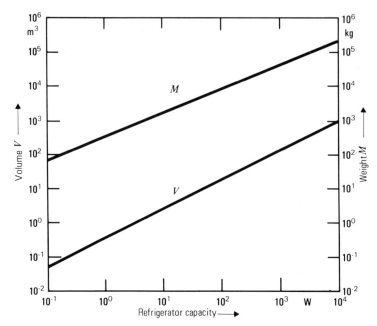

FIG. 63. Volume and Weight of Refrigerators as a Function of
Refrigeration Capacity [125].

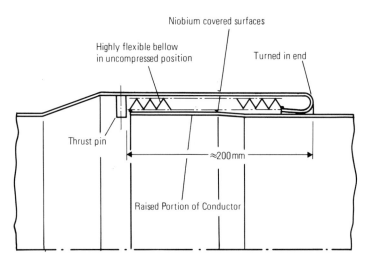

FIG. 64. Highly Flexible Bellow Joint [127].

expected even with locally higher losses. Tests on specially prepared welded joints have even shown that the increase in losses at the welded joints is negligible. In addition, the stabilizing material of the conductors has to be bridged at the joint because with an ac cable under fault conditions the contact point also becomes normally conducting. The fault current losses for a current density of 565 A/cm consequently amount to about 0.5% of the losses of a 20 m long Nb-Al conductor for each contact point and thus are without significance. This jointing technique can also be used for discrete corrugated units or pipes, should these be employed to overcome thermal contraction.

The flexible types of conductor, e.g., corrugated pipes or pipes constructed of spirally wound tapes or wires, can be made in lengths of 200 to 500 m and transported on drums so that the number of cable joints is considerably reduced and the conductor connection can be of the normally conducting type. Tests on Nb conductors with a core of very pure Al (residual resistance ratio $\geq$ 2000) have shown that it is sufficient for obtaining a low loss contact to weld the Al cores. For example, tests on Al wires (2 mm core diameter, 100 μm Nb coating) have given resistances of 3.10^{-9} Ω with butt-welded contact points. Thus losses of 5 mW per contact point can be expected for a conductor 7 cm in diameter at an effective current density of 565 A/cm. With Cu stabilized Nb_3Sn tape conductors a simple brazed connection of the overlapping tapes proves suitable. With 12.5 mm wide tapes contact resistances of 10^{-8} to 10^{-9} Ω have been obtained depending upon the length of the overlap. A prerequisite for a satisfactory performance of the normal conducting contact point is a good bond between the superconductor and the stabilizing or backing material. Combined Nb-Cu and Nb-Al conductors or even copper-stabilized Nb3Sn conductors of present-day quality fullfil this condition.

The losses at the contact points of the conductors connected normally conducting via the stabilizing material are practically insignificant compared with ac losses (or in the case of dc cables compared with thermal losses) of the individual conductor sections between the cable joints. The heat occurring locally at the contact points due to the ohmic losses is dissipated to the cooling medium with practically no increase in temperature of the conductor. The electrical insulation with flexible conductors made of overlapping wound tapes is conveniently carried out at the cable joint with special tapes, e.g., self-welding PE tapes. The winding is made correspondingly thicker to obtain the required dielectric strength.

Figure 65 shows a cable joint with a flexible conductor made of wires in a rigid thermal insulation. After connecting the phase conductors, the three helium pipes made of Invar are connected with corrugated units by welding so as to compensate for the residual longitudinal contraction during cooling. The three connecting points are axially staggered so that the increase in diameter through the corrugated units is considerably reduced. The pipes for cooling the intermediate (N_2 and He) shields are also connected with corrugated units. After applying the superinsulation the outer jacket of the cable joint is pushed over the outer pipe of the cable and welded vacuum-tight. Fig. 66 shows the rigid joint of a completely flexible single-conductor cable. This however is only the inner part of the joint up to the N_2 shield [128].

It is a general characteristic of cable joints that all inner cable components are secured against the outer pipe so as to prevent axial displacements of the same, e.g., during the cooling-down process. With rigid or semiflexible cables necessary changes in direction are carried out at the cable joints. Depending upon the design of the cable the vacuum spaces or cooling circuits are also sectionalized in the cable joints and evacuation connections are also fitted.

X. TERMINATIONS

A. Current and Voltage Leads

It is intended to use superconducting cables initially in systems in which the other important components such as generators (at least the three-phase windings of the same), transformers and switches are normally conducting. Consequently the problem arises with the termina-tions of carrying large currents at high working voltages with the lowest possible losses, from ambient temperatures to helium temperatures. As we know from other fields of applied superconductivity technology, use is made of normal conductors to introduce the current which are connected to the superconductor at a low temperature (e.g., 4.4 K). The losses occurring at the low temperature result from heat conduction and from Joule's losses in the supply lead. It is not sufficient merely to optimize the cross-section for the normal conductor because the losses then introduced of 42 mW per amp input [129] would be too high. In order to reduce supply lead losses to acceptable values, the supply leads must also be cooled. Two methods are possible: continuous cooling of the supply lead with evaporated cold helium gas or cooling the lead with the refrigerator via heat exchangers to fixed temperature levels (e.g., 4.4 K, 20 K, 80 K).

In addition to the reduction in supply-lead losses, the dielectric strength of the terminations is of great significance. Just as in con-ventional terminations, longitudinal components of the electric field occur in which would insulations exhibit a considerably reduced di-electric strength compared with the corresponding values in radial transverse fields (1:10). In addition a number of problems occur with superconductor terminations, which result from the temperature gradient in the radial and axial direction and the different thermal contraction coefficients of the materials of the terminations. Thus the

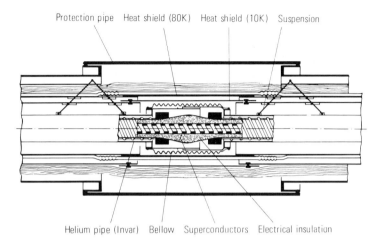

FIG. 65. Joint of AC Cable (One Phase) with Superconducting Wires or Ribbons.

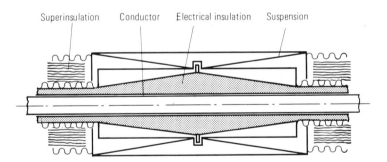

FIG. 66. Joint of DC Cable (One Conductor) with Superconducting Ribbons [128].

insulating materials to be used have a contraction which exceeds by almost one order of magnitude that of metals. If in the electrical insula-tion minute hair cracks develop, partial discharges occur, i.e., higher losses and finally disruptive discharges and destruction of the termina-tions take place.

1. Optimizing the current supply leads in relation to losses

The temperature profile of a gas-cooled current lead of length 1, cross-section F and current I is the solution of the equation

$$\frac{d}{dx}\left(F\lambda \frac{dT}{dx}\right) + \dot{m}\, c_p \frac{dT}{dx} + \frac{I^2 \rho}{F} = 0. \tag{22}$$

The first and third terms give the heat input and the Joule losses, while the second term determines the heat dissipated to the cooling gas (through put $\dot{m}$), always assuming that the heat exchange between the con-ductor and the cooling gas is ideal, i.e., that the temperature difference with respect to the cooling gas is negligibly small. The heat conductivity $\lambda(T)$ and the resistivity $\rho(T)$ are functions of the temperature while the cross-sectional area $F(x)$ in most cases is a function of the space coor-dinate x. The specific heat capacity c_p of the cooling gas He may be regarded as being approximately constant over the whole temperature range.

For design reasons it is worthwhile (e.g., to prevent freezing) to fix, in addition, to the cold end temperature T_K the warm end tem-perature T_W as well $(T(x=o) = T_W; T(x = \ell) = T_K)$. Normally the whole amount of the helium gas evaporated through the current supply lead is used for counterflow cooling (self-sustaining cooling) and the following applies

$$\dot{m}Q_L = -\lambda F\, (dT/dx)_{x = \ell}. \tag{23}$$

For every conductor material there is an optimum relation for the current supply parameter $I \cdot \ell/F$ (in the case of Cu with a residual resistance ratio of 100, this parameter, for example, is 3×10^5 A/cm [130]) at which the losses on the helium side assume a minimum value. The temperature profile of the current supply at T_W has a horizontal tangent $(dT/dx)_{x=0} = 0$, i.e., the total enthalpy of the gas is used to cool the current supply lead.

The optimization of a current supply lead and the losses of the same depend to a very great extent upon the parameters $\lambda(T)$ and $\rho(T)$ of the materials employed. For example if alloys are used for which the Wiedemann-Franz law applies $(\rho \cdot \lambda = L_n \cdot T,$ where L_n is constant) the losses in the optimum case, where T_w = 300 K, T_k = 4.2 K and with self-sustaining cooling (Eq. 23) at an ideal heat exchange are 1.131 mW/A [131]. The smaller the residual resistance of the conductor material becomes the greater are the deviations from the Wiedemann-Franz law ($\rho \cdot \lambda < L_n \cdot T$), particularly in the T_k range, which results in a reduction in the losses. For example, for Cu with a residual resistance ratio of 100 they are 1.065 mW/A [131]. These approximations for the expected losses differ to some extent depending upon what dependencies on temperature have been assumed for the individual parameters in Eq. 22. The use of metals with high residual resistance ratios causes, however, a growing tendency towards thermal instabilities [130]. Even small overcurrents result in rapid rises in temperature in the current supply lead far in excess of 300 K and the thermal destruction of the supply lead may be the consequence. In addition, the following general statements can be made.

1. In the absence of an ideal heat exchange between the surface of the conductor and the cooling gas, an increase in the minimum losses mentioned above must be expected. With heat exchange conditions which can be simply realized, this increase may be neglected in the case of pure metals while with alloys a marked increase is to be expected [132].

2. If the current supply lead is shunted at the cold end by a super-conductor with a high T_c and H_c the losses only decrease slightly [132].

3. The effect of the warm end temperature on the losses is also small and in general it does not pay to decrease T_w to the temperature of nitrogen (T = 80 K) [133, 134].

4. The ratio of the losses at full working current and no load for Wiedemann-Franz metals (alloys) is about 2.8 and decreases to about 1.3 as the residual resistance decreases [130].

5. For zero heat exchange or ideal heat exchange, the variation in the cross-section F(x) has no influence on the losses [135, 136].

6. For large refrigerator-cooled systems it may be better to use a large part of the enthalpy of the cold gases in the heat exchanger of the refrigerator and not in the current supply lead. Optimization calculations for large magnet systems have revealed that there is no great difference between this type of cooling and self-sustaining cooling [136]. Corresponding observations on cable systems are not yet available.

As our own calculations have shown, the other method of cooling the current supply leads, i.e., progressive cooling through discrete heat exchangers (e.g., at 80 K, 20 K and 4.4 K) results in the same power losses at the input of the refrigerator as the waste-gas cooling discussed in detail.

2. Resistance to fault currents

Fault current conditions are determined by the operating mode of the cable (ac or dc). The highest loads are obtained with three-phase

systems: fault currents up to 100 kA with clearing times of four cycles
(80 ms) or on the failure of the first circuit breaker   of a maximum of
0.5 s, are possible. To assess the maximum temperature increase ΔT
occurring in the current supply lead the heat exchange with the cooling
gas is disregarded. We then get for the adiabatic increase in temperature

$$\Delta T(x) = \int_{0}^{t} \frac{\rho(T(x, t))}{C_p (T(x, t)} \times \frac{I^2(t)}{\gamma \cdot F^2} \cdot dt, \qquad (24)$$

where γ is the density of the conductor material and t the clearing time
for the fault current.

At the beginning of the fault, the greatest increases in temperature
occur and also the greatest gradients in T at the cold end of the current
supply lead, this being due in particular to the specific heat capacity
which although low is highly dependent upon the temperature. But it is
at the cold end where the times for the temperature equalization processes are less than the fault current times, so that for certain ranges
of Δx (about 10 cm) a temperature equilibrium can be anticipated. During
the succeeding period any additional rise in temperature at the cold end
is considerably subdued by the large increase of $c_p \propto T^3$. At the warm
end the opposite effect can be observed, namely, a rapidly increasing
temperature. The resistivity and consequently the losses, increase
linearly with the temperature ($\rho \propto T$) while c_p remains practically constant.

For a current supply lead optimized at 10 kA ($\ell = 100$ cm, F = const)
we get roughly for Cu (1 % residual resistance) and the aluminum alloy
Al-5052-0 (0.25 Cr, 2.5 Mg, 97 Al, 40 % residual resistance) the temperature increases shown in Table 5 (to simplify the estimate the fault
current is regarded as being independent of time). As may be gathered
from this, temperature increases in the current supply lead in the case
of a normal fault (t = 80 ms) are not critical. In the rare case of a
clearing time of 0.5 s, increases in temperature have to be reckoned
with at the warm end of the current supply lead when electrolytic copper
is used as conductor material. These temperature increases while not
able to cause damage to the conductor itself are able to do so to the
adjacent electrical insulation. The favorable performance of the Al
alloy is to be attributed to the large cross-section required for this conductor which results in a high heat capacity. An improvement for current supply leads made of copper can be obtained by adding additional
thermal capacities through which no current passes.

TABLE 5. Temperature Increase of Current Lead with Fault Current

Conductor material	$\dfrac{I \cdot l}{F}$	ΔT at T_k		ΔT at T_w	
		$t = 80$ms	$t = 0{,}5$ s	$t = 80$ms	$t = 0.5$ s
Cu	3×10^5 A/cm	60 K	85 K	40 K	300 K
Al 5052	$2{:}4 \times 10^4$ A/cm	15 K	35 K	1K	6 K

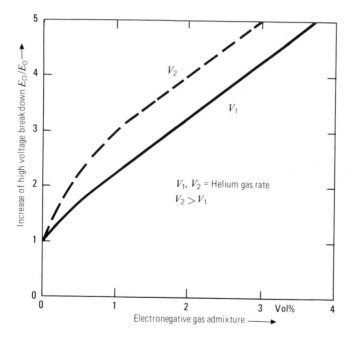

FIG. 67. High Voltage Breakdown of Gaseous Helium (300 K; 1 bar) as a Function of the Portion of Admixed Electronegative Gas [138].

3. Dielectric strength of the supply lead

As already mentioned at the beginning, a number of problems arise in connection with the insulation in the termination which must be solved. These are primarily the mechanical stresses which occur in the insulating material due to differing contraction coefficients and large temperature gradients from ambient to helium temperature and the longitudinal components of the electrical field in addition to the radial fields for which the breakdown strength is in general considerably smaller. dc voltage measurements on papers impregnated in helium (5 K; 2.9 bar) [114] have, for example, shown the dielectric strength vertical to the surface of the paper to be about 40 kV/mm and in the plane of the paper only about 4 kV/mm; this is a ratio of 10:1 which also applies approximately to oil-impregnated paper at normal temperatures. Conditions are even worse with plastic foils impregnated in helium (Tyvek) and with ac voltage where, our own measurements show the dielectric strength ratio to be about 20:1. The dielectric strength of liquid helium is comparable with that of oil at room temperature (ac voltage: about 30 kV_p/mm at 4.2 K, 1 bar and an electrode spacing of 1 mm; dc voltage: about 40 kV/mm at 4.2 K and 3 bar); thus in the termination of a superconducting cable similar design principles can be adopted for the voltage bushing as in conventional cables and transformers.

From the mechanical and electrical point of view, glass fibre reinforced cast resin materials or insulation elements wound of special paper impregnated in epoxy resin are suitable for use in the area of the radial field in the termination. In the case of an ac cable they should also be provided with potential grading capacitor elements which allow a controlled reduction of the potential both for the radial fields in the insulation unit and also for the longitudinal fields e.g., in liquid helium. The dielectric losses in a potential grading insulation element are negligibly small and are, for example, about 0.1 W at 100 kV (our own measurements). Otherwise the longitudinal components of the electrical field are reduced by stress cones which can also have potential grading in the case of ac.

A further problem arises with current supply leads cooled with waste gas owing to the exceedingly small dielectric strength of gaseous helium at room temperature which, under normal conditions, is 0.15 kV/mm. The difficulty consists in transferring to earth potential the He cooling gas being at a high potential at the warm end of the termination. Here two exceedingly elegant methods have been proposed [137, 138], which enable the breakdown strength of the warm gaseous helium to be increased so considerably that the potential can be reduced over relatively short sections. The methods are based on the idea of detaching free electrons from the gas and thus reducing its ionizing capability by the admixture of electro-negative gases or vapors or by using large area charge-carrier capture walls. Fig. 67 shows the effectiveness of the first method where the dielectric strength of a

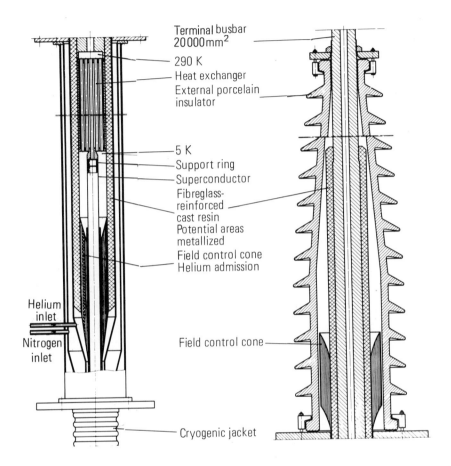

Terminal busbar
20000mm^2

290 K
Heat exchanger
External porcelain
insulator

5 K
Support ring
Superconductor
Fibreglass-
reinforced
cast resin
Potential areas
metallized
Field control cone
Helium admission

Helium
inlet
Nitrogen
inlet

Field control cone

Cryogenic jacket

FIG. 68. 200–kV Terminal for a Superconducting DC Cable (Constructed
by AEG) [114].

specific helium gas section is increased fivefold by the addition of only 3% by volume of a foreign gas. Similar improvements can also be achieved with the second method. When adding the foreign gas the same can again be removed at earth potential from the helium by freezing out, before the pure gas is again admitted to the He refrigerating plant.

B. Termination Designs

To minimize the losses in practice for an optimum current supply lead it is necessary to achieve an ideal heat exchange between the current supply lead and the cooling gas. This requirement can be fulfilled by dividing the current supply lead into thin wires (100 to 150 μm dia.) thus providing a large cooling surface [139]. For ac applications the skin effect of the current supply conductor must be taken into consideration. The skin depth, for example, for a conductor with a 1% residual resistance (Cu) is about 1 mm at 4.2 K and 50 Hz. Its influence can be suppressed by transposing the individual conductors (wires). This design feature can, however, only be realized in a simple manner when the conductor area (F) is constant since an area variation in the conductor along its length automatically causes short circuiting of the transposed individual conductors. As in the case of the phase conductor the outgoing and return conductors of the current supply should be arranged coaxially so as to prevent eddy current losses in the normal conducting termination components.

Figure 68 shows the principles of a single pole dc cable termination for 200 kV [114]. The most important part of the termination is the normal conductor designed as heat exchanger which at its lower end is connected to the cable superconductor, receives the current from this and transfers the latter from 5 K to 300 K (left half of the figure). The heat exchanger is cooled by cold helium gas rising upwards, which is introduced into the lower part of the termination. It enters the heat exchanger through a helium inlet provided in the elongated stress cone built up of wide paper tapes. The stress cone causes a soft field modification in the transition from the normal cable insulation to the glass fibre reinforced cast resin pipe with a larger diameter. This pipe constitutes the electrical insulator for the heat-exchanger conductor and is only exposed to electrical radial fields. Where the inner and outer cylindrical surfaces of this insulating pipe are potential areas of the electrical field, they are metallized to prevent cavities from occurring at the interfaces after contraction through cooling. The right hand part of Fig. 68 shows the ordinary porcelain insulator for the terminations with a stress cone at the point where the earth electrode ends.

Figure 69 shows diagrammatically the single-phase arrangement of an ac cable connection. The main feature of the design is again a normal conductor constructed as the heat exchanger, in this case

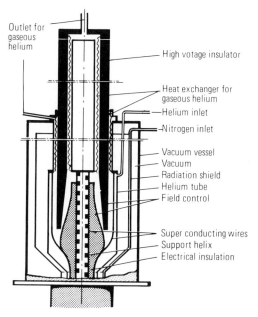

FIG. 69. Coaxial Terminal Fitting for 120 kV Superconducting AC Cable
(One Phase) (Constructed by Siemens).

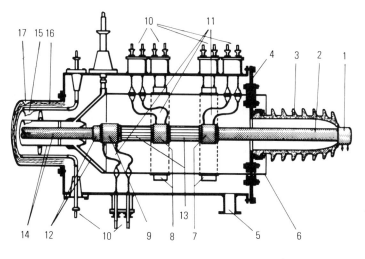

1 Terminal bolt	7 ⎧ Heat N_2, 80 K	12 Conection normal-superconductor
2 Current lead	8 ⎨ exchanger H_2, 20 K	13 Transposed Cu-conductor
3 High voltage insulation	9 ⎩ for liquid He, 4.4 K	14 Superconductor
4 Vacuum vessel	10 Coolant terminal	15 Radiation shield
5 Vacuum terminal	11 Coolant pipe electrical	16 Superinsulation
6 Return conductor	insulating	17 Protection tube

FIG. 70. Coaxial Terminal Fittings for 120 kV Superconducting AC Cable
(One Phase).

duplicated, one being for the actual phase conductor and the other for the shielding or return conductor. The longitudinal components of the electric field at the end of the conductor insulation and in the helium are reduced by means of elongated stress cones, partly by field gradient control. The inner heat exchanger, which is at a high potential, is separated from the outer one, which is at earth potential, by a cylindri-cal insulating unit which at its lower end transforms into the elongated stress cone mentioned above. The liquid helium for cooling the heat exchanger is introduced separately to the termination.

Finally, Fig. 70 shows without details the single phase arrange-ment of a three phase cable termination which is cooled in three tempera-ture stages (4.4 K, 20 K and 80 K) through discrete heat exchangers from the refrigerator and the electrical insulation of which consists of a high vacuum. The cryogenic liquids (He, H_2 and N_2) for cooling the heat exchangers which are at a high potential, are conveyed to the latter by means of pipe coils made of electrical insulating material. The heat exchangers for cooling the coaxial return conductor are at earth poten-tial and consequently no difficulties occur here. The optimized conductor cross sections between the individual temperature stages vary in size. The discrete cooling in three temperature stages necessitates altogether three cryogenic cooling media, which for electrical reasons (dielectrical strength) must be in the liquid state in the cycle, i.e., this kind of ter-mination cooling is more expensive in regard to refrigeration than cooling with evaporated helium gas.

XI. METHODS OF LAYING VARIOUS TYPES OF CABLES

In the past, cable development programs were primarily con-cerned with the development of components. The development of suit-able methods of installing superconducting cables are usually dealt with at later stages of these programs. This is why essentially only qualita-tive statements can be made on this subject.

The first step in the laying of a cable is to create a cable trench. This may be done by the usual method of manual digging or by other familiar methods, with an open trench to be preferred for certain types of cables (semiflexible and rigid). The relatively large power density of superconducting cables will naturally require smaller trench dimen-sions than with other cables of the same transmission capacity. In addition, the costly item of filling with thermal sand is not needed. The advantages are probably offset by the need to lay superconducting cables in concrete pipes, which may perhaps be formed by halves placed to-gether. The reason for this is that the thermal insulation of a super-conducting cable must not be damaged under any circumstances since this would involve troublesome consequences, as described in Section XII.

The laying methods to be used depend considerably on the kind of cable. The following is a description of the methods so far con-templated.

Fully flexible cables of the single conductor type can be laid in lengths of 200 - 250 m as conventional cables. Both the electrical connec-tions of the conductors and of the electric insulation and the mechanical connections of the thermal insulation have to be made in the joints which connect the ends of the adjoining cable sections. While the cable sections are manufactured in specially suited workshops at the factory, the connection, complicated as it already is, has to be made in the field. In order to avoid difficulties in subsequent operation, a high degree of care and cleanliness has to be applied in the manufacture of the joints. The most modern automated welding methods (Argon arc, electron beam) are required for connecting the various pipe components. Since weld seams of the highest quality are essential, sophisticated inspection methods such as involve mobile helium leakage detectors and X-ray equipment must be applied. The space required in the area of the joints is thus likely to be relatively large. What has been said with re-gard to the joints also applies to the laying methods discussed in the following for the other types of cables.

With semiflexible cables, the rigid thermal insulation has to be transported to the site in sections 15 to 20 m long and put together by welding the individual pipes. The sections have already been heated and evacuated once at the factory. The flexible cable conductor or each individual conductor is then drawn into 500 m lengths of thermal pipe sections, in the manner usual with water cooled oil filled cables. To repeat, thermal joints are required every 15 to 20 m while electric joints can be spaced at 500 m intervals.

Methods have also been discussed in which the individual sections of the thermal insulation, similar to transfer lines for cryogenic liquids, are delivered to the site and joined together as fully completed compo-nents that have been evacuated at the factory. The helium ducts at the adjoining sections are first welded together and then prefabricated joints which may be evacuated separately are fitted to the connecting points. The electric conductors are then drawn into 500 m sections of the cryo-pipe. This method would be very advantageous, since it would not be necessary to evacuate the thermal insulation before it cools down, and practically no vacuum pumps would be required. Although experience with liquid reservoirs having stationary insulation vacuum is very good, it is still doubtful whether this method is practicable considering the long service life of cables (25 to 30 years). For the present, we are assuming that the vacuum of the completed cable has to be produced and maintained by the necessary pumps.

With the rigid (pipe type) cables, there are two laying methods: thermal insulation and electrical conductor are factory -made in sections of 15 to 20 m. This requires the spacing of a common electrical and mechanical joint every 15 to 20 m. A better method is the one in which several sections of the thermal insulation are first placed together and then the rigid conductor system is run (e.g., on wheels) into the helium pipe from one end of a completed length. This method would have the advantage of allowing several electrical joints to be fabricated at a single location.

In the laying methods discussed so far, the thermal cable insulation is manufactured at the factory and delivered to the site in practically the final state. Another method calls for the rigid thermal insulation to be set up in the field [38].

This method requires a simple powder insulation instead of the more sensitive insulation using reflecting foils. Lengths of flexible cable conductors are then drawn into long sections of this rigid insulation.

The necessary buildings for the vacuum substations, as well as for the refrigerating stations in the case of major cable lengths, have to be constructed along the cable route. Electric control and supply lines laid parallel to the cable will be required. Taken as a whole, the outlay for the cable installation is considerable, and the particular cost involved represents an appreciable portion of the total cable costs.

XII. OPERATIONAL RELIABILITY AND MONITORING

Superconductor cables represent complex installations for the operation of which numerous cryogenic auxiliary units such as refrigerating and vacuum equipment are required. Sources of trouble leading to disturbances may lie either in the cable itself or in the auxiliary units. In order to detect and eliminate them in time, constant monitoring of the installations is necessary.

A requirement to be specified for the cable is practically absolute reliability, since the time needed for repairs can be months because of the long periods needed for heating up and cooling down. To avoid brief failures, this requirement also applies to the auxiliaries. Since it may not be assumed that all parts will be absolutely reliable in operation, it is planned to duplicate critical components or systems, such as moving parts. Buffer tanks for helium and nitrogen in the refrigerating stations will bridge the time following a fault in the refrigerating system until the reserve units assume full operation.

Faults in the cable may occur in the conductor system or in the thermal insulation (vacuum). Faults (short circuit) initiated by the conductor system are registered at the terminal stations and result in immediate shutdown of the cable (within the switching times of existing circuit breakers). As with conventional cables, classical methods (e.g., bridge measurements) are used for locating faults.

The main problem in operating the cable is maintaining the vacuum for the thermal insulation. This requires constant monitoring of the vacuum. The simplest method of detecting a vacuum leak would be by monitoring the helium pressure at the refrigerating stations. However, since a pressure rise in helium is only propagated at the speed of sound (approx. 100 m/s), the least favorable case of a fault halfway between two refrigerating stations (10 km apart) would involve times of about 50 sec until the fault is detected and necessary measures are taken. This period can only be reduced if discrete vacuum measuring points at relatively short intervals are spaced over the length of the cable.

No calculations exist so far on the times involved in temperature and pressure increase in the cooling circuits following the occurrence of a fault in the thermal insulation. Because of the local heat inleak occurring on a major leak, it is very likely that the superconductor will lose its superconductivity. The voltage signal received can be used to shut down the cable immediately. This is absolutely necessary as otherwise the heavy temperature rise due to the ohmic losses occurring — especially with helium or helium-impregnated tape employed as electrical insulation — will cause a reduction in the breakdown strength, making a breakdown likely.

The local temperature rise both from a breakdown in the conductor system and from a major vacuum leak can cause the pressure of the helium to rise to a point leading to local destruction of the cable. The result of this would be (in a cable with a nonsectionalized vacuum space) total collapse of the thermal insulation and the destruction of the major part of the remaining cable, as well as loss of the volume of helium in the cable.

To protect the cable against this kind of destruction, several additional measures are required during installation. The safety measures discussed here are of a purely hypothetical nature, since there is as yet no experience to draw upon in the installation and operation of a cryogenic facility of this size. As already mentioned, close to 100% reliability is required for the thermal insulation. To achieve this it is essential that:

The weld seams and their subsequent inspection are carried out with the greatest possible care;

The forces occurring during the cooling process are reduced to a minimum by appropriate design measures;

Damage to the external protective pipe is positively avoided by suitable laying of the cable, e.g., in a concrete trough.

The effect of faults which still may occur in the cable can be localized by sectionalizing the vacuum spaces at intervals of several hundred meters. This also has the benefit of facilitating regular vacuum monitoring, as well as localizing and repairing possible leaks. For small vacuum leaks, additional vacuum pumps used discretely can be employed to keep thermal losses within limits until the leak has been located and eliminated.

A pressure increase in the cooling circuits as a result of a defect in the cable may be limited by the use of safety valves set at earth potential (rupture diaphragms). The gases (He, N_2) are vented to atmosphere or, if justified by economic reasons as in the case of helium, retrieved by suitable devices (retrieval bladder, line to compressor). Using the external pipe as pressure vessel for the gaseous helium does not appear to be practicable owing to the high pressure involved (approx. 50 bar, depending on helium content). Since all of the helium coolant or part of it is at high voltage potential depending on the cable concept, it has to be reduced to earth potential at intervals of several hundred meters to allow the safety valves to reduce the pressure of the gas. This can be done in a similar manner as with the coolant feed-in points (Fig. 61b). Here it is also possible, in addition to the vacuum spaces, to sectionalize the coolant circuits by means of valves in the event of a defect, since all of the helium is at earth potential. This avoids further damage to the cable from the continued flow of coolants. The pressure rise in the coolant circuits and any helium losses occurring will be confined to the defective section.

In the adjoining unaffected parts of the cable the thermal insulation is maintained up to about 200 hours even if the coolant circuit is interrupted, due to the high heat of evaporation of liquid nitrogen. However, at a permissible pressure rise of approximately 20 bar in the helium circuit, only about 60 hours are available for draining the remaining helium from the cable via the adjacent refrigerating stations and for retrieving it in buffer vessels or preferable in road tankers.

As already stated, these proposals for protecting a superconductor cable against defects of the thermal insulation or coolant circuits are so far purely of a speculative nature. To what extent the methods discussed will be applied and whether they are economically justifiable depends exclusively on the nature and frequency of the faults occurring. Nothing definite can be said before extensive tests on long experimental cable sections have been carried out.

Irrespective of the foregoing, a capacity margin as with conventional cables has to be installed to obtain the full transmission capacity of a line, in order to cope with any failure of one of the superconductor cables of the line. An economically justifiable capacity margin can be achieved mainly where single conductor cables are used, and a fourth reserve cable is laid parallel to the three single conductor cables of a three-phase system, for example.

XIII. CHARACTERISTIC DATA OF VARIOUS CABLE TYPES

A. Data of ac Cables

As we have learned previously, cables with coaxial conductor pairs appear to have the most favorable properties of all cable configurations discussed. In this section we shall therefore confine ourselves to this kind of cable.

Superconducting coaxial cables represent typical radial field cables. The electrical processes take place mainly in the conductor and in the dielectric between the conductor and screen conductor. The fundamental relationships of such cables can be shown by means of a single cable with an earth return. The equivalent circuit shown in Fig. 71 can be used if the line length is only a fraction of the wavelength at 50 Hz ($\lambda = 6000 \, \text{km}/\sqrt{\varepsilon_r}$). U_A and U_E are the voltages at the beginning and end of the line, R' the resistance, $j\omega L'$ the series inductive impedance, G' the transverse conductance and $1/j\omega C'$ the shunt capacitive impedance between the conductor and screen. All values are referred to unit length. The following relationships exist between the voltages and currents at the beginning (U_A, I_A) and at the end of the line (U_E, I_E)

$$I_A = I_E \cosh(\gamma \ell) + \frac{U_E}{Z_W} \sinh(\gamma \ell), \text{ and}$$

$$U_A = U_E \cosh(\gamma \ell) + I_E Z_W \sinh(\gamma \ell), \tag{25}$$

where currents and voltages are vectors, Z_W is complex, and γ is the so called complex propagation constant. For this

$$\gamma = \alpha + i\beta = \sqrt{(R' + j\omega L') \cdot (G' + j\omega C')}, \tag{26}$$

where α is the attenuation constant and β the phase constant. These designations derive from the fact that the distribution of U and I along the line can be represented as a superposition of two attenuated propagating waves in opposition whose phase speed and attenuation are the same, and Z_W is the surge impedance (it is called the characteristic impedance of the transmission line). It is given by the expression

$$Z_W = \sqrt{R' + j\omega L'/G' + j\omega C'} . \tag{27}$$

The resistance R' of a superconducting ac cable is given by the hysteresis losses of the superconductor. As stated in Section V, it is,

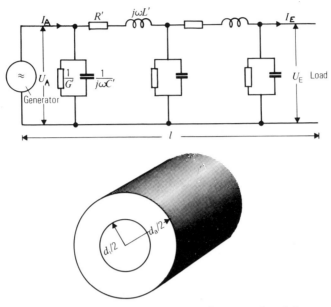

FIG. 71. Equivalent Circuit of Coaxial Cables.

for example, 1.2×10^{-10} $\Omega/$m for a 120 kV, 2500 MW ac cable where niobium is used as the superconductor, and 7×10^{-10} $\Omega/$m where Nb_3Sn is used. These resistances are about five orders of magnitude smaller than those of conventional cables. Compared with the inductive impedance, they can therefore be neglected. Even with conventional cables, the transverse conductance G' is two to three orders of magnitude smaller than the capacitive admittance. In a solid dielectric, the resistivity generally increases as the temperature decreases. In the case of superconductors with a solid dielectric, this ratio may therefore become even greater. The transverse conductance G' can thus also be neglected. Accordingly, with a superconducting cable, the following simplified relationships are obtained for the complex propagation constant and the surge impedance:

$$\gamma = j\omega\sqrt{L'C'} = j\beta; \quad \alpha \approx 0 ; \tag{28}$$

and

$$Z_w = \sqrt{L'/C'} . \tag{29}$$

The following applies for the inductance L' and the capacitance C' of superconducting coaxial conductor pairs:

$$L' = \frac{\mu_0 \mu_r}{2\pi} \ln a \quad \text{and}$$

$$C' = 2\pi \epsilon_0 \epsilon_r \cdot \frac{1}{\ln a} ,$$

where a is the ratio of the outer and inner diameters of the coaxial conductor pair, μ_0 the magnetic permeability of the vacuum, ϵ_0 the dielectric constant of free space, μ_r the relative permeability and ϵ_r the relative dielectric constant of the electrical insulation. Thus the following relations are obtained for the surge impedance and phase constant of the coaxial conductor pair:

$$Z_w = \left[\frac{\mu_0 \mu_r}{\epsilon_0 \epsilon_r}\right]^{1/2} \frac{\ln a}{2\pi} , \quad \text{and} \tag{31}$$

$$\beta = \omega \sqrt{\mu_0 \mu_r \cdot \epsilon_0 \epsilon_r} . \tag{32}$$

A further important characteristic of transmission lines, apart from Z_w and γ, is the surge impedance loading P_{nat}. This is given by

$$P_{nat} = \frac{U^2}{Z_w} = \frac{3U_0^2}{Z_w} , \tag{33}$$

where U is the line-to-line voltage and U_0 the line-to-neutral voltage.

In the case of transmission lines, it is generally important to know the relationship between the power transmitted and the surge impedance loading, since this relationship determines the behavior of the line (e.g., longitudinal voltage variation).

For the transmitted power of a cable we can write

$$P_N = 3n \cdot \xi \cdot \frac{U_0^2}{Z_w} , \tag{34}$$

where n is the number of three-phase systems used per cable, U_0 the line-to-neutral voltage and Z_w the surge impedance of the three-phase systems, and ξ the ratio of the transmitted power to the surge impedance loading of a three-phase system. ξ is given by

$$\xi = P_N/P_{nat}. \tag{35}$$

If this relationship is expressed by limit values of the super-conductor and dielectric, ξ is given by

$$\xi = \left[\frac{\mu_o \mu_r}{\epsilon_o \epsilon_r}\right]^{1/2} \frac{H_{im}}{E_{im}} = 3.75 \times 10^2 \frac{1}{(\epsilon_r)^{1/2}} \frac{H_{im}}{E_{im}}, \qquad (36)$$

where H_{im} is the amplitude of the current density of the superconductor and E_{im} the amplitude of the permissible electric field strength at the inner conductor. For ac cables $H_{im} = \sqrt{2}\ 560\ A/cm$ and $E_{im} \leq \sqrt{2}\ 100\ kV/cm$. Substituting 2 for the relative dielectric constant,

$$\xi \geq 1.5 . \qquad (37)$$

From Eq. (36) it follows that the ratio of the rated transmission power to the surge impedance loading depends only on the properties of the superconductor (H_{im}) and the dielectrics (E_{im}, ϵ_r). Eq. (37) shows that, taking account of the present state of knowledge of materials, it can be expected that the rated powers of superconducting ac cables will be larger than the surge impedance loadings, especially if the voltage test conditions for the electric insulation in Section VI D are to be conformed to. Such cables are suitable as carriers of constant load, since load fluctuations are linked with relatively large voltage variations on the load site.

In the case of pure active power transmission, the following relationship is obtained between the input voltage (U_A) and output voltage (U_E) [140]:

$$U_A = U_E \sqrt{\cos^2 \beta s + \xi^2 \sin^2 \beta s} , \qquad (38)$$

where s is the cable length, $\beta = \omega\sqrt{L'C'}$ in accordance with Eq. (28) and $\xi = P_N/P_{nat}$.

Figure 72 shows the dependence of the voltage variation ΔU on ξ, with $\beta \cdot s$ as parameter. As can be seen from Fig. 72, an inductive voltage drop must be reckoned with in the operation of superconducting cables, since $\xi > 1$. With cable lengths of up to about 100 km, this is of no importance if ξ is determined by the maximum values of the superconductor and the dielectrics. In the case of cable lengths over 200 km, however, the design of the cable will be influenced by the voltage drop. For very long cables it will be advisable to make the rated power equal to the surge impedance loading.

In constrast to superconducting cables, conventional cables are operated below their surge impedance loading. They therefore act as a capacitor and generate reactive power. To make possible long

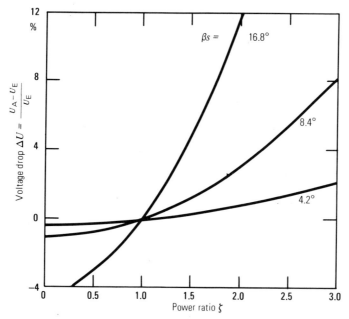

FIG. 72. Voltage Drop of AC Cables versus Ratio of Rated to Surge
Impedance Load.

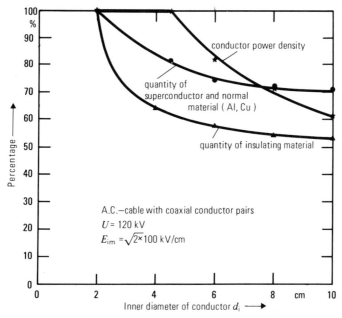

FIG. 73. Dependence of Power Density and Necessary Conductor Material
on Conductor Size.

transmission lengths (> 30 km), the cables must be provided with expensive external reactors which absorb the reactive power. In this respect, superconducting cables have clear advantages. The reactive power of superconducting cables with coaxial conductor pairs is given by

$$P_c = n \frac{3}{4} \pi \epsilon_o \epsilon_r \omega d_i^2 E_i^2 \ln a \ . \tag{39}$$

where E_i is the maximum electric field strength of the inner conductor, d_i the diameter of this conductor and n the number of three-phase systems per cable.

Attempts have frequently been made to optimize superconducting cables in respect of power density. For ac cables, the power that can be transmitted per cross-sectional area of a coaxial conductor pair is given by

$$P_N' = H_{im} E_{im} \frac{\ln a}{a^2} \ . \tag{40}$$

From Eq. (40) it can be seen that there is an optimum value for $a = \sqrt{e} = 1.65$. Taking the present maximum possible current density of $H_i = \sqrt{2} \ 560$ A/cm and a field strength $E_i = \sqrt{2} \ 100$ kV/cm, the maximum power density for a coaxial conductor pair is

$$P_N' \leq 20 \, \text{MW/cm}^2. \tag{41}$$

An interesting question for a given power and transmission voltage is that as to whether it is more favorable to employ several three-phase systems with coaxial conductors of smaller diameter or fewer systems with coaxial conductors of larger diameter. As can be seen from Fig. 73, for instance, coaxial conductors with an inner conductor diameter of 3 to 5 cm are more favorable electrically and economically than larger conductors. In the case of the small conductors, the ratio of the power that can be transmitted per conductor cross-section, referred to the expenditure for superconductor and insulating material, is more favorable than with large conductors. In all probability, however, larger conductor diameters will be necessary because of the cooling conditions required, especially in respect of minimum cool-down times. On the other hand, the diameters of flexible coaxial conductors are restricted by production limits and drum storage limits. Further advantages of several three-phase systems in a common helium tube would be higher packing density and better flexibility in operation. Systems can be switched off successively, especially on load reduction, and the problem of voltage rise at the cable end in no-load operation can be restricted to the last system in operation.

Table 6 shows the principal electrical and geometric data of a semiflexible (i.e., conductor flexible, thermal insulation rigid) 2500 MW niobium ac cable. The electric insulation consists of wound PE tapes impregnated with helium. For the sake of simplicity, ϵ_r is assumed to be 2 and tan δ 2×10^{-5}. In the light of the results obtained so far in the field of electric breakdown and the test conditions described in Section VI D, the maximum field strength amplitude in the dielectric was selected at only 78 kV/cm. As can be seen, the surge impedance is smaller by a factor of 2 to 3 and the phase constant greater by the same factor than with conventional cables. The rated power is higher than the surge impedance loading by a factor of more than 2.5. Note the low charging capacity of the cable and the very low electrical losses. The power density referred to the area of the complete cable is approximately $1.3 \, \text{MW/cm}^2$.

B. Data for dc Cables

As with ac cables, we shall also confine ourselves with dc cables to coaxial arrangements which, as stated in Section IV, ensure the best economic utilization of the superconducting material. The power of a coaxial dc cable with a voltage U_0 between the feed and return conductor and a current I is given simply by

$$P_N = U_0 I. \tag{42}$$

Introducing the current density of the inner superconductor H_i and the maximum electrical field strength at the inner conductor E_i, the power is given by

$$P_N = \frac{\pi}{2} E_i H_i \, d_a^2 \, \frac{\ln a}{a^2}, \tag{43}$$

where d_a and a again are given in XIII A.

Analogously to Section XIII A, the following then applies for the power that can be transmitted per cross-sectional area of a coaxial dc conductor pair:

$$P_N' = 2 E_{im} H_{im} \, \frac{\ln a}{a^2}, \tag{44}$$

where P becomes a maximum at $a = \sqrt{e} = 1.65$.

According to Carter [40], a current density of approximately 1500 A/cm is optimum for a dc coaxial cable. Assuming further, although this has not been proved definitely, that the dielectric strength for dc can be selected twice as high as for ac, we can, with dc cables,

TABLE 6. Tentative Characteristic Data of a Semiflexible 2500 MW
Niobium AC Cable

No. of cables per phase	1
Rated voltage, line-to-line	120 kV
Rated line current	12 kA
Linear current density (r.m.s.)	550 A/cm
Diameter of outer pipe	ca.0.5 m
Inner conductor o.d.	7 cm
Outer conductor i.d.	10 cm
No. of wires (inner conductor)	70
No. of wires (outer conductor)	105
Wire diameter	3 mm
Niobium layer thickness	30 μm
Cross-section Nb	3×49 mm^2
Cross-section Al	3×1200 mm^2
Insulation thickness	15 mm
Weight	ca. 140 000 kg/km
Peak stress in dielectric	78 kV/cm
Series inductive impedance C'	0.31 μF/km
Shunt capacitive impedance L'	71 μH/km
Surge impedance Z_W	15.1 Ω
Phase constant β per 100 km	8.4°
Surge impedance loading	950 MW
Charging power	1,4 M var/km
Losses a) cable	85 kW/km
Losses b) per terminal	50–100 kW

TABLE 7. Tentative Characteristic Data of a Semiflexible 10,000 MW
NbTi DC Cable

No of coaxial conductors	1
Rated voltage	230 kV
Rated line current	44 kA
Linear current density	1150 A/cm
Diameter of outer pipe	ca. 0.45m
Inner conductor o.d.	12.2 cm
Outer conductor i.d.	20 cm
No. of conductor strips	not defined
Nb-Ti layer thickness	
Inner conductor	130 μm
Outer conductor	100 μm
Total cross-section NbTi	115 mm^2
Cross-section Cu	1600 mm^2
Insulation thickness	3,9cm
Weight	ca. 125 000 kg/km
Peak stress in dielectric	76 kV/cm
Losses a) cable	55 kW/km
Losses b) per terminal	125–250 kW

expect a power density, referred to the coaxial conductor area, that is up to five times that with ac cables, i.e.,

$$P'_N \leq 100 \, \text{MW/cm}^2 \, . \tag{45}$$

With regard to the optimal conductor diameter, similar considerations apply as for ac cables. In addition to electrical and economic considerations, a decisive role is played by cryogenic problems.

Table 7 shows the principal electrical and geometric data of a semiflexible dc cable with a transmission capacity of 10,000 MW [40]. The hollow conductors are made of copper strips provided with a layer of NbTi. The conductors are fully stabilized for twice the operating current (= short-circuit current), which can likewise be carried by the superconductor. The operating temperature is between 4.2 and 5 K, the helium pressure approximately 4 bar. The helium flows in the inner conductor in the one direction and back again in the annular duct between the outer conductor and the helium tube (Fig. 23). The insulation between the inner conductor and the outer conductor takes the form of a wound tape of PE or nylon material ($\epsilon \approx 2$).

As shown by a comparison of Tables 6 and 7, it is expected that a coaxial dc cable will be capable of transmitting four times the power of an ac cable having the same outer diameter. The dc cable example shown is optimized in regard to superconductor expenditure. Furthermore the helium feed and return flow is separated only by the electrical insulation. As a result the dielectric is rather thick and its electrical loading is relatively low, i.e., it is not higher than the ac cable example. Optimization in respect of the utilization of the dielectric may result in a further increase in the power density of superconducting dc cables, a prerequisite naturally being higher dielectric strength.

At 34 W/km- MW for the ac cable and 5.5 W/km-MW for the dc cable, the specific cable losses have a ratio of 6:1, assuming the same type of thermal insulation for both cable designs. It should, however, be noted that the relationship indicated cannot be applied generally since here cables with different power ratings and transmission voltages were compared. With the same transmission voltage and the same overall cable diameter it is theoretically possible to obtain power density ratios of 5:1 and specific cable losses (W/km- MW) of 1:8 between Nb Ti dc cables and Nb ac cables.

The doubtlessly great advantages of dc cables are greatly restricted by the necessity of providing very expensive converter stations and more economical transmission than with ac cables will only be possible over long distances (see also Section III). In addition to restriction imposed by the converter stations, a number of other negative factors must be listed:

Suitable high power dc switchgear is not yet available; tapped lines are therefore not possible.

The power flow is not automatic and an extensive and costly control device is required.

Filters (i.e., smoothing reactors) and power factor correction capacitors are required.

In the event of a cable short circuit the large amount of energy stored in the smoothing reactors gives rise to the danger of serious damage unless counter measures are taken (see also Section IV).

Hence the distances at which dc cables become economically more attractive than ac cables still depend on fundamental questions (such as higher dielectric loading of the insulation with dc) and also on system problems. At all events they should lie above 100 km.

XIV. ECONOMICS OF SUPERCONDUCTING CABLES

Economics are the decisive criterion for the future use of superconducting cables, which would be competing with forced-cooled oil-filled cables and SF_6 pipe cables. Although the latter employ conventional engineering components, the cost data available on them from various sources are by no means identical. It is obvious that cost data regarding superconducting cables would involve much more in the way of uncertain factors. The reasons for this are obvious: the components required for the cable are still under development. The special materials for superconductor cables, such as superconductors and steel having low expansion coefficients, are produced in little quantities, meaning that their price is still comparatively high. Large refrigerating systems of the kind required for cables are still manufactured on a single-unit basis. Also, the maximum transmission voltage possible has not yet been fixed. By doubling this voltage, transmission capacity could be doubled as well, at a slight extra outlay for materials. A specific difficulty with superconductor cables exist where the likeliest transmission voltage levels vary widely, as is the case in the FRG where 400 kV is considered the next economical voltage after 120 kV.

Figure 74 shows cost estimates of various laboratories for superconducting dc and ac cables, in addition to the cost of forced-cooled oil-filled cables and cryoresistive cables, given for comparison [43, 141, 110]. The results of the estimates on the superconductor cables vary by as much as the factor 2. This is to be expected, since the estimates are based on different assumptions. Some of the prices assumed for the components and auxiliaries are based on the present cost situation, and some of the price estimates given assume mass production and correspondingly low prices.

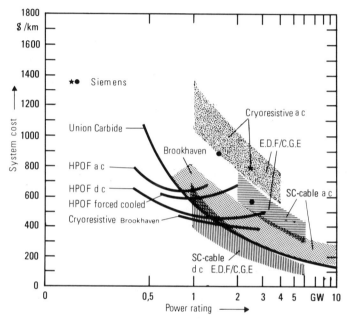

FIG. 74. Cable System Cost Breakdown.

The extrapolation of costs has definitely been taken to the extreme in the materials sector, where, for example, the prices of super-conductors are assumed to be a full order of magnitude below present prices. Another possibility of arriving at favorable cost forecasts for superconductor cables, especially with regard to high capacities, is to gradually increase the transmission voltage. Estimates assuming transmission voltages of up to 400 kV for ac cables and of 800 kV for dc cables have been published. On the basis of present know-how, there is no evidence to suggest that such transmission voltages will be at all possible.

The cost estimates made by Siemens follow present prices and apply a transmission voltage established by experiment. The estimates are close to the French figures. The discrepancies with the cryogenic resistive conductor cables are much greater than with superconductor cables. Here, again, the estimates of Siemens are in rather close agreement with the French figures. As opposed to oil-filled cables, the costs of superconductor cable systems decrease with increasing transmission capacities. There are good chances of superconducting ac cables for capacities above 2000 MW being more economical than all other transmission systems. This statement is also true, as will be seen later, with regard to SF_6 tubular conductors which are not included in Fig. 74. Fig. 74 also shows the cost estimates for super-conducting dc cables to be less by a factor of 4-5 than ac cables.

Table 8 shows the cost elements as determined by various laboratories for the major cable components. The considerable differences between some of the items can be easily explained by the way in which the costs of the components depend on the type of cable, (flexible, rigid) and the interrelation of component costs. For example, a fully flexible cable has greater thermal losses than a rigid one, meaning that the required refrigerating plant will therefore be larger and more expensive than with a rigid cable, where the cost of laying is likely to be less. The cost of superconductors increases sharply from niobium, through NbTi to Nb3Sn. If Nb3Sn conductors are used, the cost of the conductor will be relatively high, but operation at higher temperatures will permit the use of smaller refrigerating plant. Practically all cost estimates involving refrigerating equip-ment include a reserve plant.

So far, consideration has only been given to plant costs. If these are extended to include power savings from reduced transmission losses, the superconductor cable provides further advantages, as Table 9 shows. Compared with other novel types of cables, trans-mission losses are drastically reduced.

This aspect will be of increasing significance in the future, what with growing energy costs, caused by the rise in primary energy costs, and ecological requirements. Another breakthrough in the

TABLE 8. Superconducting Cable: Component Costs

	B.I.C.C.		Union Carbide		Siemens		E.D.F./ C.G.E.		CERL—Leatherhead			
Line voltage	AC 33kV		AC 138kV		AC 110kV		AC	DC	DC		DC	
Power rating	750 MW		1690 MW		2500 MW				1000 MW		10000 MW	
Spacing	10 km		8 km		10 km				10 km		10 km	
★	£	%	$	%	DM	%	%	%	£	%	£	%
Superconductor (Stabilised, +Insulation)	369	22,3	735	16,3	9500	27,2	15,3	17,6	148	14,8	520	23,9
Cryogenic Envelop	380	22,9	1300	28,8	5400	15,5	25,5	23,8	128	12,8	390	17,9
Terminals	—	—	200	4,4	1000	2,9	—	—	—	—	—	—
Refrigeration	533	32,2	955	21,2	9132	26,2	46,0	40,7	329	32,9	580	26,6
Factory assembly	160	9,6	1325	29,3	2000	5,8	13,2	17,9	395	39,5	690	31,6
Installation	215	13,0			7850	22,4						

★ x 1000

TABLE 9. Specific Installation Costs and Losses of New Underground AC High-Power Cables (Length 10 km)

Cable type	Cable data		Specific installation- costs	Specific losses
	kV	MVA	DM/MVA km	kW/km
Forced cooled oil—cable	400	1500	1740	340
SF$_6$—tube cable	400	2500	1680	270
Cryo resistive cable	400	2500	2080	350
Superconduc- ting cable	120	2500	1400	100

cost of superconducting cables would occur if in further investigations they were found to permit operation at higher voltage levels as
well (e.g., 400 kV).

XV. SUPERCONDUCTING CABLE PROJECTS

At the present time work is in progress on about ten superconducting cable projects throughout the world. In addition, a number
of laboratories or institutions are carrying out theoretical studies of
superconducting cables without having an experimental program.

In the U.S.A., for example, three laboratories are working on
the development of superconducting cables, the Brookhaven National
Laboratory, the Los Alamos Scientific Laboratory and Union Carbide.
Whereas the first two laboratories did not start their programs until
1971 and 1972 respectively, Union Carbide has been studying this
subject since 1963.

In England BICC (British Callenders Cables Ltd.) likewise
decided as far back as 1963 to build a laboratory-scale superconducting
model cable. In 1967 an alternating current of 2,080 A wa sent
through the three-meter-long test cable. Since about 1969 the program
started by BICC has been continued by the Central Electricity Research
Laboratories (CERL) at Leatherhead.

In France, work on superconducting cables has been in progress
for a relatively long time following the development of cryoresistive
cables. Concrete cable projects are being carried out in close cooperation with Compagnie Générale D'Électricité (CGE) and Air Liquid. The
Électricité de France is carrying out intensive studies of this subject
and at the Laboratoire Centrale des Industries Electrique (LCIE) measurements have been carried out on the dielectric strength of insulating
media at low temperatures.

In Japan intensive studies of superconducting cables have been
made since 1968. Concrete cable projects are being pursued by the
Furukawa Electric Company. Economic feasibility studies have been
made so far by the Electrical Laboratory, Ministry of International
Trade and Industry.

In Germany work on the development of superconducting cables
likewise started at a relatively early stage. Siemens commenced
work in this field at the end of 1968 and at about the same time a
working group was formed by AEG-Telefunken, Kabelmetal und Linde
for the joint development of superconducting cables. Development
work in this field has also been carried out since 1965 by the Institute
for Low Temperature Research (ATF) at Graz in Austria. Finally,
mention must be made of the Institute for Power Economy,

Krshishanowski, in Moscow, which is likewise developing superconducting cables. The state of development of the various cable projects is
shown in Table 10. The word "rigid" in this table means that both
the conductor system and the thermal insulation are rigid; "semiflexible" indicates a flexible conductor system and a rigid thermal
insulation, and "totally flexible" means that both the conductor system
and the thermal insulation are flexible.

In conclusion allow me to briefly mention the plans for further
development in this field insofar as they are known. The CERL in
England intends to check by means of further theoretical and experimental work to be completed by the end of 1975 whether the bulk transmission of electric power in and in the proximity of conurbations is
possible technically and economically. In France (CGE, Air Liquid,
LCIE) it is proposed to carry out further intensive studies of conductors
and dielectrics by the end of 1975. In Germany, the approved research
and development programs expire in 1973 and 1974. After that, it is
proposed in a second stage of the program to study problems of cable
installation, service monitoring and service reliability. It is proposed
to install 100 m long test lines parallel to existing conventional transmission lines. Here, the AEG-Kabelmetal-Linde Group intend to
change over from dc to ac cables. The plans have, however, not yet
been approved. The ATF in Graz is likewise planning to install a
long (50 m) experimental cable run parallel to an existing overhead
line.

In the U.S.A. the most detailed program for the further
development of superconducting cables is that of Union Carbide
which extends over the next ten years. In the plans approved for up
to 1975, the emphasis will be on conductor problems and the electrical
and thermal insulation with the aid of long model sections. By the
end of 1973, Brookhaven National Laboratory and Los Alamos Scientific
Laboratory propose to carry out tests on long experimental arrangements. BNL, for example, will use a two times 20 m long completely
flexible cryogenic envelope of corrugated tube design which will be
supplied by Kabelmetal in Germany. Little information is available
from Russia. It is known, however, that a long test section is to go
into operation in 1975. In Japan it appears that the development of
cryogenic cables is to be taken up in great style in the future. In
April 1973 the "Research Committee on New Power Transmission
Systems" was founded whose function is to work out the objectives
for a large national program. This program, which is to start in
1975, will extend over a period of nine years and have a budget of
ten thousand million yen. Approximately half this is to be spent
on the research and development of superconducting cables.

TABLE 10. Cable Projects in the World

	SIEMENS AG	AEG—Kabelmetal—Linde	ATF—Graz	BICC	CERL—Leatherhead
Company or Laboratory					
Kind of transmission	A.C.	D.C.	A.C.	A.C.	A.C.
Conceptual design	coaxial conductor pairs semi-flexible	parallel single conductors totally flexible	multiple three phase conductor system coaxial conductor pairs totally flexible	concentric tubes rigid	coaxial conductor pairs rigid, semi-flexible
First aimed voltage, kV	120	$\pm$ 200	$\geqslant$20	35	132 275
Current, kA	12	12.5	$\geqslant$5	13	
Power capacity, MW	2500	5000	$\geqslant$100	750	1400 4000
Thermal envelope	Superinsulation +LN$_2$ shield	Superinsulation +LN$_2$-shield	Superinsulation +LN$_2$-shield	Superinsulation +LN$_2$-shield	Superinsulation +LN$_2$-shield
Conductor S.C.-material	Nb-wire or ribbon (Nb$_3$Sn)	Nb$_3$Sn-ribbon	Pb-wire Nb-corrugated tube	Nb	Nb
Stabilizing material	Al Cu	Cu	Cu	Al	Cu,Al
Configuration	Helically wound hollow conductor	Helically wound hollow conductor	straight multiple wires corrugated tubes	rigid tubes	Helically wound hollow conductor
Operation temperature,K	4.4–5.6	ca.6–8	ca.4.4–5	4.4–5.5	4.4–5
Operation pressure,bar	$\geqslant$4	ca. 6	$\geqslant$4	$\geqslant$4	4–6
Electrical insulation	wrapped plastic foil	wrapped paper	wrapped plastic foil	Vacuum/Helium	Helium wrapped plastic foil
Terminal designs	under construction	under construction	under development	?	under development
Joint designs	under construction	constructed	under development	?	under development
Cable models	30m—model cable one phase 120kV 12kA in preparation	16m current tests 20m voltage test in preparation	50m totally flexible multiple conductors cryogenic test	3m—test cable 2080A 1967	8m-long one phase loss—measurements dielectric measurement in pr.
End of approved program	1974	1973	?	1969	1975

TABLE 10. (Continued)

Company or Laboratory	CGE—Air Liquide		Brookhaven Nat.Lab	Union Carbide	Los Alamos	Furukawa		Krshishanowski-i-Inst. Moskau
Kind of transmission	A.C.	D.C.	A.C.	A.C.	D.C.	A.C.	D.C.	A.C.
Conceptual design	coaxial pairs parallel conductors semi-flexible		coaxial conductor pairs semi-flexible	coaxial conductor pairs rigid	parallel and coaxial conductors semi-flexible	coaxial conductor pairs multiple conductors		?
First aimed voltage, kV	125	±100	132	138/230/345	100	154	33	35
Current, kA	14	20	13.8	7.1/11.8/17.75	50	3	33	10
Power capacity, MW	3000	4000	3000	1690/4710/10590	5000	1000	5000	600
Thermal envelope	Powder insulation +LN$_2$-shield		Superinsulation +LN$_2$-shield	Superinsulation He-gas cooled Cu-shield	Superinsulation +LN$_2$-shield	Superinsulation +LN$_2$-shield		?
Conductor S.C.-material	Nb	NbTi	Nb$_3$Sn-ribbon	Nb	not yet defined	Nb	NbTi	Nb
Stabilizing material	Al	Al	Al or Cu	Cu	Al or Cu	Cu		
Configuration	helically wd hollow conduct.	helically wound	Helically wound hollow conductor	Rigid tubes of Invar-Copper-Nb composite				
Operation temperature, K	4.4–5	4.4–5	6–8	4.2–5	?	4.2–5		?
Operation pressure, bar	≥4		6	≥4		≥4		?
Electrical insulation	wrapped foil insulation		wrapped plastic foil	supercritical He	wrapped plastic foil	wrapped plastic foil		?
Terminal designs	under development		under development	under development	?	?		?
Joint designs	under development		under development	under development	?	?		?
Cablemodels	18m full scale cryogenic envelope		2×20m long flexible cryogenic envelopes in preparation	7m - long test facility for a.c. measurements		7m - long test section coaxial pair current tests		12 - long one phase test model
End of approved program	1975		?	1975	?	see text		?

XVI. SUMMARY AND CONCLUSIONS

It has been shown that today the bulk of electrical power is transmitted by overhead lines and only a small percentage by underground
cables. One of the main reasons for this is the relatively lower costs of
overhead lines. In isolated cases, the transmission ratings per route
have reached the range of 1 to 2 GW. They can be raised to the order of
10,000 MW by increasing the transmission voltages to the range of
1000 kV and by employing several circuits per tower. Therefore, as the
transmission power ratings increase, overhead lines will continue to play
an important role in future transmission systems.

On the other hand, it is an irrefutable fact that in the future a
steadily increasing percentage of transmission systems will take the form
of underground cables, especially in urban areas where the given conditions favor this practice (e.g., right-of-way problems, environmental
considerations). Conventional high voltage cables — their power limit is
approximately 500 MW per system — will not be able to solve future power
transmission problems economically. The designs that will, in principle,
offer economical solutions are: forced-cooled oil-filled cables with an
estimated power limit of 4000 MW; SF_6 insulated tubular conductors, whose
power limits are even higher depending on the intensity of cooling employed;
and superconducting cables which have a practically unlimited transmission
capacity. Hence, intensive development work is in progress on these new
types of transmission systems at many establishments throughout the world.
Work on superconducting cables is justified by the fact that they will, in all
probability, on the basis of present knowledge represent the most
economical solution for power ratings of 2000 MW or more. Transmission
powers of this magnitude and per system will become topical in about 10
to 15 years.

At the present time, research and development work on superconducting cables is still mainly directed toward the individual components. The
concepts being studied at various establishments throughout the world
include rigid, semiflexible and fully flexible cable types. The latter two
must clearly be given preference. They can be wound on drums and can
therefore be transported and installed in long lengths (in the case of the
semiflexible type this applies for the cable cores). Moreover, they
permit a simpler solution of the problem of differing thermal contraction
of the cable components. The cable conductors primarily under consideration are hollow conductors of circular cross-section which offer good flow
conditions for helium, the medium used for cooling. In the majority of
three-phase ac cable designs each phase of the conductor is provided with
a coaxial screening conductor. In the case of dc cables, coaxial conductor
arrangements also afford advantages. With the rigid cable types, rigid
tubular conductors in a rigid thermal insulation are used, while, with the
semiflexible types, corrugated tubular conductors of helically wound tapes
or wires in a rigid thermal insulation are employed. In the fully flexible
types, the thermal insulation is also made flexible by means of corrugated
tubes.

For the conductors of dc cables preference is given to NbTi and Nb$_3$Sn at the present state of the art. However, attention is also being paid to other materials, especially those with high current densities at low magnetic flux densities and low costs. The main materials under consideration for three-phase ac superconducting cables are Nb and Nb$_3$Sn. Niobium is distinguished by particularly low losses at power frequencies. At an effective linear conductor current density of 560 A/cm and an operating temperature of 4.4 to 6K, these attain a maximum loss of 5 μW/cm^2. The losses of Nb$_3$Sn are appreciably higher, e.g., 50 μW/cm^2, but this material permits a higher operating temperature, which leads to an improved operating efficiency. However, new methods of manufacture are being studied with the aim of reducing the losses of Nb$_3$Sn by a factor of 2 to 3.

Like the superconductors for other applications, the cable conductors must be stabilized. Highly pure copper and aluminum are used as stabilizing materials. The optimal solution for cable conductors is considered to be cryostatic stabilization of the maximum short-circuit current. While this presents no problems with dc cable conductors, difficulties are encountered with three-phase ac conductors owing to the high short-circuit currents and the skin effect. Experiments and calculations carried out so far have not provided any clear conclusion as to the short-circuit performance of stabilized Nb conductors. The same applies for Nb$_3$Sn three-phase ac conductors for which efforts are being made to carry the short-circuit current in the superconductors by means of a large number of superconducting layers. A three-component conductor of Cu-NbTi-Nb has also been proposed where the short-circuit current would flow in the NbTi. These two cable conductors are by no means the most economical solution.

A very important cable component, which is not without its problems, is the electrical insulation. In all cable concepts undergoing development, the insulation is maintained at the temperature of the helium. For rigid cable types consideration is being given mainly to insulation by vacuum or by pressurized helium (supercritical He). For the electrode spacings usual in cables, a maximum dielectric strength of 200 kV/cm can be expected at 4K in the vacuum (including spacers) and of 100 to 150 kV/cm for helium (including spacers). For fully flexible or semiflexible types, preference is being given to insulation consisting of wound tapes.

In order to keep the dielectric losses within acceptable limits, it is necessary to use materials having a low dissipation factor (tan $\delta \approx 10^{-5}$ at 4 K) for three-phase ac cables, i.e., preferably nonpolar plastics, such as polyethylene. In the case of dc cables, on the other hand, a selection can be made purely on the basis of the maximum dielectric strength. Results obtained previously with materials used in conventional designs indicate that it will be possible to obtain dielectric strengths of 100 to 150 kV/cm for three-phase ac cables and of about

200 kV/cm for dc cables. Apart from electrical characteristics, an important role is also played by mechanical properties (such as flexibility, embrittlement and thermal contraction at low temperatures). Paper-like plastic tapes have better mechanical properties than compact tapes, but poorer electrical properties. Unsolved problems that will still have to be studied intensively include corona phenomena in tape insulation. At the present state of the art, it is expected that transmission voltages of 100 to 200 kV will be possible for three-phase ac cables and 200 to 400 kV for dc cables.

Thermal insulation of the helium-cooled conductor system to prevent the inflow of heat from the outside can be provided by methods similar to those employed for the insulation of cryogenic containers. The generation and maintenance of an insulating vacuum in long cable sections is best carried out by vacuum pumping stations located at points not too far apart (e.g., 1 km). Evacuating periods of a few days are required. However, proposals have also been made to construct the insulation from sealed sections evacuated during the manufacture. In addition to the super-insulation for reduction of radiation, most cable concepts incorporate a heat sink in the form of a liquid-nitrogen-cooled shield placed between the outer cable tube and the helium tube. Other designs only provide super-insulation in the vacuum space between the helium tube and the outer tube or several intermediate shields cooled by helium. If the thermal insulation is built up from rigid tubes, the insulating capacity is better than that of flexible insulation constructed from corrugated tubes. Where an N_2 intermediate circuit is provided, the thermal losses are about $0.1 \, W/m^2$ at $4 \, K$ and $2 \, W/m^2$ at $80 \, K$ for the first design, and $0.5 \, W/m^2$ at $4 \, K$ and $4 \, W/m^2$ at $80 \, K$ for the second design.

With stationary cooling of the cable, helium and nitrogen are used in single-phase form in order to avoid the well known hydrodynamic problems of two-phase flow. Subcooled nitrogen is the most suitable medium for nitrogen cooled circuits, and supercritical or pseudo-liquid helium are most suitable for the helium circuit. The only problem arising with a stationary helium cooling circuit is the possible occurrence of helium oscillations which should be avoidable. The required mass flow rates, and thus the pressure drops of the cooling media, are relatively small and thus permit the cooling stations to be arranged at large distances ($\geq 10 \, km$). The cooling down of long cable sections is more problematic. The cooling down periods may be of the order of a few weeks, and large quantities of helium must be made available for filling the cable. The filling operation should preferably be carried out with the aid of large helium road tankers. The same applies to the warming up of cables in the event of repair work becoming necessary. Because of extremely low losses, the cooling expenditure for superconducting cables is very small. The power input to the refrigerator for a 10 km section of a 120 kV, 2500 MW three-phase ac cable is, for example, less than 1 MW. The specific losses of dc cables are smaller by a factor of 3 to 6. Refrigerators of the type required for superconducting cables have already been built for other applications.

In previous studies, careful attention has been paid to cable joints and terminals. Suitable joints and terminals have already been designed and manufactured. The main problem of joints is the connection of the conductor and the electrical insulation. However, this can be solved satisfactorily. With terminals, the principal problems are minimizing of the losses of the normal conducting current leads and the heavy reduction of the dielectric strength of wound insulation where the electric fields are axial. The first problem can best be solved by providing helium gas-cooling for the normal conducting leads. The second problem can be overcome by employing methods such as are used for conventional high voltage bushings for cables and transformers (e.g., potential grading cones, potential dividers). The conditions are of course made more difficult by the large thermal contraction of the insulating materials and by the temperature gradients which may lead to high mechanical stresses. The inlets and outlets for the helium are likewise provided at the cable joints or terminals, the changeover from earth potential to high voltage potential and vice versa likewise being carried out with the aid of potential dividers.

The data available at the present time on cable laying, operating reliability and supervision can only be of a qualitative and speculative nature, since such points are invariably included in the later development programs of almost all cable projects. In regard to cable laying, it is, however, possible to say that the cable trenches of superconducting cables will be smaller than those of forced-cooled oil-filled cables or SF_6 tubular conductors owing to the higher power density. On the other hand, it would be advisable to lay superconducting cables in a concrete pipe to protect the thermal insulation, which is of course vitally important. The laying of fully flexible or semiflexible cables is a relatively simple matter. It will be possible to employ methods similar to those used for conventional oil-filled cables or forced-cooled oil-filled cables installed in pipes. The laying of rigid tubular cables is likely to be much more complicated, electrically and mechanically, owing to the large number of cable joints.

Faults in superconducting cable systems may in principle occur in any of the numerous auxiliary devices or in the cable itself. In view of this, consideration has been given to the duplicating of vital auxiliary components. Feasible cable faults include electrical breakdown and local collapse of the thermal insulation. By sectionalizing the longitudinal vacuum space, it is possible to confine vacuum failures and any cable defect resulting from these to a restricted area. In principle, it would also be possible to sectionalize the helium circuit so that only a small fraction of the helium filling of the cable would be lost in the event of a cable defect. It will not be possible to decide whether this solution is expedient or not until more experience has been gained on the frequency and type of faults occurring and more definite information is available on helium price trends.

With the aid of the measurements so far available for the various cable components, projections and comparisons were made for the main properties of superconducting coaxial three-phase ac cables and dc cables. In the case of three-phase ac cables, it was found that it is necessary to operate them at a power level appreciably above their normal load for optimal utilization of the superconductor, in contrast to thermally limited conventional cables which have to be operated at values well below their normal load. This offers the advantage that superconducting cables do not require expensive reactors for the compensation of reactive power where power is transmitted over long distances. On the other hand, the maximum possible transmission distances are limited to values between 100 and 200 km by an inductive voltage drop. Transmission over longer distances is only possible if the power to be transmitted is closer to the natural load, i.e. higher transmission voltages are available than has been assumed, or the superconductors are not utilized optimally.

On the assumption that a maximum linear effective superconductor current density of 560 A/cm and a maximum effective field strength of 100 kV/cm can be realized for a three-phase ac cable, the maximum power density obtained for a coaxial conductor pair is about 20 MW/cm^2, if the cooling ducts are also included. The power density obtained per coaxial conductor pair for dc cables is about 100 MW/cm^2 owing to the higher current-carrying capacity of the superconductors and the higher dielectric strength of the insulation. In addition to a higher power density, dc cables have much lower cable losses. Typical values are 34 W/km MW for a three-phase ac cable and 5.5 W/km MW for a dc cable. Estimates have shown, however, that despite these tremendous advantages, dc cables are unlikely to be able to compete over distances of less than 200 km owing to the high costs of the converter stations required.

The importance attached to the development of superconducting cables is demonstrated by the fact that no less than twelve cable projects are being carried out at ten different laboratories throughout the world. Cable studies that have no experimental program are excluded from this figure. The ratio of three-phase ac to dc cable projects is 2:1. The projects are generally in the stage of component development. Short model sections that permit a study of combining of components have been tested or are being prepared. In some cases these also serve for the testing of cable fittings and presuppose a certain level of development of cable manufacturing techniques. With such test sections, completion of the first section of the development programs is in most cases scheduled until the end of 1975.

Although component development work has so far produced quite good results, it cannot be looked upon as having been completed. Problems that must be studied even more intensively in the near future include the short-circuit current behavior in three-phase ac cable conductors and electrical insulation. For the first problem special importance is attached to the fixing of the performance limit of aluminum stabilized Nb conductors, the development of manufacturing methods for low-loss and favorably priced Nb_3Sn conductors and of three-component conductors, such as Cu-NbTi-Nb conductors. As already mentioned, the Nb_3Sn conductor is particularly interesting owing to its higher possible operating temperatures. In the case of the electrical insulation, studies of wound synthetic insulation must be continued, especially with regard to the occurrence of corona phenomena.

The next stages in the development of the cables will primarily be devoted to problems of cable installation, supervision of operation and operating reliability. Above and beyond the laboratory experiments, it will be necessary to install test sections of about 100 m in parallel to existing conventional transmission lines and to test them under normal power supply conditions. Not until the results of such experiments are available will it be possible to provide more definite data on the power capability, reliability and economic efficiency of superconducting cables and their auxiliary devices. The plans for cable development work extend over a period of about 10 years. It is expected that superconducting cables will be available for practical applications by about 1985.

ACKNOWLEDGEMENTS

This paper was only possible by the intense help of my colleagues P. Penczynski and F. Schmidt. P. Penczynski made very valuable contributions to Sections IV, V, IX, X, and XII and F. Schmidt to Sections II, III, VII and XIV. The cable program of Siemens AG Germany is supported at a 50% level by the Bundesministerium für Forschung and Technologie.

REFERENCES

1. A. Buch, Krausskopf-Verlag GmbH, Mainz, (1973).
2. A. Hofmann, ETZ A 92, 663 (1971).
3. G. Bogner, Fachausschuss "Low Temperatures" Deutsche Physik.
   Gesell., Münster, 19, Mar. 24, 1973.
4. E. Weghaupt, Part 1: Techn. Rundschau 20, 33 (1972); Part 2:
   Techn. Rundschau 24, 43 (1972).
5. D.P. Gregory, Sci. Am. 228, 13 (1973).
6. Underground Power Transmission Report to the Federal Power
   Commission by the Commisions' Advisory Committee on Underground Transmission, April 1966, by Arthur D. Little.
7. G.G. Sauve and R.H.P. Thom, Electrical World, Sept. 1,
   44 (1972).
8. A. Hofmann, ETZ A 91, 65 (1970).
9. Industrie Elektrik und Elektronik 18, No. 4, 86 (1973).
10. M. Erche, Energietechnik 99, (1971).
11. "Underground Power Transmission", a study of the Electric
    Research Council by Arthur D. Little, Inc. October 1971.
12. E.F. Pescke, Siemens Forschungs und Entwickl. Ber. 2, 46 (1973).
13. F. Winkler, ETZ A 92, 131 (1971).
14. J.D. Endacott, et al., Postgraduate course on high voltage power
    engineering, University of Manchester, 1969.
15. H.W. Graybill and J.A. Williams, IEEE Trans. PAS 89, 17 (1970).
16. P. Brückner, ETZ A 92, 733 (1971).
17. Report to Conférence Internationale des Grands Réseaux
    Electriques á Haute Tension (CIGRE) Stud. Com. 21, No. 11 (1971).
18. R. Jocteur and M. Osty, CIGRE Report Stud. Comm. 21, No. 7
    Part I (1972).
19. H. Gähler: Private communication, Siemens-Kabelwerk, Berlin.
20. A. v. Weiss, Energie-Ubertragung; ETZ A 88, 521 (1967).
21. H. Paul, Int. Elek. Rundschau, 24, 87 (1970); H. Paul, Electronics
    and Power 16, 87 (1970).
22. H. Meinke and K. Range, Nachrichtentechn. Z. 16, 161 (1964).
23. A.V. Pastukov and F.E. Ruccia, Conference on low temperatures
    and electric power, London, March 24, 1969.
24. S.H. Minnich and G.R. Fox, IEEE Winter Power Meeting, 1970.
25. H. Nagano, M. Fukasawa, S. Kuma and K. Sugiyama, Cryogenics
    13, 219 (1973).
26. K. Hosokawa, Report to CIGRE Study Committee 21, No. 6 (1972).
27. P. Graneau, Cryogenic Eng. News 4, 19 (1969).
28. B.C. Belanger and M.F. Jefferies, Cryogenics and Ind. Gases, 7,
    17, (1972).
29. M.J. Jefferies, S.M. Minnich and B.C. Belanger, IEEE Under-
    ground Transmission Conference, Pittsburgh, May 22, 1972, p. 77.
30. R. McFee, Power Engr. 65, 80 (1961).
31. R. McFee, Elec. Engr. 81, 122 (1962).

32. P. Grassmann, Techn. Rundschau 54, 9 (1962).
33. G. Bogner and W. Heinzel, Solid State Elec. 7, 93 (1964).
34. G. Bogner and W. Heinzel, Spring Meeting of the Institute of Metals in London, March 18, 1964.
35. W. Kafka, US Patent No. 3,292,016.
36. P.A. Klaudy, Advances in Cryogenic Engr., 11, New York (Plenum Press), 1966, p. 684, K.D. Timmerhaus (Ed.).
37. E. Bochenek et al. CIGRE Study Committee 31, Melbourne, March 1973.
38. P. Dubois et al., IEEE Pub. No. 72 CHO 682-5-TABSC, 1972, p.173.
39. G. Bogner and F. Schmidt, ETZ A 92, 740 (1971).
40. C.N. Carter, Cryogenics 13, 207 (1973).
41. H. Engelhard and E. Bochenek, Colloquium in Heidelberg, Oct. 21, 1971.
42. T. Horigome, N. Ito, S. Ihara and S. Sekine, CIGRE Study Committee 31 (1972), London, August 1972.
43. E.B. Forsyth, Brookhaven National Laboratory, March 1972, Contr. No. AT(30-1)-16 United States Atomic Energy Commission.
44. J. Nicol, A Study for the Electric Research Council by Arthur D. Little Inc., Case No. 73411, October 1971.
45. E. Scheffler, International Symposium Hochspannungstechnik, München 1972, p.579.
46. E.C. Rogers and D.R. Edwards, Electr. Rev. 181, 348 (1967).
47. W. Kafka, Elek. Zeit. A 90, 89 (1969).
48. R.W. Meyerhoff, Cryogenics 11, 91 (1971).
49. P.A. Klaudy, ETZ A 89, 325 (1968).
50. R.L. Garwin and J. Matisoo, Proc. IEEE 55, 538 (1967).
51. G.L. Guthrie, J. Appl. Phys. 42, 5719 (1971).
52. W.F. Gauster, D.C. Freeman and H.M. Long, Paper 56 II E (USA), World Power Conference (1964), p. 1954.
53. P.A. Klaudy, Titisee (1972), p. U1-U74.
54. D.R. Edwards, R.J. Slaughter, Electrical Times 3, 166 (1967).
55. E.F. Hammel, Los Alamos Scientific Laboratory, LASL (1972), p.94, proposal for dc sc power transmission lines.
56. P. Klaudy, Elektrotechn. und Maschinenbau 89, 93 (1971).
57. Y.B. Kim, C.F. Hempstead and A.R. Strnad, Phys. Rev. Letters 9, 306 (1962).
58. R. Hampshire, J. Sutton and M.T. Taylor, Conference on Low Temperatures and Electric Power, London 1969, p.69.
59. P.R. Aron, and G.W. Ahlgren, Adv. in Cryogenic Engr. 13, 21 (1967).
60. H.A. Ullmaier, Phys. Stat. Sol. 17, 631 (1966).
61. W.J. Dunn and P. Hlawiczka, Brit. J. Appl. Phys. Ser. 2, 1, 1469 (1968).
62. T.A. Buchhold, Cryogenics 3, 141 (1963).
63. H. London, Phys. Letters 6, 162 (1963).
64. C.P. Bean, Rev. Mod. Phys. 36, 31 (1964).
65. C.P. Bean and J.D. Livingston, Phys. Rev. Letters 12, 14 (1964).
66. D. St. James and P.G. de Gennes, Phys. Letters 7, 306 (1963).

67. H.J. Fink, Phys. Rev. Letters 14, 309 (1965).
68. G. Fournet and A. Mailfert, J. Phys. (Paris) 31, 357 (1970).
69. P. Penczynski, Siemens Forschungs und Entwicklungsberichte No.5 2, 296 (1973).
70. P.H. Melville, J. Phys. C 4, 2833 (1971).
71. R.A. French, Cryogenics 8, 301 (1968).
72. P. Penczynski, DPG-Frühjahrstagung Münster (1973).
73. R. Grigsby and R.J. Slaughter, J. Phys. D 3, 898 (1970).
74. T.A. Buchhold and R.L. Rhodenizer, IEEE Trans. Mag. MAG 5, 429 (1969).
75. J.C. Male, Cryogenics 10, 381 (1970).
76. R.W. Meyerhoff, IEEE Pub. No. 72 CHO 682-5-TABSC, 1972, p. 194.
77. Brookhaven National Laboratory, Power Transmission Project, Semiannual Report, Technical Note No. 11, March 28, 1973, p. 43.
78. R.G. Rhodes, E.C. Rogers and R.J.A. Seebold, Cryogenics 4, 206 (1964).
79. T.A. Buchhold and P.J. Molenda, Cryogenics 2, 344 (1962).
80. C.H. Meyer, D.P. Snowden and S.A. Sterling, Rev. Sci. Instr. 42, 1584 (1971).
81. M.T. Taylor, Conference on Low Temperatures and Electric Power, London (1969), p.61.
82. W. Kafka, Deutsche Patent No. 1, 250, 526.
83. E. Massar, ETZ A 89, 335 (1968).
84. G. Bogner, Proc. of the ICEC 3-Conference, Berlin 1970, p.35.
85. J.A. Baylis, IEEE Pub. No. 72 CHO 682-5-TABSC, 1972, p. 182.
86. M.J. Chant, Cryogenics 7, 351 (1967).
87. P.S. Vincett, Brit. J. Appl. Phys. (J. Phys. D.) Ser. 2, 2, 699 (1969).
88. L. Jedynak, J. Appl. Phys. 35, 1727 (1964).
89. A.H. Powell et al. Int. J. Electronics, 21, 393 (1966).
90. J.S.T. Looms et al., Brit. J. Appl. Phys. (J. Phys. D.) Ser. 2, 1, 377 (1968).
91. R.P. Little and S.T. Smith, IEEE Trans. on Electron Devices E.D.12, 77 (1965).
92. R. Rohrbach, CERN Report 64-50 (1964).
93. P. Graneau and J. Jeanmonod, IEEE Trans. on Electr. Insul. EI 6, 39 (1971).
94. R. Hawley, Vacuum, 18, 383 (1968).
95. O. Milton, IEEE Trans. on Electr. Insul. EI 7, 9 (1972).
96. J. Shannon et al., J. Vac. Sci. Tech. 2, 234 (1965).
97. D.A.Swift, Vacuum 18, 583 (1968).
98. J. Thoris et al., Cryogenics 10, 147 (1970).
99. B. Fallou et al., Cryogenics 10, 142 (1970).
100. J. Gerhold, Cryogenics 12, 370 (1972).
101. G. Matthäus and P. Massek, Siemens Forschungslaboratorium, Erlangen (1973).
102. B. Fallou, et al., Low Temperatures and Electric Power, London 1969, Proc. Conf. Int. Inst. Refrig., Comm. 1, Pergamon Press, New York 1970, p.377.

103. R.J. Meats, (to be published).
104. K.N. Mathes, IEEE Transactions on Electr. Insul. EI 2, 24 (1967).
105. J.A. Gardner, AIAA Journal 7, 1639 (1969).
106. R.N. Allan and E. Kuffel, Proc. IEE 115, 432 (1968).
107. G. Matthäus and P. Massek, Siemens Forschungslaboratorium, Erlangen (1973).
108. E.B. Forsyth et al., Cryogenic Engineering Conference 1973, Paper No. J-2 Atlanta, August 8, 1973 (to be published).
109. K.G. Müller, Vakuumtechnische Berechnungsgrundlagen Verlag Chemie, Weinheim 1961.
110. L. Deschamps, Y. Jegou and A.M. Schwab, Report on Electricity in France, April 1972.
111. P.E. Glaser, Machine Design 39, 146 (1967).
112. P.E. Glaser, J.A. Black, R.S. Lindstrom, F.E. Ruccia and A.E. Wechsel, NASA Sp-5027 (1967).
113. G. Bogner and F. Schmidt, Naturwissenschaften 57, 414 (1970).
114. H. Heumann, Mitteilungen der Kabelwerke der AEG-Telefunken-Gruppe 4/1972.
115. G. Bogner, Autumn School entitled, Herbstschule über Anwendung der Supraleitung in der Elektrotechnik und Hochenergiephysik, Titisee (1972), p. V1-V40.
116. H. Morihara et al., Cryogenic Engineering Conference 1973, Paper No. J-4 Atlanta, Aug. 8, 1973 (to be published).
117. R.V. Smith, Cryogenics, 9, 11 (1969).
118. V. Arp, et al., NBS-Report No. 10, p.703, July 1971.
119. K. Edney et al., Cryogenics 7, 355 (1967).
120. W.G. Steward et al., Developments in Mechanics 4, 1513 (1967), Johnson Publishing Co., New York.
121. K. Kellner et al., International Institute of Refrigeration 1970, p. 195 (Pergamon Press Ltd.).
122. D.N.H. Cairns et al., International Institute of Refrigeration 1970, p. 155 (Pergamon Press Ltd.).
123. C. Laverick, Cryogenics 11, 442 (1971).
124. G. Bogner, Cryogenic Engineering Conference 1973, Paper No. I-1, Atlanta, USA (to be published).
125. T.R. Strobridge, IEEE Transactions on Nuclear Science 16, 1104, (1969).
126. W.H. Hogan, Cryogenic Engineering Conference 1973, Paper No. A-2 Atlanta, USA (to be published).
127. E.C. Rogers, R.J. Slaughter and D.A. Swift, Proc. IEE, 118, 1493 (1971).
128. E. Bochenek, HGÜ-Colloquium in Heidelberg, October 1971.
129. R. McGee, Rev. Sci. Instr. 30, 98 (1959).
130. J.M. Lock, Cryogenics 9, 438 (1969).
131. R. Agsten, Cryogenics 13, 141 (1973).
132. J.W.L. Köhler, G. Prast and A.K. DeJonge, Proceedings of the Third International Cryogenic Engineering Conference, Berlin, May 25, 1970, p. 192.

133. M. Rauh, Dissertation 1971, Juris-Druck and Verlag, Zurich.

134. J.E.C. Williams, Cryogenics $\underline{3}$, 234 (1963).

135. F. Lange, Cryogenics $\underline{10}$, 398 (1970).

136. D. Güsewell and E.U. Haebel, Proceedings of the Third International Cryogenic Engineering Conference, Berlin, May 25, 1970, p. 187.

137. F. Schmidt and P. Massek, Deutsche Auslegeschrift 2,163,270.

138. F. Schmidt, G. Matthäus and P. Massek, Deutsche Auslegeschrift 2,164,706.

139. K.R. Efferson, Rev. Sci. Instr. 38, 1776 (1967).

140. P. Denzel, Grundlagen der Übertragung Elektrischer Energie, Springer Verlag, Berlin-Heidelberg-New York, 1966.

141. H.M. Long, W.T. Beall, L.K. Eigenbrod, R.W. Meyerhoff, and J. Notara, Edison Elec. Inst. Project RP 78-7 (1969).

Chapter 8

THE USE OF HYDROGEN AS AN ENERGY CARRIER

Cesare Marchetti

Head Materials Division

Euratom CCR - Ispra - Italy

I. INTRODUCTION

The logical motivation to find an energy carrier, or energy intermediate, able to satisfy the requirements of the energy market not covered
or coverable by electricity, has been that of extending the use of nuclear
reactors as primary energy sources. This may seem a premature pre-
occupation as nuclear plants are still striving to penetrate the fringes of
the electric market. However, one should keep in mind that energy
markets have very long relaxation times in respect to the introduction of
new fuels as shown in Fig. 1. Furthermore energy systems take very
long to develop technically; e.g., nuclear power plants took 15-20 years
to reach the prototype stage, and will take another 20-30 years to make
a dent in the electric market. So we have to start right now if we want to
be ready with a new system in the year 2000, the converging point of many
forecasts of doom, concerning particularly energy.

From a thermodynamical point of view, nuclear reactors are very
rudimentary machines, e.g., a water reactor simply throws away 50%
of the practically pure free energy of the fission by degrading it into
300 C heat. And reactors are bound to safety regulation and economies
of scale that make them improbable as small localized sources to warm a
kettle or run a car. The two inherent characteristics of nuclear reactors:
relative low grade heat produced and very large size, make the use of an
intermediate energy vector a necessity, particularly for that part of the
energy market not covered by electricity.

When we first examined this problem we did set the following conditions for this energy vector. It should be:

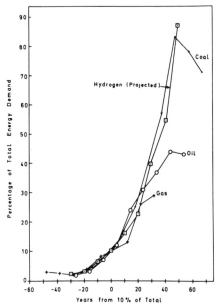

FIG. 1. Normalized Consumption Curve for Hydrogen Based on the Growth History for Gas, Oil and Coal.

1. easily transportable
2. storable
3. very flexible in use
4. pollution free

Hydrogen, produced by water decomposition, clearly satisfies all four requirements.

II. HYDROGEN AS AN ENERGY VECTOR

A. Hydrogen Transportation

Hydrogen pipelines already exist, Fig. 2, and they permit the transportation of hydrogen at a cost substantially analogous to that of natural gas, on a caloric basis [1]. The pipeline is an unobtrusive and very efficient system to transport and distribute energy. To give some "rule of thumb" figures: a 58" pipeline can carry between 15 and 20 GW depending on pressure and optimization conditions of the line, at a cost of 10-20% the current value of hydrogen per 1000 km, with an energy expenditure per 1000 km of 1-2% of the energy transported.

Hydrogen can also be transported as a liquid, in trucks, railway cars and barges. The technology developed to build large spherical containers used in the U.S. space program [2], is adequate for building

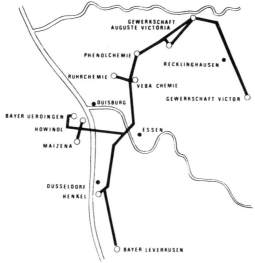

Transportation costs for hydrogen and natural gas

Energy Gcal/sec	Type of gas	Optimal distance between stations km	Capital costs 10⁴$/Gcal/sec		Transportation cost $/Gcal/ 1000km	H₂/CH₂	
			Pipelines	Pumping stations		Capital	Transp. cost
1	H₂	500	242	10	1. 23	1. 44	1. 36
	CH₄	150	166	10	0. 90		
2	H₂	500	190	10	0. 98	1. 43	1. 40
	CH₄	150	131	8	0. 70		
3	H₂	500	163	10	0. 85	1. 42	1. 39
	CH₄	150	114	8	0. 61		
4	H₂	500	145	10	0. 77	1. 39	1. 38
	CH₄	200	104	8	0. 56		

Note: Length of the pipeline: 100ₒkm - Max. pressure: 65 atm.

FIG. 2. H_2 Transport by Pipeline (Northern Germany)

large sea tankers analogous to LNG tankers. But liquefaction is thermodynamically and technically a costly operation and can be justified, at least
on a short term basis, only when the final use of hydrogen is in liquid form,
e.g., for fueling vehicles.

In the wake of the present revival of interest for airships one may add
that liquid hydrogen being so light, less than 1/10 the density of water,
floating liquid hydrogen in the air appears a logical proposition. It would
be enough for the airship to carry about 10% of the cargo in gaseous form
at room temperature to float the ship and the cargo. These airships
could be quite large affairs, 1000 m. length and 250 m. diameter can be
a reference size [3].

B.   Hydrogen Storage

Hydrogen is easily and cheaply stored at various level of stored
volumes and relaxation times. The pipeline is the obvious small size,

short relaxation time system. The storage is made by changing the pressure in the pipeline (but keeping it constant at the user level). The incremental costs of a pipeline vs. tube diameter is low, and if this diameter is chosen with a view toward an inevitable increase in traffic, the storage capacity is inexpensive. The relaxation time depends on the structure of the grid, but can be typically hours.

Aquifiers are the next best solution. They are currently used in France to store "bubbles" of gas injected in an aquifier protected by an impermeable roof [4]. If the roof is domed, then very large stable bubbles can be formed, with capacities and relaxation times of days. A variant is to leach a hole in a salt dome and use it as a gas bottle. In that case relaxation times can be much shorter [3].

Exhausted gas fields (or structures that may have contained them) have a wide range of capacity, e.g., the Groningen field could contain the equivalent of a couple years of energy in the form of hydrogen for Europe. The gas can be drawn typically at the rate of 1% per day, so this kind of storage is especially indicated for seasonal peak shaving.

LH2 storage is expensive and probably liquifaction and regasification will not pay for the reduction in volume even with improved technologies. If liquid hydrogen will be required as such, for surface and air transportation, then LNG technology is indicated.

III.   HYDROGEN USES

To check the flexibility of usage, we dissect the nonelectrical energy market into a dozen sectors, analyzing for eacn of them the potential use of hydrogen. A literature search led us to the discovery that in each sector, a study of the use of hydrogen was being considered.

As examples we describe some of the current and potential uses of hydrogen.

a. Transportation: H_2 cars to eliminate pollution [6],[7], LH_2 planes to improve performance [8], [9],

b. Household: Catalytic heaters and burners [10], Catalytic H_2 lighting [11] (To realize the all gas house as counterpart to the all electric house),

c. Ore Reduction: Steel making with H_2, the main driving force being pollution control and the reduced availability of proper coking coals [12],

d. Basic Chemistry: Here hydrogen is familiar and all sorts of syntheses can start from it. Ammonia is an obvious example, but $3 H_2 + CO_2 \rightarrow CH_3OH + H_2O$ opens the way to petrochemistry.

A detailed analysis shows that practically all the energy market not covered by electricity can be covered by hydrogen, with certain conditions about the price [13].

IV. THE HYDROGEN ENERGY SYSTEM

Let us try to see, with a rudimentary system analysis, what are the consequences of the properties of hydrogen on the structure of a system producing and distributing hydrogen just as electricity is produced and distributed. Its storability has a profound effect on the operation of the generating facilities: reserve capacity and peaking capacity are no longer necessary, and hydrogen stations can work like chemical plants, e.g., 8000 hrs/year and consequently be dimensioned on the mean yearly load. The importance of that point on capital investments can be better appreciated by observing that in an electrical system the installed power (generation and distribution) is about double of the mean power, i.e., on the average 50% of the equipment is laying idle. Inexpensive transport and flexibility in use, have a profound effect on the size of the generating units. We took the electrical system as a model and we found that the size of the generators did double every 6-7 years from Edison's time up to now. The interpretation of this extraordinarily constant growth rate during such a long period of time is very simple, and very illuminating about the mechanisms of interaction between technology and the market. With electricity consumption doubling every 9-10 years, the intensity of the distribution nets increases roughly at the same rate. On the other hand the improvements in electricity transport make the (linear) extension of a net double every 20-25 years. If we assume that the power of a new machine is a constant fraction, e.g. 10% of the power of the energy net at that moment, we find the 6-7 years doubling time for the size of the generators. Electricity covers about 10% of the energy market (but it consumes 25% of the primary energy). Thus, the market left for hydrogen is about an order of magnitude larger than for electricity, and the same is true for the distance at which hydrogen can be economically transported by pipelines. Consequently the mean energy handled by an hydrogen net can be three orders of magnitude larger than for an electrical net.

To speak of power stations in TW instead of GW may sound shocking for specialists in reactor technology, but they should not be surprised. After all the size of electric generators has increased by five orders of magnitude since Edison's time.

This large concentration in power has very important ecological and economical consequences:

1. The reactors can be located in a very few isolated sites, and much more money can be spent in improving their intrinsic safety.

2. The economics of scale can fully display their beneficial effects. With a mature technology three orders of magnitude in size may mean a factor of five in unit cost (e.g., assuming that Cost = (Size) 0.75, a potential liable to eliminate any competition. Incidentally Edison's generators were five orders of magnitude smaller than the present ones and did cost essentially the same per kW (in 1900 dollars!). Technological improvements and economies of scale have completely compensated for inflation and devaluation.

Furthermore, a hydrogen energy system can solve the chemical pollution problems linked to the use of energy. With classical fuels antipollution measures may become impossible if pollution has to stay at least constant with an exponential increase both in total energy consumption, and in its localization. The inevitable CO_2 production would also have climatic effects on a global scale.

V.   EFFECT OF A HYDROGEN ENERGY SYSTEM ON THE ELECTRICAL SYSTEM

In a nutshell, the large market, the easy transportability and storability make hydrogen production and distribution a practical operation on a continental scale. It may become worldwide if transportation of LH_2 in tankers appears viable technically and economically. This would mean that primary energy production could be concentrated on a dozen sites serving all the world, with very great benefits for reactor location and safety. The economies of scale would clear the market from competition and may provoke a revolution in the structure of the electrical power system. A model of the things to come has been sketched in a study made by TEMPO General Electric.

In this study [19] the following scheme (ECO-Energy) is proposed (Fig. 3):

1. Reactors will produce only hydrogen (and oxygen),
2. Hydrogen (and oxygen) will be transported as gas by pipelines,
3. Electricity will be generated in small units (e.g., 200 MW) located in the consuming areas,
4. Reserve and "mutual help" is provided through a chicken-net grid connection between the units.

These units are of a particular type taking the best advantage of the fact the combustion product of hydrogen and oxygen is water. A steam turbine is operated like a gas turbine in the high temperature region by gradually injecting hydrogen and oxygen into the steam. The blades face

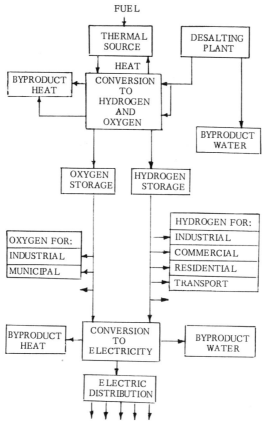

FIG. 3. "ECO-Energy" According to TEMPO - GE.

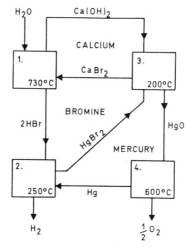

FIG. 4. Mark I - The First ISPRA Thermochemical Water Splitting Process.

only steam and with foreseeable progress in materials the maximum steam temperature can be 1400 C. These hybrid internal combustion gas-steam turbines may reach efficiencies of 60-65% which, coupled with a similar efficiency in the water splitting process, would compare favorably with the usual 40% efficiency of conversion from fission to electricity. The factor of five reduction in the cost of heat from a X1000 nuclear reactor, due to the economy of scale would make a local reactor almost certainly non-competitive.

How can electrical systems compete? Let's be optimistic and assume that in the future a high temperature superconductor may be discovered. Then, on a continental scale, electricity would compete with hydrogen, with a disadvantage of only a factor of 10 due to the size of the market. A world-level system would be perhaps too daring for electricity. Apart from the factor of ten, the problem of storage remains. This simple fact means doubling the generating capacity, i.e., capital per unit of energy sold. If we add to that the above factor of ten, defense against hydrogen in the boundary markets (e.g. house comfort) will be unreasonable unless super-conducting coils will be able to store, e.g. 10 KWh/liter, at a marginal cost by using inexpensive high temperature superconducting windings.

VI.   ISPRA WAY TO HYDROGEN

A certain number of assumptions underlie the presentation of hydrogen as the unique energy carrier. The most important one is that nuclear heat can be converted into hydrogen in some way. Obviously, one can make electricity and then electrolyze water. The technique exists, but unfortunately, the economics are awful, and it is easy to see why. Heat from the reactor has to be transformed into steam, then into mechanical energy, then electricity, then to hydrogen in the electrolyzer. Capital costs soar at each step and efficiency obviously goes down (to 25%). On the other hand, to reach another factor of 1000 in size is difficult for electromechanical systems.

At Ispra we have started a more down to earth route to the single step transformation: The thermochemical water splitting [14] [15] in Figure 4 shows an example of a possible process. The temperature at which a reactor can release its heat is insufficient to thermally disassociate a water molecule (2000-3000°C would be necessary). The decomposition, therefore, is performed in two, three or more steps, each of them thermally operated. The example in Fig. 4 is the first process we discovered, in 1969, so it was christened Mark-1. In the meantime many other processes were found. About 20 more at Ispra, a dozen elsewhere, essentially in Germany and the USA, where the fad of inventing water splitting process has developed very rapidly.

Theoretical efficiencies can be easily calculated remembering three characteristic numbers for hydrogen: 68.3 kcal/mole high heating value,

n°	Mark		Elements	max T (°C)	Number of Reactions
1	Mark	1	Hg,Ca,Br	730–780	4
2	"	1B	Hg,Ca,Br	730–780	5
3	"	1C	Cu,Ca,Br	900	4
4	"	1S	Hg,Sr,Br	800	3
5	"	2	Mn,Na(K)	800	3
6	"	2C	Mn,Na(K),C	850	4
7	"	3	V,Cl,O	800	4
8	"	4	Fe,Cl,S	800	4
9	"	5	Hg,Ca,Br,C	900	5
10	"	6	Cr,Cl,Fe(V)	800	4
11	"	6C	Cr,Cl,Fe,(V),Cu	800	5
12	"	7	Fe,Cl	800	5
13	"	7A	Fe,Cl	1000	5
14	"	7B	Fe,Cl	1000	5
15	"	8	Mn,Cl	900	3
16	"	9	Fe,Cl	650	3

FIG. 5. Summary Table of the Chemical Cycles to Produce Hydrogen.

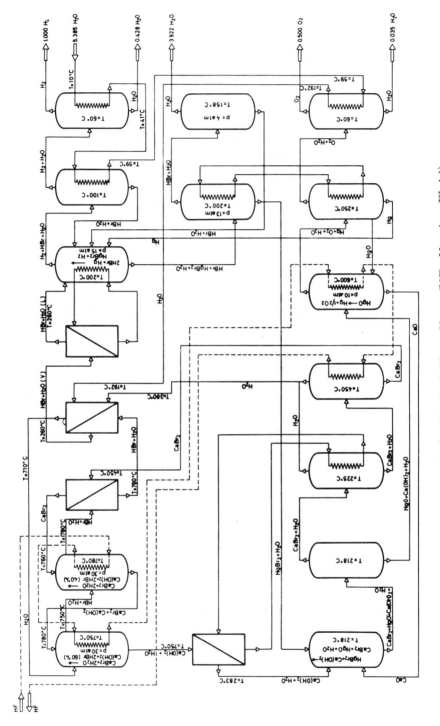

FIG. 6. Mark-1 Flow-Sheet (HF, Heating Fluid)

58 kcal/mole low heating value, 56.7 kcal/mole free energy. The maximum amount of hydrogen that can be produced can be calculated using Carnot efficiency and free energy [16]. But hydrogen is a fuel, and efficiency can also be defined as the ratio between the heat produced in burning hydrogen and the heat used to produce it. So Carnot efficiency has to be multiplied by $58/56.7 = 1.02$ or $68.3/56.7 = 1.2$ depending on what heating value we choose, and that depends in what way we burn the hydrogen. However, there is no ambiguity in calculating the waste heat from the plant: one has to use the high heating value, so waste heat will be proportional to $1-1.2\mu_c$ (μ_c = Carnot efficiency). It is interesting to note that with a source temperature at about 1400°C the theoretical waste heat is zero.

A number of processes found in Ispra are reported in Fig. 5. To get a more precise feeling about the working of those processes we shall describe Mark-1 more in detail.

The reactions are the following:

1. $CaBr_2+2H_2O \xrightarrow{\text{730 C}} Ca(OH)_2+2HBr$ (water splitting),

2. $Hg+2HBr \xrightarrow{\text{250 C}} HgBr_2+H_2$ (hydrogen switch),

3. $HgBr_2+Ca(OH)_2 \xrightarrow{\text{200 C}} CaBr_2+H_gO+H_2O$ (oxygen shift),

4. $HgO \xrightarrow{\text{600 C}} Hg+(1/2)O_2$ (oxygen switch).

The sum is $H_2O \longrightarrow H_2+(1/2)O_2$.

In the first reaction, the water molecule is split in two, the two halves becoming attached to Ca and Br. In the second reaction hydrogen is separated from Br by reacting the HBr formed in the first reaction with Hg. In the third reaction oxygen is transferred from Ca to Hg, and Br viceversa, so that the original chemical $CaBr_2$ is reconstituted. In the fourth reaction O_2 is separated from Hg by thermal decomposition, and Hg is reconstituted closing the cycle. The final net result is the decomposition of a water molecule.

Figure 6 gives the flowsheet of a plant and Fig. 7 shows how heat is drawn from the hot helium produced by a HTG reactor.

Two questions come to mind: What is the efficiency one can expect from a real process? What is the cost of this hydrogen? Both questions are very difficult to answer precisely at this stage because they require at least a pilot plant, but some intermediate approximation can be illuminating.

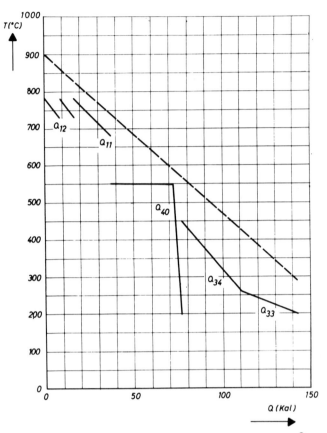

FIG. 7. Example of Process Thermal Coupling Using 900° C Helium Gas
as Primary Heat Carrier.

Assumptions	
Reactor investment/kWth	$50
Fuel cycle cost/kWth h	0. 5mils
Market price for H_2	15mils/m³
″ O_2	12mils/m³
″ D_2O	$60/kg
Operation	8, 200hr/year
Efficiency	50%
Gross product value, about	$40
Running costs	
Fuel	$ 4
Operation	$ 2
Capital charges	
Reactor (int. & depr. 16%/year)	$ 8
Chemical plant (int. & depr. 25%/year)	$26
Breakeven investment for the chemical plant	100$KWth

Note: The site is chosen in order to sell the by-products at
current prices.

FIG. 8. Orientative Economic Analysis for the First Water Splitting Plant.

For the efficiencies we have prepared computer models of a plant, obviously very simplified, but containing realistic values for the key parameters, e.g., the Δt in the heat exchangers. Such models give for Mark-1 practical efficiencies in the range of 50-55% with the relatively crude flow-sheets we have indicated.

To estimate the costs, we worked backward, assuming a breakeven cost for the hydrogen produced, i.e, a production cost equal to the market price, and we calculated the investment, i.e., the amount of money we can spend for the plant. The very encouraging result is that this chemical plant can cost between one and two times the cost of the associated reactor (Fig. 8). Comparing the construction of a reactor with the relative straightforwardness of a chemical plant, the chances of reaching a competitive cost appear high.

At Ispra, we are limited in our research to the exploration of new cycles, laboratory work and bibliographies to collect all the data necessary for the realization of a pilot plant. The funding, from the origin of the project to the end of 1976 amounts to about \$10M.

Other laboratories have recently joined the research effort. In particular the University of Aachen, the Julich Center, the Institute of Gas Technology (Chicago), General Electric Center in Schenectady, Oak Ridge, Los Alamos and Argonne Centers, have considered cycles, each of them having at least one process different from ours. At the University of Aachen in particular, Prof. Knoche has developed a computer program to "invent" processes by selecting thermodynamically self-consistent systems out of systematic combinations of chemical elements.

Apart from water splitting using nuclear heat, other processes to decompose water should be mentioned. A quite original concept consists in using the plasma of a fusion reactor, doped with Al to produce u.v. light of the proper wavelength to decompose steam photochemically [17]. Others have tried to imitate the working of chlorophyll where two photons are used in a cascade to decompose one water molecule. A very interesting solution to this problem, avoiding the intricacies of organic or bio-chemistry, has been outlined by A. Fujishima and K. Honda [18]. They use a semiconductor, TiO_2, to absorb the two photons and make the energy available for water decomposition. The sun would be the obvious source of light. Another method proposed is to concentrate solar light to obtain temperatures of 3000 - 4000K, and crack water by brute force. Actually hydrogen is the perfect medium to store sunlight and to ship it long distance between large scale sun stations and consumer points. Hydrogen is probably the only solution if the power sources are floated on the ocean.

Thus, if we are able to solve the many technological problems, the world is ready for a revolution in the structure of the energy system.

REFERENCES

1. G. Beghi et al., Transport of Natural Gas and Hydrogen in Pipelines, Ispra 1550, May 1972.
2. J.R. Bartlit et al. in Proc. of the Intersociety Energy Conver. Eng. Conf. 1972, p. 1312.
3. N.P. Biederman, Pipeline & Gas Journal, 197, 62 (1970).
4. J. Colonna, Annales des Mines, 10, 7 (1969).
5. Anon, The Oil and Gas Journal, 69, 67 (1971).
6. R.J. Schoeppel et al., Proc. of the Intersociety Energy Conver. Eng. Conf. (1972), p. 1375.
7. K.H. Weil, Proc. of the Intersociety Energy Conver. Eng. Conf. 1972, p. 1355.
8. F.E. Jarlett, Aviation and Space Conf. (ASME), June 16, 1968.
9. G.D. Brewer, Lockheed Laboratory Report, May 16, 1973.
10. E.R. Kweller, R.B. Rosenberg, Gas Industries Natural Gas, Edition 15, 11 (1971).
11. J.M. Reid et al., Luminescent gas lamp U.S. Pat. 3.582.252, June 1, 1971.
12. R. Wild, Chemical and Process Eng. Vol. 50, No. 2, 55 (1969).
13. C. Marchetti, Chemical Economy and Engineering Rev., 5, 7 (1973).
14. Proceedings Round Table on Direct Hydrogen Production, Ispra 12, Dec. 1969 CCR EURATOM, Ispra, EUR/C-IS/1062/1/69 e.
15. Hydrogen Production from Water Using Nuclear Heat: Progress Report No. 1 ending December 1970, Report EUR 4776 e. Progress Report No. 2 ending December 1971, EUR 4955 e, Progress Report No. 3 ending December 1972, EUR 5059e.
16. J.E. Funk, R.M. Reinstrom, I & EC Process Design Develop. 5, 336 (1966).
17. B. Eastlund, W.C. Gough, Generation of Hydrogen by U.V. High Produced by the Fusion Torch - 163 Nat. Meeting Am. Chem. Soc., Boston, 1972.
18. A. Fujihima and K. Honda, Nature, 238, 37 (1972).
19. W. Hausz et al. ECO-Energy, Proc. of the Intersociety Energy Conver. Eng. Conf. (1972), p. 1316.
20. J.E. Johnson, "The Economics of Liquid Hydrogen Supply for Air Transportation," Cryogenic Eng. Conf., Atlanta, Aug. 10, 1973.

Chapter 9

TUNNEL JUNCTIONS FOR COMPUTER APPLICATIONS

J. Matisoo

IBM Thomas J. Watson Research Center

Yorktown Heights, New York 10598

I. INTRODUCTION

The main focus of this NATO Advanced Study Institute is on large scale applications of superconductivity. Frequently, as here, this means applications involving high power levels and large size. In computer applications of superconductivity, just the opposite is true. One is interested in low power and compact structures. Large scale enters only from the point of view of the number of units.

This paper describes the use of Josephson junctions as computer devices and indicates what motivates this work. Since the idea of superconducting devices is not new, we first review the background and then go into technical matters. We present a discussion of how the devices function, how they are made, what is known about them so far, and why they continue to look extremely interesting.

For the sake of those not acquainted with computers, Fig. 1 indicates the large functional blocks. A machine consists of a central processing unit, or CPU, which controls the operations which are performed and does the arithmetic. Control operations are things like instructions to get a given piece of information, and so on. The arithmetic operations consist of the usual simple ones such as addition, subtraction, multiplication, and so on, and include logical operations as well. Information, both programs and data, are stored in a box labeled "memory and storage". The distinction between memory and storage is somewhat arbitrary. Storage usually means a box which can contain a great deal of information but to which one does not have particularly rapid access. These large units of a computer are made up of relatively simple circuits and devices. Figure 2 lists a number of devices which are in common use. For

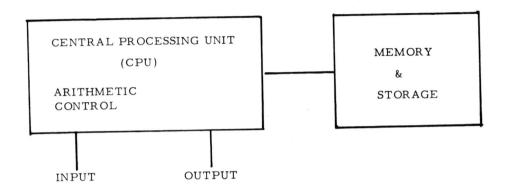

FIG. 1. Block Diagram of Computer System.

ELEMENTARY BUILDING BLOCKS

ARITHMETIC & CONTROL

(CIRCUITS FOR AND, OR, INVERT
FUNCTIONS)

Vacuum Tubes

Transistors

MEMORY & STORAGE

(STORE BINARY INFORMATION)

Magnetic Cores

Transistors

Tapes

Disks

FIG. 2. Table of Common Elementary Building Blocks.

arithmetic and control, transistors are used. The functions which are performed by the transistor circuits on the simplest level are AND, OR, and INVERT functions. The function performed by memory and storage devices is, of course, to store binary information. Magnetic cores, transistors, tapes and disks are commonly used. In the magnetic devices, storage of information is achieved by orienting the magnetization in a specific direction for a zero and in another for a one. In transistor circuits, usually some voltage level defines a zero and another voltage level defines a one. Logical operations can then be performed by having a device or circuit switch from one level to another when the appropriate logical conditions are met. An ideal device is one which switches between states very rapidly and is very small so that many can be placed close to one another. These two properties enable the computer to process data rapidly. A very important factor in placing devices close to one another is the amount of power the devices dissipate. This power shows up as heat and must be removed, lest the temperature of the devices gets to be too high. To assure adequate heat removal, the device spacing may have to be larger than the device size itself. The larger the spacing between devices, the longer the signal propagation time, and the slower the data processing rate. Apart from these simple considerations, many other engineering aspects, of course, must be taken into account when weighing the relative merits of different computer devices.

Figure 3 illustrates the sort of speed power values for present day semiconductor technology as compiled by Keyes[1], and the corresponding value for Josephson tunneling devices. The speed power advantage for these devices is such that effort in solving the considerable problems involved in the fabrication and use of such devices is justified.

The first superconducting switch was a piece of superconducting wire switched between the superconducting and normal states [2]. The form adopted by Buck is shown in Fig. 4 [3]. Here a piece of superconducting wire is driven normal by means of a magnetic field generated in the control winding. The resistance of the "gate" wire is zero if there is no control current present, and it has its normal state value when the full control current is present. The speed of operation of such a device is quite slow, on the order of milliseconds. This is primarily due to the large size of the device. Note, however, that the power dissipation is small, being zero when the gate wire is superconducting.

An improved version of the cryotron is illustrated in Fig. 5. It was realized that by reducing the device dimensions, by utilizing thin film techniques, improved performance would be obtained. The thin film cryotron contained a thin film strip of tin or indium which could be driven normal by means of a magnetic field. The magnetic field is obtained by passing current through an insulated control line. Initially, such devices were made by sequential vacuum deposition through metal masks within the same pumpdown cycle [4]. This technique limits linewidths to 5 to 10 mils. Photoprocessing techniques were introduced leading to much narrower lines [5] .

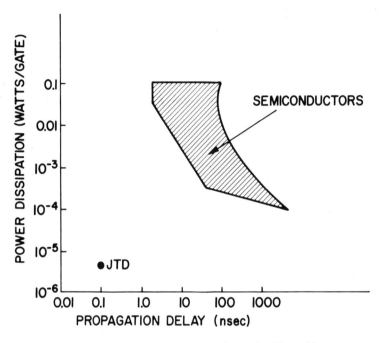

FIG. 3. Speed–Power Range of Logic Circuits.

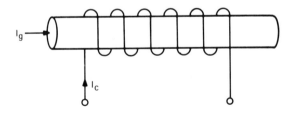

FIG. 4. Wire Wound Cryotron.

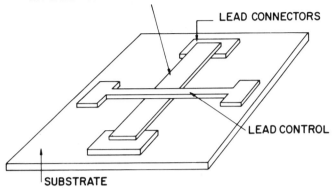

FIG. 5. Thin Film Cryotron.

Considerable amount of effort was devoted to the development of superconducting cryotron circuitry. By mid-1960's, however, these devices had lost the competition to room temperature transistor devices. There were two basic reasons for this. The makers of thin film cryotron circuits introduced integrated circuit concepts. In this pioneering effort, they were ahead of their time and encountered a sizable number of problems. The semiconductor industry has today solved many, but not all, of these problems. The second, and perhaps more important reason, is that the device itself did not turn out to be competitive in performance with improved versions of the transistor.

The discovery of the Josephson effect and progress in integrated circuit processing techniques which are independent of specific device technology have changed this picture. In the following paragraphs, we describe some principles and results of our work in the area of Josephson tunneling devices for computer applications.

II. BASIC PRINCIPLES

Figure 6 illustrates some basic principles of superconducting circuits; namely, current steering and persistent currents. Consider a superconducting loop as illustrated. Initial application of current I_0 leads to an equal division in the two branches (this is true because the inductances of the two branches are equal). If we have a switch in the lefthand branch which can be opened, the total current will divert to the right hand branch.

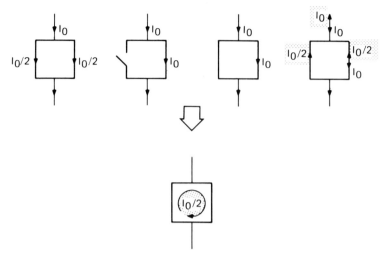

FIG. 6. Diagram Indicating Basic Current Steering and Persistent Current
         Ideas.

The switch may now be closed and the current will remain in the righthand
branch. This is what we mean by current steering. It constitutes an
elementary logic operation. To obtain persistent current, one further
step is added; the incoming current I_0 is terminated. It is easy to see that
this leads to a persistent current $I_0/2$ in magnitude circulating in
this case clockwise. As long as the loop remains wholly superconducting,
this current will persist. It is obvious that the circulating current could
have been stored in a counter-clockwise direction. These two circulating
directions can then constitute two states of a memory cell.

    In principle, any superconducting device can be connected as the
active element in the loop. However, from the point of view of the current
transfer times, it matters a great deal which element is used. Use of
Josephson tunneling devices leads to extremely rapid current transfer
times.

    The experimental conditions which must be met in order to have the
Josephson effects observable in tunneling structures are illustrated in
Fig. 7 [6]. This shows a simple circuit for measuring the I-V characteristics of two pieces of lead separated in space by a distance D. Under conditions 2 current flows because electrons are tunneling from one metal to
the other. The I-V characteristics are linear for voltage drops of interest,
a few millivolts. The current for a given voltage depends on the potential
barrier; in particular, the current level is set by the spacing D. As discovered by Giaever, the I-V characteristic becomes sharply non-linear
when the metals become superconducting [7]. Little current flows for

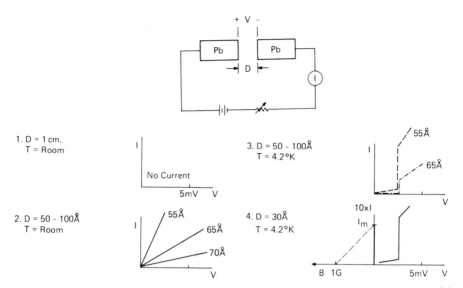

FIG. 7. Experimental Conditions for Josephson Effects to be Observable.

voltages below the energy gap; current increases sharply at the gap voltage
and approaches the normal state character asymptotically. This I-V
characteristic reflects the density of states in a superconductor, and the fact
that the spacing D is sufficiently large so that the superconducting correla-
tions across the space are washed out. When they are not then as theorized
by Josephson and experimentally verified by Rowell and Anderson [8], a
zero voltage current of limited magnitude (I_m) appears in the I-V characteristic. Furthermore, the magnitude of this zero voltage current
is strongly magnetic field dependent. Given a structure in which current
can be carried either at zero voltage, or a voltage corresponding to the
energy gap (2.5 millivolts for lead at 4 K), and a threshold which is controllable, an interesting switching device may be possible.

III. DEVICES AND TECHNOLOGY

    The possibility which a Josephson tunnel junction offers is that of a
very fast switch. This was already true in the very first device investigations made by Matisoo [9] and is even more true today. As with all
superconducting devices, power dissipation is low. These properties
motivate the development of a technology for fabrication of such devices
and circuits. The basic difficulty lies in the fact that the tunnel barrier
must be very thin, on the order of 30 Å or less. The device characteristics should also be stable with respect to thermal cycling and storage. The
development of such a technology satisfactory for fabrication of experimental Josephson tunneling device circuits has been described by Greiner

et al[10]. Figure 8 illustrates their basic gate configuration. It consists
of a base electrode M_2, which is separated from the counterelectrode M_3,
by a native oxide of M_2 grown to the desired thickness by an RF
oxidation technique [11]. Above, but insulated from, the junction thus
formed is a control line M_4. The entire structure rests on an insulated
superconducting ground plane. The ground plane is niobium; the ground
plane insulation is Nb_2O_5 formed by liquid anodization. The vacuum deposited
lead alloy films are used as junction electrodes, controls and interconnection lines. Patterning of layers is accomplished by means of a photore-
sist lift-off technique. In addition to the gate element itself, other components of the technology are insulated crossings, superconducting contacts, and, for the purposes of a logic circuit, terminating resistors. All
of these are described in more detail by Greiner et al. Figure 9 is a
photograph of a gate (the configuration is the same as that of Fig. 8). The
linewidths are on the order of mils.

Actual I-V characteristics for a tunnel junction are illustrated in
Fig. 10. This figure shows the major device parameters I_{max}, the gap
voltage, the resistance below the gap R_j. Also identified is the resistance of the junction when both electrodes are normal. Finally, it indicates
the current value, I_{min}, at which the junction returns to the zero voltage
state. That is, the current voltage characteristic for typical junctions
is hysteretic. I_{min} is a dynamic property of the junction [12, 13]. As a
result, the value of I_{min} depends upon the environment in which the
junction finds itself. This question has been discussed at length by
Zappe [14].

The relevance of these parameters becomes clear when one examines
the circuit model for the junction (Fig. 11) [15, 16]. The tunnel junction
clearly has a capacitance which, because of the thinness of the oxide is
quite large; it has a resistive current voltage characteristic, as illustrated
in Fig. 10. Finally, there is the supercurrent term. In cases in which the
spatial variations are of no interest, the lumped model can be used successfully. However, generally spatial variations are of interest. In that case,
the junction can be viewed as a distributed parallel plate transmission
line which has a readily calculable inductance per unit length, capacitance
per unit length, and a resistance per unit length. The supercurrent in
this case is also per unit length. We have implicitly assumed a one dimensional configuration. All of the circuit elements, with the exception of the
supercurrent term, are familiar and their terminal behavior can be readily
characterized. In 1962 Josephson provided us with a prescription for the
behavior of this element [17]; namely, what the value of a supercurrent is
at any given instant of time at some point in the junction depends on the
magnetic field in the junction and what the value of the supercurrent is at
any given point in the junction as a function of time depends upon the voltage
across the junction. The actual functional forms involved integrals of the
magnetic field, or flux, and integrals over the voltage. The second equation, which relates the frequency of a supercurrent to the voltage via e/h
is the basis for the well known fundamental constant experiments undertaken
by the University of Pennsylvania group [18]. This model, once the

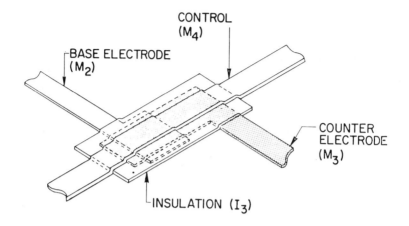

GATE

FIG. 8. Basic Device Configuration.

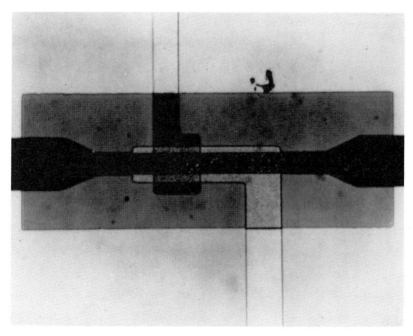

FIG. 9. Photograph of Typical Device.

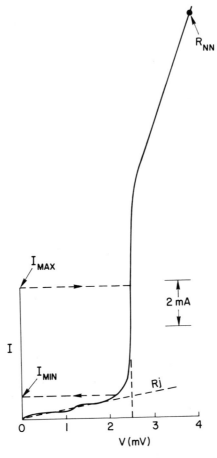

FIG. 10. I-V Characteristic Identifying Common Device Parameters.

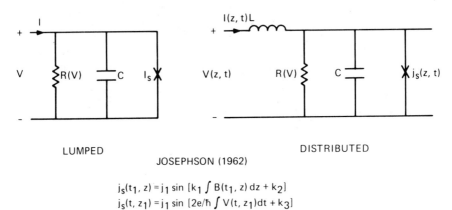

$$j_s(t_1, z) = j_1 \sin [k_1 \int B(t_1, z) dz + k_2]$$
$$j_s(t, z_1) = j_1 \sin [2e/\hbar \int V(t, z_1) dt + k_3]$$

FIG. 11. Equivalent circuit for Josephson Tunnel Junctions.

appropriate parameters have been measured, can be extensively tested. The measurement of these parameters is particularly simple, as nearly all can be determined from the dc I-V characteristic. How the capacitance can be derived from the dc I-V characteristic is not obvious. The characteristic impedance of a tunnel junction is low. The phase velocity of electromagnetic waves in the junction interior is typically 1/20th of the speed of light. Therefore, the junction acts like an open-ended resonator, even for dimensions as small as a 10th of a millimeter. The junction resonances manifest themselves as steps of constant voltage in the dc I-V characteristic. The resonant frequency can be obtained by measuring the voltage and then using the 2e/h relationship. Knowing the resonant frequency, mode, dimensions, and the inductance, which is readily calculable, one can infer the junction capacitance.

This device model has been extensively tested by quite a number of people working in various aspects of Joseph tunnel junctions. We show several illustrations of particular interest to us. The first of these is a static device property. How the zero-voltage current threshold I_m, varies with the control current in a specific in-line configuration is illustrated in Fig. 12. This shows a comparison between measured and calculated I_m dependence for a junction of specific length and current density, so that the L/λ_j ratio is equal to 3 [19]. The agreement between the calculation and experiment is excellent. Earlier comparisons in cases with $\lambda_j > L$ have also yielded good agreement [16].

More interesting than static comparisons is the dynamic large signal behavior of the device. Figure 13 shows a point junction in a superconducting loop. This is a case of interest in current steering. The corresponding equation, which describes the dynamics of this system, is also shown [14]. Were it not for the I_m sine, etc. term, we could solve the equation trivially. The same is true of the switching from the zero voltage state to the gap along some load line of an isolated junction. Zappe has shown that excellent approximations can in fact be obtained by neglecting the non-linear term [14]. It is of importance only when the voltage becomes small. We illustrate what is meant in a subsequent figure.

Consider now the solution of the equation for an isolated junction and compare it with experiments. Figure 14 shows the results of such a comparison carried out by Zappe and Grebe for a junction with dimensions roughly 0.1 millimeter by 0.1 millimeter [20]. Again, the agreement is such that the model description must be considered valid. The figure also illustrates a switching time for the device of 60 psec   (1 ps = 10^{-12} sec), an extremely short switching time. Jutzi and Mohr of IBM Zurich have established upper bound of 38 psec for switching in some smaller devices [21]. More accurate, direct measurements are almost impossible to obtain. They calculate that their devices were switching in 5-10 psec. These examples illustrate that these devices can be modeled with considerable confidence.

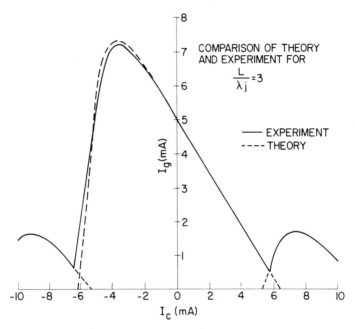

FIG. 12.   Comparison of Calculated and Experimental I_m vs. I_c for $L/\lambda_j = 3$.

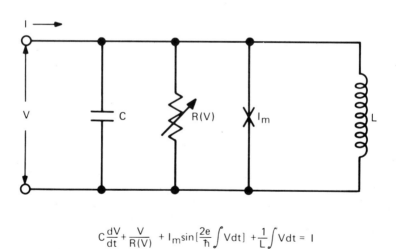

$$C\frac{dV}{dt} + \frac{V}{R(V)} + I_m \sin\left[\frac{2e}{\hbar}\int V dt\right] + \frac{1}{L}\int V dt = I$$

FIG. 13.   Point Junction Model of Device in a Loop and Corresponding Differential Equation.

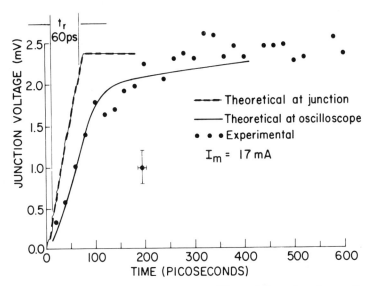

FIG. 14. Switching of a large Josephson Tunnel Junction from the Zero Voltage State to the Gap.

IV. LOGIC AND MEMORY CIRCUITS

How one can perform the logical operations of OR, AND, and INVERT using a Josephson tunneling gate is illustrated in Fig. 15 [22]. The circuit consists of a tunneling gate to which is being supplied a current $I_g < I_m(0)$. The gate has three equivalent control lines. The gate output consists of a transmission line of characteristic impedance Z_L and is properly terminated. The impedance levels are chosen such that when the gate switches, a current i, is generated in the output line. This output line in turn acts as a control for a number of other logic gates. The reason for proper termination is also illustrated in Fig. 15. The output current level i, is established in a single pass of the wavefront down the output line, rather than suffering multiple reflections if the line is terminated, for example, in a short. It is clear that proper termination gives the fastest operation. The OR, AND, and INVERT functions are performed by suitably selecting the gate and current levels so that the gate switches when any one of the controls is present (OR), or arranging the situation so that the gate switches only when all inputs are present (AND). Note that power dissipation occurs only during the portion of time that the gate is on. Figure 16 is a photograph of such a logic gate. The dynamic operation of such circuits has been experimentally investigated by Henkels [23].

The principle of storage or memory is that of persistent currents, as illustrated in Fig. 6. Figure 17 is a photograph of such a cell proposed by Anacker [24], and experimentally investigated by Zappe [25]. The

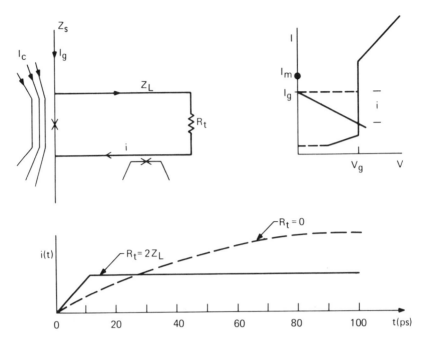

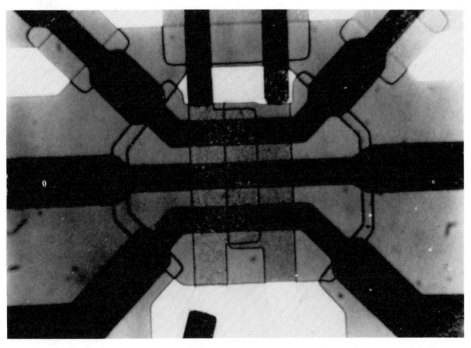

FIG. 15. Diagram Illustrating Logic Circuit Principle.

FIG. 16. Photograph of Typical Logic Gate.

idea is to store in the superconducting loop a circulating current of either clockwise or counterclockwise polarity. A sense gate below the super-conducting loop can distinguish between the presence of a clockwise cir-culating current or counter-clockwise circulating current. In any memory array in which one has a number of such cells, it is necessary to select by some means the desired cell. In writing or storing information, a cell is selected by coincident current along two X and Y lines. The Y line is called the bit line, and the X line is called the word line. Depend-ing on the information present and the polarity of the bit current, either the lefthand or righthand gate will switch, transferring current to the opposite side of the loop. When the writing currents are removed, a cir-culating current of roughly 1/2 the X line current is stored in the loop. The sensing of the information is done in a similar coincident current mode and is non-destructive. In sensing, the coincidence is between a sense gate current and the X line current.

An important design consideration in such memory cells is that the current transfer be stable. This is achieved in essence by assuring that the circuit is critically damped. Figures 18 and 19 illustrate Zappe's experimental results and calculations of the current transfer process. Figure 18 shows the solution to the equations of motion under design con-ditions (see Fig. 13). The points are experimentally determined points. Although there is obviously some scatter, the agreement is reasonable. Figure 18 shows the corresponding voltage waveform in which the end of the current transfer process is much more readily identifiable.

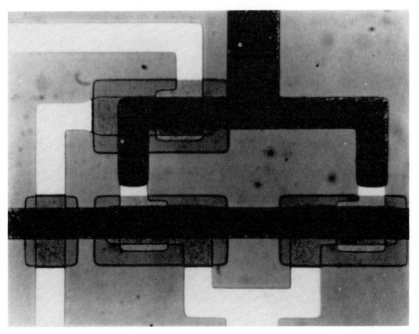

FIG. 17. Photograph of a Loop Memory Cell.

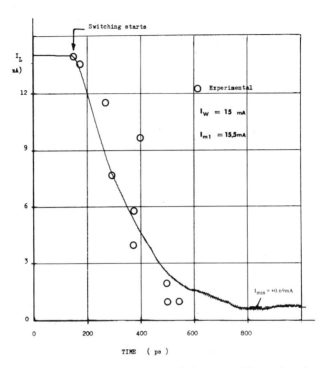

FIG. 18. Experimental and Theoretical Current Transfer in a Memory Cell.

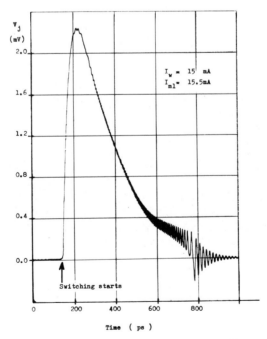

FIG. 19. The Corresponding Voltage Waveform.

Incidentally, the figure illustrates that even though the device size is quite large, approximately 0.1, 1 millimeter by 0.1 millimeter, and the loop length is also quite large, small switching times of about 600 psec for the cell have been obtained.

Finally, Fig. 20 shows one such memory cell being operated quasistatically under realistic memory test conditions, in which the cell is alternatively written, disturbed and read continuously.

V.   SUMMARY

The devices of the form that have been described are, I believe, quite well understood. The elementary circuits which have been experimented with to a degree are also well understood and have a speed power advantage over semiconductor circuits. Provided that this technology can remain competitive with respect to linewidth resolution, the performance advantage at the device level should carry on into the computer system. Since the patterning techniques are essentially the same as those used in the semiconductor technology, improvements in that technology in linewidth resolution should carry over to Josephson tunneling devices as well. Also because of this similarity (although there are obvious differences such as the RF oxidation, for example), the cost of making a wafer in this technology should not be appreciably different from the costs involved in the semiconductor technologies. The need to refrigerate to 4 K is

WORD CURRENT

BIT CURRENT

SENSE CURRENT

SENSE SIGNAL

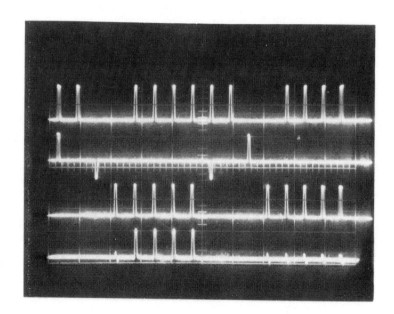

FIG. 20.   Photograph Illustrating Typical Memory Cell Test Pattern.

an added cost. Because power dissipation levels in the circuits are reasonably low, we anticipate needing refrigeration capacities on the order of a few watts at 4 K. Hopefully, the cost of such refrigerators is not prohibitive.

Because the electrical device properties are radically new, a great deal of work in that area remains to be done. Similarly, as Greiner et al [10] have pointed out, the present state of the technology is suitable only for experimental circuit work. Nevertheless, my assessment of the technology potential is optimistic.

REFERENCES

1.   R.W. Keyes, IEEE Spectrum 6, 36 (1969).
2.   J.M. Casimir-Jonker and W.J. deHaas, Physika 2, 935 (1935).
3.   D.A. Buck, Proc. IRE 44, 482 (1956).
4.   I. Ames, M.F. Gendron and H. Seki, Trans. 9th Nat. Vac. Symp. pp. 133, McMillan, New York, 1962.
5.   J.P. Pritchard, J.P. Pierce, and B.G. Slay, Proc. IEEE 52, 1207 (1964).
6.   B.D. Josephson, Adv. Phys. 14, 419 (1965).
7.   I. Giaever, Phys. Rev. Letters 5, 147 (1960).
8.   P.W. Anderson and J.M. Rowell, Phys. Rev. Letters 10, 230 (1963).
9.   J. Matisoo, Appl. Phys. Letters 9, 167 (1966); Proc. IEEE 55, 172 (1967).
10.   J.H. Greiner, S. Basavaiah, I. Ames, Vac, Science and Technology (to be published).
11.   J.H. Greiner, J. Appl. Phys. 42, 5151 (1971).
12.   W.C. Stewart, Appl. Phys. Letters, 12, 277 (1968).
13.   D.E. McCumber, J. Appl. Phys. 39, 3113 (1968).
14.   H.H. Zappe, J. Appl. Phys. 44, 1371 (1973).
15.   B.D. Josephsen, Advan. Phys. 14, 419 (1965).
16.   D.N. Langenberg, D.J. Scalapino and B.N. Taylor, Proc. IEEE 54, 560 (1966).
17.   B.D. Josephson, Phys. Letters 1, 251 (1962).
18.   See for example. B.N Taylor, W.H. Parker, and D.N. Langenberg, Rev. Mod. Phys. 41, 375 (1969).
19.   S. Basavaiah (to be published).
20.   H.H. Zappe and K.R. Grebe, IBM J. Res. Dev. 15, 405 (1971); J. Appl. Phys. 44, 865 (1973).
21.   W. Jutzi, Th. O. Mohr, M. Gasser and H.P. Gschwind, Elect. Letters 8, 589 (1972).
22.   W. Anacker, Proc. of the 1972 Fall Joint Computer Conference, Anaheim, California, 1972.
23.   W.H. Henkels (to be published).
24.   W. Anacker, IEEE Trans. Mag., MAG-5, 968 (1969).
25.   H.H. Zappe, Device Research Conference, Boulder, Colorado (1973) (to be published).

Chapter 10

HIGH GRADIENT MAGNETIC SEPARATION: AN INDUSTRIAL APPLICA-

TION OF MAGNETISM

D. Kelland, H. Kolm, C. deLatour, E. Maxwell,
and J. Oberteuffer

Francis Bitter National Magnet Laboratory, MIT

Cambridge, Massachusetts 02139

I.   INTRODUCTION

A widespread commercial application of magnetism is the separation of mixed materials with different magnetic properties. Most magnetic separators use permanent magnets or electromagnets which remove ferromagnetic impurities such as bolts from non-magnetic products. Flour and cereal, among other common foodstuffs are subjected to cleaning by electromagnets. The grade of scrap iron is improved while loading it for transportation by lifting it magnetically which reduces the non-ferromagnetic content.

Less well known but of great economic importance are the magnetic separators used to improve low grade mineral ores. The reduction to metal requires a certain percentage of mineral in the ore which is not always found occurring naturally. About 23 minerals are improved or beneficiated by magnetic separation [1]. In some low grade taconite, iron is found as magnetite which is separated from a background of silica by conventional magnetic separators.

These devices have a variety of physical forms utilizing drums, belts, and grates, to accomplish the separation. The most familiar of the conventional separation devices is probably the magnetic drum separator. One of these is shown schematically in Fig. 1. It consists of a revolving drum in which a set of stationary magnets provide a magnetic force on the surface of the drum. The force holds the magnetic particles against the revolving drum carrying them over to the right while allowing the non-magnetic particles to fall off the drum to the left. This device is useful for the separation of strongly magnetic particles of sizes somewhat larger than 100 microns. The relatively low magnetic fields and field gradients at the surface of the drum are produced by either electromagnets or permanent magnets within the drum.

For separations of materials with smaller particles, having weaker magnetic properties, machines of this kind do not work. Several attempts have been made in the past decade or two to develop separators which would treat non-ferromagnetic ores which contain paramagnetic or weakly ferromagnetic components. These typically have magnetizations several orders of magnitude less than that of magnetite. For example, α-Fe_2O_3, at 10 kG has a magnetization of 0.58 emu/gm whereas that of magnetite is 96 emu/gm. Also, the development of machines with higher magnetic field gradients, to trap smaller particles with smaller magnetic moments, has been the object of many efforts, as there exist large deposits of ores requiring very fine grinding to liberate the mineral components.

In the past, advancements have been made in available magnetic-field strength by using larger coils and currents, improved magnetic circuits, and smaller circuit gaps. Higher field gradients in the working volume of the separators have been produced by the use of sharp pointed elements, grooved plates, and spheres of various diameter. All of these methods shunt the field through large volumes of ferromagnetic material, reducing the effective open field-volume and the throughput capacity of the device. The optimum system would have a small material-volume, a large field-value throughout the open volume, a large number of gradient sites, and the highest possible field gradient at these sites.

A magnetic separator with these features was developed at the Francis Bitter National Magnet Laboratory about five years ago [2]. Dr. Henry Kolm, working on the problem of removing iron-stained titanium dioxide from kaolin clay, designed a magnetic separator with a finely-divided ferromagnetic matrix which occupies about 5% of the field volume. The first material used for the matrix was steel wool. It has many small fibers with points and sharp edges which have radii of curvature on the order of a few microns. When placed in a region of magnetic field, there are field gradients created at these points and edges which extend into the space between the fibers. Because the steel wool is distributed throughout the volume of a pipe or can forming the separator, but does not fill the space with material, high magnetic field gradients are thus provided all through the effective open volume. These gradients attract and hold micron-size particles which have been magnetized by the background field. If the steel wool is magnetically saturated, gradients of the order of 1 kG/micron can be obtained. The strong magnetic fields are produced by efficient magnet design using normal magnets. They could be produced for large volumes and high fields by superconducting magnets.

A schematic of a simple high gradient magnetic separator is shown in Fig. 2. The feed, usually in the form of a liquid slurry, is passed down through the cannister in which the ferromagnetic matrix is packed. In the presence of the strong magnetic field, the magnetic particles are trapped on the matrix; the liquid and non-magnetic particles pass easily

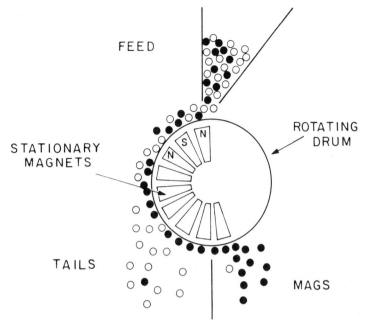

FIG. 1.   A Conventional Drum-Type Magnetic Separator.

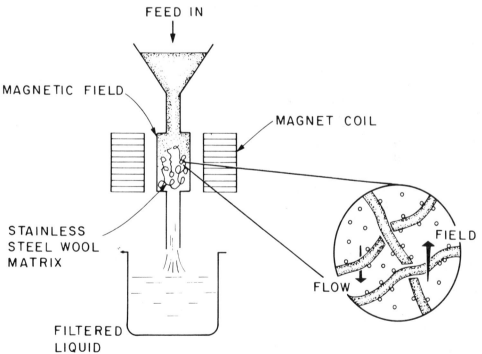

FIG. 2.   Schematic Diagram   of a High Gradient Magnetic Separator
          (HGMS).

through the relatively open structure of the matrix, and are collected
below. Magnetic components are retrieved by reducing the applied magnetic
field to zero and backwashing the matrix.

The basic features of this static separator have been incorporated in
a device designed for continuous operation for the treatment of
materials in which the magnetic component is large. In iron ore this magnetic component may be 50% or more of the solid content. To stop the
operation of a static machine periodically in order to clean out the magnetically-trapped ore would not be practical.

In 1971 the RANN (Research Applied to National Needs) Division of
the National Science Foundation awarded a contract to the National Magnet
Laboratory at MIT to develop a continuous process based on high gradient
magnetic separation for beneficiation of certain iron ores [3]. We are now
engaged in a pilot plant test of a continuous device in the laboratory of a
major ore producer in Minnesota. Other research at the National Magnet
Laboratory on high gradient magnetic separation [4] has included the ap-
plicability of HGMS to the removal of pyritic sulfur from coal [5] purification of water [6] and an investigation of the basic characteristics of
HGMS devices and processes [7].

The potential applications for high gradient magnetic separation devices and techniques are numerous. In addition to the beneficiation of a
variety of minerals by high gradient magnetic separation, water purification
and liquid-solid separations in chemical processing can be accomplished.
Through its application for the beneficiation of low grade minerals,
HGMS may be useful in recovering value from minerals tailing piles as well
as from coal and oil ash. Water purifications include removal of suspended
magnetic solids such as those found in steel mill effluents. The removal
of dissolved heavy metal constituents which are present in many industrial
water effluents is also possible. Although superconducting magnets are
not necessary for many present applications of high gradient magnetic
separation, it is anticipated that future large-scale high-field applications
will use superconducting magnets.

II.   PRINCIPLES OF HIGH GRADIENT MAGNETIC SEPARATION

A simple expression for attractive magnetic force on a body
is given by $F = VM(dH/dx)$, where V is the volume of the body,
M is its magnetization in the field H, and dH/dx is the field gradient [8].
High gradient magnetic separation devices seek to maximize this force
maximizing the magnetic field variation across the volume of the particle
to be trapped. Field gradients as large as 1 kG/micron are produced by
applying strong magnetic fields to ten micron diameter fibers of ferromagnetic materials as found in the matrices of HGMS devices. Since the
magnetization of a particle is in general a function of the ambient magnetic
field, maximizing the magnetic field in a HGMS device also increases the

magnetic attractive force. For ferromagnetic materials this effect
generally saturates at fields of the order of 10 kilogauss, but for paramagnetic particles, the magnetization increases with increasing magnetic
field beyond 10 kilogauss. For these materials larger magnetic fields
enhance the magnetic trapping forces.

It maybe seen intuitively that the range of the magnetic force implied
by the magnetic field gradient dH/dx must be matched to the size of the
particle to be trapped. If the particle is much larger than the range of
the gradient force then only a small portion of the particle will feel the
effect of the force. On the other hand, if particle is small compared to the
range of the gradient, then the difference of magnetization across the
particle, which accounts for this dipole force, will not be large. For a
given magnetic field gradient, then, the magnetic force as a function of
increasing particle size may be thought of as first increasing as the volume
of the particle, for particles small compared to the magnetic field
gradient range, and finally saturating at some upper limit for a particle
which is large compared to the range of the magnetic field gradient.

In a magnetic separation device it is the competition between the
force of magnetic attraction on a particle and the gravitational or hydrodynamic drag forces which determines the efficiency of the separation.
This is indicated schematically in Fig. 3. It may be shown that the ratio
of the magnetic force on a particle to the competing gravitational and

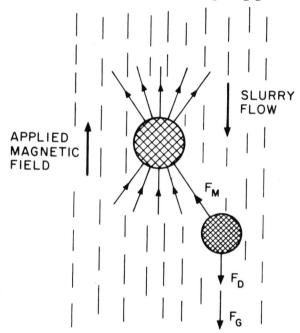

FIG. 3.   Schematic Representation of a Fiber in a High Gradient
          Magnetic Separator.

hydrodynamic drag forces, is a maximum for a magnetic field gradient range which approximately matches the size of a magnetic particle. In HGMS separators this characteristic distance can be very small and this is why the process is more effective for small particles than with conventional magnetic separators.

The hydrodynamic drag force depends on either the first or second power of the diameter of the particle whereas the gravitational force depends on the third power. If the characteristic distance associated with the magnetic field gradient is adjusted to the particle diameter so as to maximize the magnetic force, the magnetic force will depend on the second power of the diameter. Because of this difference in the dependence of these three forces on the particle diameter, the drag force becomes dominant for very small particles and the gravitational force for very large particles. In between,the magnetic force dominates and exhibits a maximum. For cupric oxide particles (density = 4, magnetic susceptibility = 2.4×10^{-4} emu/g.) this occurs for a diameter of approximately 0.5 mm. in a field of 10 kG. For higher fields, or particles of higher susceptibility or lower density, the maximum occurs at smaller diameters and vice-versa. In the earlier conventional magnetic separation art, operating principally with ferromagnetic or ferritic particles (like magnetite), these considerations are not critical as the high intrinsic magnetization of the particle, even in weak fields, insures that the magnetic force exceeds the competing forces for a wide range of particle sizes.

III.   THE CHARACTERISTICS OF HIGH GRADIENT MAGNETIC SEPARATION

The efficiency of the separation accomplished by a magnetic separation device may be expressed in two ways. First, the amount of magnetic material recovered in the magnetic component (mags) may be measured relative to the amount of magnetic material in the feeds. This percentage is called "recovery". Second, the separation may be characterized by the purity of the magnetic component separated; that is, the percentage of magnetic material in the mags relative to the total mass recovered in the mags. These two measures are independent quantities and together determine the efficiency of the separation.

For most separation devices, both conventional and high gradient devices, there is a relation between the grade and the recovery of the desired component, the product of the separator. As the operating variables of the separator are changed,the recovery can be increased at the expense of the grade, and vice versa.

For high gradient magnetic separation devices,the recovery as a function of the two principle operating variables of the separator is shown in Fig. 4. The material used in this test was a mixture of 10 micron cupric oxide powder and aluminum oxide power slurried in water.

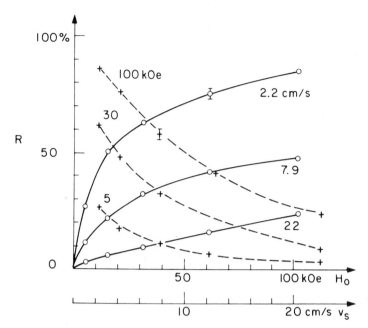

FIG. 4.   Recovery Percent of Desired Component in Mags Relative to
Feed, vs. Applied Magnetic Field (Solid Curves) and Slurry
Flow Rate (Dashed Curves).

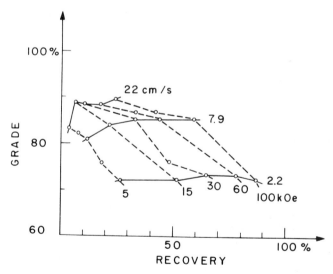

FIG. 5.   The Grade, Percent Desired Component in Mass Relative to
Total Mass of the Mags, vs. Recovery under Various Conditions
of Applied Field and Slurry Flow Rate.

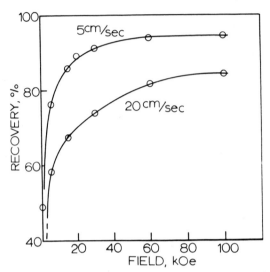

FIG. 6.   Recovery of a Non-Magnetic Iron Ore vs. Applied Magnetic
Field at Two Different Values of Slurry Flow Rate.

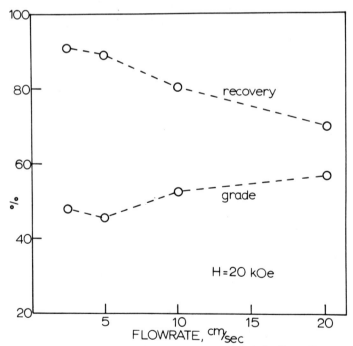

FIG. 7.   Grade and Recovery of a Non-Magnetic Iron Ore vs. Slurry
Flow Velocity in a HGMS device.

It may be seen that the recovery of cupric oxide increases with increasing magnetic field at all slurry velocity rates, and decreases with increasing slurry velocity for all values of applied magnetic fields. By contrast the grade varies very little with increasing magnetic field and increases with increasing velocity. If these points are plotted on a grade recovery curve, as shown in Fig.5, it may be seen that increasing both the magnetic field and the slurry flow velocity can increase simultaneously the recovery of the desired component in the mags as well as the grade.

Similar results are obtained with taconite ores. Fig. 6 shows the recovery of iron for field values up to 100 kG with two flow rates indicated. The recovery is higher for slower flows (for a given field) at which more material is trapped. The increase in recovery at higher fields is more pronounced for higher flow rates (lower curve). Tradeoff between field and flow capacity can be sought on the basis of other parameters and ultimately on economics.

Figure 7 shows the grade (and recovery) of the magnetically-trapped product for a field of 20 kG plotted against flow rate. The grade increases with flow rate because less non-magnetic material is trapped mechanically. The stripping forces are higher in relation to the magnetic trapping forces at higher flow rates, indicated by lower recovery. These figures indicate the importance of having a high field available throughout the volume of the separator and show that grade improves with increasing flow rate--a vital consideration in the economics of separation.

Figure 8 contains two grade-recovery curves, one for constant H and the other for constant flow rate. The former shows that higher grades may be attained at the expense of recovery by choosing higher flow rates. The same result can be gotten by decreasing the field, but for a combination of good grade and recovery it would be better, except for very low grades, to work at higher fields. It may be concluded that operating high gradient magnetic separation devices at high throughput velocities and high fields will yield a more efficient separation. Depending on the economics, it may be desirable to use more costly but stronger magnetic fields in order to increase the efficiency of the separation process. The use of superconducting magnets to produce these fields may be indicated even when a less efficient separation is possible using normal magnets.

IV. HIGH GRADIENT MAGNETIC SEPARATION DEVICES

Two basic types of high gradient magnetic separation devices have been developed at the National Magnet Laboratory, and at Magnetic Engineering Associates of Cambridge, Massachusetts. These are batch type devices and a continuous "Carousel" device whose particular usefulness depends on the application.

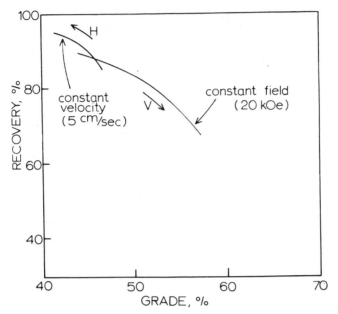

FIG. 8.   Effect of Increasing Applied Field or Slurry Velocity on
Grade-Recovery Point for an HGMS Device.

A batch device, such as that shown schematically in Fig. 2 is useful
when the feed contains a relatively small amount of magnetic material
to be trapped in the separator. This allows a convenient duty cycle for
the separator; for example, in the removal of fine weakly magnetic
particles which stain kaolin. The slurry is allowed to pass through the
device for approximately 20 minutes during which time the impurity
particles are trapped. With the field reduced, the wash cycle takes place
in about 5 minutes. The use of slurry valves to switch between units
during the wash portion of the cycle or a simple surge tank makes the batch
device useful in this application where the material to be trapped represents perhaps one percent of the total volume of the feed material. The
capacity of the device for trapped material is approximately equal to 5
percent of the volume. In the case of water treatment where magnetic
impurities might constitute 2,000 parts per million of the total feed
volume, it is indicated that a batch device would be adequate.

When a large proportion of the feed material is to be trapped by the
separator, the duty cycle of a batch type device may become inconveniently
short. For these applications such as beneficiation of iron ores, a continuous HGMS device is indicated. Such a device is shown in Fig. 9. A
wheel or "Carousel" containing cells which are packed with a filamentary
ferromagnetic material such as stainless steel wool is driven by an
external motor and allows the matrix cells to rotate into the region of
strong magnetic field. Here the feed slurry stream passes through the matrix

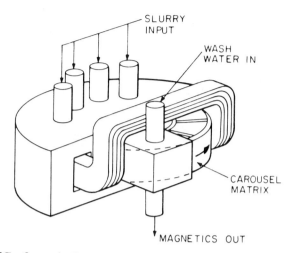

FIG. 9. A Continuous "Carousel" HGMS Device.

and out below the device. In this region magnetic particles are trapped in the cells. They are washed by the wash water which was introduced by the pipes to the right of the feed pipe and finally, when the cell has rotated around the opposite side of the device where the field is small, the particles are washed out. The wheel turns continuously with new cells constantly being introduced into the slurry stream. The device shown in Fig. 9 is similar to the pilot plant device referred to in the introduction which will be used for the beneficiation of non-magnetic taconite ores. The wheel or Carousel in this device is approximately 45 centimeters in diameter and has a capacity of 3 tons of ore per hour.

Some particles, either by their small size or magnetic properties cannot be trapped directly even in the matrix of a high gradient separator. This is the case with some suspended solids, bacteria, and color particles with contaminate water systems. It is possible, however, to associate such particles with a magnetic seed. Finely divided iron oxide is coagulated together with the contaminants into flocs which are then collected in the separator. The removal of the flocs from the water stream is accomplished within a very short time after the flocculant is added. Very fast flow rates are possible because of the large open volume of the separator and because the amount of material to be removed from a water supply or effluent stream is small compared to the volume of the stream.

In Table 1, data is given for surface and bottom samples from the Charles River in Boston. Coliform bacteria and other solids have been removed with the seeding technique. Approximately 1000 parts per million iron oxide is used as the seed and the flocculant is aluminum sulphate.

TABLE 1. EFFECT OF MAGNETIC TREATMENT ON WATER QUALITY

| | CHARLES RIVER SAMPLE | | | | DEER ISLAND SEWAGE SAMPLE | |
| | SURFACE | | BOTTOM | | | |
	control	treated	control	treated	control	treated
Coliform Bacteria (/100ml)	16,000	0	16,000	300	2.8×10^6	18,000
Turbidity JTU units	20	2	1,700	1	50	3
Color (color units)	105	3	3,700	1	150	20
Suspended Solids (mg/liter)	7	5	690	5	45	9

TABLE 2. EFFECT OF MAGNETIC TREATMENT ON PHOSPHATE CONTENT OF WATER

| | ORTHOPHOSPHATE parts per billion P | |
	Control	Treated
CHARLES RIVER SAMPLE		
no clay added	330	180
bentonite clay added	330	60
DEER ISLAND SAMPLE		
no clay added	3000	300
bentonite clay added	3000	⟨100

Dissolved ions can be separated from water systems in similar ways. The orthophosphate ion is a nutrient for algal growth and its removal from water supplies can limit growth.

Table 2 shows orthophosphate removal from Charles River water and from sewage samples taken at a bottom sewage treatment facility on Deer Island. Here, magnetite is the seed and added aluminum ions bind the phosphate ions into flocs. Included in this table are data for the same process but with added colloidal clay. The clay serves to absorb orthophosphate ions thereby aiding in the coagulation. More than one stage of treatment can reduce the phosphate content to very low levels.

V. CONCLUSIONS

Since magnetic separation represents a large commercial use of magnetism, the potential use of superconducting magnets for this application may represent its largest commercial potential. The use of superconducting magnets in conventional existing magnetic separation devices is probably quite limited. However, for the large-scale high-gradient type magnetic separation devices the use of superconducting coils is very likely to be significant. It should be noted that the introduction of any new technology is not a simple process; high gradient magnetic separation has not been accepted in many areas of potential application. For this reason it has not been thought wise to try to introduce the additional concept and technology of superconductivity in these areas. Once the technology of high gradient magnetic separation can be accepted using normal magnets it is likely that the economics of superconducting magnets for these devices can prevail over the resistance to that new technology. Indeed,it appears that for most high gradient magnetic separation devices larger than pilot scale, that is with bores larger than about 2 feet in diameter and matrix volumes of the order of a few cubic feet, superconducting magnets would have favorable economics. These devices would be assumed to operate in the range of 10 kilogauss. Naturally, even for pilot scale devices which were required to operate at fields of 100 kilogauss, the use of superconducting magnets would be indicated. Nearly all of the commercial-industrial-large-scale applications for high gradient magnetic separation would involve devices whose capacity is somewhat larger than a few cubic feet in volume. This is especially true in the case of water treatment. A large steel mill may discharge on the order of 100 million gallons of water per day. At the flow rates which may be achieved in high gradient magnetic separation devices this implies a magnetized volume on the order of 1,000 cubic feet, probably in several units. Such devices may be made with normal magnets but it would appear that the favorable economics of superconducting magnets would prevail, at this point. Thus, use of superconductivity in future magnetic separation devices appears assured, even though the time scale is uncertain.

REFERENCES

1.   Bureau of Mines Staff, Mineral Facts and Problems, Bureau of
      Mines Bulletin 650, U.S. Department of the Interior, 1970.
2.   H. Kolm, E. Maxwell, J. Oberteuffer, D. Kelland, C. de Latour,
      and P. Marston, Magnetism and Magnetic Materials Conference
      Proceedings 1971 (abstract only).
3.   D.R. Kelland, IEEE Trans. on Magnetism Vo. Mag-9,
      September 1973.
4.   J.A. Oberteuffer and D.R. Kelland, eds., Proceedings of the High
      Gradient Magnetic Separation Symposium, MIT Francis Bitter
      National Magnet Laboratory, Cambridge, Massachusetts, 1973.
5.   S.C. Trindade and H.H. Kolm, IEEE Trans. on Magnetics, Vol.
      Mag-9, September 1973.
6.   C. de Latour, IEEE Trans. on Magnetics, Vol. Mag-9, September
      1973.
7.   J.A. Oberteuffer, IEEE Trans. on Magnetism, Vol. Mag-9,
      September 1973.
8.   E.C.Stoner p. 37 in Magnetism and Matter, London: Metheun
      and Co. (1934).

Chapter 11

PROGRAMS ON LARGE SCALE APPLICATIONS OF SUPERCONDUCTIVITY
IN FRANCE

G. Bronca

Commissariat a l'Energie Atomique and Delegation Generale

a la Recherche Scientique et Technique

Saclay, France

I. INTRODUCTION

The research in the field of large scale application of superconductivity is supported in FRANCE by three different sources.

1. Private sources corresponding to universities, national laboratories and laboratories belonging to the national utilities company (EDF) or private industry.

2. Military program funded by the Ministry of Defense.

3. The government agency for research (Delegation Generale a la Recherche Scientifique et Technique).

In that last case funding is very often based on 50% support by the government. These different sources may be combined for any particular program. There is little information at present on the military program, however, we review how the civil applications program is organized for the coming years.

II. THE SITUATION IN 1971

In 1971, while the national utilities company (EDF) was doing a study on applications in its own field, the government agency (DGRST) decided to make a general study which will allow the government to approve and support a coherent program for the coming years. This study was made in 3 parts:

1.   To obtain general information on the work done and the state of the
      art in the field of superconductivity and connected applications in
      France and elsewhere.

2.   Transmit this information to the responsible leaders of
      all the laboratories and companies involved in possible applications.
      Using this information we ask them what they will want to do and
      what support they will need.

3.   Define the government program. A very large theoretical and
      experimental effort has been made a few years ago mainly in university
      laboratories on general superconductivity problems. This effort
      was progressively slowed down and today small groups are doing
      research on this general subject. In the field of magnet applications,
      high field coils were built for different laboratories but it was
      restricted to high energy physics. However, there were also various
      constructions of magnets and tests with homopolar machine proto-
      types, rotating machine components, cryogenic transformers and
      transmission lines, superconducting materials, and low temperature
      materials behavior.

In the five years (1966-1971), the D.G.R.S.T. spent about 4 million
French francs per year in different fields of research and a total of 10
million francs to support many projects. It means that the same amount of
money was funded directly by the laboratories doing the work. It is very
likely that about 10 million francs was spent during the same 5 year
period from private sources.

III.   FUTURE PROGRAMS

      The conclusion of the study and the corresponding program are as
follows:

A.   Basic Research

      This is mainly carried out by university-type laboratories (ORSAY,
GRENOBLE, ECOLE NORMALE SUPERIEURE, LCIE, SACLAY,...) and
except for some particular areas it will not be supported through this
program. This includes studies like high temperature superconductors,
pinning, flux penetration, and conductor behavior.

B.   Applications to Research Apparatus

      There is a small market for high field coils for research in different
fields as solid state physics, chemistry and biology. The order of magnitude

of this market is about 1 million francs per year. This work does not need any government support. At present, the main application concerns high energy physics, fusion reactors and radio frequency cavities. With the linear accelerator and RF separator it seems that the progress is slow and only a small group is continuing a modest effort at the university level.

Plasma physics devices and fusion reactors seem to be promising for the future. However, there is at present no project incorporating a superconducting magnet and only one industrial group intends to start a small effort (on the basis of about 1 million francs per year). However, this is not included in their first part of the program.

Many magnets were built and tested in the past for high energy physics experiments. Large detector models, magnets for polarized protons targets, dipole and quadrupoles were successfully used in the laboratories and in experimental areas. All this effort was directly paid for by the high energy laboratory. The future of such applications, even if the market is not very large, is very important for all the projects being planned for beams in experimental research with national accelerators and at CERN. The design, tests and operational use of dipoles, quadrupoles and large detectors give experience both to the industries which build them and to the laboratories which use them. This will be fruitful for any other applications. Moreover, these applications will continue for a period of 5 to 8 years and may be the only application during that time.

Following this conclusion it was decided to support a 3-year program (1972-1974) for the design the construction and the test of a pulsed dipole which could be considered as a good prototype for many lenses. This program is lead by the SACLAY group sharing the work with ALSTHOM, Laboratories of MARCOUSSIS and THOMSON-BRANDT for the conductor. The total cost of this program is 11.2 millions francs, half of it being supported by the D.G.R.S.T. The tests of this dipole will start by mid 1974.

As a final remark I would like to mention the collaboration existing in Europe for both pulsed and dc fields between RUTHERFORD High Energy Physics (Great Britain), Kernforschungszenstrum/Karlsruhe (Germany) and the CEA/SACLAY (France) known as the GESSS.

C. Superconducting Alternators

This application is the most promising for the future in France. The need for very large alternators using a superconducting inductor will appear in about 15 years. EDF is interested in promoting the development of such machines for its network and the companies are interested in the market which could be very large. The general scheme would be to spend 5 years for general studies, 5 years to build a large prototype and 5 years to get operational experience.

In the 5 year period (1966-1971), paper work on the problems raised by superconductivity was made in many places. The COMPANY ELECTRO-MECANIQUE (CEM) will continue its present small effort and intends to participate to the study and the construction of a very large alternator made by the swiss company BBC.

The Societe ALSTROM decided late 1971, with the agreement of EDF to start a four year program divided in two phases. Phase A (ending mid 1974) is concerned with feasibility and economical interest. It will include the study of a 1 to 3 GW alternator and the construction and the test of model at the proper scale for parts of the alternator. Phase B (end of 1975) will be devoted to the design of a significant size prototype (order of magnitude 100 MW). The decision to build a prototype will be made at the end of phase B. This program is presently being followed. ALSTHOM and EDF are leading the work with the collaboration of laboratories of MARCOUSSIS, THOMSON-BRANDT and other laboratories and the financial support of D.G.R.S.T.

D.   Transmission Lines

It is clear now that very high power transmission lines will not be needed in FRANCE for 15 to 20 years. Much work was done in the past on cryoresistive and superconducting line and both ac and dc current. This research at the laboratories of MARCOUSSIS in connection with EDF was reported at the ANNAPOLIS Conference. Since that time the effort has slowed down due to the remoteness of the application. Nevertheless, everybody agrees that the insulation problem at low temperature is important. In the next few years, this will be the only effort in collaboration with EDF, Laboratoire Central des Industries Electriques and Laboratories de MARCOUSSIS.

E.   Homopolar Machines

C.E.M. and Laboratoires de Recherches de l'ALSTHOM conducted a series of studies in the past and LCIE designed and tested successfully a model of a homopolar machine (60 kW at 600 rpm). The application of homopolar machines for use in industry (Aluminum), even though it looked interesting, did not meet large support from the users. This is true also for civil naval applications. Finally there are the military applications of this work which is supported by the Ministry of Defense.

F.   Other Applications

Many other applications have been looked at, however, there is not enough interest to justify a development program. Electrical switching needs material properties which are not available today and no effort can

be made before such properties are reached. Energy storage had some
very successful tests in Laboratoires de MARCOUSSIS and C.E.M. both
working in collaboration with CEA. Coils storing many hundred kilojoules
were built and performed well. However, this is again a field which has no
civil support in the near future and the work is done now only for military
applications. Superconducting levitated trains use are not forecast in
France by the national transport company before a very long time.

G.   Conductor and Cryogenic Apparatus

Both short and long range applications will need development of
superconducting conductor and cryogenic apparatus. Almost all the work
mentioned previously was made using the collaboration of THOMSON-BRANDT
(conductor) and AIR LIQUIDE (cryogenic apparatus). The work in this
field will continue when necessary in close connection with the groups in
charge of the particular program.

For the conductor this effort is supported by high energy physics
magnets and by the superconducting alternator studies. In addition, but
on a small scale, work is done on Nb_3Sn and new materials. For the
cryogenic problem, it seems today that refrigerators and liquifiers are
available for the needs which are foreseen. A small effort is continuing
to improve the present situation. A large program will be started only
if connected to a well defined application program and appears necessary.

IV.   CONCLUSIONS

The program which developed in 1972 from the conclusion of the
1971 study is progressing well along the 2 main directions: in the near
future, magnets for applications in research especially high energy physics
and for the far future generators for electric power systems. Currently
there is no new information which would lead to a change in this program.
It is very likely that in 1975 a new study must be made concerning the work
for the period 1975-1980. In Table 1 we have listed the main research
institutions, group leaders and the projects in France.

TABLE 1.   Superconductivity Research in France

INSTITUTION	GROUP LEADERS	PROGRAM
Centre de Recherches des Tres Basses Temp-eratures (GRENOBLE).	LACAZE	Basic Research.
Laboratoire National des Champs Intenses.	PAUTHENET VALLIER	Magnetism research Magnet construction.
Universite d'ORSAY	FRIEDEL	Basic Research.
Ecole Normale Supe-rieure	KLEIN LEGER	Basic Research.
CEA/SACLAY-Departe-ment SATURNE.	BRONCA	Pulsed and dc magnets for HE physics and nuclear physics .
CEA/SACLAY - Departe-ment des Particules elementaires.	DESPORTES	dc magnets for HE Physics.
CEA/SACLAY ALSTHOM/BELFORT Lab. De MARCOUSSIS	BRONCA GOYER DUBOIS	Construction of a prototype a pulsed dipole 1972-1974.
ALSTHOM (BELFORT);	RUELLE	Study of an alternator.
EDF Etudes et Recher-ches (CLAMART).	MALAVAL CROITORU	General applications in the field of production and transmission of energy.
Laboratoires de MARCOUSSIS	DUBOIS MALDY	Collaboration in alternator design . Trans-mission line (insulation). Stored energy.
Laboratoire Central des Industries Elec-triques (FONTENAY-aux-ROSES).	FOURNET MAILFERT	Basic Research. Basic study of alternator design insulation study. Homopolar machine.
Compagnie Electro-Mecanique (Le Bourget)	MALANDAIN	Preliminary alternator study. Stor-age energy. Magnet design.
THOMSON-BRANDT	CALAVAS	Nb-Ti production. New material studies.
AIR LIQUIDE (Sassenage)	CHAPERON CARBONEL	Refrigeration and liquefier, general cryogenic studies.

Chapter 12

PROGRAMS ON LARGE SCALE APPLICATIONS OF SUPERCONDUCTIVITY

IN THE FEDERAL REPUBLIC OF GERMANY

G. Bogner

Research Laboratories

SIEMENS AG Erlangen, Germany

I. INTRODUCTION

The work on applied superconductivity in Germany was and is mainly concentrated in the industrial laboratories such as: SIEMENS AG, VACUUMSCHMELZE GMBH, KRAFTWERK UNION (a common daughter of SIEMENS and AEG), a working group of AEG-TELEFUNKEN, KABELMETAL and LINDE AG and also in the large governmental laboratories, such as: KERNFORSCHUNGS-ANSTALT Jülich (KFA) and INSTITUT FUR EXPERIMENTELLE KERNPHYSIK (IEKP) of the Gesellschaft fur Kernforschung in Karlsruhe. The superconductivity program of the KFA will be finished at the end of 1973. Part of its activities has already been moved and the remainder will be moved to the IEKP Karlsruhe. Fundamental research on superconductivity is done in a series of technical universities, but this work will not be mentioned in the scope of this report.

The research and development work on applied superconductivity is partly financed by the research budgets of the industries or governmental laboratories and partly by the BUNDESMINISTERIUM FÜR FORSCHUNG UND TECHNOLOGIE (BMFT) within the so called New Technology Program. In one case research will probably also be supported by the DEUTSCHE FORSCHUNGSGEMEINSCHAFT (DFG). For a later phase of the German land transportation program, significant fundings can be expected from the BUNDESMINISTERIUM FÜR VERKEHR (BMV).

The major applied programs which will be treated in this report are:

1.   Superconducting materials
2.   Magnets for high energy physics and plasma physics
3.   Magnets for industrial or special applications
4.   Rotating electrical machinery
5.   Electric power transmission
6.   Land transportation systems
7.   High frequency cavities and Josephson-elements

II.   SUPERCONDUCTING MATERIALS

The development of practical superconducting materials, especially high field superconductors, is concentrated in the industrial laboratories. VACUUMSCHMELZE GMBH, Hanau, had started research and development on superconducting materials in 1962. The first commercial superconductors were single core conductors of NbZr 25. The present standard program includes different types of fully stabilized, partially stabilized and intrinsically stable copper coated superconductors with circular and rectangular cross sections. The most used alloy is NbTi 50% with current densities up to 2.5×10^5 A/cm^2 at 5 T. For special applications, intrinsically stabilized superconductors have been fabricated in single pieces up to 200 kG.

The current research program of this company consists of basic work on pinning, stabilization, fluxjumps, technological and metallurgical work on different types of superconductors. Besides NbTi 50%, other alloys are included, 40, 45 and up to 65% Ti and third components. A special research program concerns filament conductors for pulse and ac applications (including conductors with copper-nickel and copper-copper-nickel substrates having thin filaments of 10μ m in diameter) and different techniques of transposition. The program on pulsed or ac materials was started in 1969 in connection with the original plans of CERN to convert the new large European proton accelerator to a superconducting version.

The commercial superconductors of the VACUUMSCHMELZE are known as VACRYFLUX. They have been successfully tested in a large number of laboratory scale magnets and the following bigger magnet projects contain this material:

1.   Testing magnet for BEBC-bubble chamber at CERN "Braracourcix".
2.   BEBC-bubble chamber magnet at CERN (about 50 tons supercond. mat.)
3.   "Pluto"-magnet at DESY-Hamburg
4.   Stellarator prototype magnet "Wendelstein"-Siemens, MPI-Garching.

The general development work on superconducting materials, especially on dc superconductors, was and is financed by money of the company. The pulse materials program is supported by a 50% subsidy of the BMFT. The present program section will be finished at the end of 1973. But there are plans to continue the materials program on a broad

basis after this date. The programs are under the leadership of Dr.
Assmus and Mr. Hillman.

A development program on superconducting alloys is also carried
out by KRUPP-FORSCHUNG, Essen. The work is concentrated on
optimization of NbTi-alloys with respect to proper manufacturing and
superconducting properties. Commercial materials are not yet available,
as far as known. The program is also supported at a 50% level by the
BMFT and will be finished in 1974. The leaders of the program are Dr.
Hillenhagen and Mr. Willbrand.

At the Research Laboratories of the SIEMENS AG in Erlangen a
program on developing vapor deposited Nb_3Sn-ribbons has been carried
on for several years. The material was successfully used in some high
field test magnets (B > 10T) for their own laboratory. It is not available
commercially at the moment. The program was financed 100% by the
company. It was and is conducted under the leadership of Dr. R.
Gremmelmaier and Dr. H. Pfister.

At the same laboratory research work on the A15-superconductors
Nb_3Al and $Nb_3(AlGe)$ is carried out. The materials are produced by de-
positing Al or AlGe melts on Nb wires and a subsequent heat treatment at
1800 C. The critical temperatures attained so far are 18.5K for
Nb_3Al and 19.5K for $Nb_3(AlGe)$. The current densities in the superconducting layers reached values of about 10^5 A/cm^2 at 5T.

Besides this program at SIEMENS AG, intensive work is concentrated
on the development of multifilament conductors of Nb_3Sn and V_3Ga in a matrix
of copper-based alloy, produced by the known solid state diffusion process.
Within this program there is a tight collaboration between SIEMENS AG
and VACUUMSCHMELZE. The former being especially engaged in the
metallurgical part of the work, the latter in the problems of reliable and
optimum manufacturing processes for this material. Conductors with
circular and rectangular cross-sections with lengths up to 100 m have
already been fabricated. The current densities attained so far are about
10^6 A/cm^2 (at 5 T) with respect to the superconducting layer and about
10^5 A/cm^2 with respect to the total conductor. The critical temperatures
for these materials were 14.5 K (V_3Ga) and 17.5 K (Nb_3Sn). The investigations started in 1971 and are planned to be finished in 1974. The work
is also sponsored at a 50% level by the BMFT. Program-leaders are again
Dr. R. Gremmelmaier and Dr. H. Pfister.

Important work is also done on niobium conductors at the SIEMENS-
Research Laboratories at Erlangen. This work will be treated in more
detail in connection with high-frequency cavities and power transmission
lines.

III.   MAGNETS FOR HIGH ENERGY- AND PLASMA PHYSICS

A.   DC Magnets for High Energy Physics

At present there is only one large dc magnet project in Germany. In May 1973 DESY Hamburg invited bids for constructing the large super-conducting magnet system "Oktopus". Oktopus is a system of eight superconducting windings of rectangular shape, arranged around a hollow space like the driving paddles of a (Mississippi) paddle-steamer around the axis. The single magnets are of remarkable size (1.8 x 4 m) and they have to be cooled by forced flowing helium. The system is to be used as a detector magnet at the collision point of the Electron-Positron-Storage-Ring, the system axis coinciding with the beam axis. First a single prototype segment will be built with the schedule for fabrication and test taking about 1.5 years. It is planned that two years later the complete Oktopus will begin its operation.

The responsible leader for this project is Dr. G. Horlitz, DEUTSCHES ELEKTRONEN SYNCHROTRON (DESY), Hamburg, Dr. G. Horlitz is also leader of the DESY-group which is engaged in superconducting magnets for high energy physics.

At the INSTITUT FÜR EXPERIMENTELLE KERNPHYSIK (IEKP) Karlsruhe conceptual design work was started on beam bending and detector magnets for new high energy physics experiments in different European laboratories. This work will be accomplished within a common program of the GESSS with the aim of standardizing and simplifying magnet types for high energy physics. (GESSS = Group European Superconducting System Studies, a collaboration of the laboratories C.E.N. Saclay France, Rutherford Lab. U.K. and IEKP Karlsruhe, Germany). Responsible for the IEKP part are Prof. W. Heinz and Dr. P. Turowski.

B.   Pulsed Magnets for High Energy Physics

In 1970 plans have been discussed at CERN, according to which the new large proton-synchrotron (2.2 km diameter) should be designed to an energy level of 150 GeV with conventional magnets, but later on should be completed in several steps by adding superconducting magnets units leading to a 650 GeV energy level. The decision, whether the super-conducting version or not, was originally envisaged for the end of 1973.

To prove the feasibility of superconducting synchrotron magnets the above mentioned GESSS-Group was formed in 1970, with IEKP Karlsruhe as German partner. After successful tests of some laboratory-scale magnets, a first large prototype magnet with 1.4 m length, 8 cm Ø bore and a rated flux density of 4.5 T has successfully been operated at the IEKP Karlsruhe in July 1973. The next step in this program will be an improved

prototype with low losses and low training effects. The development of
superconducting prototype magnets is supported by parallel investigations
on superconducting composites, cables and tapes, testing of structural
materials at low temperatures and studying the influence of radiation on
material properties. The program is headed by Prof. W. Heinz and
Dr. P. Turowski (the latter as far as the prototypes are concerned). One
third of the budget comes directly from the New Technology Program of the
BMFT. The other two thirds from the normal budget of the Institute.
It is intended to continue the program until 1976 and complete it by constructing a 1:1 scale magnet which fulfills all specifications of HEP.

The part conversion of CERN II requires the large scale fabrication
of superconducting synchrotron magnets. Therefore in 1971 SIEMENS
AG started a program of developing superconducting synchrotron magnets.
The program was well synchronized with the corresponding program of
IEKP Karlsruhe. Its special concern was developing and observing
fabrication methods for the magnets which are close to industrial ones.

In 1972 it was realized that the superconducting version of CERN II
was more and more unlikely. In coordination with the BMFT, who supports
the efforts of SIEMENS AG by 50% it was decided to reduce the research
staff, the number of subprojects and to stretch the planning from the end
of 1973 to July 1975. Currently a prototype dipole magnet with a magnet
length of 0.5 m, an aperture of 6 cm and a rated flux density of 5 T is at
assembly and will be tested in November 1973. The magnet program of
SIEMENS Erlangen is headed by Dr. R. Gremmelmaier and Dr. G.
Bogner, project leader is Mr. H. Salzburger.

C.   Magnets for Plasma Physics

The work on plasma physics with respect to controlled nuclear fusion
and magneto hydrodynamic generators is concentrated at the two large
governmental laboratories MAX PLANCK INSTITUT FÜR PLASMAPHYSIK,
Garching near Munich and KERNFORSCHUNGSANSTALT Jülich near
Aachen. As far as the application of superconductivity to plasma physics
is considered the activities of KFA Julich have been moved to IEKP
Karlsruhe, as already mentioned above.

At the MPI for Plasmaphysics Garching two larger magnet projects
for controlled fusion experiments have been developed: A multipole
(quadrupole) system, which consisted of two concentric Nb_3Sn rings which
were to be freely suspended in a magnetic field of external conventional
coils. This project has been stopped a little while ago since it was not
expected to yield new results in comparison with more advanced experiments
(especially Tokamaks) elsewhere. In 1970, there were plans to put a
superconducting stellarator arrangement into operation in 1974. It was
originally intended to produce the toroidal magnetic field of 4 T and a mean
diameter of 4 m by means of 40 intrinsically stable superconducting

magnet units in separate cryostats. For this purpose, in an intensive collaboration with SIEMENS Research Laboratories in Erlangen, a prototype magnet with a free bore of 0.8 m was developed and successfully tested at the SIEMENS labs at the end of 1972. The results obtained with this magnet were considerably above the state of the art at the end of 1972 (maximum flux density: 5.7 T, current density in conductor: 187×10^6 A/m^2, stored energy 1.7 MJ).

During the construction of the prototype magnet, it turned out that a conventional pulsed stellarator arrangement with watercooled magnets would be much more flexible with respect to other experiments and it was therefore decided to cancel the superconducting project. The experiments with the prototype magnet, however, will be continued at Garching. In these tests especially,the mechanical behavior of this type of magnet will be investigated by simulating stellarator arrangements. The tests will be headed by Mr. Schmitter of MPI Garching. In accordance with other laboratories in the world MPI Garching doesn't provide any larger superconducting experimental machine within this generation of experiments. The next generation of experiments will be carried out on an European basis, most probably also using pulsed conventional magnet arrangements.

At the IEKP Karlsruhe an experimental and theoretical study on the feasibility of superconducting energy storage devices with short pulse times has just been started. The first step of this program is the construction and test of a 100 KJ coil with a superconducting switch, designed for a discharge time of ≤ 10 ms and a maximum discharge voltage of 50 KV. In the second step, a storage device for some MJ and a maximal voltage of 100 KV will be constructed. It should be ready in 1976. Within this research area, a program has been started for the development of optimum superconducting switches for storage devices. The latter work is a continuation of the investigations already started in 1972 at KFA Julich. It includes research work for the selection of suitable superconducting materials and construction work to yield a switch system with short cycling times. The work is harmonized with theoretical studies, which have been carried out at the DEUTSCHE FORSCHUNGS- UND VERSUCHSANSTALT FÜR LUFT- UND RAUMFAHRT (DFVLR) in Stuttgart under the leadership of Prof. W. Peschka. It is hoped that within the scope of this work additional important data will be obtained, for instance on high voltage behavior of materials and cryogenic fluids, high voltage-high current feed throughs and on optimization of cooling with helium at various states.

With respect to system studies, studies on energy storage systems and fusion magnet systems (in collaboration with MPI Garching) will be emphasized in 1973 to 1975. The program mentioned above is under the leaderships of Prof. W. Heinz and Dr. P. Komarek. It is financed at two thirds by the budget of the Institute and at one third directly by the BMFT.

At the KFA Jülich a large superconducting MHD-magnet was operated together with the largest noble gas MHD-installation "Argas". Conceptual designs and technical and economical studies on MHD-magnets for power plants have been carried out. The work managed by Dr. P. Komarek was finished. A superconducting 10 MW-MHD-project was planned at the MPI Garching. Design studies carried out together with the SIEMENS Research Laboratories in Erlangen were concluded in early 1972. Meanwhile, this project, which was under the leadership of Dr. Mundenbruch MPI Garching was also stopped.

IV.   MAGNETS FOR INDUSTRIAL AND SPECIAL APPLICATIONS

The application of superconducting magnets for industrial use is limited to some small test facilities for magnetizing new conventional persistent magnetic materials, such as SmCo, up to higher flux densities. Preliminary studies on magnetic separation of minerals and water-purification have been carried out, but at present there is no active experimental work in this field.

Research and development work on a special but very interesting application of superconducting magnets, namely as lenses in electron microscopy, is performed at the Research Laboratories of SIEMENS AG in Munich. The main advantages of superconducting lenses in comparison with conventional ones are: small volume, small weight and small focal length. (At 5 MV beam voltage for instance the weight of a superconducting, lens, including cryostat, is 100 times smaller than that of a conventional one.) A further advantage is its high resolution. An objective lens without any iron was developed which has an especially high peak induction. It consists of a coil made of NbTi multicore wire, coaxially aligned shielding cylinders made of Nb_3Sn sintermaterial in the center of the coil and a shielding casing of the same material around the system. Such a lens was used in a 400 KV-electron microscope. A resolution of 5Å was achieved.

The plans for the future include the improvement of the components and the design of a high voltage microscope column consisting of a cryostat with all lenses superconducting. The work is under the responsibility of Dr. Pfisterer and Dr. Fuchs. The project leader is Dr. I. Dietrich. The work started in 1966 was first entirely financed by the company. In the past years it is supported by the BMFT at a 50 percent level. The program is funded until the end of 1974.

V.   ROTATING ELECTRICAL MACHINERY

Within this area two topics are investigated at present in the FRG, large turbo alternators for power stations and dc machines for propulsion.

A.   Superconducting Turbo-Alternators

The power output units of the largest conventional alternators built at present by the KRAFTWERK UNION AG (KWU) are 1500 MW. They are achieved by 4-pole watercooled units. The estimated power limits of conventional alternators are: about 2500 MW for two-pole alternators and about 3000 MW for four-pole machines. It is expected that there will be need for such power units in the late eighties. After some preliminary studies on superconducting alternators KWU and SIEMENS started at the beginning of 1973 a three year program in which it shall be elaborated whether it is technically feasible and economically attractive to construct and use superconducting alternators with power ratings ≥ 3000 MW. Within this program no model generator will be built. This decision will be made after the end of the study program. The program is an intensive collaboration of the KWU and the SIEMENS Research Labs. in Erlangen. KWU is responsible for the complicated magnetic field computations, and the electrical and mechanical design of the rotor and stator. SIEMENS AG Research Labs are responsible for measurements on superconductors, mechanical measurements on structural materials, the development and construction of rotating current leads, and rotating He-transfer- and vacuum-coupling. The program is headed by Dipl. Ing. Heinrichs and Abolins of KWU and Dr. R. Gremmelmaier and Dr. G. Bogner of SIEMENS AG. It is supported at a 50% level by the BMFT.

For the past two years the INSTITUT FÜR ELEKTRISCHE MASCHINEN UND GERÄTE of the Technical University of Munich has carried out work on large superconducting alternators. The investigations were concentrated on theoretical and experimental determinations of the magnetic fields and reactances of such generators. The experimental work was supported by BBC/Switzerland. The further aim of this program is to construct within the next year a 300-400 KW model alternator with rotating armature and stationary superconducting field winding. The program is headed by Prof. H.W. Lorenzen and Dr. J. Sergel. The Institute has applied for supporting this one year program at the DEUTSCHE FORSCHUNGSGEMEINSCHAFT (DFG).

B.   Rotating Machinery for Propulsion Systems

In this field work is carried out by SIEMENS AG which is especially concerned with dc machinery of the heteropolar type, with a superconducting quadrupole field winding and a normal conducting rotating armature with commutator. There is a sharing of the work between the NÜRNBERGER MASCHINEN UND APPARATE WERK (NMA) of SIEMENS AG which is responsible for the electrical and mechanical design of a motor with a power of some megawatts, and the SIEMENS Research Laboratories in Erlangen which are responsible for the problems concerning superconductivity and cryogenics. The leading personnel at NMA are Dipl. Ing. Benecke and Dr. Thum; for the Research Laboratories again Dr. R. Gremmelmaier and Dr. G. Bogner. The program is carried out parallel to the alternator program and is also supported by the BMFT at a 50% level.

VI. ELECTRIC POWER TRANSMISSION

Relatively large efforts are being put into the development of superconducting high power cables. There are two programs, one by SIEMENS AG with the aim of developing an ac cable (120 kV; 2500 MW) and one by the working group of AEG-TELEFUNKEN-KABELMETAL and the German LINDE with the task of developing a dc cable ($\pm$ 200 kV; 5000 MW). In the working group AEG is responsible for the elctrical conductor and dielectrics, KABELMETAL for the thermal insulation and LINDE for the cooling cycle. The work on cables began during the second half of the sixties and was financed at first only by the companies. Since 1970 the programs are sponsored by the BMFT at a 50% level.

The favored design of the working group is a totally flexible single conductor cable type with a helically wound hollow conductor and a thermal insulation consisting of corrugated tubes. The concept preferred by SIEMENS is semiflexible i.e. with flexible conductors in a rigid thermal insulation. The conductor used for the ac cable most probably will be especially developed Nb, and that for the dc cable commercial Nb_3Sn. Extended measurements on thermal insulations, conductors, and dielectrics were carried out by both parties. SIEMENS emphasized the development of copper and aluminum conductors clad with thin layers of low lossy Nb. Design and construction of joints and terminals are under way.

At present at the Research Laboratories of AEG, Frankfurt, a 15 m long cable section bent into an oval loop is being tested. This piece of cable contains a joint and the current is fed in by induction with the aid of four transformers. A high voltage test is in preparation for which a dc voltage of 200 kV to ground will be applied to a 20 m long test section. The test will be performed at the KABELWERKE RHEYDT of AEG.

At the Research Laboratories of SIEMENS in Erlangen a 30 m long, one phase model cable is under preparation. It shall be tested synthetically for an ac operation voltage of 120 kV and a current of 12 kA. The flexible conductor for this model cable will be fabricated at the SIEMENS KABELWERK Berlin. The test is planned for the second half of 1974.

The first sections of the development programs which were timed to the development of the cable components will be finished at the end of

1973 (AEG, KABELMETAL, LINDE) and at the end of 1974 (SIEMENS).
Future plans include improvements of the components and designs,
intensified work on field installation techniques, operational control and
safety. Also test lines with lengths of the order of 100 meters will be
operated parallel to existing conventional lines. The working group
(AEG-KBELMETAL-LINDE) has decided to change to ac cables for this
second part of the program. It has to be pointed out, however, that the
second part of the program has not yet been approved, neither by the
companies nor by the government. The responsible persons for the
programs are: Dr. H. Heumann and Dr. H. Franke (AEG), Prof.
Scheffler KABELMETAL and Dr. Baldus LINDE AG; Dr. R. Gremmelmaier
and Dr. G. Bogner for the SIEMENS program.

VII. LAND TRANSPORTATION SYSTEMS

     In Europe, the installation of a nonconventional high speed ground
transportation system is considered to be necessary for the nineties. It
should supplement the further development of conventional railways and
it should replace the medium range air traffic (1300 km) which is running
into more and more problems. The Ministries of Research and Technology
(BMFT) and of Transportation (BMV) have combined their efforts with
the Federal Railways (DB), some universities, and some interested
industrial companies, to prepare for these tasks in the near future.

     The phase of preliminary studies has ended with the presentation of
the "Study on a High Speed Ground Transportation System" of the HSB-
Studiengesellschaft in December 1971.

A.   The Main Topics to be Worked on in Western Germany:

1.   Advanced railway techniques
      Here the automation of the transportation of goods as
      well as the increase of the effective travelling speed for
      passengers are of prime interest. Experimental runs of
      up to 400 km/h on a track of the Large Experimental
      Facility Donauried in the late seventies will lead to basic
      technical and economical knowledge on the upper speed limit.

2.   Air Cushion techniques
      An eleven ton vehicle has been successfully operated by the
      end of 1972 at more than 130 km/h by KRAUSS-MAFFEI (KM).
      The goal is to get comparative data to a similar vehicle
      with ferromagnetic lift and guide components on the same
      track.

3.   Magnetic levitation techniques
      Preliminary studies on permanent magnet levitation (AEG; KRUPP)
      being completed, the inductive and the ferromagnetic systems
      are studied now with great priority. KM and MBB (Messerschmidt-
      Bölkow-Blohm) put into operation experimental vehicles (of the
      ferromagnetic suspension type) as early as 1971. In 1972 they
      achieved,
      180 km/h on a 700 m track (MBB)
      164 km/h on a 1000 m track (KM)

      In these months MBB is constructing a straight track of 1200 m
      for 400 km/h and KM is completing a double curved track of 2500 m
      for 350 km/h.

      All these efforts aim at the layout of realistic vehicles (100 tons)
to be operated in the Donauried facility (see below), the design of which
is beginning now. The projects incorporating inductive means of suspension are conducted by the three electrical companies AEG, BBC and
SIEMENS. Their work will be presented in more detail now.

B.   The Inductive Levitation Technique

First Step: R&D-on Components: AEG, BBC and SIEMENS have put their
efforts together in a cooperative basic program on transportation components, the main objects being:

1.   Linear Motors (Induction and synchronous types,frequency converters,
      coordination: BBC, Dr. Hochbruck).

2.   Magnetic suspension (coordination: SIEMENS, M. Marten; for the
      superconducting part subcoordination: Dr. Bogner, C. Albrecht)

3.   Power pick up systems (coordination: AEG, Dr. Bopp).

      The goal of the program is the realization of prototypes of the
components. The main concept is inductive suspension with superconducting magnets (SIEMENS). BBC is investigating the problems of conventional magnets for ferromagnetic suspension. The basic program is
50% supported by the BMFT. The first step is not yet completed.

Second Step: Test of Components on Experimental Tracks: The BMFT
sponsors application studies, tests of components and the construction of
a test track in Erlangen at a funding level of more than 50%. The "Erlangen
Circular Test Track" will be described below. For closer cooperation
in planning and execution of the tests at Erlangen (and for the following
steps) AEG, BBC and SIEMENS founded the "Project Group Magnet
Levitation" (PM); central office in Erlangen, project manager Mr. A.
Lichtenberg.

Third Step: Large Test Facility DONAURIED: 40 km north of Augsburg
a large test facility will be built for high speed experiments. Final
extension is 70 km of tracks forming a closed double loop. Tracks for
advanced railway investigations (400 km/h) and for magnetic suspension
techniques   (500k m/h) are provided separately. Several organizations
share in the realization of this big project:

1.   GBI: (Gesellschaft fur Bahntechnische Innovation = Company for
        innovation of tracked transportation techniques)

        Concern: Collection and harmonizing of all requirements as
                          being demanded by the different appliers and
                          partners.
        Begin: 1971
        Sponsored: by BMFT
        Manager: Dr. Straimer
        Associates: AEG, BBC, KM, KRUPP, MBB, SIEMENS, STRABAG
        Further
        Associates: Dyckerhoff and Widmann, LINDE, MAN, SEL

2.   BAUSTAB: (construction staff)

        Concern: Planning and execution of the construction
        Begin: 1972
        Sponsored: by BMV
        Manager: Dr. Töpfer
        Participants: working staff of the DB, now being subordinated
                          directly to the BMV.

3.   PK: (Planungskonferenz = planning conference)

        Concern: Supervision of the planning of the facility,
                          harmonization of the different proposals
                          between GBI, DB and BMV, preparations for an
                          organization which operates the facility
        Begin: 1972
        Leader: Dr. Hoch, BMV
        Participants: BMV, GBI and a project group of the DB

        The start of construction of the first section is envisaged for 1974,
with the beginning of tests on the tracks scheduled for 1975/76. The
first separate sections shall enable the technical comparison of the two
competitive magnetic levitation systems under equivalent conditions and
thus lead to a final decision by the end of the seventies on the system
to be realized.

        The experimental runs on the completed tracks will provide for know-
how on speed limits and operational behavior of realistic vehicles which
will be prototypes of the final high speed coaches of a public transportation

system for advanced railways as well as the selected magnetic suspension technique.

C.   The Erlangen Facilities

A circular test track has been built in Erlangen in order to test several components in realistic continuous operation. On this circle of 280 m diameter an experimental vehicle weighing about 18 tons will be operated at 200 km/h maximum. In the meantime this vehicle is nearing completion. The main components of the system are:

- the concrete track (Dyckerhoff and Widmann)
- the power pickup system (AEG-Teldix)
- the vehicle itself (MAN)
- the linear induction motor (AEG)
- the cryogenic supply (LINDE)
- the superconducting magnets (SIEMENS)
- the command and control system

VIII. HIGH FREQUENCY CAVITIES AND JOSEPHSON ELEMENTS

A.   High Frequency Cavities

The development of superconducting Nb cavities with high Q-values and high critical HF-fields is carried out in close collaboration with IEKP Karlsruhe and the Research Laboratories of SIEMENS in Erlangen. The work is primarily concerned with the construction of HF beam separators and proton linacs.

At SIEMENS special surface treatments (electropolishing and anodizing) of Nb were developed which lead to X-band Nb cavities with quality factors of 1×10^{10} and critical RF fields of 0.16 T (for a TM_{010} mode) for instance. Also manufacturing techniques for high precise resonator structures such as separators were successfully developed. At present SIEMENS is manufacturing two 3 m long particle deflectors by order of IEKP Karlsruhe. They will be finished in 1974 and will be applied to the CERN 300 GeV proton synchroton in 1976.

At the IEKP Karlsruhe a prototype of a superconducting proton linear accelerator has been constructed and tested as a first stage of a projected accelerator with a duty factor unity, current 1 mA and energy 0.5 - 1 GeV. The first segment consists of helix structures. In the meantime, two half meter sections each of five $\lambda/2$ spirals have been built. The spirals are made of electropolished, anodized niobium tubes (6.3 mm diameter) and cooled with superfluid helium and are housed in a lead or niobium resonator. Such sections have been successfully tested; protons were accelerated with a maximum gradient of 1.4 MV/m. The research work of SIEMENS is sponsored at a 50% level by the BMFT. The leading personnel at IEKP Karlsruhe are: Prof. Citron, Mr. Lengeler (separator) and Dr. Kuntze (Linac); at SIEMENS: Dr. R. Gremmelmaier and Dr. H. Pfister.

B.   Josephson Elements

Research work on Josephson elements is carried out at the Physi-kalisch-Technische Bundesanstalt in Berlin (H.D. Hahlbohm) and in Braunschweig (V. Kose). In Braunschweig a measuring apparatus with niobium point contact elements is constructed for the determination of e/h and for voltage standards. At the SIEMENS Research Laboratories in Erlangen Josephson elements consisting of tunneling films and weak link bridges are being developed. Niobium films prepared by evaporation or sputtering techniques are preferred as superconducting films, and niobium oxide as insulating layer. Photo etching as well as electron beam techniques are provided for preparing the weak links. The work of SIEMENS is sponsored by the BMFT at a 50% level. It started in 1972 and will continue up to 1974. The leading personnel are: Dr. R. Gremmelmaier and Dr. H. Pfister.

IX.   CONCLUSIONS

In the Federal Republic of Germany a well balanced program on the application of superconductivity exists. Besides magnets for high energy physics the programs are especially designed for applications for the eighties and nineties. The main efforts at present are in the land transportation systems including levitation by superconducting magnets. The total per years funding of the applied superconductivity program is quite significant. Nevertheless, it has to be increased during the coming years for the construction and testing of large scale superconducting models.

ACKNOWLEDGMENTS

I especially want to thank Dr. Komarek from IEKP Karlsruhe and Dipl. Ing. Albrecht, SIEMENS AG, for their very valuable contributions. Also the kind support from Mr. Hillmann, VAC, Dr. Isolde Dietrich and Dr. H. Pfister, SIEMENS, has to be mentioned.

Chapter 13

PROGRAMS ON LARGE SCALE APPLICATIONS OF SUPERCONDUCTIVITY
IN ITALY

C. Rizzuto

G.N.S.M. del C.N.R. e Istituto di Scienze Fisiche

Genova - ITALY

I.   INTRODUCTION

The development of applied superconductivity in Italy is less advanced than in other more industrially developed countries. A start, however, has been made. We give a brief outlook of what is being done, including also a few details on the research laboratories and low temperature facilities existing in Italy. The details are shown in Tables 1 to 4.

II.   CRYOGENICS

The low temperature research laboratories are mainly organized and financed by C.N.R. (National Research Council), either directly, as is the case of the Metrology laboratory in Turin, or the laboratory of Cold Technology in Padova, or through the national Group of Structure of Matter (G.N.S.M.) which has sections operating in several cities, of which those operating at cryogenic temperatures are in Genova, Padova, Roma, Catania, Palermo and Cagliari or through two industries: S.I.O. (Italian Oxygen Company, connected to Air Liquide, France) Milano, and Rivoira (which has an agreement with Gardner, USA) Torino. Both of them have facilities either for importing the liquid helium or for liquefying it (Table 2).

III.   APPLIED SUPERCONDUCTIVITY

Applied superconductivity studies of interest to large scale applications are being made in:

A)   Frascati (C.N.E.N.): Superconducting materials: development of
       Nb_3Sn strips and wires and construction of superconducting magnets.

615

This research is being done also in collaboration with other laboratories and industries. This research was started about five years ago.

B) Genova (ASGEN): A 85 KW homopolar motor with superconducting field windings, and two series rotating disks is being built. This work is in cooperation with the C.N.R./G.N.S.M. section of Genova, which is presently performing measurements on the low temperature properties of fiberglass epoxies and of low temperature instrumentation.

C) Palermo (University, Faculty of Engineering): A research program has started this year on superconducting levitation for transportation. This laboratory has already been working on conventional levitated transport and linear motors.

D) Another large scale application (although not conventional) can be defined the levitation of a five ton heavy aluminum cylinder in the gravitational wave experiment being studied by the Solid State Electronics Laboratory of C.N.R./G.N.S.M. in Rome (in collaboration with other laboratories in Rome).

Some research work preliminary to large scale applications is being or has been conducted, mainly on matters relating to superconducting cables at Pirelli (Milano), Genova (engineering faculty) and SNAM (Monterotondo). Other smaller scale programs are also listed in Tables 3 and 4.

IV. FUTURE RESEARCH

The above is the picture of applied superconductivity in Italy. What are the near future prospects?

A program based on the development of superconducting materials and on the preliminary design of a 100 MW superconducting generator which should be financed with about 2 million US $ is presently being considered by C.N.R. While there seems to be a warm support by everybody to its scientific content, the main difficulties are being experienced to find how to overcome the bureaucratic situation which exists in Italy.

TABLE 1. LIST OF ORGANIZATIONS APPEARING IN THE TABLES

ORGANIZATION

C.N.R.	- National Research Council
G.N.S.M./C.N.R.	- National Structure of Matter Group
C.N.E.N.	- Nuclear Energy Committee
I.S.M.L.	- Light Metals Experimental Laboratory (Novara)
E.N.E.L.	- National Electrical State Industry
A.S.G.E.N.	- Ansaldo S. Giorgio Generale (Genova)
S.I.O.	- Italian Oxygen Company (Milano)
S.N.A.M. connected to E.N.I.	- National Petroleum State Industry (Technical Branch)
I.E.N.G.F.	- Electrical and Frequency Standards Institute (Torino)
I.N.F.N.	- National Institute of Nuclear Physics
L.T.M.	- Laboratory for Non Common Metals (Milano)

Funding of programs is made in most cases: a) from Government to C.N.R.,
b) through University (Ministry of Education), c) C.N.E.N. (Ministry of Industry)
d) I.M.I. (Industrial Development Fund), e) E.N.E.L.

TABLE 2.   CRYOGENIC LABORATORIES IN ITALY

Laboratories	He Consumption Liters/year	Type of Programs
CNEN (Frascati)	~10,000 (Collins)	High Energy Physics - Liquid Helium- Applied Superconductivity
CNR/GNSM (Genova)	~ 6,000 (CTI 1400)	Solid State - Cryopumping - Applied Superconductivity
SNAM (Rome)	Starting (CTI 1400)	Cryopumping - Solid State
Metrology (Torino)	~ 500 (Collins)	Temperature - Metrology
CNR/GNSM (Palermo)	(Collins)	Solid State
University (CNR/GNSM &Catania)	~ 500 (Philips)	Tunnelling
University (Trieste)	(Philips)	Transport in Metals
SIO (Milano)	~ 5,000 (TBT)	Sale (+ Air Liquide)
Rivoira (Torino)	~ 500 (Collins)	Sale (+ Gardner)
Euratom (Ispra)	(Linde)	
C.N.R. (Naples)		Josephson tunnelling
CNR/GNSM (Cagliari)		Solid State
CNR/GNSM (Bologna)		Electron microscopy
IENGF (Torino)		Voltage and frequency standards
ASGEN (Genova)		Superconducting motor
C.N.R. (Rome)		Cryogenics
C.N.R./G.N.S.M. (Rome)		He properties, applied s/c.
C.N.R./G.N.S.M. (Padova)		He properties, applied s/c.
I.N.F.N. (Padova)		High energy physics

TABLE 3. LARGE SCALE APPLICATIONS OF SUPERCONDUCTIVITY IN ITALY: DEVICES

Organization	Contact at school (or manager)	Device	Type of Program	Funding K$ and source	Status & Future Plans
ASGEN-Genova	Stopiglia Vivaldi	DC Homopolar motor /generator, twin disks 90 KW	Design and Construction and test	400 for 3 years IMI (Ind. Development funds)	Design completed Construction started in 1973
Aeronautics Institute, University of Palermo	Prof. Mattioli	Train levitation	Design preliminary tests		Preliminary
Laboratorio Elettronica Stato Solido CRN/GNSM Rome	Modena	Gravitational waves detector (static levitation)	Construction testing	~100	In construction
Pirelli/Dunlop (Italy)	Lombardi	S/C cables	Feasibility estimate		Completed, no future plans in Italy.
ENEL		S/C cables	Feasibility est.		Feasibility study in progress
INFN/V Genova	Manuzio	S/C cavities (Pb + Nb)	Small accelerators	INFN	Pb tested Nb in construction
Electrotechnic Institute Univ. of Genova	Centurioni Molinari	S/C Cables and generators	Electro-dynamic forces in cables S/c	CNR+ENEL	In Progress
Univ. of Rome CNEN Rome + CNR Naples	Cerdonio Barone Paterno	Josephson instrumentation	Construction and testing	CNR	In Progress
Univ. of Genova CNR/GNSM	Vaccarone			CNR + CNEN CNR	
IENGF (Torino)	Andreone	Josephson voltage standard	Construction		Completed. Josephson frequency standard in program.

TABLE 4. MATERIALS

Organization	Contact at school (or manager)	Type of Material	Type of Program	Funding K$ and source	Status and Future Plans
CNEN	Sacerdoti Sacchetti Spadoni Ricci	Nb_3Sn) Nb_3Al) Ribbons + stabilized cables and Cu stabil.	Metallurgical reaction and feasibility and H_c and I_c tests	~200 over 4 years CNR	Tested effect of defects on Nb_3Sn formation + I_c (Zr + cold working)
ISML	Ceresara	Nb_3Al Nb_3Sn	Preparation and physico - chemical investigations	CNR	Formation in thin filaments + ribbon + multifilaments (tested at CNEN)
LTM (Milano)	Donolato Strocchi	Nb + Nb alloys + ribbons	Metallurgical	~70 CNR	Alloys for tests at CNEN
University of Genova Electrotechnics	Centurioni Molinari	Dielectric	Testing	~20 CNR + ENEL	LN_2 in progress LHe_4 in program
SNAM		Dielectric	Testing	ENEL	Solid insulation in LN_2
University of Genova CNR/GNSM	Rizzuto Pizzo Fassino Vaccarone	He^4 Fiber glass	Test of rotary joints and behavior Thermal Conductivity and diffusivity Mech. strains	~10 CNR	Rotary dewar in construction Measurements in progress. Programs on single thin wires, fluxons etc. Basic research on s/c alloys
University of Bologna CNR/GNSM	Vittori Antisari	V_3Si	Electr. microscopy of fluxon lines	30 ~ 50	Testing penetration of fluxons at H_{c1}

PROGRAMS ON LARGE SCALE APPLICATIONS OF SUPERCONDUCTIVITY IN JAPAN

K. Yasukochi and T. Ogasawara

Department of Physics, College of Science and Engineering

Nihon University, Chiyoda-ku, Tokyo

I. INTRODUCTION

The following notes give a short survey of the present, large scale application of superconductivity in Japan. Since in many aspects, research and development have taken a different way in Japan compared with other countries, we thought it necessary to begin with a short historical account of the progress (Section 2). Section 3 gives a review of the present state, where we have divided the major projects into the following lines: 1) MHD power generation systems, 2) superconducting rotating machines, 3) cryogenic power transmission systems, 4) magnetically levitated transportation, and 5) application for high energy physics. Programs on basic superconductivity science, superconducting materials and refrigeration systems are not included. This survey covers data up to August 1973. A tabulated version can be found at the end of these notes.

II. HISTORY OF THE LARGE SCALE APPLICATION OF SUPERCONDUCTIVITY IN JAPAN

The High Field Magnets Conference held at the Massachusetts Institute of Technology in 1961 was the starting point of the development of superconducting magnet technology. Up to that time, the research activities in superconductivity had been very low in our country. The state of affairs may be conceived from the fact that the number of helium liquefiers installed in Japan at that time was not more than ten and basic studies on superconductivity were made in only a few laboratories. Largely from the impetus due to the discovery of high field superconductors, several companies started the development of superconducting materials,

621

and an increased effort was paid to basic studies in universities and
governmental institutions.

1966 is the year   worth mentioning for cryogenic engineering in
Japan. Firstly, the Cryogenic Association of Japan was founded in order
to organize and promote research and development of applied superconductivity and refrigeration systems. The second event was the start
of a big project on magnetohydrodynamic power generation systems at
the "Electrotechnical Laboratory", which belongs to the "Agency of
Industrial Sciences and Technology". In this project some emphases was
given to the development of a large scale superconducting magnet for an
MHD generator. This target, together with the first International
Cryogenic Engineering Conference held in the next year (1967) at Kyoto,
stimulated the research on the application of superconductivity. As one
of these results, the production of superconducting wires and cables was
started by several companies such as Hitachi Ltd., Mitsubishi Electric
Co., Toshiba Electric Co., Ltd., Japan Vacuum Engineering Co., Ltd.,
and the Furukawa Electric Co., and small superconducting magnets found
their way into many laboratories. With the advance of the big MHD project,
main contractors, Hitachi, Mitsubishi Electric and Toshiba Electric
constructed relatively large superconducting magnets of fully stabilized
mode in cooperation with the Electrotechnical Laboratory; Mitsubishi
operated a 75 kG solenoid type magnet with a stored energy of 2 MJ in
1968 and Hitachi developed   a saddle shaped magnet of 4.5 MJ in the
next year (1969).

This MHD project also has led to the development of other types of
large scale application of superconductivity, for example: the Toshiba
team paid special attention to the superconducting rotating machine and
constructed a small scale dc homopolar generator (1969). This was
succeeded in 1970 by a 3000 kW dc homopolar generator by the association
for machinery promotion. With regard to cryogenic power transmission
systems, a study group was organized as early as 1964 under the Central
Research Institute of the Electric Power Industry, and a survey of information on cryogenic power transmission was made together with a few
members from some universities and the Electrotechnical Laboratory;
later four cable companies and Tokyo Electric Power Company, Inc. joined
this group. Among these companies, Furukawa Electric engaged in a
model test on superconducting cables from 1970 on.

1970 is also the year that the Japanese National Railways and the
Ministry of Transportation made the decision to start research and
development efforts for a high-speed ground transportation system. This,
we think, as a replacement of the MHD project, will become the research
activity   in large scale application of superconductivity in the late 1970's.
At present, special efforts have been given to a system of linear motor
propulsion and superconducting magnet levitation for suspension and
guidance. And the hardware development by JNR is progressing smoothly
in cooperation with industrial companies like Toshiba Electric, Hitachi,

Mitsubishi Electric and Fuji Electric, and a demonstration vehicle has been constructed recently. The final confirmation of the feasibility is expected in 1977.

In European countries and in the United States, high energy physics laboratories have taken the leadership in the development of large scale application of superconductivity. But, very unfortunately, this is not the case in Japan. After many turns and twists, the National Laboratory for High Energy Physics made a start in 1971 at a reduced scale compared to the original expectations. Experimental works on superconducting magnet systems for high energy physics have just begun but, for the time being, are not progressing too rapidly.

III. CURRENT STATUS OF THE LARGE SCALE APPLICATION OF SUPERCONDUCTIVITY IN JAPAN

In the following, we intend to give a survey on the current status of large scale application of superconductivity in Japan. For convenience, the programs have been divided into five categories; 1) MHD power generation systems, 2) superconducting rotating machines, 3) cryogenic power transmission systems, 4) magnetically levitated transportation, and 5) applications in high energy physics experiments.

A. MHD Power Generation System

The research and development of MHD power generation systems was started in 1966 at the Electrotechnical Laboratory, the Agency of Industrial Sciences and Technology, Ministry of International Trade and Industry (MITI). Here, it seems to be necessary to explain briefly this typically Japanese concept of a "big project". There exists a considerable amount of new and innovative technology which is important for future Japanese industries. If the technology in question is of high risk for private investment at the current state of art, the government supports the development of that technology through the so called Agency of Industrial Sciences and Technology, and entrusts development to industrial companies. In the case of the "MHD big project", as it is formally called, the Electrotechnical Laboratory took a leading part and the main contractors were as follows; Hitachi, Mitsubishi Electric, Toshiba Electric, the Japan Oxygen Company and Toyoda Machine Works, Ltd. At the start of this project, namely in 1966, Stekly of AVCO Everett Research Lab had just finished the construction of the first large scale saddle shaped superconducting magnet for MHD. Consequently, in the Japanese MHD project, the development of a big superconducting magnet for a 1000 kW class generator was set up as a target of the first seven years plan. It must be noticed that emphasis was given to the combined operation of a superconducting magnet and a combustion system. As the first step of the project, a 45 kG saddle shaped magnet (AVCO type) with a stored energy of 4.5 MJ was constructed.

The next stage was started about 1970 and aimed at the construction of a big superconducting magnet for a 1000 kW MHD generator. The magnet is of the race-track type, with a uniform (90%) field space at room temperature of 10 x 20 x 120cm. According to the design, the central field is 50 kG with a stored energy of 70 MJ, which is the largest of its kind in the world. A successful test was just completed and the combined operation with a combustion system is going to be carried out by the end of this year. The total budget for 1966 to 1972 was about 5 billion yen.

The first phase of the MHD big project came to an end by 1972, and continues for three years with a budget of 900 Million yen from 1973 on. In the second phase plan of this project, a prototype MHD generator has been proposed, but the approval of the government is not obtained yet.

B.   Superconducting Rotating Machines

Toshiba Electric has commenced the research and development of a dc homopolar generator in 1967, and completed the test of a 10kW machine in 1969. Encouraged by this success, the "Japan Society for Promotion of Machine Industry" adopted in 1970 a plan of the construction of a 3000 kW dc homopolar generator to meet civilian technological needs. This society has a branch which promotes the extensive application of new machines, and funds are supplied by the Japanese Professional Bicycle Racing Association. Toshiba Electric was entrusted with the construction in cooperation with Furukawa Electric, The Japan Oxygen Company, and Toyoda Machine Works, Ltd. The total budget is about 180 million yen for 1971 - 1973. The superconducting magnet of this machine generates an average of 40 kG with the stored energy of 26 MJ. The armature consists of two discs in series connection and the output voltage is 150 V at 20 kA. Carbon brushes are adopted as current collectors. After completion of the test operation, which is now going on, the generator is arranged to be transported to Furukawa Nikko Works for a practical trial use for a period of six months as a power source for a copper electrolyzing plant.

On the other hand, the application of superconductivity to synchronous machines is relatively new, only one or two years old. Recently Fuji Electric Co. has constructed a 30 kVA synchronous generator with an outer stationary superconducting field of 4 poles, and an inner rotating armature. The test was completed in 1972. As a next step, they plan to develop a synchronous generator of inner rotating field. Although there is a growing tendency for research and development of synchronous machines in Japan, the project has not yet come to a large scale stage.

C.   Cryogenic Power Transmission System

In Japan, recent increment of electric power demand is drastic and amounts to about 10 percent per year, which must be compared with 7 percent per year in the United States. Therefore, power transmission systems of high capacity is a matter of urgent concern. Under this situation, a study group on superconducting power transmission was organized in the Central Research Laboratory of the Electric Power Industry in 1964. This group made studies from various points of view. With the participation of some cable manufacturers, the group was reorganized in 1968 into the present "Committee on Superconducting Power Transmission". The efforts of this committee include actual designs of experimental model cables, conceptual designs, and cost estimates of some power transmission systems. An interim report of these results appeared in 1973. On the other hand, the Electrotechnical Laboratory has been conducting experimental studies on ac losses and economy analyses of superconducting power transmission.

Since 1969 or so, active programs have been started by some cable manufacturers. Furukawa Electric Co. engaged in the development of superconducting materials and their application to superconducting power cable. In 1971, the group has made an electrification test of 154 kV class model cable; a composite conductor, 25 micron Nb lined to copper tube of 40mm inner diameter. The electrification test was carried out on a coaxial core of 8m length up to 3,000 Amp ac and 5,000 Amp dc. A model test of cryoresistive cable with liquid nitrogen as coolant was made by the Furukawa group and both electrification and electric breakdown tests were completed in 1972. Similar studies have been pursued by Hitachi Cable, The Fujikura Cable Works, and Showa Electric Wire & Cable. These model cables are 20 ~ 30 m in length.

As for future plans of the superconducting power transmission system, the Agency of Industrial Sciences and Technology made a proposal to the government as a big project. In this plan, the total budget amounts to 20 billion yen for the first period of 8 ~ 9 years. Although this proposal has not yet been approved by the government, the plan is expected to start in 1975. For this purpose, the MITI organized the Committee on New Power Transmission Systems in April 1973 and feasibility studies and discussions in research and development plan have been started. Cable makers are awaiting the take-off of this big project for the substantial scaling up of their works, and the heavy electric industry groups are considering their future plan of related superconducting electric machines in connection with this project.

D.   Magnetically Levitated Transportation

The Japanese National Railways started the operation of Tokaido Shinkansen (New Tokaido Line) between Tokyo and Osaka in 1964, and today the number of passengers amounts to 250,000 per day. Due to the steady increase of passengers, the capacity of New Tokaido Line is expected to be saturated by 1983 or so, and therefore an additional line must be constructed by 1985. In order to meet this growing demand, a new high-speed ground transportation system is needed, which can connect

Tokyo and Osaka (515 km) in about one hour. This system not only would become important in connection with a nationwide network expanded from Tokyo and Osaka, but also would release the rush on domestic airlines. In view of the above situation, JNR and the Ministry of Transportation made a decision in 1970 to start research and development of a high-speed ground transportation system. The Tokaido zone between Tokyo and Osaka is the so-called "Tokaido Megaloporis". The consideration of factors such as safety, noise and pollution has led to the following conclusion: The most desirable system is the combination of a linear motor propulsion with the active side on ground, and a superconducting magnet levitation for suspension and guidance. Along this line, JNR has been engaged in research and development by their own investment. It seems that JNR intends to have this project designated as a national project, after the feasibility is confirmed. In parallel with JNR's efforts, the Ministry of Transportation is supporting corresponding research projects of industrial companies such as Hitachi, Toshiba Electric, Mitsubishi Electric, Fuji Electric, Vacuum Metallurgical Co., Ltd., and Sumitomo Shipbuilding & Machinery Co., Ltd. The funding support, however, is relatively low and active researches in these industrial companies are maintained rather by their own corporate efforts.

In the beginning of a series of tests for superconducting magnet levitation, JNR constructed a test facility of a rotating disc type, 1.4 m in diameter. A round cryostat containing two semi-circular superconducting magnets is suspended above the disc with normal conductor loops. Dynamic tests on levitation can be made up to a speed of 100 km/hr. The rotating disc experiment was succeeded by a testing facility with a linear track of 220 m in length. The dynamic characteristics of magnetic levitation in combination with a linear synchronous motor drive were studied up to the speed of 50 km/hr. The next step was the construction of a demonstration vehicle with superconducting magnet levitation and linear induction drive, which can carry four passengers. The track is 480 m in length. The demonstration was performed in public in 1972. This vehicle is of course a test car of the magnetic levitation and also demonstration for 100th railway anniversary commemoration of JNR. In JNR's research and development efforts, we can pick up several characteristic points; (i) linear motor drive of active side on ground, (ii) loop track, (iii) combined test run of maglev and LSM. It should be emphasized that the operation of the demonstration vehicle has a considerable effect on the reduction of psychological and sociological barriers against this mode of transportation. A field test at the speed of 500 km/hr on a 7km track is scheduled to be performed in 1975. The research and development for the confirmation of the feasibility is expected to be completed by 1977 and if the results are positive, the target date for the line to be operational is 1985.

E.   Application for High Energy Physics

Since 1958, discussions arose among Japanese nuclear scientists on the necessity of a large scale proton synchrotron. It was 1971 that the National Laboratory for High Energy Physics (KEK) was established. After

a long decade of intensive efforts of nuclear scientists   the Science
Council of Japan and the Ministry of Education gave a total budget of
8 billion Yen. The 12 GeV proton synchrotron is now under construction.
Unlike both Europe and the United States, where high energy
physics laboratories have been progressive centers in the development of
superconducting magnet technology, the KEK is not playing an important
role at present. Recently some work on superconducting magnets for
high energy physics has been started; a superconducting Q-magnet of
20 cm bore and 50 cm length for the beam transportation, and a superconducting magnet for polarized targets. There are also some discussions
of future plans for scaling up the energy of the accelerator. One of them
is a plan nicknamed "TRISTAN", i.e., Three Ring Intersecting Storage
Accelerator in Nippon.

TABLE 1A.   PROJECTS OF LARGE SCALE APPLICATION OF SUPERCONDUCTIVITY
           - PUBLIC AGENCIES, GOVERNMENT, UNIVERSITIES

Organization	Person in Charge	Project Categories	Current Status, Funding
Electrotechnical Laboratory Tanashi-shi, Tokyo	K. Fushimi Y. Aiyama H. Ishii T. Akiyama	MHD Power Generation System	a) 1,000 kW class MHD power generation system - ETL Mark V, race track type SC magnet with uniform (90% up) field space, 100 x 250 x 1,200mm, central field 50 kG, stored energy 70 MJ, completed 1972. b) Refrigerator for Mark V, refrigeration capacity 3 kW at 20 K, liquefaction rate 250 ℓ/hr, completed 1972.
	T. Horigome	SC Power Transmission System	a) Feasibility studies and conceptual design of superconductive and cryoresistive power transmission systems. b) Analysis and model test of 10m long, 1cm diam. supercritical helium passage. c) Studies on thermal and electrical insulators. d) Large scale refrigeration systems; turbo-compressor operating at low temperature. e) AC-losses in superconductors.
	Y. Aiyama S. Todoriki	SC Magnet for Fusion Reactor	a) Design studies. b) Nb3Sn fine-multi conductors. c) SC hollow conductors.
National Laboratory for High Energy Physics (KEK) Oho-machi, Tsukuba-gun, Ibaraki-Ken	H. Hirabayashi M. Kobayashi Y. Kimura T. Nishikawa	SC Magnets for High Energy Phys.	a) Superconducting Q-Magnet for KEK beam channel, Panofsky type, 20cm bore, 50cm length, G = 1.6 G/cm, 1972-74, under construction. b) Basic studies on AC losses in SC magnets.
		RF Cavity	a) Preliminary studies on SC RF-Cavities.
		Design Studies on SC Accelerator	a) Design studies on future superconducting synchroton; nickname "TRISTAN".

TABLE 1A. PROJECTS OF LARGE SCALE APPLICATION OF SUPERCONDUCTIVITY –
PUBLIC AGENCIES, GOVERNMENT, UNIVERSITIES (continued)

Organization	Person in Charge	Project Categories, Current Status, Funding	
Musashino Electrical Communic. Lab. Midori-cho, Musashino-shi, Tokyo	Y. Hoshiko	SC Coaxial Transmission Line	a) Miniature SC coaxial cable, 1.5mm outer diam. with Pb-Cu conductor and FEP dielectric; attenuation: 0.6 db/km at 1 GHz.
Japanese National Railways (JNR)			
Technical Development Department Marunouchi, Chiyoda-ku, Tokyo	Y. Kyotani	Magnetically Levitated Transportation	a) Extensive studies of economy and technical aspects of high-speed ground transportation systems. b) Feasibility studies and conceptual design of linear motor systems with active side on ground and SC magnet levitation for suspension and guidance, from 1970 on. c) Test facilities of rotating disk type for maglev, completed 1970. d) Model vehicle for testing of dynamic characteristics of maglev in combination with LSM drive, completed 1971. e) Construction of the demonstration vehicle ML-100, completed 1972. f) Future plans: vehicle field test at 500 km/h on 7 km track, 1975.
Tohoku University:			
Research Institute for Electr. Commun., Katahira, Sendai,	Y. Onodera	SC Transmission Line	a) Basic study on superconducting active lines.
Laboratory for Nuclear Science, Kanayama, Tomizawa, Sendai,	Y. Kojima	SC RF-Cavity	a) Basic studies on superconducting RF-cavities for high energy physics experiments.

TABLE 1A. PROJECTS OF LARGE SCALE APPLICATION OF SUPERCONDUCTIVITY –
PUBLIC AGENCIES, GOVERNMENT, UNIVERSITIES (continued)

Organization	Person in Charge	Project Categories,	Current Status, Funding
(Tohoku Univ. continued) Faculty of Science, Katahira, Sendai,	T. Ohtsuka	Magnetically Levitated Transportation	a) Analysis of dynamic behaviour of magnetically levitated trains.
University of Tokyo:			
Department of Engineering, Bunkyo-ku, Tokyo	K. Oshima	Magnetically Levitated Transportation	a) Technical analysis and feasibility study on cryogenic systems for maglev.
Instit. for Nuclear Study, Tanashi-shi, Tokyo	T. Katayama	SC RF-Cavities	a) Basic studies on superconducting RF-cavities for high energy physics experiments.
Nihon University College of Science & Engineering, Kanda, Chiyoda-ku, Tokyo	K. Yasukōchi	SC Magnet Technology	a) Basic studies on SC magnet technology. b) Technical analysis of various applications of SC magnet including maglev.
Nagoya University Faculty of Science, Chigusa-ku, Nagoya	R. Kajikawa	SC Magnet for High Energy Physics	a) Superconducting solenoid, 50mm inner diam with polarized targets.
Osaka University Faculty of Engin., Suita-shi, Osaka	K. Ura	SC RF-Cavities	a) Experiments on Pb RF-Cavities for Electron microscopes.
Kyushu University Faculty of Engin., Hakozaki-cho, Fukuoka	F. Irie	SC Power Transmission systems	a) Basic studies on superconducting power transmission systems.

TABLE 1B. PROJECTS OF LARGE SCALE APPLICATION OF SUPERCONDUCTIVITY – COMPANIES

Organization	Person in Charge	Project Categories	Current Status, Funding
Hitachi Ltd.:			
Hitachi Works Hitachi-shi, Ibaraki-Ken Central Research Lab. Kokubunji-shi, Tokyo	T. Kasahara H. Kimura	SC Magnet for MHD	a) 45 kG saddle shaped magnet, stored energy 4.5 MJ, joint work with Electrotechnical Laboratory, supported by Agency of Industrial Sciences and Technology, completed 1969. b) SC magnet for ETL-Mark V 1,000 kW MHD generator, race track type, 50 kG, stored energy 70 MJ, joint work with Electrotechnical Laboratory. c) Design analysis of MHD magnet.
		Magnetically Levitated Transportation	a) Test facility of linear synchronous motor, partially supported by Ministry of Transportation, 7 M¥, FY 1972, completed 1972. b) Basic research on SC magnet for maglev. c) Technical analyses.
Mitsubishi Electric Co.:			
Central Research Lab. Amagasaki-shi, Hyogo-Ken	K. Akashi	SC Magnet for MHD	a) 75 kG SC magnet (solenoid), stored energy 2 MJ, joint work with Electrotechnical Laboratory, supported by Agency of Industrial Sciences and Technology, completed 1968. b) Saddle type SC magnet of CSI type, 20 kG, supported by Agency of Industrial Sciences and Technology, completed 1970.
	E. Ono M. Iwamoto	Magnetically Levitated Transportation	a) Test facility for dynamic characteristics of maglev, rotating drum type, 360 km/hr, partially supported by Ministry of Transportation, FY 1973, completed 1973. b) Development of lightweight cryostat. c) Technical analyses of LSM.
	M. Iwamoto	SC Magnet for Plasma Physics	a) 40 kG cusp type SC magnet for generation and confinement of laser plasma, jointly used by Tokyo Univ. and Nagoya Univ., completed 1967.

TABLE 1B.   PROJECTS OF LARGE SCALE APPLICATION OF SUPERCONDUCTIVITY - COMPANIES (continued)

Organization	Person in Charge	Project Categories,	Current Status, Funding
Tokyo Shibaura Electric Co., Ltd.:			
Toshiba Research and Development Center, Power and Control Lab. Kawasaki-shi, Kanagawa-Ken	H. Ogiwara	Magnetically Levitated Transportation	a) Facilities for static and dynamic test of maglev, large superconducting magnet (0.3m x 1.2m), dynamic test using a disc track of 3.1m in diameter, partially supported by Ministry of Transportation, 20 M¥, FY 1971, completed 1971.
			b) SC magnet and cryostat for null-flux type suspension, partially supported by Ministry of Transportation, 9 M¥, FY 1972, completed 1972.
			c) SC magnet and cryostat for suspension and guidance, 4m long, partially supported by Ministry of Transportation, 12 M¥, 1973.
Toshiba Heavy Electric Lab. Tsurumi-ku, Yokohama-shi, Kanagawa-Ken	M. Yamamoto S. Nakamura	DC Homopolar Generator	a) 3,000 kW disc type DC homopolar generator, ordered by Japan Society for Promotion of Machine Industry, 180 M¥, for 1971-1973, cooperation of three companies; after completion, the generator will be put to practical trial use at a copper electrolyzing plant.
The Furukawa Electric Co.:			
Central Research Lab. Shinagawa-ku, Tokyo	Y. Furuto	SC Power Transmission System	a) Electrification test of SC power cable, 10m long, AC 3,000 Amps, DC 5,000 Amps, completed 1972.
			b) Conceptual design of SC (AC and DC) power cables.
			c) Research on materials (electric insulation and thermal insulation).
			d) Basic research on SC transformers.

TABLE 1B.   PROJECTS OF LARGE SCALE APPLICATION OF SUPERCONDUCTIVITY - COMPANIES   (continued)

Organization	Person in Charge	Project Categories	Current Status, Funding
(Furukawa, continued) Central Research Lab.	Y. Kashiwayanagi	SC Coaxial Cable for Telecommunication	a) Miniature SC coaxial cable, 1.5mm outer diameter, Pb-Cu conductor and FEP dielectric; attenuation: 0.6 dB/km at 1GHz: joint study with the Electrical Communication Laboratory, 1972.
High Tension Lab. Yokosuka-shi, Kanagawa-Ken	K. Kikuchi	CR Power Transmission System	a) Electrification and electrical breakdown test on 20m long CR cable, 1972. b) Research on electric insulation at low temperature.
Fuji Electric Co., Ltd.: Central Research Lab. Yokosuka-shi, Kanagawa-Ken	S. Akiyama H. Fujino	SC Synchronous Generator	a) 30 kVA, outer stationary SC field and rotating inner armature, horizontal axis, completed 1972. b) Scaled-up version, inner rotating SC field, horizontal axis, 1973-1974.
	S. Akiyama	Magnetically Lavitated Transportation	a) Lightweight superconducting magnet for a test facility of maglev, completed 1972. b) Refrigeration system. c) Subsidiary equipment, partially supported by Ministry of Transportation, 3.1 M¥, FY 1973. d) Rotating drum type test facility for maglev characteristics, 1973.
		Others	Power transmission, feasibility studies on SC Transformer and circuit breaker.
Hitachi Cable, Ltd. Hidaka-cho, Hitachi-shi, Ibaraki-Ken	K. Hidaka	CR Power Transmission System	a) Electrification and electrical breakdown test on 30m long CR cable, completed 1972. b) Research on electrical insulation at low temperature. c) Conceptual design and feasibility studies.

TABLE 1B.   PROJECTS OF LARGE SCALE APPLICATION OF SUPERCONDUCTIVITY - COMPANIES (continued)

Organization	Person in Charge	Project Categories	Current Status, Funding
The Fujikura Cable Works, Ltd. Kiba, Koto-ku, Tokyo	I. Kubo	CR Power Transmission System	a) Conceptual design and feasibility studies. b) Electrification and electrical breakdown test on 20m long CR cable, 1972.
Sumitomo Electric Industries, Ltd. Kitahama, Higashi-ku, Osaka	K. Kojima	CR Power Transmission System	a) Conceptual design and feasibility studies.
Showa Electric Wire & Cable Co., Ltd. Kawasaki-shi, Kanagawa-Ken	K. Haga	CR Power Transmission System	a) Conceptual design and feasibility studies. b) Electrification and electrical breakdown test on 20m long CR cable, 1972.
Central Research Institute of Electric Power Industry Komae-shi, Tokyo	N. Yamada H. Okamoto	Cryogenic Power Transmission System	a) Conceptual design, feasibility studies, economy and technical analyses by the Committee on Superconducting Power Transmission (a research organization on advanced power transmission).

PROGRAMS ON LARGE SCALE APPLICATIONS OF SUPERCONDUCTIVITY IN SWITZERLAND

T. Bratoljic, Brown, Boveri and Cie., Baden

G. Meyer, Brown, Boveri and Cie., Zurich

Switzerland

I. INTRODUCTION

The activities in Switzerland are relatively few because of the small market in this land, the confined export possibilities and the restricted means in government support. Table I gives a short survey of the institutions involved and their essential activities.

II. PROGRAMS

At SIN a 5T superconducting solenoid of 8m length with an ID of 12cm is under construction. The fully impregnated coils are cooled by supercritical He circulating in a copper pipe winding layer at the center surface of the coils. The new concept combines the advantages of high current density and closed circuit cooling at low He capacity. A 1m model was successfully tested at Karlsruhe last year.

Brown, Boveri and Company, Limited (BBC), together with three other Swiss companies Métaux Précieux, Neuchâtel; Metallwerke Dornach A.G. Dornach and Schweizerische Metallwerke Selve & Cie, Thun, develop and fabricate superconductors on the basis NbTi. The focal points are hollow and compact conductor cables, impregnated with solder and filament conductors for dc and ac magnets. The working group has constructed the first big magnet cooled with supercritical helium (OMEGA). In the low temperature laboratory of BBC in Zurich-Oerlikon studies and tests on impregnated coils and superconducting rotors are in progress.

After developing its water cooled rotor winding BBC began at the end of the 60's to consider the possibility of the application of superconductivity as a further step in the progress of ac turbogenerators. An important encouragement in this field resulted from the success of the first MIT experimental machine having a rotating superconducting field winding. From 1970 to the beginning of 1972 BBC carried out technical and economic feasibility studies. The points of particular interest were the possible maximum rating of turbines and superconducting generators and a comparison of performances and costs of both superconducting and conventional generators at a level of 1000 MVA, just achieved by conventional machines. The design was based on materials already commercially available, and it was important to consider the superconductor and the structural materials, both at low and at room temperatures. It is significant, that there is a considerable lack of data on material behavior, especially regarding its mechanical performance at He temperature. This would be a very worthwhile field of research for many universities and institutes.

The design concept corresponds to a machine with rotating cryostat, like the first MIT machine. This concept avoids the large diameter rotating vacuum seals as required by a machine with a stationary cryostat, and makes it easier to overcome big forces acting on the outer surface of the rotor in the case of a sudden short circuit of the stator winding. The superconducting field winding is in slots in the cold inner rotor and the warm outer rotor-cylinder, at room temperature, serves as a damper. The stator winding is an air-gap winding built into a structure of insulating material. The outer part of the stator is a laminated iron cylinder, which protects the surroundings from the strong magnetic field and also increases the useful magnetic flux in the region of the stator winding.

The considered design gives a possible rating of about 3500 MVA at 3000 r.p.m. This rating is roughly twice the maximum possible rating of conventional machines. The total weight of the superconducting machine of approximately 850 tons corresponds to a specific weight of 0.25 kg/kVA. The weight of the rotor is about 10% of the total weight. The utilization coefficient is very near to 100 kVA min/m^3. The losses amount to 15 MW so that the efficiency is better than 99.5%.

It should be also possible to build large 4-pole superconducting generators with a rating twice that of conventional machines. Comparing 1000 MVA superconducting and conventional machines shows the superconducting generator to have 1/3 rotor volume, 1/2 total volume and 2/3 weight. The losses are considerably smaller (about half). However, the first estimation of production costs does not indicate any advantage of the superconducting machines at this level of rating. At higher ratings one can expect that the superconducting generator would be competitive also in this respect. The results of the feasibility studies being satisfactory, BBC plan to continue with development work in this field.

TABLE I. SUPERCONDUCTIVITY PROGRAMS IN SWITZERLAND

UNIVERSITIES:

1.   Université de Genève, Dept. de physique de la matière condensée.
        Magnetic ions in A-15 superconductors
        Intrinsic properties of A-15 superconductors and their
            function of composition
        Influences on T_c in A-15 superconductors

2.   Université de Lausanne, Institut de physique expérimentale
        Mechanical and physical properties of high field superconductors

3.   Universität Bern, Physikalisches Institut
        Design of a superconducting split coil spectrometer to be
            connected with the Muon channel at SIN

4.   Eidgenössische Technische Hochschule Zürich (ETHZ)

     A. Low Temperature Laboratory:

             ac-losses in superconducting compound conductors
             Pressure effects in unisotopic conductors
             Periodic pinning structures (matching effect)
             Thermal conductivity in periodic proximity-effect-
             arrangements

     B. Dept. Electrical Engineering (in initial stage):

             Special problems on generators and motors
             Superconducting switches, fast and slow energy extraction
             Energy storage
             Water cleaning

INSTITUTIONS:

1.   Swiss Institute for Nuclear Research Villigen (SIN)
        Superconducting Muon Channel

COMPANIES:

1.   International Business Machines (IBM), Laboratory Zürich
        Josephson Computers (see Matisoo, Chapter 9).

2.   A.G. Brown Boveri and Cie. Baden (BBC) and the Working Group on
     Superconductivity
        Development and fabrication of NbTi Superconductors
        ac losses in superconducting compound conductors (in collaboration
        with ETHZ, Low Temperature Laboratory)
        Components of superconducting magents
        Design of superconducting ac machines .

PROGRAMS ON LARGE SCALE APPLICATIONS OF SUPERCONDUCTIVITY

IN THE UNITED KINGDOM

A.D. Appleton

International Research and Development Co. Ltd.

Newcastle Upon Tyne, England

I. INTRODUCTION

There has been an increase in the level of activity in superconductivity in the United Kingdom during the past year or so and the indications are that this trend will continue. One of the more significant developments has been the decision of the Science Research Council to fund work programs in universities on topics associated with superconducting ac generators; this work will be carried out in close liaison with industry and is described in this survey.

In the research laboratories of the Central Electricity Generating Board at Leatherhead (CERL), a test rig has been commissioned to provide data for the development of superconducting ac power transmission systems and a program of experiements is underway; a decision will be taken within the next two years on the future scale of this work.

At International Research and Development Co. Ltd. the major emphasis has been on the development of a superconducting dc marine propulsion system on behalf of the Ministry of Defense (Navy), although the industrial applications of superconducting dc motors and generators have not been neglected. Also at IRD work continues on the development of superconducting ac generators and it is hoped to step up the level of funding in this most important area.

The high standard of work which we have come to expect from the Rutherford High Energy Laboratory continues with the development of the prototype quadrupole magnet to meet a possible CERN requirement together with some more basic studies including an evaluation of filamentary niobium-tin superconductors. The grapevine suggests that we may be

639

hearing a great deal more on the subject of filamentary niobium-tin in the not too distant future.

II.   THE ACTIVITIES OF THE SCIENCE RESEARCH COUNCIL

In the light of the growing interest in superconducting ac generators, the Science Research Council, after discussions with industry, universities and government laboratories, decided to promote further activities in this area. A working party was set up to hold a number of meetings between October 1971 and July 1972 and the recommendations were as indicated in the following abstracts from the report.

"The working party takes the view that although the timescale for the development of superconducting alternators is necessarily long, and the justification for such machines at present uncertain, their potential is sufficiently great to warrant research being undertaken directly towards this objective. It is thought that while a large research effort should not be committed at this stage, it is nevertheless important that an early start should be made. This is for two main reasons. Firstly, although production of large superconducting ac generators is not envisaged before the 1990s research work should start as soon as possible if this date is to be met. The working party considers that the likely timescale for the development of the technology is illustrated in Table 1. Secondly, in view of the leading position of the UK in several relevant aspects of superconductivity, including materials, magnets, dc machines and helium refrigerators, it is particularly desirable to maintain that position in ac machines bearing in mind the growing worldwide activity in this area."

"For although much of the impetus towards the development of large superconducting ac machines has come from industry and much of the development work will necessarily be carried out by the manufacturers, there exists within a number of university departments a considerable body of expertise which would be of immediate relevance to such a project. Effective use of these resources can be made in circumstances where research workers are able to interact both with development work in industry and with research work in other university departments.

Therefore the way ahead recommended by the working party is that university research should start as soon as possible in obtaining basic data and the criteria necessary for determining design parameters directed towards superconducting ac generators, and relevant to both the near-conventional and the iron-less stator concepts. It is important that close contact should be maintained with industry if the research is to be fully effective. Indeed it is considered that this field offers scope to university departments of engineering for a realistic involvement with industry with the possibility of collaborative research projects."

TABLE 1. Likely Timescale for the Development of Superconducting ac Generators

Year	19'	73	74	75	76	77	78	79	80	81	82	83	84	85	86	87	88	89	90	91	92
Research		————————————																			
Rig work and construction of 60 MW prototype				————————————																	
Running of prototype								———————													
Design and construction of 600 MW machine										————————————————											
Design and construction of 1000 MW machine														————————————————							

"The working party therefore recommends the following areas to be particularly appropriate for research now in university departments.

1.   The thermodynamic, hydrodynamic and heat transfer properties of helium in a rotating frame of reference.
      Information and experience should be obtained on the behavior of helium in a high speed rotating frame with special reference to rotational inertia forces and pulsating force fields. The effect of these forces on the heat transfer properties of the refrigerant also needs study.

2.   Impedances, eddy-current losses and forces. Terminal performance at least by the conventional two-axis approach. It is envisaged that the work on impedances, eddy-currents and forces will be essentially a computing study, together with the use of some scale models. These studies are closely interconnected, and will proceed in parallel. There is also a good case for initiating study on terminal performance leading in due course to the development of suitable control techniques.

3.   Fatigue and hysteretic heating of composite materials. The cyclic forces which the non-conducting support structure will experience can be estimated fairly accurately; the parameters affecting the consequent fatigue and hysteretic heating demand study."

III.   RESEARCH GRANTS TO UNIVERSITIES AND LABORATORIES

A.   Oxford University

A research grant has been awarded to Oxford University (Department of Engineering Science) "to carry out a fundamental experimental program on heat transfer and fluid flow in a rotating frame of reference which will produce basic data of direct relevance to the superconducting ac generator program." The value of the grant is £33,161 over a period of three years and the work will be under the direction of Dr. C.A. Bailey. The study of heat transfer to cryogenic fluids has been in progress at Oxford for a number of years and the new grant supplements this work with special reference to superconducting ac generators. It is proposed to investigate the thermodynamic behavior of helium contained within flow passages inside a rotating body. A theoretical study of the flow systems shown in Fig. 1 started in 1972 and this will now be supplemented by experimental work. The latter will involve flowing helium fluid along various shaped tubes at different Reynolds numbers and heat flux inputs, initially stationary and then at speeds increasing to 3600 rev/min. Initial work will be confined to single phase supercritical or sub-cooled helium but two-phase flow will be considered later. A feature of the Oxford program is the availability of a set of computer programs "HEPROP" which have been developed to calculate the helium properties necessary for the analysis and interpretation of experimental heat transfer results. "HEPROP" covers the temperature range 2K to 1700K and pressures up to 1000 bar and is stored in the University's ICL 1906A computer.

B.   Leeds University

A research grant has been awarded to Leeds University (Department of Electrical and Electronic Engineering) "to study, in a general way, basic aspects of the terminal performance of superconducting ac machines. Suitable circuit models of the machine will be established and used, with digital and analog computing techniques, to examine in particular: short circuit, load changing, negative sequence and stability behavior. The results will be used (a) in the determination of desired parameters and character-istics for successful operation in the overall system, and (b) in calculating those torques and forces within the machine which are crucial for proper mechanical design. The essential associated study of machine impedances will be carried out using both experimental and computational techniques." The value of the grant is £44,883 over a period of three years and will be under the direction of Professor P.J. Lawrenson.

The work proposed at Leeds will be for a machine without iron in the magnetic circuit, and is based to some extent on earlier work at International Research and Development Co., Ltd., with whom there has been close collaboration for a number of years. As a starting point the 2-axis model incorporating seven coils (two each for the rotor screen, environ-

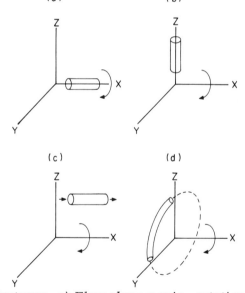

Fig. 1. Flow Systems: a) Flow along x axis, rotation about x axis. b) Flow along z axis, rotation about x axis. c) Flow para-llel to x axis, rotation about x axis. d) Flow around x axis, rotation about x axis.

mental screen and stator circuits, and one for the superconducting field) and computer programs developed at IRD will be employed. Improved models will be developed and experimental work on parameter determination will be employed to support and guide the theoretical approaches. The analyses of the machine performance will be carried out with increasing accuracy as the mathematical and experimental models are developed. Attention will be given particularly to end effects, and later work will allow for actual eddy-current distribution in the screens and supports.

C.   Nottingham University

A research grant has been awarded to Nottingham University (Department of Mechanical Engineering) to consider the problems of hysteretic heating and fatigue in non-metallic composite materials which may be employed to support the stator winding in superconducting ac generators. Problems arise because of the large cyclic forces which appear on the stator conductors and which are, therefore, transmitted to the support structure. There is considerable expertise on this type of problem at Nottingham University. The work program will include the determination of the elastic constants of a wide range of fibre reinforced materials, the evaluation of thermal conductivities, the performance of fatigue tests at 100 Hz, the measurement of hysteretic heating, and stress calculations. The analytical and experimental tools required for this program of work are already well advanced at Nottingham University. The value of the grant is £14,812 over three years and the work is under the direction of Dr. M.J. Owen.

D.   Other Universities

The above research grants to the Universities of Oxford, Leeds and Nottingham have already been approved by the SRC, and a number of others are expected to be awarded in the near future. Among the other Universities expected to take a leading role in the program of work on ac generators is Southampton University with electrical aspects under the direction of Professor P. Hammond and cryogenic work under the direction of Dr. R. G. Schurlock.

E.   Central Electricity Research Laboratories

1.   Superconducting power transmission

Work on superconducting power transmission continues at the CERL laboratory of the CEGB at Leatherhead as described in a recent article [1] by Dr. J.A. Baylis, Head of the Section for the Power Application of Superconductors.

The work at CERL is maintained at a level which will enable a decision to be made within the next two years on whether superconducting cables might find application in certain special situations such as bulk transmission to and around large cities. To be economical it is estimated that a rating of about 4GVA is required and this is the likely magnitude of power requirements.

A new facility constructed for CERL by BOC has recently been brought into commission for testing conductors and dielectrics for superconducting cables (it is worth noting that the new rig is not a prototype transmission line). The test facility is 8 m long and 300 mm bore and is supplied with helium from the refrigerator which has been in operation for a number of years at CERL. Tests have been carried out on a co-axial pair of conductors of 100 mm and 63 mm diameter, the length being 5.5 m; the conductors are niobium on copper and were developed and produced by IMI to specifications approved by CERL. The tests have included operation at currents up to 16 kA and it is now proposed to undertake experiments at high voltage to test dielectric materials; the test voltages will be 100 kV at 50 Hz and 300 kV impulse. It is not possible to conduct the current voltage tests simultaneously but this is not of particular consequence for the results required from the rig.

The co-axial pair of niobium-clad copper tubes have been described in a joint paper [2] by CERL and IMI. Although a niobium layer 1 μm thick will carry the required current, a 50 μm layer has been used tor ease of fabrication and improved reliability. The tests have shown that such a conductor can carry the required current, but two problems remain: thermal contraction, and circuit conditions. The latter may be solved by employing a substitute of high purity aluminum or perhaps a type 2 superconductor (probably Nb-Zr) and work is continuing. The problem of thermal contraction has been greatly eased by the development of a flexible conductor described by Baylis; the concept of the latter is to replace the solid tube by a set of helically wound strips (wound into a suitable plastic former) a dielectric of polyethylene tape and a further set of helical conductors. It is planned to test short lengths of this cable in the new test rig rotor later this year after the tests on dielectric losses have been completed. The work on the cables is directed by Dr. J.A. Baylis with support on the material side from Mr. B.J. Maddock.

F.   International Research and Development

1.   Superconducting dc machines

A high level of activity on the development of superconducting dc motors and generators is being maintained at IRD with particular

emphasis at the present time on a superconducting marine propulsion
system on behalf of MOD(N). The generator is completed, the motor
is nearing completion and cooldown is expected to take place before
the end of September 1973. The entire system will then undergo extensive land trials before it is installed in a ship for sea trials. In parallel
with the manufacture and testing of this first superconducting propulsion
system, detailed studies are proceeding with larger systems [3],
including those suitable for civil marine propulsion systems and land
based applications. Significant design improvements have been made
in the last year or so and the status of the machine is at the pre-
production level. Considerable attention has been given to effecting
simplifications to the machines and attractive designs are now available for a range of applications. It is hoped to accelerate the
introduction of these new machines into industry.

To take one important example, superconducting marine propulsion
systems are particularly attractive where space is at a premium as in
container ships for example, and these conditions will apply to
tankers if large clean water ballast capacity becomes mandatory.
The location of a gas turbine and generator power unit near the deck
level can result in a 10% or so increase in cargo volume. There are
numerous other advantages for marine propulsion systems.

2.   Superconducting ac generators

The substantial level of work on the development of superconducting
ac generators of two years ago has been reduced for the last year,
but it is hoped to expand again to at least the previous level in the
reasonably near future. Work has continued on consolidating the
earlier work [4] and improving the design, particularly in respect to the
rotor. This work at IRD is carried out in close collaboration with
C.A. Parsons & Co. Ltd. Close liaison has been maintained with
a number of universities and the new SRC program will assist the
development of these machines. It continues to be the case for
superconducting ac generators (as indeed it is for superconducting dc
machines) that when designs are made, in greater depth improvements
present themselves, resulting in benefits both technical and economic.   Further developments on the IRD work on ac generators are not
planned for a year or two but in this order of time it is hoped to make
substantial progress.

G.   General Electric Company

The GEC group of companies is giving consideration to the development of superconducting ac generators but work to date is limited to design
studies.

H.   Rutherford High Energy Laboratory

Work continues at a high level at RHEL on the development of dipole and quadrupole magnets and also the further development of NbTi and Nb_3Sn superconductors [5-7].

The program may be summarized as follows:

1.   Objectives

Accelerators: a) CERN II Conversion. b) EPIC (Electron Proton Intersecting Complex). A new e-p storage run idea is being worked on at RHEL. This run dipole/quadrupole magnet has a field ~ 5T with ~10cm bore and several meters long and approximately 1000 off.

Beam Handling Magnets: For general use and special experiments requiring high fields. Dipoles and quadrupoles 5-15 cm bore approximately 1 meter long with 5 ~7 T dc fields.

Special Purpose Magnet: An example is a polarized target magnet with very uniform field and wide angle access.

2.   Magnet Construction

CERN Magnets: Dipole magnets AC3 and AC4 are completed. AC5 is a prototype to fit the CERN requirements but not as long as the actual CERN magnets. It has an effective length of 1.3m; overall length, 1.5 m; bore, 10cm; field, 4.5T; current, 3100A; rise time, 2 seconds; field uniformity, better than 10^{-3} within a 70mm diameter cylinder at all field levels; coding, liquid helium channels, laminated iron yoke. The magnet will be completed in the summer of 1974.

A prototype bending magnet is being constructed which will be loaned to Imperial College for optical spectroscopy work. It has an effective length of 1.1m; overall length, 1.6 m; coil bore, 140 cm; warm bore, 10 cm; field, 5T; current, 720A; conductor, 2.2 x 1.1 mm rectangular composite, 361 filaments. When inside its support structure, the winding will be impregnated with resin at high pressure, ~ 5000 lb/in^2, in an attempt to overcome mechanical degradation and turning. The magnet will be completed approximately June 1974.

A polarized target magnet (PT55) is being constructed. It has an outer diameter of 60 cm; inner diameter, 27 cm; central field, 2.5 T; peak field, 5.6T along the surface of the large coil; uniformity of field, 1 x 10^{-4} in a target cylinder 3 cm diameter x 5 cm length; scatter angle, 120º (core); stored energy, 430 kJ; current 105 A, conductor Cu:NbTi. The magnet will be completed in mid 1974.

Under an AERE/UK Universities/Canadian Universities collaboration a solenoid for use as a Triumf accelerator is being constructed. It has an effective length of 1m; coil bore, 14 cm; warm bore, 10 cm; field, 6T; current, 230A; conductor IMI C361/1000 1 mm diameter. The magnet will be completed by the end of 1973.

A model is being constructed of the energy transfer device developed by Dr. P. F. Smith. It has an overall diameter of $\sim$1 meter; field, $\sim$4T; current, 4000 A; energy transfer, 2×10^5 J.

3.   Conductor Development

Niobium titanium materials in collaboration with IMI, Ltd. a) 8917 filament 3 component composite now in satisfactory use for the AC5 magnet. b) 14701 filament composite with "spider" to reduce ac losses. c) 177241 filaments. Some problems remain to be overcome.

For cables, flat tapes work well and are used for all ac magnets and energy storage magnets.

In collaboration with AERE, Harwell, filamentary Nb_3Sn is being developed. Some promising samples have been tested and details will be published.

I.   Other Activities

Magnetic Levitation: A contract has been awarded to Warwick University (Dr. Rhodes) by the Wolfson Foundation to study superconducting magnetic levitated transport. The contract is for £150,000 over a period of five years and work is expected to commence in the autumn of 1973.

Culham Laboratory: The LEVITRON experiment at Culham is working well; this was developed by Mr. D. N. Cornish to study magnetic confinement stability. The decision not to employ superconducting magnets for the next phase of TOKAMAK experiments may result in a reduction of the level of activity on superconductivity at Culham. On the other hand the level of activity on CTR at Culham is being increased.

Oxford Instrument Co. Ltd: A hybrid magnet has been constructed for the University of Oxford. a) Superconducting (outer): Nb Ti: 240mm bore (ambient) 6.6T; b) Copper (inner): water cooled: 50mm bore (ambient) 9.4T:2MW.

Also the Oxford Instrument Co. is concerned with the construction of the cryostat for the RHEL, AC5 dipole magnet.

REFERENCES

1. Baylis, J.A., Electrical Times, Issue 4232, 9, (1973).
2. Graeme-Barber, C. Maddock, B.J., Popley, R.A. and Smedley, P.N. Cryogenics, 12, 317, (1972).
3. Appleton, A.D. and Bartram, T.C., Third Ship Control Systems Symposium, Ministry of Defense, Foxhill, Bath, Somerset, (1972).
4. Appleton, A.D. and Anderson, A.F., Proc. 1972 Applied Superconductivity Conference, Annapolis, p. 136, IEE Publ. 72CH0682-5-TABSC (1972).
5. Wilson, M.N., RHEL/M/A26.
6. Thomas, D.B. and Wilson, M.N., RPP/A93 (RHEL).
7. Wilson, M.N., Cryogenics, 13, 361 (1973).

PROGRAMS ON LARGE SCALE APPLICATIONS OF SUPERCONDUCTIVITY IN THE UNITED STATES

R.L. Powell, F.R. Fickett and B.W. Birmingham

National Bureau of Standards, Institute for Basic Standards

Boulder, Colorado 80302

I. INTRODUCTION

A. Summary

A brief overview is given of U.S. programs on large-scale applications of superconductivity. For each of the projects that was separable,
we have given the organization, the manager or key contact man, a brief
description of the device and its application, comments on the type of
program, the source and amount of funding whenever possible, the current
status of the project, and projected plans for the next few months to a year.
For convenience, the programs have been divided into seven categories
based upon their intended application 1) generators for electric power
systems, 2) electric power transmission, 3) machinery for propulsion
systems and auxiliary power supplies, 4) magnets for energy storage,
MHD, and CTR systems, 5) magnets and cavities for high energy physics
experiments, 6) magnets for industrial and medical applications, and 7)
land transportation systems. For each category there are summary
tables and a brief interpretive and descriptive text. Programs on basic
superconductivity science and general background materials studies have
not been included unless they are directly supported as part of an applied
project. The total level of funding for the programs reviewed here is
about $26 M per year, with most of that coming from the federal govern-
ment, especially the Atomic Energy Commission. The main non-govern-
ment support in the field comes from the Electric Power Research Institute,
though a significant amount also comes from manufacturers.

Although there have been many changes in support for particular
projects, the total funding for all large-scale applications of superconductivity has remained fairly constant over the last two years. Projected

funding for the next year appears to maintain this constancy. The main exceptions to this generally level trend are the expansion this year in the naval propulsion program and the increased corporate funding last year for superconducting generators. However, there has been a substantial expansion, by a factor between 2 and 3, in funding for large-scale super-conducting devices since the overview written in 1967 by one of the present authors (BWB).

There has been substantial progress on superconducting generators during the last year: two multi-megawatt machines have been built and tested. At present, because of lack of high-level funding, work in this area is confined mainly to applied research and component development, but not to construction of larger generators. Development of transmission lines is progressing smoothly across a broad front, several different types of lines are being developed at various laboratories. During the next year, some large-scale prototypes should be completed and tested; substantial materials studies and component development projects are presently under-way. There is a very strong program on naval propulsion systems, this program is very well funded and there has been substantial progress. During the next year a 400 hp motor/generator system should be tested and plans formalized for much larger systems. Energy magnets for energy storage, MHD, and CTR experiments are being developed fairly aggres-sively. There is a large variety of systems under development at various laboratories; several units have been built already and many more are under construction. There could be a strong effort in this area if it is decided to incorporate superconducting magnets in the next generation of fusion experiments. Magnet systems for high-energy physics experiments continue to be heavily supported, with the newer magnets fitting into two classes a) routine, low-innovative, and b) highly sophisticated. Both types are useful for making present accelerator experiments more effective and for making future accelerators more powerful. With some favorable decisions on upgrading present storage rings and accelerators there could be a rapid expansion of superconducting devices in this area. Support of magnets for industrial and medical applications continues at a relatively low level. Especially for industrial magnets, large demonstration systems are necessary before superconducting devices will be included in many applications. Utilization of superconducting magnets in land transportation systems is highly problematical at present. Many general decisions, both technical and societal, must be made before superconducting levitation pads will become common features of high-speed ground transportation vehicles.

B. Summary Tables

In Tables 2 through 8 we have attempted to present, in a succinct manner, an overview of the large-scale applications of superconductivity in the U.S.A. For each project we have listed: the institution or company actually doing the work; a person who was our main contact point (not necessarily the project leader); a description of the magnet, device or system; an indication of the type of program, whether a paper study or an experimental test, etc.; the source and, when available, the amount of funding at the present; the status of the project as of summer 1973; and any future plans and proposals, complete with the requested funding when this information was available.

Table 1 is a listing of abbreviations used in the summary tables. These are primarily abbreviations of funding agencies and contractors although some fiscal terms also appear. The tables are inserted into corresponding summaries.

We have not listed small supporting research projects nor have we listed any of the very large number of general projects studying the properties of normal materials at low temperatures or studying new superconducting materials and configurations, unless they are directly funded as part of an applied project. We have also not discussed refrigeration, although in most large-scale applications it may well be one of the more important factors. We have made every attempt, within our limitations, to discover all the existing research. It is likely that we have managed to overlook some projects, particularly in-house efforts by funding agencies. Needless to say, we would appreciate receiving any updating material that is available.

A special note is called for on our reporting of the funding situation. In many instances the current funding is uncertain, usually for one of two reasons: it is corporate funding, which is often considered proprietary information, or it is a part of a much larger program such that the fiscal data on the superconducting devices is not immediately available. The situation on future funding is, of course, always uncertain and any of the listed dollar amounts should be considered with some caution.

II. GENERATORS FOR ELECTRIC POWER SYSTEMS (Table 2)

This application of superconductivity has been relatively slow in developing. Small toy-like models were built as early as the mid- and late-1960's with the research being sponsored mainly by the Army and Navy. A pioneer group at AVCO Everett, now mainly dispersed, also carried out important conceptual designs, feasibility studies, and technical analyses, primarily for military applications.

Present progress in this field can be attributed, both directly and indirectly, to the farsighted December 1968 decision of the Edison Electric Institute (a cooperative organization funded by privately owned electric utilities) to sponsor research on large superconducting synchronous machines at the Cryogenic Engineering Laboratory of MIT. That research group, presently headed by Professors Smith and Wilson, has led in proposing innovative machines, in giving consultative advice, and in promoting advanced training for engineers now working for the government (especially the Navy) and for equipment manufacturers. Their main project in the last few years has been to design, construct, and test a 2 MVA, 2-pole synchronous generator. That machine has been constructed and many preliminary tests have been completed. Although there have been a few small operational problems, the generator has worked very well and has proven the technical feasibility of superconducting ac machinery. The MIT group hopes to have the generator connected to the Cambridge power system this fall, with it acting as a synchronous condenser rather than as a generator.

Building upon the experience and concepts of the MIT group, the Westinghouse team has designed, constructed, and tested a more sophisticated 5 MVA, 2-pole generator which has obtained wide publicity and was well described in the proceedings of the Winter 1973 IEEE meeting. The generator has run at full speed, full voltage, and full current, but not all three simultaneously. Full load tests in an industrial test stand are scheduled for late this year.

At about the time that the MIT/EEI and Westinghouse machines were being constructed, the Cryogenics Division of the National Bureau of Standards sponsored market analyses, feasibility studies, and technical analyses for much larger, up to several gigawatt, machinery. Their program to support the second generation of machinery, in the 10 to 100 megawatt range, was discontinued by the fund impoundments which occurred in late 1972.

At present there are no active programs for larger generators. That activity will have to await high-level decisions to fund substantial machinery development by either the Electric Power Research Institute (a successor to the Research Division of EEI and the Electric Research Council) or by the federal government directly. Until that time, the present level of funding, about 3/4 million dollars, will support continued applied research and development of special components, but not construction of much larger generators.

III. ELECTRIC POWER TRANSMISSION (Table 3)

Funding for this application is relatively high, about 3-1/2 to 4 million dollars per year, and research is progressing across a broad front: cryoresistive and superconducting, ac and dc, rigid and flexible, vacuum and fluid electrical insulation, and background materials studies. Cryoresistive systems have been included in this section because most of the technical problems are very similar and many of the results, especially on insulation and refrigeration, are interchangeable. The main funding for these developments comes from the Atomic Energy Commission and from the Electric Power Research Inst., which receive funding support from the Dept. of Interior as well as from private utilities. Formerly the National Science Foundation supported a substantial amount of the research, but now their funding is restricted to the cryoresistive program at the National Magnet Laboratory and the multilayer program at Stanford University.

Cryoresistive applied research and component testing are being carried out at General Electric and at the National Magnet Laboratory (NML). Most of the work at both places has been limited to materials studies, particularly insulators, and to small components, but the NML group has tested model 8 ft. long sections and terminals and expects to test more components at full voltage later this year.

The Los Alamos Scientific Laboratory has concentrated on dc transmission lines with much of the work up to the present being on technical analyses and short sample tests. Later this year they expect to have a large racetrack test bed ready for testing of full scale transmission line components.

Both the Brookhaven National Laboratory and the Linde Division of Union Carbide are working on ac transmission lines. Neither group has built and tested a large transmission line model, though both have carried out much applied research and some short sample tests. Both groups will assemble and test large components later this year and early next year.

Strong supporting research for these programs is also being done now at the National Bureau of Standards and the Oak Ridge National Laboratory.

In general the programs on transmission lines are progressing well. Since most of them started within the last year or two, they are still in early exploratory stages, but substantial components and model lines should be completed within the next year at several laboratories.

IV. MACHINERY FOR PROPULSION SYSTEMS AND AUXILIARY POWER SUPPLIES (Table 4)

As mentioned in Section II the Army and Navy sponsored research on rotating machinery for their special applications in the mid-1960's. A few small kilowatt motors were actually constructed as part of those early projects. The programs continued at a relatively low level until the recent strong efforts initiated by the Naval Ship Research and Development Center, Annapolis branch. The Navy program alone is supported at about the 4 million dollars per year level. As is typical of large dynamic development programs, a multi-faceted approach is being used. There is a thorough in-house effort to build a 400 hp dc homopolar motor and a 300 kW dc homopolar generator and to study designs and concepts for more advanced, higher-powered, propulsion systems. They also support large programs at Garrett and General Electric to design advanced systems, both dc and ac-dc hybrid, for machinery in the 20 to 40 thousand horsepower range.

The Air Force has a much smaller program to develop a model rotor, and eventually a generator, for a 400 cycle airborne auxiliary power supply. The Army program is now at a considerably reduced level.

The Advanced Research Projects Agency is strongly supporting the various Dept. of Defense machinery programs with research projects on materials and auxiliary engineering systems. This program has just begun, no significant results have been obtained yet. They are also supporting research and development programs on alternative, non-superconducting, machinery as potential competitors to superconducting systems.

The total overall effort on this application of superconductivity is funded at about 6 million dollars per year, including background materials and systems research. Again the programs are relatively new, only one or two years old, and no total systems have yet been constructed and tested. Within the next year or two, however, there should be substantial achievements and actual propulsion systems in operation.

V. MAGNETS FOR ENERGY STORAGE, MAGNETOHYDRODYNAMIC
   (MHD), AND CONTROLLED THERMONUCLEAR REACTION (CTR)
   SYSTEMS (Table 5)

Research and development on these types of magnets is quite varied, many laboratories have active projects, and the funding sources are more numerous than usual. The Air Force, Atomic Energy Comm. (AEC), Electric Power Research Institute, the Office of Coal Research and others contribute about 6-1/2 million dollars per year for the total programs; 2-1/2 million for MHD, a little less than 3 million for CTR, and a little more than 1 million for energy storage. Many of the magnets and systems for these programs are natural successors to previous beam handling and other magnet systems developed for high energy physics experiments. As a consequence several of the design teams working on these programs gained considerable experience on earlier projects that were quite similar. There is a much greater data and experience base for these programs than there was for rotating machinery or transmission lines, for example.

Energy storage magnets have been designed for both slow discharge and rapid pulse systems. The slow discharge toroidal magnets are designed for peak load shaving of central power systems. To be useful they would need to be in the 10^{12} - 10^{14} J, 1-10 GW hour, capacity range; one 50 kJ coil has been started at the Los Alamos Scientific Laboratory. However, most of the efforts in this area have been confined to conceptual designs and feasibility studies. The pulsed energy storage magnets are being developed for special military applications, either airborne or land mobile, and for pulsed energy supplies to fusion experiments. They generally are solenoidal. Two magnets, 7 and 30 kJ, have been built and two more, 100 and 300 kJ, are under construction.

Progress on pulsed energy storage magnets is good; several moderate size systems will be in operation during the next year, but development work on slow discharge, peak-shaving magnets is quite limited.

Research and development on controlled thermonuclear reaction magnets is, quite naturally, supported primarily by the CTR Division of the AEC. Many different designs and laboratories are being supported, though none of them have been supported for a long enough time to have large working magnet systems in operation. Many short sample tests and small model magnets have been completed, but large scale funding and the resultant development awaits the high-level decision whether to use superconducting or normal magnets for the next generation of CTR feasibility experiments. At present it appears that CTR program managers are unwilling to gamble their feasibility experiments on large scale superconducting magnet systems, especially the pulsed ones. However, superconducting magnet systems are being supported in parallel efforts so that they will be available for future, larger CTR systems where they are probably necessary, rather than merely more efficient.

The Office of Coal Research is especially strong in supporting MHD magnet research for potential future energy conversion plants although the Air Force is also supporting projects for airborne applications. A few small MHD magnets have been constructed, especially for the Air Force, and many more larger ones are being designed at present. The new ones are large, both in size and in field, up to almost 2 meters I.D. (room temperature access of 20 by 40 centimeters) and up to 8 tesla. These magnets are often sophisticated in design, with cold iron and tapered solenoids or split coils for field shaping. By mid- or late-1974, there should be at least two large scale MHD magnet systems in operation.

VI. MAGNETS AND CAVITIES FOR HIGH ENERGY PHYSICS EXPERIMENTS (Table 6)

Magnet systems for high energy physics (HEP) experiments are supported exclusively by the High Energy Physics Program of the AEC, at a level of about 7-1/2 million dollars per year. In the past, this program has led in the development of new magnets and has contributed substantially to research on new materials and to studies on stabilities in superconductors. In addition to the projects listed in Table 6, the Division of Physical Research of the AEC supports about 2-1/2 million dollars of basic research on superconducting materials each year.

The age of designing and constructing big superconducting magnets for bubble chambers is probably over; the emphasis has changed to much smaller, but much more complex, magnets. Present efforts on HEP magnets can be roughly divided into two categories: 1) low-innovative, relatively well-known beam handling magnets and 2) highly sophisticated magnets with complicated field shapes, high fields, and difficult stability requirements. In the first category are many dipole and quadrupole beam-

line magnets designed for more effective use of present primary or secondary beams or for upgrading the energy of present storage rings. In the
second category fall some very interesting magnets that tax the designing
skills of the laboratory technical staffs. As examples of the latter there
are beam splitters, analyzing spectrometers, flux trappers and flux
shielders.

Because of the many different projects there is no simple way to
describe their current status and progress. Many have been completed,
many are being constructed, and even more are being designed for future
systems, especially for upgrading or energy doubling of present accelerators.

In contrast, the research and development on RF cavities for linear
accelerators has decreased considerably, with only small efforts at Brookhaven National Laboratory and Stanford Linear Accelerator Center. Most
of their efforts are now devoted to materials studies rather than hardware
development.

VII. MAGNETS FOR INDUSTRIAL AND MEDICAL APPLICATIONS
(Table 7)

As is obvious from Table 7, industrial uses of superconducting
magnets are essentially nonexistent. One major reason is the reluctance
of the rather conservative mineral industries to get deeply involved with
a new and "untried" technology. This attitude appears to be changing and
a symposium held at the Massachusetts Institute of Technology in May of
1973 on magnetic techniques in the mineral industry was well attended.
Another reason for the rather low interest is that the problems that superconducting magnets are best able to handle haven't reached crisis propor-
tions yet. This is especially true of the many prospective applications of
the high gradient magnetic separation techniques such as coal desulfuriza-
tion and water purification.

The first generation of large scale demonstration projects will
almost certainly use normal magnets. Superconducting magnet applications
appear to be several years in the future and no significant amount of funding
is likely to be available in the near future.

The situation with medical applications is much more encouraging.
Here the jobs require the high fields and small sizes unique to superconducting magnets, and the applications are of a sort that capture the public
imagination and interest. For most applications to date, the magnets are
not complex in design and little is needed in the way of new magnet technology. What is needed are methods of reducing the size and complexity
of operation of the magnet systems and increasing their reliability. Beyond
that, the major problems lie in education of hospital staff members to the
peculiarities of working near high magnetic fields and in gaining acceptance
of the techniques for general use on humans. The pion channel experiment
is an obvious exception to almost all of the above statements. There the

magnet problems are very complex and the entire system is going to require some innovative work to prepare it for use on humans.

The funding situation for medical applications appears reasonably good. This is particularly so because the clinical experiments tend to be funded from medical sources such as the National Institutes of Health, whereas the magnet development funds come from physical science sources such as National Science Foundation. The immense public interest in this field makes it a prime candidate for creating acceptance of superconducting magnet technology.

VIII.   LAND TRANSPORTATION SYSTEMS (Table 8)

Compared to some other countries, the existing U.S. commitment to magnetically levitated (Maglev) transportation is not large. It has yet to be determined whether the commitment will be to conventional electro-magnet (attractive) or a superconducting magnet (repulsive) system. It seems clear that only one of these systems will be pursued to the large scale test vehicle stage. There is a widespread feeling that the economic and sociological problems of this mode of transportation are at least as difficult to solve as the strictly technical ones.

Not only is the type of suspension in debate, so also is the propulsion system, the main problem being to find the relatively low weight, large gap, reasonably silent motive system necessary for high speed propulsion through populous areas.

The funding for Maglev research is almost exclusively from the Department of Transportation (DOT) with the exception of a few corporate efforts and the National Science Foundation funding of the magneplane project. DOT has indicated that their projected total funding for Maglev research and development will be near that of the Tracked Air Cushion Vehicle -- around $25 M.

TABLE 1.   ABBREVIATIONS USED IN THE SUMMARY TABLES

CY	Calendar year
FY	U.S. Government fiscal year, e.g., FY 73 began on July 1, 1972 and ended on June 30, 1973
pa	per annum
AEC	Atomic Energy Commission, Germantown, MD
AFAPL	Air Force Aero Propulsion Lab., Wright-Patterson AFB, OH
ARPA	Advanced Research Projects Agency, Dept. of Defense, Arlington, VA
BNL	Brookhaven National Lab., Upton, NY
DOT	Dept. of Transportation, Washington, D.C.
EPRI	Electric Power Research Inst., temporarily headquartered in New York City. Incorporated in 1972, it is continuing research programs initiated by the Edison Electric Inst. and the Electric Research Council.
LASL	Los Alamos Sci. Lab., Los Alamos, NM
MIT	Mass. Inst. of Technology, Cambridge, MA
NBS	National Bureau of Standards, Boulder, CO
NSF	National Science Foundation, Washington, D.C.
NIH	National Institutes of Health, Bethesda, MD
NML	National Magnet Lab. at Mass. Inst. of Technology, Cambridge, MA
NSRDC	Naval Ship Research and Development Center, Carderock, MD
NSRDC(A)	Naval Ship Research and Development Center, Annapolis, MD
NSRDC(A)	Naval Ship Research and Development Center, Carderock, MD
OCR	Office of Coal Research, Dept. of Interior, Washington, D.C.
ONR	Office of Naval Research, Arlington, VA
USAMERDC	U.S. Army Mobile Equipment Research and Development Center, Ft. Belvoir, VA

TABLE 2. SUMMARY OF PROGRAMS ON GENERATORS FOR ELECTRIC POWER SYSTEMS

Organization	Manager	Device	Type of Program	Funding (K $) Source	Status	Future Plans
Arthur D. Little, Cambridge, Mass.	John Bishop, Jr.	Large rotating electrical machinery	Market analysis	33 in FY 73 from NBS	Completed Feb. 1973	-
General Electric Co.-Corporate Research and Development, Schenectady, N.Y.	G. R. Fox	S.C. generators	Market analysis, conceptual designs, feasibility studies, technical analyses	25 in FY 73 from NBS	Completed Mar. 1973	-
		S.C. generators	Market analysis, feasibility studies, conceptual designs, experimental research, reviews, component design, construction, and test	Corporate, estimated to be between 250 and 500 pa	Continuing	Applied research and component development continuing
Magnetic Corp. of America, Waltham, Mass.	Z. J. J. Stekly	S.C. generators	Feasibility studies	Corporate	Completed analysis for Off. of Sci. and Tech. study	-
Mass. Inst. of Tech.-Cryogenic Engineering Lab. and Electric Power Systems Engineering Lab., Cambridge, Mass.	J. L. Smith, Jr., G. L. Wilson	(a) 2 MVA generator; (b) large S.C. generators	(a) Design, construction, and test of components and unit; (b) Conceptual designs and feasibility studies	168 in CY 73 and uncertain in CY 74 from EPRI, 16 in C72-73 from NSF	Generator constructed, and preliminary, non full-load, tests completed	Component redesign, additional tests and use as synchronous condenser scheduled later in 1973
		S.C. rotating electrical machinery	Training	Equivalent to about 200 pa from Navy, NSF, etc.	Advanced training of engineers from industry, government, and academia	Continuing for an average 7 to 11 students
National Bureau of Standards-Cryogenics Div., Boulder, Colo.	B. W. Birmingham	(a) S.C. rotating electrical machinery; (b) helium properties	(a) Feasibility studies, reviews; (b) Applied research and technical analyses	90 in FY 73 for in-house studies	Studies also contracted out to A.D. Little, General Electric, and Westinghouse	Continuing at lower funding level
Univ. of Texas, Austin	H. H. Woodson	S.C. rotating electrical machinery	Training	-	-	Planned for initiation in CY 74
Westinghouse Electric Corp.-Electromechanical Div., Large Rotating Apparatus Div., and Research Lab., Pittsburgh, Penn.	C. J. Mole	S.C. generators	Conceptual designs, feasibility studies, and technical analyses	25 in FY 73 from NBS	Completed Mar. 1973	-
		(a) 5 MVA generator; (b) large S.C. generators	(a) Unit design, construction, and test; (b) Conceptual designs and feasibility studies, reviews, advanced component design, construction, and test	(a) Corporate, estimated to be between 1000 and 2000 for developing generator; (b) 250 to 500 pa continuing	Generator constructed, and preliminary, non full-load, tests completed	Additional tests and installation in full-load test stand scheduled later in 1973. Applied research and component development continuing

TABLE 3. SUMMARY OF PROGRAMS ON TRANSMISSION OF ELECTRIC POWER

Organization	Manager	Device	Type of Program	Funding (K$) Source	Status	Future Plans
Atomic Energy Commission-Div. of Appl. Technology, Germantown, Md.	B. C. Belanger	S. C. transmission systems	Funding support of programs at Brookhaven Nat. Lab., Los Alamos Sci. Lab., and Oak Ridge Nat. Lab.	700 in FY 73, 1000 in FY 74	See specific programs	Continuing
Brookhaven Nat. Lab., Upton, N.Y.	E. Forsyth	Flexible ac S.C. transmission line	Conceptual designs, feasibility studies, technical analyses, applied experimental research, design, construction and test of components. Support research at Nat. Bur. Standards-Cryogenics Div.	440 and 637 in FY 73 from NSF, 450 in FY 74 from AEC	Three main studies in progress: 1) ac losses in S.C. cable materials, 2) high-voltage insulation studies, 3) refrigeration	Research continuing. Flexible 20 meter dewars (outer part of transmission line) will be tested in Fall 1973. Ultimate goal is 1/2 mile long, 69 KV transmission line in 4 to 5 years
Electric Power Research Inst., temporarily NY, NY; Dept of Interior, Washington, D.C.	C. Starr for EPRI and F.F. Parry for DOI	Transmission systems, S.C. and cryoresistive	Cooperative funding support of programs at General Electric, Mass. Inst. of Tech.-Nat. Magnet Lab., Stanford Univ., Union Carbide, and Westinghouse. The DOI share of the total funding varies, but is about 20%	About 1,500 pa	See specific programs	Continuing
General Electric Co.-Corporate Research, Schenectady, NY	G.R. Fox	Cryoresistive, liquid nitrogen cooled transmission line components	Applied experimental research, design, construction, and test of components	687 for 2 years from EPRI/DOI	Have designed improved dielectric insulation systems	Construction and tests on 1000 ft transmission system
General Electric Co.-Electric Utility Engr. Operation, Schenectady, NY	G.D. Brewer	dc S.C. transmission lines	Feasibility studies, economic and technical analyses	25 in FY 73 from LASL	Completed June, 1973	-
Los Alamos Sci. Lab., Los Alamos, New Mexico	W. E. Keller	dc S.C. transmission line	Conceptual designs, feasibility studies, technical analyses, applied experimental research, and design, construction and test of components and systems	400 in FY 73 and 300 in FY 74 from AEC	Have built and tested many short sample models for critical currents	Large 30 meter racetrack test bed with switchgear will be completed late summer 1973. Best preliminary test specimens will be constructed in long lengths and tested
Stanford Univ., Stanford, Calif.	T. H. Geballe, R. H. Hammond	ac S.C. transmission line	E experimental research on critical currents, losses, and techniques for thin film superconductors and dielectrics. Subcontracts to Systems Controls and Nat. Bureau Standards.	365 in FY 73 and FY 74 from N.S.F.	Testing techniques and materials, some short test-sections	Continuing research

TABLE 3. SUMMARY OF PROGRAMS ON TRANSMISSION OF ELECTRIC POWER (continued)

Organization	Manager	Device	Type of Program	Funding (K$) Source	Status	Future Plans
Mass. Inst. of Tech.-Nat. Magnet Lab., Cambridge, Mass	P. Graneau	Cryoresistive, liquid nitrogen cooled transmission line	Applied experimental research, and design, construction, and tests of components and unit	350 for two years, Mar. 72-Mar. 74, from EPRI/DOI. Funded jointly to Underground Power Corp., but most work is at MIT. Also 150K in FY 73 from NSF	Have designed, constructed, and tested single terminal system and 8' long cable components	Full 138 KV tests at Walz Mill scheduled for Fall 1973 on single terminal. Design of the second phase, 2 terminal and conductor system, is nearly complete. Third phase, 1000' long, complex system is being designed
National Bureau of Standards-Cryogenics Div., Boulder, Colo	M.C. Jones, V. D. Arp	Heat transfer and flow research in support of S.C. transmission line development	Applied experimental research and technical analyses on heat transfer to subcritical He I and He II under forced circulation, transient heat transfer computations, and flow instabilities in long lines. Technical analyses of refrigeration systems and operating parameters.	In FY 73, 50 from BNL, 27 from AEC, 30 from AFAPL, and 30 from NBS. In FY 74, 20 from BNL, 30 from AFAPL and 20 from NBS	He I research complete, others still in progress	Continuing, except He II research which is postponed
Oak Ridge Nat. Lab., Oak Ridge, Tenn	H.M. Long	Materials research in support of S.C. transmission line development	Applied experimental research on cryogenic dielectrics, ac loss mechanisms in high field S.C. and dispersion hardening of aluminum	300 in FY 73 and 150 in FY 74 from AEC	Some research on liquid nitrogen and solids in liquid nitrogen completed. Equipment for second and third projects completed	Continuing mainly on first project, dielectric studies, breakdown and flash-over, in liquid nitrogen and helium up to 130 KV
Stanford Univ., Stanford, Calif	T. H. Geballe and W.A. Phillips	Materials research in support of S.C. transmission line development	Applied experimental research on thin film superconductors and cryogenic dielectrics	80 in FY73 and possible 200 in FY74 from EPRI/DoI	Some research completed and published	Continue at higher funding level
Systems Control Inc., Palo Alto, Calif	N.L. Badertscher	dc S.C. transmission lines and ac lines.	Feasibility studies, economic and technical analyses	25 in FY 73 and 33 in FY 74 from Stanford.	Completed dc in June, 1973.	Complete ac studies.
Union Carbide Corp.-Linde Div., Tarrytown, NY	D. A. Haid	ac S.C. transmission line	Conceptual design, feasibility studies, technical analyses, and design, construction, and test of components and systems	2,100 for 3-1/2 years through April 75 from EPRI/DOI	Have built small working model, have tested several components, and completed some experimental research	Large header will be built soon, large scale tests scheduled for Spring 1974
Westinghouse Electric Corp.-Advanced System Tech. Group, Madison, Penn "Walz Mill"	L. Kilar	Test stand development and testing in support of S.C. and cryoresistive transmission lines	Development of test facility for special lines	200 in FY 74 from EPRI/DOI	Modifications begun	Ready for tests in Fall 1973 and Spring 1974

TABLE 4A. SUMMARY OF PROGRAMS ON MARINE ELECTRIC PROPULSION SYSTEMS

Organization	Manager	Device and Application	Type of Program	Funding (K$) Source	Status	Future Plans
Garrett Corp.-AiResearch Manuf. Co., Torrance, Calif	J.H. Dannan	Electric propulsion systems for (a) 40 000 hp/shaft twin screw Small Waterplane Twin Hull (SWATH), (b) two 20 000 hp/shaft contra-rotating SWATH, (c) 20 000 hp/shaft hydrofoil, and (d) 30 000 hp/shaft surface effect naval vessels	Conceptual designs and feasibility studies in first year; design, construction, and tests of components and units in succeeding years. Two concepts for propulsion: S.C. dc homopolar generator/S.C. dc homopolar motor and normal ac generator/rectifiers/S.C. dc homopolar motor	Corporate during FY 73; about 1,000 in FY 74 from NSRDC(A)	Begun July 1973	If preliminary studies are successful, hardware components will be started in FY 75
General Electric Co.-Corporate Research, Schenectady, NY	G.R. Fox	Same as above	Conceptual designs and feasibility studies in first year; design, construction and test of components and units in succeeding years. One concept for propulsion: S.C. dc homopolar generator/S.C. dc homopolar motor	Same as above	Same as above	Same as above
Maritime Adm.-Off. of Asst. Adm. for R and D	C.W. Parker	Marine electric propulsion systems for large vessels	Research and development	Funding requested but not received, for FY 74	-	Await further development of naval propulsion systems
Mass. Inst. of Tech.-Cryogenic Engr. Lab., Cambridge, Mass	J.L. Smith	S.C. naval electric propulsion systems	Conceptual designs, feasibility studies, primarily of ac systems (also some support of 2 MVA generator listed in Table 2). Also postgraduate training of Naval Officers	100 in FY 73 and 74 from ONR	Various innovative concepts have been proposed	Continuing
Naval Ship Research and Development Center, Carderock, Md	S.T.W. Liang	S.C. naval electric propulsion systems	Conceptual designs and feasibility studies	Estimated 50 in FY 72	Completed 1972	-
Naval Ship Research and Development Center (Annapolis), Annapolis, Md	L.G. Adams	S.C. electric propulsion systems for Arctic Surface Effect Vehicles (SEV)	Feasibility studies and technical analyses	Estimated 50 from ARPA	Completed June 1973	-

TABLE 4A. SUMMARY OF PROGRAMS ON MARINE ELECTRIC PROPULSION SYSTEMS (continued)

Organization	Manager	Device and Application	Type of Program	Funding (K$) Source	Status	Future Plans
Naval Ship Research and Development Center (Annapolis), Annapolis, Md	W.J. Levedahl	Electric propulsion systems for (a) 40 000 hp/shaft twin screw Small Waterplane Twin Hull (SWATH), (b) two 20 000 hp/shaft contra-rotating SWATH, (c) 20 000 hp/shaft hydrofoil, and (d) 30 000 hp/shaft surface effect naval vessels	Conceptual designs and feasibility studies. Support research and development at Garrett Corp. and General Electric Co. and conduct similar studies in-house on various advanced concepts	About 2,000 in FY 74, mainly contracted out	Begun Jan. 1973	Continuing
		400 hp S.C. homopolar motor	Design, construction and test of motor and its auxiliary systems	About 1,500 pa for this and the following three projects	Motor completed, undergoing tests on liquid metal brushes	Continuing tests, install in complete test bed Fall 1973
		300 KW S.C. homopolar generator	Design, construction and test of generator and its auxiliary systems	-	S.C. magnets complete	Complete construction in October 1973 and test during 1974 in large test facility
		Test facility for S.C. propulsion systems	Design, construction and use of test facility with 10 000 amp rectifiers, water brakes, switchgear, and 1000 hp marine turbine	-	Under construction	To be completed in Fall 1973
		Switchgear and refrigeration systems for S.C. naval electric propulsion systems	Design, construction, and test of lightweight, 30,000 amp, switchgear and lightweight, reliable refrigeration systems	-	Under development	Continuing
Westinghouse Electric Corp.- Electro mechanical Div., Pittsburgh, Penn	C.J. Mole	Naval electric propulsion systems, for 20 000, 30,000, 40 000 hp motors, ac or dc., normal or S.C.	Conceptual designs, feasibility studies and technical analyses for normal and superconducting homopolar dc machinery and synchronous /asynchronous ac machinery	100 in FY 73 from ONR	Completed March 1973	-
		Same as the one at top of this column	Conceptual designs and feasibility studies	Corporate	Completed Feb. 1973	-
		dc, electric propulsion system, homopolar generator, normal room temperature system, about 4 MW	Design, construction and test of generator and its components, especially the current collection system which is similar in technology to S.C. homopolar machinery	810 in FY 73 and 800 in FY 74 from ARPA	In progress	To be tested late 1973 or early 1974

TABLE 4B. SUMMARY OF PROGRAMS ON AIRBORNE AND MILITARY GENERATORS

Organization	Manager	Device and Application	Type of Program	Funding (K$) Source	Status	Future Plans
Air Force Aero Propulsion Lab., Wright-Patterson Air Force Base, Ohio	E.L. Boyer	Airborne 5 MVA 400 Cycle S.C. generator	Funding support of research and development on airborne S.C. generators at Westinghouse and background research at Nat. Bur. of Standards and in-house	55 in FY 73 for in-house research	In progress	Continuing
U.S. Army Mobile Equipment Research and Development Center, Ft. Belvoir, Va	J.H. Ferrick	Mobile S.C. ac generators	Design and test of S.C. generators. Small in-house awareness program only at present	20 in FY 73 and 74	Now dormant	–
	L.I. Arnstutz	Mobile S.C. machinery	Funding support of research and development on auxiliary components of S.C. machinery, such as refrigeration systems at General Electric	About 125 pa	In progress	Continuing
Magnetic Corp. of America, Cambridge, Mass	Z.J.J. Stekly	Airborne 5MVA 400 cycle S.C. generator	Conceptual designs, feasibility studies and design construction, and test of model S.C. coil for rotor of generator. Subcontract from Westinghouse	–	Completed model coil in 1972	
Westinghouse Aerospace Electrical Div. and Research Lab., Pittsburgh, Penn	J. H. Parker	Airborne 5MVA 400 cycle S.C. generator	Applied experimental research and design, construction, and test of rotor and possibly later, complete generator	223 in FY 73, 1,400 originally budgeted. FY 74 funding uncertain	Rotor coils completed and tested, rotor under construction.	Rotor tests to be finished in 1973 and complete generator to be built if rotor works satisfactorily

TABLE 4C. SUMMARY OF PROGRAMS IN DIRECT SUPPORT OF MILITARY MACHINERY

Organization	Manager	Type of Program	Funding (K$) Source	Status	Future Plans
Argonne Nat. Lab., Argonne, Ill	E. Fisher	Applied research on fatigue damage in S.C. composites and their stabilizing metals	FY 74 from ARPA through NBS	To begin late Summer 1973	Continuing
Advanced Research Projects Agency, Arlington, Va	E.C. von Reuth	Funding support of background research for various Dept. of Defense programs on S.C. rotating machinery, including those at about 12 universities and at General Electric and the National Bureau of Standards	FY 74 to NBS and about 500 pa to universities	In progress at universities, to begin late Summer 1973 at NBS	Continuing
Battelle Memorial Inst.-Columbus Labs., Columbus, Ohio	F.J. Jelinek	Applied research on specific heat and thermal expansion of selected S.C. and structural materials	FY 74 from ARPA through NBS	To begin late Summer 1973	-
	J.E. Campbell	Compilation and review of handbook data for designing S.C. machinery	FY 74 from ARPA through NBS	To begin late Summer 1973	-
General Electric Co.-Research and Development Center, Schenectady, NY	B.D. Hatch	Applied research on referigeration requirements and equipment for S.C. machinery, current collection systems, and S.C. coil tests	About 250 pa from ARPA	Preliminary studies completed	Experimental equipment for current collection and coils to be completed Fall 1973
	D.B. Colyer	Development of miniature turbine refrigerators	About 125 pa from USAMERDC	Preliminary tests completed	Continuing
	W.B. Hillig	Applied research on thermal expansion and tensile properties of S.C. composites and structural materials	FY 74 from ARPA through NBS	To begin late Summer 1973	Continuing
Martin Marietta-Denver Div., Denver, Colo	F. Schwartzberg	Applied research on large scale fracture properties of structural materials	FY 74 from ARPA through NBS	To begin late Summer 1973	Continuing
National Bureau of Standards-Cryogenics Div., Boulder, Colo	V.D. Arp	Applied technical analyses of refrigeration cycles and properties of helium	85 in FY 73 and 48 in FY 74 from ARPA	Completed refrigeration cycle analyses, properties of helium in progress	Complete equation of state for helium in various useful forms
	R.P. Reed	Technical program management and applied research on (a) zero-field and magneto-thermal conductivity of structural materials, (b) dynamic elastic properties of structural materials, (c) fracture toughness and fatigue of structural materials, and (d) exploratory research on low-thermal conductivity, high-strength composites	FY 74 from ARPA	To begin late Summer 1973	Continuing
Westinghouse Electric Corp.-Research and Development Center, Pittsburgh, Penn	G.G. Lessmann	Applied research on mechanical properties of welds, materials processing, and characterization of structural materials	FY 74 from ARPA through NBS	To begin late Summer 1973	Continuing

TABLE 5A. SUMMARY OF PROGRAMS ON ENERGY STORAGE MAGNETS

Organization	Manager	Device and Application	Type of Program	Funding (K$) Source	Status	Future Plans
Air Force Aero Propulsion Lab., Wright-Patterson AFB, Ohio	J.M. Turner	Pulsed energy storage S.C. solenoid for airborne usage	Conceptual designs and feasibility studies of lightweight, high repetition rate, low energy storage solenoid. To be contracted out. Presently fund magnet development at Magnetic Corp. of America	Estimated 100 in FY 74		Probably start late in CY 73
Los Alamos Sci. Lab., Los Alamos, New Mexico	W. E. Keller	Long term energy storage S.C. toroids for peak load generation. Codename SMES	Conceptual designs and feasibility studies for 10^{14} J toroid. Applied research and design construction and test of 50 kJ bumpy toroid prototype	100 in FY 73 and in FY 74 from AEC	Have started construction of 50 KJ prototype	Complete and test prototype in FY 74
	H.L. Laquer, J.D. Rogers	Pulsed energy storage S.C. solenoids for compression coils of theta pinch experiment. Codename METS	Conceptual design and feasibility studies for 1-1/2 MJ system followed by applied research and system design and construction of prototype 30 kJ and 300 kJ solenoids	368 in FY 73 and 545 (or more) in FY 74 from AEC	Have built and tested 30 KJ solenoid, beginning assembly of 300 KJ solenoid	Complete 300 KJ solenoid
Magnetic Corp. of America, Waltham, Mass	E.J. Lucas	Pulsed energy storage S.C. solenoids for airborne usage	Design, construction and test of 7 kJ and 100 kJ prototype solenoids and applied research on S.C. materials, switches, and non-conductive dewars	201 in FY 73 and 53 in FY 74 from AFAPL	Tests on 7 KJ solenoid are complete, components for 100 KJ solenoid are nearly finished	Complete 100 KJ solenoid in 1973
	Z.J.J. Stekly	Long term energy storage S.C. system for peak load generation	Conceptual designs, feasibility studies on 10 GW hour energy storage system for Off. of Sci. and Tech.	Corporate	Completed 1972	
Nat. Bur. of Standards-Cryogenics Div., Boulder, Colo	J. Hord	Fluid cooling systems for storage magnets	Applied research and development of fluid helium, including He II, forced circulation systems	30 in FY 73 and 20 in FY 74 from AFAPL	Apparatus built, undergoing tests	Complete pumping tests
U.S. Army Mobile Equipment Research and Development Center, Ft. Belvoir, Va	J.H. Ferrick	Pulsed energy S.C. magnet for mobile applications	Conceptual design, feasibility studies of 100 kJ S.C. coil; design, construction and test of 10 kJ prototype S.C. coil	about 25 in FY 73, 100 in FY 74	Preliminary studies finished	Complete prototype coil in FY 74
Univ. of Wisconsin, Madison, Wisc	R.W. Boom, H.A. Peterson	Long term energy storage S.C. system for peak load generation	Conceptual designs and feasibility studies of 1000 MW, 10 000 MW hr system and applied research on ac losses in superconductors	200 in CY 73 from General Electric, NSF, Wisconsin Alumni Res. Fund, and Wisconsin Utilities Assoc.	In progress	Continuing
Los Alamos Sci. Lab., Los Alamos, New Mexico	K. Thomassen	Rotating variable S.C. inductor	Conceptual designs. Code name RETS	20 in FY 73 from AEC	Completed	

TABLE 5B. SUMMARY OF PROGRAMS ON MAGNETS FOR CONTROLLED THERMONUCLEAR REACTIONS

Organization	Manager	Device	Type of Program	Funding (K$) Source	Status	Future Plans
Atomic Energy Comm.-Div. of Controlled Thermonuclear Research, Germantown, Md	W.C. Gough	S.C. systems for developing CTR programs	Funding support for programs at Argonne Nat. Lab., Lawrence Livermore Lab., Los Alamos Sci. Lab., Oak Ridge National Lab.	About 1,850 in FY 74	See individual programs	–
Argonne Nat. Lab., Argonne, Ill	C. Laverick	S.C. systems for developing CTR programs	Review and recommendations on various projects	About 25 in FY 73 and 25 FY 74 from AEC	Continuing	–
Brookhaven Nat. Lab., Upton, NY	J.R. Powell	S.C. system for developing CTR programs	Review and recommendations on various projects, especially on applications of present technology to steady state devices	About 60 in FY 73 and 50 in FY 74 from AEC	Continuing	–
Intermagnetics General Corp., Guilderland, NY	C.A. Rosner	Magnet for laser fusion research	Design and construction of solenoid magnet	Constructed for KMS Fusion, Inc.	Completed Spring 1973	–
Lawrence Livermore Lab., Livermore, Calif	C.E. Taylor C.D. Henning	Applied research for CTR systems	Applied research on Nb_3Sn tapes, bonding techniques, multilayer composite tapes, and mechanical properties of composite tapes. Also design, construction, and test of small demonstration magnets	About 200 in FY 73 and 100 in FY 74 from AEC	Baseball II magnet operating successfully. Small demonstration magnets have been built	May expand program substantially if prototypes prove feasible
Los Alamos Sci. Lab., Los Alamos, New Mexico	H.L. Laquer, J.D. Rogers, and K. Thomassen	Magnets and inductors for energy storage in CTR systems	See Table 5A for description of programs	–	–	–
Oak Ridge Nat. Lab., Oak Ridge, Tenn	H.M. Long	Applied research and prototype magnets for CTR systems	Applied research on materials, especially multifilamentary NbTi wires and V_3Ga ribbons, potting techniques, and stability conditions. Design, construction, and test of prototype magnets	About 150 pa from AEC	Mirror quadropole magnets operating successfully. Many short sample tests completed in 10 tesla magnet	Nb_3Sn insert in magnet for 15 tesla short sample tests
Princeton Univ.-Plasma Physics Lab., Princeton, N.J.	J. File	Discrete-coil toroidal magnet for fusion research	Conceptual designs and feasibility studies for large S.C. magnet systems. Design, construction, and test of toroidal magnet components	100 in FY 73 and 50 in FY 74 from AEC	One discrete coil is being constructed and will be tested Dec. 1974	Uncertain, continuation proposal has been submitted
Univ. of Wisconsin, Madison, Wisc	R.W. Boom	S.C. magnets for Tokomak reactors	Conceptual designs and feasibility studies of constant tension magnet windings for Tokomak reactors and other magnet systems	About 30-50 pa for S.C. portion of program from AEC, Northern States Power Co., and Wisconsin Electric Utilities	Feasibility studies for some toroidal systems completed	Extend studies to other systems such as mirror magnets

TABLE 5C.   SUMMARY OF PROGRAMS ON MAGNETS FOR STEADY STATE PLASMA EXPERIMENTS

Organization	Manager	Device and Application	Type of Program	Funding (K$) Source	Status	Future Plans
NASA Lewis Research Center-Electrophysics Div., Cleveland, Ohio	W.D. Coles	Four solenoid magnets for dc, steady-state plasma heating experiments code name: SUMMA	Design and test of 4-solenoid, mirror compound magnets with design fields of 90-50-90 kG, 43" long	700 in FY 74	3 of 4 magnets operated to nearly full field, 80 kG on one end, 50 on other	Carry out experiments on dc steady-state plasma heating
		Bumpy toroid for dc steady state plasma heating and containment experiments	Design and test of 12 magnet, 30 kG bumpy toroid, 5 ft major diameter	475 in FY 74	Began operations during FY 73	Same as above

TABLE 5D. SUMMARY OF PROGRAMS ON MAGNETS FOR MAGNETOHYDRODYNAMIC GENERATORS

Organization	Manager	Device and Application	Type of Program	Funding (K$) Source	Status	Future Plans
Air Force Aero Propulsion Lab., Wright-Patterson AFB, Ohio	R. F. Cooper	Airborne MHD magnets system	Design, construction and test of S. C. magnet system	Possibly 200 later in FY 74	-	-
Argonne Nat. Lab., Argonne, Ill	J. P. Purcell	S. C. magnet and cryogenic system for central station electric power generation, prototype, 8 tesla	Design, construction and test of a split solenoid, untapered, with cold-iron field shaping	1,100 in FY 73 and FY 74 from OCR (DOI) through Stanford Univ.	Magnet and cryogenic design nearly finished	S. C. materials will be ordered soon, magnet to be operating in Fall 1974
AVCO Everett Research Lab., Everett, Mass	A. M. Hatch	S. C. magnet and cryogenic system for central station electric power generation, prototype, 5 tesla	Design, construction and test of a solenoid pair, tapered, with cold-iron field shaping. Magnet construction subcontracted to Magnetic Engineering Assoc.	750 pa for S. C. magnet and cryogenic portion of total program from OCR(DOI), EPRI, some private utilities, and corporate	Magnet design finished cryogenic and vacuum systems designed	Start construction and assembly in Oct. 1973
Ferranti Packard Ltd., Toronto, Ontario	D. L. Atherton	Airborne MHD magnet 1-1/2 MJoule	Design and construction of a saddle coil dipole magnet	400 for total project from AFAPL and Canadian government	Construction completed	Testing will be carried out by Magnetic Corp. of America
Magnetic Corp. of America, Waltham, Mass	Z. J. J. Stekly	Airborne MHD magnet, 1 MJoule	Design, construction and test of saddle coil magnet	About 400 from AFAPL for this system and testing of the above unit	Magnet completed Summer 1973, most tests complete	More tests scheduled
		S. C. magnet and cryogenic system for central station electric power generation	Conceptual designs and feasibility studies for 8 tesla system	Subcontract from Stanford Univ.	Designs completed	-
Magnetic Engineering Assoc., Inc., Cambridge, Mass	P. Marston	S. C. dipole magnet for AVCO Everett Program	Design and construction of dipole magnet	300 from AVCO Everett	Magnet design finished	Assembly of magnet to begin Fall, 1973
Stanford Univ.- Dept. of Mech. Engr., Stanford, Calif	R. H. Eustis	MHD power generation system	Conceptual designs and design, construction, and test of total MHD generation system. S. C. magnet designs and construction subcontracted to Magnetic Corp. of America and Argonne Nat. Lab.	About 1,500 for magnet design and construction	Conceptual design completed. Final design of magnet and cryogenic system nearly finished	Magnet system to be operating in Fall, 1974
Stanford Linear Accelerator Center, Stanford Calif	S. J. St Lorant	Split solenoid for MHD magnet model tests	Design, construction, and test of small, 60 kG split solenoid triaxial access	54 in FY 72 from AEC, lesser in FY 73	Completed modifications in early summer 1972. Used at SLAC in 1973, now at Stanford Univ.	Use in occasional special experiments

TABLE 6. SUMMARY OF PROGRAMS ON MAGNETS AND CAVITIES FOR HIGH ENERGY PHYSICS EXPERIMENTS

Organization	Manager	Device and Application	Type of Program	Funding (K$) Source	Status	Future Plans
Atomic Energy Comm.-High Energy Physics Program, Germantown, Md	C.R. Richardson	S.C. systems for acceleration and control of high energy particle beams	Funding support for programs at Argonne Nat. Lab., Brookhaven Nat. Lab., Lawrence Berkeley Lab., National Accel. Lab., and Stanford Linear Accel. Center	About 7,600 in FY 74 for operating expenses and equipment construction	See individual programs	-
Argonne Nat. Lab., Argonne, Ill	J.R. Purcell	S.C. magnet for 15 ft. bubble chamber at Nat. Accel. Lab.	Design, construction, and test of 400 MJ, 30 kG (central) bubble chamber magnet	About 2,100 for completion from Nat. Accel. Lab.	Magnet completed Fall 1972	-
		Beam line dipole and quadrupole magnets	Design and construction of working beam line magnets for ZGS system		Continuing	-
			Design S.C. magnets for new storage ring for present ZGS and design, construct, and test prototype magnets	100 in FY 73, 150 in FY 74 from AEC	One quadrupole and one dipole model magnet tested, another dipole near completion	Continue constructing prototype dipoles and quadrupoles
Brookhaven Nat. Lab.-Accelerator Dept., Upton, NY	J.P. Blewett, G.T. Danley, W.B. Sampson	Niobium RF cavities	Mainly applied research on materials rather than design and construction of cavities	Reduced from previous years, about 300 from AEC and NAL	Research on metallurgy continuing	Continuing research, hardware design postponed
		Beam handling dipole and quadrupole magnets for present AGS accelerator. Also primary beam splitter	Applied research on S.C. materials and design, construction and test on beam magnets, both secondary and primary	About 575 for operating and equipment expenses in FY 74 from AEC	Two dipoles nearly completed for 30 GEV bubble chamber beam. Splitter being installed after life tests. Some materials research completed	Construct 4 more dipoles and 4 quadrupoles for another beam, continue materials studies
		Beam handling dipole and quadrupole model magnets for future ISABELLE accelerator	Conceptual designs, feasibility studies, applied research on S.C. materials and design, construction and test of slow-pulse dipole and quadrupole model magnets	About 275 in FY 74 from AEC	Two 1 meter dipoles built and tested. Quadrupoles designed. Some materials studies completed	Continue materials studies and designs, build 3 meter dipoles
Lawrence Berkeley Lab., Berkeley, Calif	W.S. Gilbert	Beam handling dipole and quadrupole magnets for present Bevatron	Design, construction, test, and continued usage of beam handling magnets for Bevatron	350 in FY 74 from AEC	dc dipole and two dc quadrupole magnets on line in March, 1973	Continuing
		Pulsed storage ring dipole magnets for future accelerators	Design, construction and test of model dipole magnets		Magnet #8 tested at 1 hz., #9 tested at 60 and 1000 hz, #10 tested soon	Continuing

TABLE 6. SUMMARY OF PROGRAMS ON MAGNETS AND CAVITIES FOR HIGH ENERGY PHYSICS EXPERIMENTS (continued)

Organization	Manager	Device and Application	Type of Program	Funding (K$) Source	Status	Future Plans
National Accelerator Laboratory, Batavia, Ill	R. W. Fast	Dipole and quadrupole magnets for present system	Design, construct, and test beam line dipole and quadrupole magnets	200 in FY 73 and 250 in FY 74 from AEC	Final design of two large analyzing dipoles	Design of additional dipole and quadrupole magnets. Run two dipoles Spring 1974
	B.P. Strauss	Dipole and quadrupole magnets for energy doubler for S.C. synchrotron	Design of large scale systems, and design, construction, and test of model and prototype magnets	1,500 in FY 73 and FY 74 from AEC	Six slow cycle model, 5 dipole, 1 quadrupole constructed. Eight full size magnets under construction	Complete and test full size prototype magnets. Complete design report and two prototypes Spring 1974
		Dipole and quadrupole magnets for muon external beam line	Design, construct, and test beam line dipole and quadrupole magnets for muon beam line	300 in FY 74 from AEC	First magnet is under construction	First to be installed in Spring 1974
Stanford Linear Accelerator Center, Stanford, Calif	E. L. Garwin	Dipole for flux trapping of dipole, quadrupole, and sextapole fields	Design, construction, and test of hollow cylinder, 17 kG dipole for flux trapping	50 in FY 74 from AEC	In progress	-
	P. Wilson	S.C. RF cavities for linear accelerator	Design, construction, and test of cavity sections	300 in FY 73 now discontinued	Reproducibly demonstrated 10^9-10^{10} Q, high breakdown field 1 kG	Postponed construction mainly materials problems at present
	S.J. St Lorant	S.C. spectrometer for spark chamber	Design, construction and test of spectrometer solenoid having four sections	900 in FY 73 and FY 74 from AEC	First coil is being tested, all four will be tested by end of 1973	Complete system in 1973
		Internal magnet for rapid cycle bubble chamber	Design, construction and test of 22 kG magnet inserted into bubble chamber	25 in FY 73 from AEC	Completed Nov. 1973. 2×10^5 pictures taken	Continue operating
		Magnet for polarized target	Test and use of magnet designed at Yale Univ., 60 kG at target	65 in FY 73 from AEC	Completed in 1972	Continue using in polarized target. experiments
		S.C. magnetic shield for primary electron beam	Design, construction, and test of Mark II model flux excluding tubes for cleaner scattered particle experiments	Orig. model funded for 15-20. Mark II 15-20 in FY 73 and FY 74	Final assembly in progress	Complete assembly and test of flux-excluding tubes

TABLE 7. SUMMARY OF PROGRAMS ON MAGNETS FOR INDUSTRIAL AND MEDICAL APPLICATIONS

Organization	Manager	Device and Application	Type of Program	Funding (K$) Source	Status	Future Plans
Intermagnetics General Corp., Schenectady, NY	C. Rosner	Industrial magnets for ore separation and beneficiation	Feasibility studies and technical analyses	Less than 50 in FY 74 from NSF	Study ends late 1973	Large scale demonstration facility possible later
Magnetic Corp. of America, Waltham, Mass	Z.J.J. Stekly	50 kOe, 5" ID high gradient separator magnet system; industrial Kaolin purification	Applied experimental research, system design, construction and test	Corporate	One magnet delivered to "a major Kaolin producer" for field tests	A second, identical, separator unit is being constructed for laboratory research
Mass. Inst. of Tech.-National Magnet Lab., Cambridge, Mass	H. Kolm, E. Maxwell and others	Industrial high gradient magnetic separation for water treatment, ore beneficiation, coal desulfurization, etc	Feasibility studies, laboratory tests with normal magnets	About 250 over last 2-1/2 yrs from NSF	All work to date is with normal magnets. No S.C. magnets have been proposed. Many laboratory demonstrations	Field tests in taconite beneficiations planned. If industry can be convinced of economics of high fields, later, large volume magnets must be superconducting
Vanderbilt Univ., Nashville, Tenn	C.E. Roos	Magnet system for scrap separation by eddy current interactions	Feasibility studies, laboratory tests	300 from EPA over 3 yr period ended Aug. 72. Also funds from Vanderbilt and Steiner-Liff	Most work with normal magnets. S.C. magnet tried but was not successful	Questionable whether size effects can be handled. Applications of S.C. magnets appear limited
Magnetic Corp. of America, Waltham, Mass	Z.J.J. Stekly	Toroidal magnets and cryogenic system for Stanford Univ. π meson accelerator	Conceptual designs, component design, construction and test	Contract from Stanford Univ. project listed below	Magnet pancakes nearly completed	-
Mass. Inst. of Tech.-National Magnet Lab., Cambridge, Mass and Mass General Hospital, Boston, Mass	D.B. Montgomery	Medical, high gradient 20 kOe magnet. Magnetic guidance for catheter through intercranial blood vessels. Primary proposed use is in repair of aneurysm	Applied experimental research; magnet design, construction and test	Hardware development from NSF, medical application funded separately by NIH	Tests have been made with normal magnets. S.C. magnet is being wound	New proposal for $150 K for 2 yr has been submitted to NSF. Use of present system for clinical trials has been approved.
Stanford Univ.-Physics Department, Stanford, Calif	H.A. Schwettman	Medical collection and focusing of π mesons for cancer therapy	Applied experimental research, magnet design, construction and test	500 from NSF over past 2-1/2 yr, ends Sept. 73. Some medical being carried out at about same level.	Some preliminary tests have been made on prototype coils	System should be mostly completed in 1973. Human studies may start in 1975, clinical trials 1978. Renewal proposal to be submitted for about $250 K for pion channel installations and improvements
Stanford Linear Accelerator Center, Stanford, Calif	S. St. Lorant	Medical tapered solenoid ~1800 Oe at 12 cm. Positioning of silicone containing magnetic particles within tumors-cancer therapy	Applied experimental research, magnet design, construction and test	25 for magnet from UCLA Medical Foundation	Magnet and cryogenic system built and tested for about 10 operations	Proposals will be submitted for construction of an optimized magnet system. Probably for ~$50-60 K

TABLE 8. SUMMARY OF PROGRAMS ON LAND TRANSPORTATION SYSTEMS

Organization	Manager	Device	Type of Program	Funding (K$) Source	Status	Future Plans
Ford Scientific Laboratories, Dearborn, Mich	J. Reitz	Magnetically levitated (Maglev) suspension, propulsion system	Technical analyses, experimental modeling	Corporate; also 370 in past 2-1/2 yr and 75 in first half of FY 74 from DOT	Completed good mathematical model of S.C. magnet moving over track, some experimental verification with magnet wheel, and evaluation of S.C. magnet paddlewheel and helical devices for propulsion	Renewal proposal to DOT to study both S.C. and ferromagnetic systems of suspension. Small, high speed vehicles to be built and tested
General Motors Research Lab., Warren, Mich	T.C. Wang	Maglev suspension propulsion system	Technical analyses, experimental modeling	Corporate, estimated 250	S.C. magnet built for wheel tests. Several studies of dynamics completed	Indications are that this program is growing. Further experimental and theoretical studies of dynamics and propulsion planned
Magnetic Corp., of America, Waltham, Mass	Z.J.J. Stekly	Levitation pads and dewar system	Technical analyses	Subcontract from Ford	Completed	–
Mass. Inst. of Tech.-Nat. Magnet Lab., Cambridge, Mass	H. Kolm	Magneplane system including vehicle, guideway and linear synchronous motor	Feasibility studies, conceptual design, and technical analyses of magneplane system, some experimental modeling. Operational 1/25 scale total system with cryogenic vehicle.	300 in CY 71 and 72 from Avco-Everett Mass. Inst. of Tech., and Raytheon. 95 in CY 72 and 115 in CY 73 from NSF	Completed extensive studies of all aspects of levitation, propulsion and control as well as economic analyses. 1/25th scale system constructed.	Renewal proposal for $160 K for a 1 yr program submitted to NSF. Outside contracts proposed for guideway economics and propulsion studies.
Stanford Research Institute, Menlo Park, Calif	H.T. Coffey	Maglev vehicle and guideway	Technical analyses of suspension system using a test vehicle	80 for first half of FY 74 from DOT	Towed four-magnet vehicle built and tested at low speed. Passive damping system tested	Increased instrumentation and installation of active damping system

SUBJECT INDEX

MAR 10 '76